MATHEMATICAL METHODS IN PHYSICS

MATHEMATICAL METHODS IN PHYSICS

Samuel D. Lindenbaum
Israel

World Scientific
Singapore • New Jersey • London • Hong Kong

Published by

World Scientific Publishing Co. Pte. Ltd.
5 Toh Tuck Link, Singapore 596224
USA office: 27 Warren Street, Suite 401-402, Hackensack, NJ 07601
UK office: 57 Shelton Street, Covent Garden, London WC2H 9HE

Library of Congress Cataloging-in-Publication Data
Lindenbaum, Samuel D.
Mathematical methods in physics / Samuel D. Lindenbaum.
p. cm.
ISBN-13 978-981-02-2760-9 -- ISBN-10 981-02-2760-4
I. Mathematical physics. I. Title.
QC20.L498 1996
530.1'5--dc20 96-31604
CIP

British Library Cataloguing-in-Publication Data
A catalogue record for this book is available from the British Library.

Preface

This book comprises a set of class notes on a Columbia University (New York) 2-semester graduate course in Mathematical Methods in Physics. Although the course is normally taken by first-year MA or PhD physics majors, it can be used by senior undergraduates having the requisite mathematical background and majoring in physics, engineering or other technically related fields.

Some studies, not usually found in standard mathematical physics texts, are presented to complement the basic theories. These include: the *theory of curves in space* in Chapter 1; *analytic continuation* in Chapter 6; and *retarded and advanced D-functions* in Chapter 11.

In addition, applications of the theories are inserted throughout the text where deemed appropriate. These include: moment of inertia in "Tensor Analysis"; Maxwell's equations, magnetostatics, stress tensor, continuity equation, electrostatics, and heat flow in "Tensor Fields"; special relativity in "Matrix and Vector Algebra in N-Dimensional Space"; Fourier series, Hermitian operators, Legendre polynomials and spherical harmonics in "Hilbert Space"; electrostatics, hydrodynamics and Gamma function in the study of analytic functions in "Theory of Functions of a Complex Variable;" vibrating string, vibrating membrane and harmonic oscillator in "Theory of Ordinary Differential Equations"; age of the earth and temperature variation of the earth's surface in "Heat Conduction"; and field due to a moving point charge (Liénard-Wiechert potentials) in "Wave Equations".

Lastly, Appendix A (Part I) and Appendix B (Part II) comprise approximately 150 pages of problems, with their solutions, which represent applications as well as extensions to the theory.

Contents

Chapter 11 Wave Equations

Appendix B. Problems and Solutions

PART I

Chapter 1

Vector Analysis

1.1 Vector Algebra

1.1.1 Rotation of Coordinate Axes

Following a rotation of coordinate axes (Fig. 1-1), the new coordinates x_1' x_2' x_3' can be expressed most generally as

$$x_1' = u_{11}x_1 + u_{12}x_2 + u_{13}x_3,$$

$$x_2' = u_{21}x_1 + u_{22}x_2 + u_{23}x_3,$$

$$x_3' = u_{31}x_1 + u_{32}x_2 + u_{33}x_3,$$

or, in summation form,

$$x_i' = \sum_{j=1}^{3} u_{ij}x_j.$$

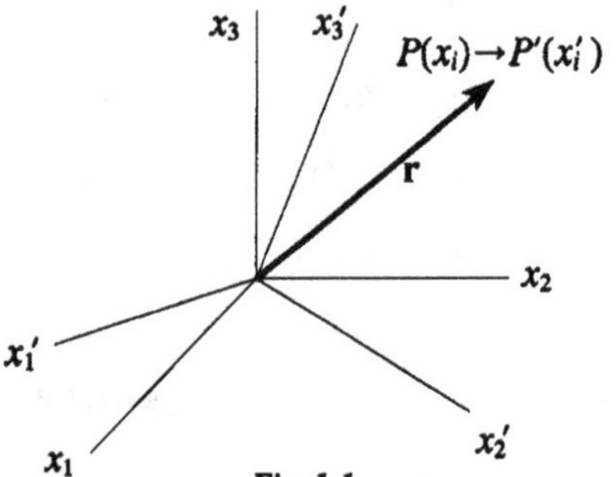

Fig. 1-1

Since the length of a vector remains unchanged under a coordinate rotation, i.e. $|\mathbf{r}| = |\mathbf{r}'|^2$,

$$\begin{aligned} x_1^2 + x_2^2 + x_3^2 &= x_1'^2 + x_2'^2 + x_3'^2 \\ &= x_1^2(u_{11}^2 + u_{21}^2 + u_{31}^2) + x_2^2(u_{12}^2 + u_{22}^2 + u_{32}^2) \\ &+ x_3^2(u_{13}^2 + u_{13}^2 + u_{33}^2) + x_1x_2(u_{11}u_{12} + u_{21}u_{22} + u_{31}u_{32}) \\ &+ x_1x_3(u_{11}u_{13} + u_{21}u_{13} + u_{21}u_{23} + u_{31}u_{33}) \\ &+ x_2x_3(u_{12}u_{13} + u_{22}u_{23} + u_{32}u_{33}). \end{aligned}$$

Hence,

$$(u_{11}^2 + u_{21}^2 + u_{31}^2) = 1. \quad (u_{11}u_{12} + u_{21}u_{22} + u_{31}u_{32}) = 0.$$

$$(u_{12}^2 + u_{22}^2 + u_{32}^2) = 1. \quad (u_{11}u_{13} + u_{21}u_{23} + u_{31}u_{33}) = 0.$$

$$(u_{13}^2 + u_{23}^2 + u_{33}^2) = 1. \quad (u_{12}u_{13} + u_{22}u_{23} + u_{32}u_{33}) = 0.$$

or, in summation form,

$$u_{ij}u_{ik} = \delta_{jk},$$

where summation over i is implied, and where $\delta_{jk} = 0$ for $j \neq k$, and $\delta_{jk} = 1$ for $j = k$.

1.1.2 The Inverse Rotation

We had for the coordinate system rotation $\Sigma \to \Sigma'$,

$$x_1' = u_{11}x_1 + u_{12}x_2 + u_{13}x_3.$$

$$x_2' = u_{21}x_1 + u_{22}x_2 + u_{23}x_3.$$

$$x_3' = u_{31}x_1 + u_{32}x_2 + u_{33}x_3.$$

(1) $$x_i' = \sum_{j=1}^{3} u_{ij}x_j.$$

For the inverse coordinate system rotation $\Sigma' \to \Sigma$,

$$x_1 = \upsilon_{11}x_1' + \upsilon_{12}x_2' + \upsilon_{13}x_3'.$$

$$x_2 = \upsilon_{21}x_1' + \upsilon_{22}x_2' + \upsilon_{23}x_3'.$$

$$x_3 = \upsilon_{31}x_1' + \upsilon_{32}x_2' + \upsilon_{33}x_3'.$$

(2) $$x_i = \sum_{j=1}^{3} \upsilon_{ij}x_j'.$$

However, since $x_1^2 + x_2^2 + x_3^2 = x_1'^2 + x_2'^2 + x_2'^2$,

(3) $$\sum_{i=1}^{3} u_{ij}u_{ik} = \delta_{jk}, \quad \text{and} \quad \sum_{i=1}^{3} \upsilon_{ij}\upsilon_{ik} = \delta_{jk}.$$

From (1) and (2), we have

$$
\begin{aligned}
x_1' &= u_{11}(\upsilon_{11}x_1' + \upsilon_{12}x_2' + \upsilon_{13}x_3') + u_{12}(\upsilon_{21}x_1' + \upsilon_{22}x_2' + \upsilon_{23}x_3') \\
&\quad + u_{13}(\upsilon_{31}x_1' + \upsilon_{32}x_2' + \upsilon_{33}x_3') \\
&= x_1'(u_{11}\upsilon_{11} + u_{12}\upsilon_{21} + u_{13}\upsilon_{31}) + x_2'(u_{11}\upsilon_{12} + u_{12}\upsilon_{22} + u_{13}\upsilon_{32}) \\
&\quad + x_3'(u_{11}\upsilon_{13} + u_{12}\upsilon_{23} + u_{13}\upsilon_{33}).
\end{aligned}
$$

Employing a similar procedure for x_2' and x_3', we have the result:

$$
\begin{aligned}
u_{11}\upsilon_{11} + u_{12}\upsilon_{21} + u_{13}\upsilon_{31} &= 1. \\
u_{21}\upsilon_{12} + u_{22}\upsilon_{22} + u_{23}\upsilon_{32} &= 1. \\
u_{31}\upsilon_{13} + u_{32}\upsilon_{23} + u_{33}\upsilon_{33} &= 1.
\end{aligned}
$$

$$
\begin{aligned}
u_{11}\upsilon_{12} + u_{12}\upsilon_{22} + u_{13}\upsilon_{32} &= 0. \\
u_{21}\upsilon_{11} + u_{22}\upsilon_{21} + u_{23}\upsilon_{31} &= 0. \\
u_{31}\upsilon_{12} + u_{32}\upsilon_{22} + u_{33}\upsilon_{32} &= 0.
\end{aligned}
$$

$$
\begin{aligned}
u_{31}\upsilon_{11} + u_{32}\upsilon_{21} + u_{33}\upsilon_{31} &= 0. \\
u_{11}\upsilon_{13} + u_{12}\upsilon_{23} + u_{13}\upsilon_{33} &= 0. \\
u_{21}\upsilon_{13} + u_{22}\upsilon_{23} + u_{23}\upsilon_{33} &= 0.
\end{aligned}
$$

or, in summation form (summation over i implied):

(4) $$u_{ji}\upsilon_{ik} = \delta_{jk}.$$

Comparing (4) with (3), we see that $u_{ji} = \upsilon_{ij}$.

Hence,

$$x_i' = \sum_{j=1}^{3} u_{ij}x_j. \qquad x_i = \sum_{j=1}^{3} u_{ji}x_j'.$$

1.1.3 Definition of a Vector

A vector $\mathbf{A}$ is defined as the totality of 3 quantities (A_1, A_2, A_3) which, under a rotation of axes

$$x_i' = \sum_{j=1}^{3} u_{ij} x_j,$$

transform as do the coordinates themselves, i.e.

$$(A_1, A_2, A_3) \rightarrow (A_1', A_2', A_3'),$$

where

$$A_i' = \sum_{j=1}^{3} u_{ij} A_j.$$

Examples of Vectors

(1) Position vector $\mathbf{r}(x_1, x_2, x_3)$: since the x_i are the coordinates themselves.

(2) Velocity vector $\mathbf{v} = \dot{\mathbf{r}} = \frac{d\mathbf{r}}{dt}\left(\frac{dx_1}{dt}, \frac{dx_2}{dt}, \frac{dx_3}{dt}\right)$: since time is invariant to a rotation.

(3) Acceleration vector $\mathbf{a} = \frac{d\mathbf{v}}{dt} = \frac{d^2\mathbf{r}}{dt^2}\left(\frac{d^2x_1}{dt^2}, \frac{d^2x_2}{dt^2}, \frac{d^2x_3}{dt^2}\right)$.

1.1.4 Equality of Vectors

$\mathbf{A} = \mathbf{B}$ if and only if $A_i = B_i$ $(i = 1, 2, 3)$. If this is true in one coordinate system, then it is true in any other coordinate system, since

$$A_i' = \sum_{j=1}^{3} u_{ij} A_j = \sum_{j=1}^{3} u_{ij} B_j = B_j'.$$

Hence, to prove the equality of two vectors, one need only show it to be true in any convenient coordinate system. It then follows that all vector equations are invariant to rotations.

1.1.5 Addition (Subtraction) of Vectors

If **A** and **B** are separately vectors, then

$$\sum_{j=1}^{3} u_{ij}(A_j + B_j) = \sum_{j=1}^{3} u_{ij}A_j + \sum_{j=1}^{3} u_{ij}B_j = A_i' + B_i',$$

which defines the transformation of a new vector formed from

$$(\mathbf{A} + \mathbf{B}): (A_1 + B_1, A_2 + B_2, A_3 + B_3).$$

It follows from the linearity of the transformation that $(\mathbf{A} + \mathbf{B}) = (\mathbf{B} + \mathbf{A})$.

1.1.6 Geometrical Interpretation of Vectors

(a) A vector $\mathbf{A}:(A_1, A_2, A_3)$ can be represented by a line segment and a direction.

Define $\mathbf{r} \equiv \mathbf{r}_P - \mathbf{r}_Q$, where $\mathbf{r}_P$ and $\mathbf{r}_Q$ are coordinate vectors of points P and Q, respectively (Fig. 1-2).

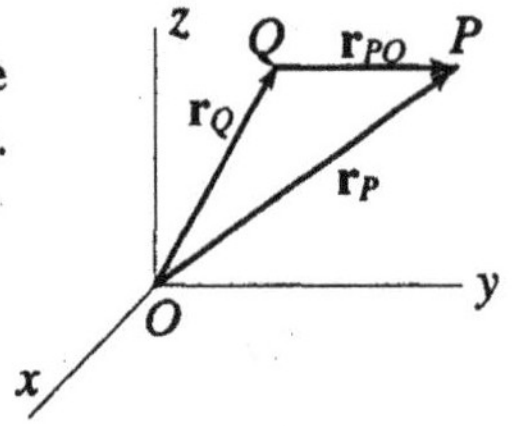

Fig. 1-2

Define

$$A_1 \equiv x_P - x_Q,$$
$$A_2 \equiv y_P - y_Q,$$
$$A_3 \equiv z_P - z_Q,$$

so that $\mathbf{r}_{PQ} = \mathbf{A}:(A_1, A_2, A_3)$, where the set (A_1, A_2, A_3) transforms as a vector. Hence, if a quantity is a vector (A_1, A_2, A_3), it can be represented by a directed line segment.

(b) Converse: if a quantity can be represented by a directed line segment, then it is a vector.

1.1.7 Scalar Product

The scalar product is defined as: $\mathbf{A}\cdot\mathbf{B} = \sum_{i=1}^{3} A_i B_i$.

Theorem: The scalar product of 2 vectors is a scalar (i.e. the same in a rotated frame).

Proof: Under a rotation $A_i \to A_i'$, $B_i \to B_i'$, and $\sum_{i=1}^{3} A_i B_i \to \sum_{i=1}^{3} A_i' B_i'$.

$$\sum_{i=1}^{3} A_i' B_i' = \sum_{ijk=1}^{3} u_{ij} A_j u_{ik} B_k = \sum_{jk=1}^{3} A_j B_k \sum_{i=1}^{3} u_{ij} u_{ik} = \sum_{jk=1}^{3} A_j B_k \delta_{jk} = \sum_{k=1}^{3} A_k B_k .$$

1.1.8 Example of Scalar Product

In Mechanics: $\mathbf{F} = m\mathbf{a} = m \dfrac{d\mathbf{v}}{dt}$.

$$\mathbf{F}\cdot\mathbf{v} = m \frac{d\mathbf{v}}{dt}\cdot\mathbf{v} = m \sum_{i=1}^{3} \upsilon_i \frac{d\upsilon_i}{dt} = m \sum_{i=1}^{3} \frac{d}{dt}\left(\frac{\upsilon_i^2}{2}\right) = m \frac{d}{dt} \sum_{i=1}^{3} \frac{\upsilon_i^2}{2}$$

$$= m \frac{d}{dt} \frac{\upsilon^2}{2} = \frac{d}{dt}\left(\frac{m\upsilon^2}{2}\right).$$

$$\int_1^2 \mathbf{F}\cdot\mathbf{v}\,dt = \int_1^2 \mathbf{F}\cdot d\mathbf{s} = W_{12} = \int_1^2 d\left(\frac{m\upsilon^2}{2}\right) = \mathrm{KE}_2 - \mathrm{KE}_1 .$$

and the rate of doing work equals the rate of change of kinetic energy.

1.1.9 Definition of a Scalar

A scalar is defined to be a quantity which does not change under a rotation of coordinates.

Let $\mathbf{B} = \mathbf{A}$, so that $\mathbf{B}\cdot\mathbf{A} = \mathbf{A}\cdot\mathbf{A} = \sum_{i=1}^{3} A_i^2 = A^2$, where A is the magnitude of $\mathbf{A}$.

Example: If $\mathbf{A} = \mathbf{r}$, $r = \sqrt{x^2 + y^2 + z^2}$, the magnitude of the position vector $\mathbf{r}$.

Since $\mathbf{A}\cdot\mathbf{B}$ does not change under a rotation, choose axes as shown in Fig. 1-3.

Then $\mathbf{A}:(A, 0, 0)$, $\mathbf{B}:(B\cos\theta, B\sin\theta, 0)$, and $\mathbf{A}\cdot\mathbf{B} = AB\cos\theta$, where $\theta = (\mathbf{A}, \mathbf{B})$, the angle between $\mathbf{A}$ and $\mathbf{B}$.

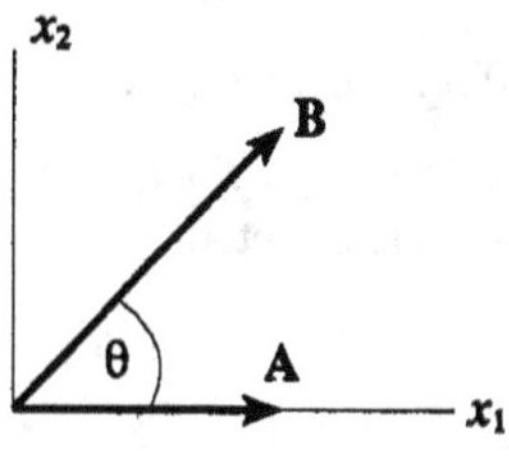

Fig. 1-3

Since $\mathbf{A}\cdot\mathbf{B} = AB\cos\theta$ in this coordinate system, and since $\mathbf{A}\cdot\mathbf{B}$ is invariant, $\mathbf{A}\cdot\mathbf{B} = AB\cos\theta$ in all coordinate systems.

If $\mathbf{A}\cdot\mathbf{B} = 0$ and both $\mathbf{A}$ and $\mathbf{B} \neq 0$, then $\theta = 90°$ and $\mathbf{A}$ is perpendicular to $\mathbf{B}$.

(Scalar Product) Theorem Converse: If the quantities (A_1, A_2, A_3) satisfy $\sum_{i=1}^{3} A_i B_i$ = a scalar for any arbitrary vector **B**, then **A**:(A_1, A_2, A_3) is a vector.

Proof: Since **B** is a vector, $B_i' = \sum_{j=1}^{3} u_{ij} B_j$.

$$\sum_{i=1}^{3} A_i B_i = \sum_{i=1}^{3} A_i' B_i' \text{ (by hypothesis)}$$

$$= \sum_{i=1}^{3} A_i' \sum_{j=1}^{3} u_{ij} B_j = \sum_{ij=1}^{3} u_{ij} A_i' B_j = \sum_{ij=1}^{3} u_{ji} A_j' B_i$$

by interchanging j and i (i.e. employing the dummy index rule). Hence, $\sum_{i=1}^{3}\left[A_i - \sum_{j=1}^{3} u_{ji} A_j'\right] B_i = 0$. But since **B** is an arbitrary vector, $A_i = \sum_{j=1}^{3} u_{ji} A_j'$, which is the law for the inverse transformation of the components of a vector.

1.1.10 Unit Vectors

In the Σ system, the unit vectors $\boldsymbol{i}_o, \boldsymbol{j}_o, \boldsymbol{k}_o$ have the following properties:

$$\boldsymbol{i}_o{:}(1,0,0) \qquad \boldsymbol{j}_o{:}(0,1,0) \qquad \boldsymbol{k}_o{:}(0,0,1)$$

$$\boldsymbol{i}_o \cdot \boldsymbol{i}_o = \boldsymbol{j}_o \cdot \boldsymbol{j}_o = \boldsymbol{k}_o \cdot \boldsymbol{k}_o = 1. \qquad \boldsymbol{i}_o \cdot \boldsymbol{j}_o = \boldsymbol{j}_o \cdot \boldsymbol{k}_o = \boldsymbol{k}_o \cdot \boldsymbol{i}_o = 0.$$

If we now rotate coordinates to the Σ' system (Fig. 1-4), as a result of which vector components transform as

$$x_i' = \sum_{j=1}^{3} u_{ij} x_j,$$

we can find the components of $\boldsymbol{i}_o', \boldsymbol{j}_o', \boldsymbol{k}_o'$ in the Σ system. Similarly, from the inverse transformation

$$x_i = \sum_{j=1}^{3} u_{ji} x_j',$$

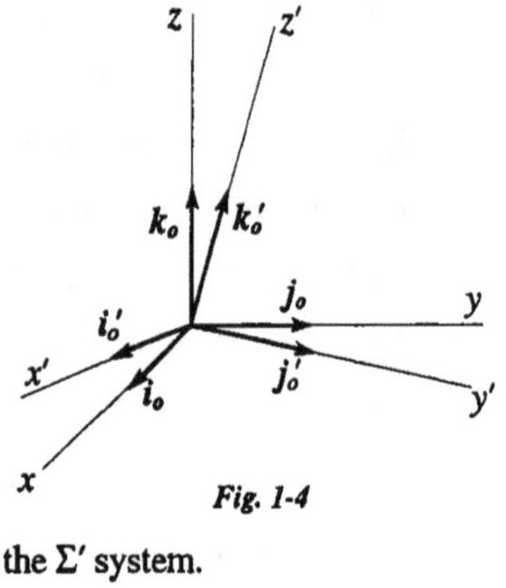

Fig. 1-4

we can find the components of $\boldsymbol{i}_o, \boldsymbol{j}_o, \boldsymbol{k}_o$ in the Σ' system.

In the Σ' system, $\boldsymbol{i}_o'$:(1,0,0), $\boldsymbol{j}_o'$:(0,1,0), $\boldsymbol{k}_o'$:(0,0,1). Then from $x_i = \sum_{j=1}^{3} u_{ji} x_j'$, it follows that

$$\boldsymbol{i}_o'\cdot\boldsymbol{i}_o = u_{11}\cdot 1 + u_{21}\cdot 0 + u_{31}\cdot 0 = u_{11}.$$

$$\boldsymbol{i}_o'\cdot\boldsymbol{j}_o = u_{12}\cdot 1 + u_{22}\cdot 0 + u_{32}\cdot 0 = u_{12}.$$

$$\boldsymbol{i}_o'\cdot\boldsymbol{k}_o = u_{13}\cdot 1 + u_{23}\cdot 0 + u_{33}\cdot 0 = u_{13}.$$

$$\boldsymbol{j}_o'\cdot\boldsymbol{i}_o = u_{11}\cdot 0 + u_{21} 1 + u_{31}\cdot 0 = u_{21}.$$

$$\boldsymbol{j}_o'\cdot\boldsymbol{j}_o = u_{12}\cdot 0 + u_{22}\cdot 1 + u_{32}\cdot 0 = u_{22}.$$

$$\boldsymbol{j}_o'\cdot\boldsymbol{k}_o = u_{13}\cdot 0 + u_{23}\cdot 1 + u_{33}\cdot 0 = u_{23}.$$

$$\boldsymbol{k}_o'\cdot\boldsymbol{i}_o = u_{11}\cdot 0 + u_{21}\cdot 0 + u_{31}\cdot 1 = u_{31}.$$

$$\boldsymbol{k}_o'\cdot\boldsymbol{j}_o = u_{12}\cdot 0 + u_{22}\cdot 0 + u_{32}\cdot 1 = u_{32}.$$

$$\boldsymbol{k}_o'\cdot\boldsymbol{k}_o = u_{13}\cdot 0 + u_{23}\cdot 0 + u_{33}\cdot 1 = u_{33}.$$

Hence, $\boldsymbol{i}_o' = u_{11}\boldsymbol{i}_o + u_{12}\boldsymbol{j}_o + u_{13}\boldsymbol{k}_o.$

$$\boldsymbol{j}_o' = u_{21}\boldsymbol{i}_o + u_{22}\boldsymbol{j}_o + u_{23}\boldsymbol{k}_o.$$

$$\boldsymbol{k}_o' = u_{31}\boldsymbol{i}_o + u_{32}\boldsymbol{j}_o + u_{33}\boldsymbol{k}_o.$$

Using the transformation $x_i' = \sum_{j=1}^{3} u_{ij} x_j$, we get

$$\boldsymbol{i}_o\cdot\boldsymbol{i}_o' = u_{11}\cdot 1 + u_{12}\cdot 0 + u_{13}\cdot 0 = u_{11}.$$

$$\boldsymbol{i}_o\cdot\boldsymbol{j}_o' = u_{21}\cdot 1 + u_{22}\cdot 0 + u_{23}\cdot 0 = u_{21}.$$

$$\boldsymbol{i}_o\cdot\boldsymbol{k}_o' = u_{31}\cdot 1 + u_{32}\cdot 0 + u_{33}\cdot 0 = u_{31}.$$

$$\boldsymbol{j}_o\cdot\boldsymbol{i}_o' = u_{11}\cdot 0 + u_{12} 1 + u_{13}\cdot 0 = u_{12}.$$

$$\boldsymbol{j}_o\cdot\boldsymbol{j}_o' = u_{21}\cdot 0 + u_{22}\cdot 1 + u_{23}\cdot 0 = u_{22}.$$

$$\boldsymbol{j}_o\cdot\boldsymbol{k}_o' = u_{31}\cdot 0 + u_{32}\cdot 1 + u_{33}\cdot 0 = u_{32}.$$

$$\boldsymbol{k}_o\cdot\boldsymbol{i}_o' = u_{11}\cdot 0 + u_{12}\cdot 0 + u_{13}\cdot 1 = u_{13}.$$

$$\boldsymbol{k}_o\cdot\boldsymbol{j}_o' = u_{21}\cdot 0 + u_{22}\cdot 0 + u_{23}\cdot 1 = u_{23}.$$

$$\boldsymbol{k}_o\cdot\boldsymbol{k}_o' = u_{31}\cdot 0 + u_{32}\cdot 0 + u_{33}\cdot 1 = u_{33}.$$

Hence,
$$\begin{aligned} \boldsymbol{i}_o &= u_{11}\boldsymbol{i}_o' + u_{21}\boldsymbol{j}_o' + u_{31}\boldsymbol{k}_o'. \\ \boldsymbol{j}_o &= u_{12}\boldsymbol{i}_o' + u_{22}\boldsymbol{j}_o' + u_{32}\boldsymbol{k}_o'. \\ \boldsymbol{k}_o &= u_{13}\boldsymbol{i}_o' + u_{23}\boldsymbol{j}_o' + u_{33}\boldsymbol{k}_o'. \end{aligned}$$

1.1.11 Vector Product

If **A** and **B** are vectors, we define a quantity $(\mathbf{A}\times\mathbf{B}):(\varepsilon_1, \varepsilon_2, \varepsilon_3)$, where

$$\varepsilon_1 = A_2B_3 - A_3B_2. \quad \varepsilon_2 = A_3B_1 - A_1B_3. \quad \varepsilon_3 = A_1B_2 - A_2B_1.$$

Problem: (a) Show that **A×B** is a vector, and (b) Find the magnitude and direction of (**A×B**).

Solution:

(a) To solve this, we need merely show that the three new quantities $\varepsilon_1', \varepsilon_2', \varepsilon_3'$ are related to the old quantities $\varepsilon_1, \varepsilon_2, \varepsilon_3$ by the rule for the transformation of vector components.

Take any arbitrary vector **C** and construct

$$(\mathbf{A}\times\mathbf{B})\cdot\mathbf{C} = \begin{vmatrix} C_i & C_2 & C_3 \\ A_1 & A_2 & A_3 \\ B_1 & B_2 & B_3 \end{vmatrix} = \sum_{i=1}^{3} C_i(\mathbf{A}\times\mathbf{B})_i .$$

If we can show the determinant to be a scalar, then **A×B** is a vector. In other words, we have to show that the determinant does not change under a rotation. The determinant, however, is merely the volume of a parallelopiped formed on the vectors **A, B** and **C**, and this volume does not change under the rotation.

(b) Choosing the system of coordinates shown in Fig. 1-5:

$$\mathbf{A}:(A,0,0). \quad \mathbf{B}:(B\cos\theta, B\sin\theta, 0).$$

$$\mathbf{C}:(0,0,AB\sin\theta).$$

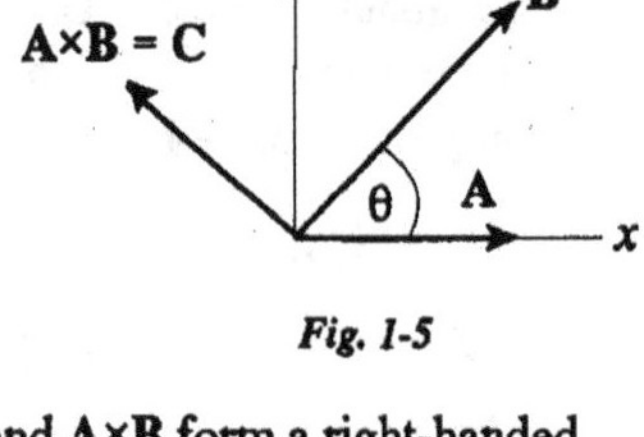

Fig. 1-5

Hence $|(\mathbf{A}\times\mathbf{B})| = |\mathbf{C}| = AB\sin\theta$ and **A, B** and **A×B** form a right-handed system.

1.2 Examples and Applications

1.2.1 Central Forces (Refer to Fig. 1-6)

By Newton's law, $\mathbf{F} = m\ddot{\mathbf{r}}$, where $\frac{d^2\mathbf{r}}{dt^2} = \ddot{\mathbf{r}}$.

When **F** is parallel to **r**, we have the case of central forces (e.g., electrons in an atom, planets about the sun, etc.).

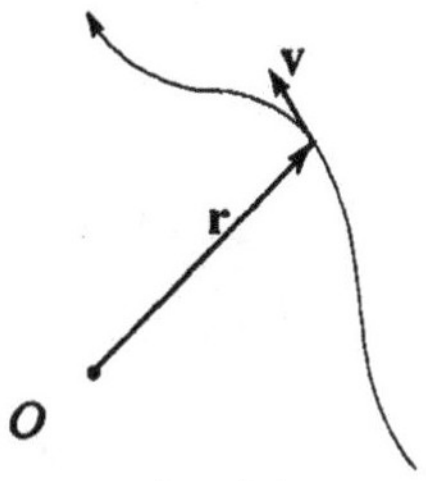

Fig. 1-6

We now show that in the case of central forces, the momentum vector $\mathbf{p} = m\ddot{\mathbf{r}} = \mathbf{v}$ always lies in a plane.

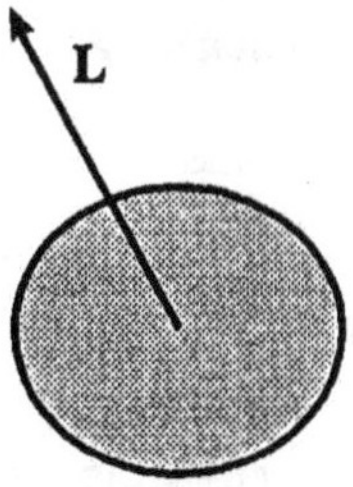

Fig. 1-7

Define a new vector, $\mathbf{L} = \mathbf{r}\times\mathbf{p}$, the angular momentum vector. Then $\dot{\mathbf{L}} = \dot{\mathbf{r}}\times\mathbf{p} + \mathbf{r}\times\dot{\mathbf{p}}$; but $\dot{\mathbf{r}}\times\mathbf{p} = 0$ and $\mathbf{r}\times\dot{\mathbf{p}} = \mathbf{r}\times(m\ddot{\mathbf{r}}) = \mathbf{r}\times\mathbf{F} = 0$ for **F** parallel to **r**; hence $\dot{\mathbf{L}} = 0$ or **L** is a constant of the motion. But **L** is perpendicular to **p** by definition, so that **p** must lie in a plane about **L** (Fig. 1-7).

Note: Given a problem wherein a particle is under the influence of central forces, it can always be reduced from a three-dimensional one to a two-dimensional one, i.e. its motion lies in a plane.

1.2.2 Infinitesimal Rotation (Refer to Fig. 1-8)

Consider a rotation about the z-axis:

At time $t = 0$, $\mathbf{r} = \mathbf{r}(0)$; at time $t = \delta t$, $\mathbf{r} = \mathbf{r}(\delta t)$, where $\mathbf{r}(\delta t) - \mathbf{r}(0)$ is generated by a rotation of amount $\delta\varphi$ about the z-axis. But this is the same as a rotation of coordinate system by an amount $\delta\varphi$ in the opposite direction (Fig. 1-9).

$$x(\delta t) = x(0)\cos\delta\varphi - y(0)\sin\delta\varphi.$$

$$y(\delta t) = y(0)\cos\delta\varphi + x(0)\sin\delta\varphi.$$

$$z(\delta t) = z(0).$$

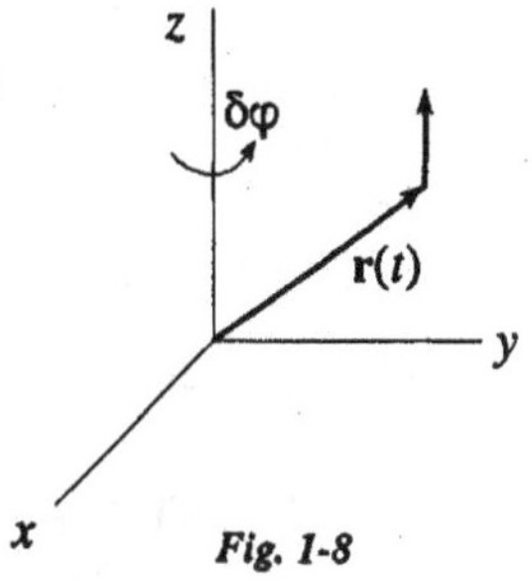

Fig. 1-8

For infinitesimal rotations, $\sin\delta\varphi \approx \delta\varphi$; $\cos\delta\varphi \approx 1$. Hence,

$$x(\delta t) = x(0) - y(0)(\delta\varphi) + O(\delta\varphi)^2.$$

$$y(\delta t) = y(0) + x(0)(\delta\varphi) + O(\delta\varphi)^2.$$

$$z(\delta t) = z(0).$$

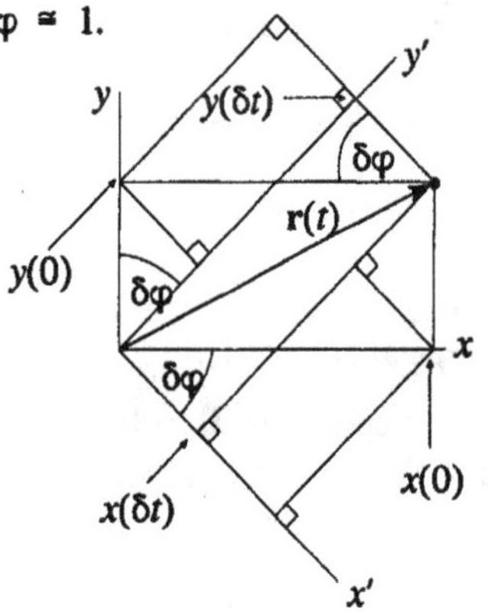

Fig. 1-9

$O(\delta\varphi)^2$ denotes terms "of the order of" $(\delta\varphi)^2$.

Let $\delta t \to 0$ and expand about $x(0)$ and $y(0)$.

$$x(\delta t) = x(0) + \dot{x}(0)\,\delta t + \cdots$$

$$y(\delta t) = y(0) + \dot{y}(0)\,\delta t + \cdots$$

Comparing the above expressions with those derived from Fig. 1-9, we have:

$$\dot{x} = -y\left(\frac{\delta\varphi}{\delta t}\right). \quad \dot{y} = x\left(\frac{\delta\varphi}{\delta t}\right). \quad \dot{z} = 0.$$

We define a vector $\overline{\omega}$ having direction parallel to the axis of rotation and having magnitude $|\overline{\omega}| = \delta\varphi/\delta t$. In this system, $\overline{\omega} = (0, 0, \delta\varphi/\delta t)$.

Note that $\dot{\mathbf{r}} = \overline{\omega}\times\mathbf{r}$ yields:

$$\dot{x} = \dot{\mathbf{r}}_x = -\omega_z y = -y(\delta\varphi/\delta t).$$

$$\dot{y} = \dot{\mathbf{r}}_y = \omega_z x = x(\delta\varphi/\delta t).$$

$$\dot{z} = 0.$$

If now $\overline{\omega}$ is constant in time, what is the acceleration of the particle?

$$\mathbf{a} = \ddot{\mathbf{r}} = \frac{d}{dt}(\overline{\omega}\times\mathbf{r}) = \overline{\omega}\times\dot{\mathbf{r}} + \dot{\overline{\omega}}\times\mathbf{r} = \overline{\omega}\times(\overline{\omega}\times\mathbf{r}) = (\overline{\omega}\cdot\mathbf{r})\overline{\omega} - \omega^2\mathbf{r}.$$

(The last step involves a vector identity which is proven later — see Problems.)

If we now hypothesize that, in addition, $\mathbf{r}\cdot\overline{\omega} = 0$, (rotation perpendicular to the plane of motion), then $\mathbf{a} = -\omega^2\mathbf{r}$, and is opposite in direction to $\mathbf{r}$.

1.3 Theory of Curves in Space (Refer to Fig. 1-10)

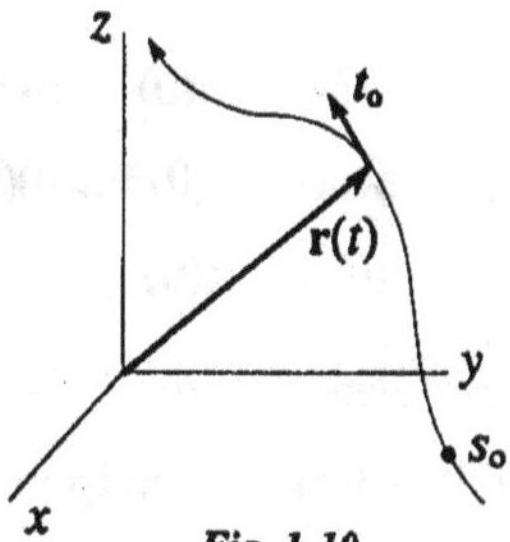

Fig. 1-10

Define a unit tangent vector $t_o = \mathbf{v}/\upsilon$. Then

$$\mathbf{a} = \dot{\mathbf{v}} = \frac{d}{dt}(\upsilon t_o) = \dot{\upsilon} t_o + \upsilon \dot{t}_o .$$

But $t_o \cdot t_o = 1$. Hence, $t_o \cdot \dot{t}_o = 0$ and $\dot{t}_o$ is perpendicular to t_o.

The acceleration, therefore, has 2 components, one along the tangent vector t_o and one perpendicular to t_o.

Define a unit normal vector n_o

$$n_o = \frac{\dot{t}_o}{|\dot{t}_o|} .$$

Let s be defined as the length of the curve from some fiducial point s_o, so that $\upsilon = \dot{s}$. Then, since $t_o = f(t)$,

$$\dot{t}_o = \frac{dt_o}{ds}\frac{ds}{dt} = \upsilon \frac{dt_o}{ds}, \qquad |\dot{t}_o| = \upsilon \left|\frac{dt_o}{ds}\right| ,$$

so that

$$n_o = \frac{\dot{t}_o}{\upsilon \left|\frac{dt_o}{ds}\right|} .$$

and

$$\mathbf{a} = \dot{\upsilon} t_o + \upsilon \dot{t}_o = \dot{\upsilon} t_o + \upsilon^2 \left|\frac{dt_o}{ds}\right| n_o .$$

Define Radius of Curvature

$$R = \frac{1}{\left|\frac{dt_o}{ds}\right|} ,$$

so that

$$\mathbf{a} = \frac{\upsilon^2}{R} n_o + \dot{\upsilon} t_o .$$

Physical Meaning of Radius of Curvature R (Figs. 1-11, 1-12, 1-13)

Since t_o and t_o+dt_o are unit vectors, $|dt_o| = |t_o|d\theta = d\theta$. If ds is approximated by the arc of a circle, what will be the radius of curvature of the arc, R_{arc}?

$$\frac{1}{R_{arc}} = \frac{d\theta}{ds} = \left|\frac{dt_o}{ds}\right| = \frac{1}{R}.$$

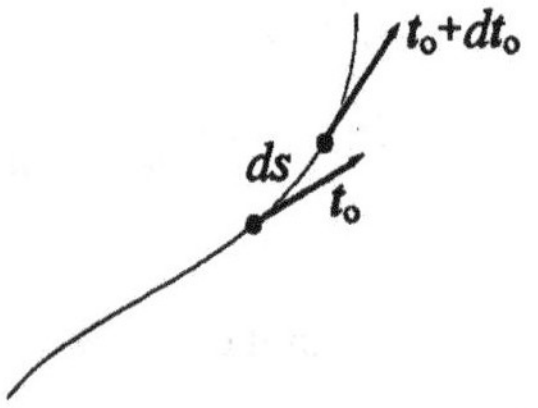

Fig. 1-11

Hence the Radius of Curvature is the value of the equivalent radius of a circle having an arc length ds.

Define the binomial unit vector $b_o \equiv t_o \times n_o$. If a particle moves along a path such that it always lies in a plane, b_o will never change direction, being always perpendicular to that plane. If the curve of motion should leave the plane, b_o will change direction, and we have torsion.

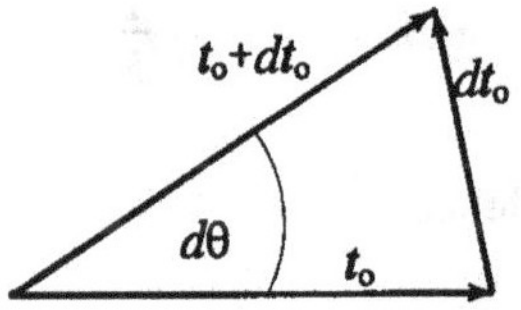

Fig. 1-12

Define torsion $T \equiv \left|\frac{db_o}{ds}\right|$. As the particle moves a distance ds along the curve, b_o turns through an angle $d\theta'$, where $d\theta' = |db_o|$. Hence torsion is the rate of change of the $t_o \times n_o$ plane with respect to distance along the curve.

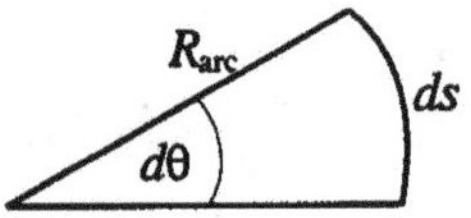

Fig. 1-13

Define

Curvature (1st Curvature)	$C_1 \equiv \frac{1}{R} = \left\lvert\frac{dt_o}{ds}\right\rvert.$
Torsion (2nd Curvature)	$C_2 \equiv \left\lvert\frac{db_o}{ds}\right\rvert.$
Total Curvature	$C^2 \equiv C_1^2 + C_2^2.$

From the definition of the binomial unit vector b_o, $\dot{b}_o = \dot{t}_o \times n_o + t_o \times \dot{n}_o$. But $\dot{t}_o$ is parallel to n_o; hence $\dot{b}_o = t_o \times \dot{n}_o$, and is perpendicular to t_o, and since $b_o \times b_o = 0$, $\dot{b}_o$ is perpendicular to b_o. Therefore, $\dot{b}_o$ is perpendicular to t_o and to b_o, and lies in the n_o direction.

Problem: Show that $C = \left|\frac{d\boldsymbol{n}_o}{ds}\right|$.

Solution:

$$\boldsymbol{n}_o = \boldsymbol{b}_o \times \boldsymbol{t}_o.$$

$$\frac{d\boldsymbol{n}_o}{ds} = \frac{d\boldsymbol{b}_o}{ds}\times\boldsymbol{t}_o + \boldsymbol{b}_o\times\frac{d\boldsymbol{t}_o}{ds}.$$

Since $d\boldsymbol{b}_o$ is in the $\boldsymbol{n}_o$ direction, $\frac{d\boldsymbol{b}_o}{ds}\times\boldsymbol{t}_o$ is parallel to $\boldsymbol{b}_o$ with magnitude $\left|\frac{d\boldsymbol{b}_o}{ds}\right| = C_2$, and since $d\boldsymbol{t}_o$ is in the $\boldsymbol{n}_o$ direction, $\boldsymbol{b}_o\times\frac{d\boldsymbol{t}_o}{ds}$ is parallel to $\boldsymbol{t}_o$ with magnitude $\left|\frac{d\boldsymbol{t}_o}{ds}\right| = \frac{1}{R} = C_1$.

Hence, $\frac{d\boldsymbol{n}_o}{ds}$ is a vector lying in the $\boldsymbol{b}_o$, $\boldsymbol{t}_o$ plane, and

$$\left|\frac{d\boldsymbol{n}_o}{ds}\right|^2 = \left|\frac{d\boldsymbol{b}_o}{ds}\right|^2 + \left|\frac{d\boldsymbol{t}_o}{ds}\right|^2.$$

or,

$$C^2 = C_1^2 + C_2^2.$$

Chapter 2

Tensor Analysis

2.1 *n*th Rank Tensor

A tensor of rank n, $T_{i_1 i_2 \ldots i_n}$ is defined to be the totality of 3^n quantities which, under a rotation of axes, $x_i' = \sum_{j=1}^{3} u_{ij} x_j$, transforms as

$$T'_{i_1 i_2 \ldots i_n} = \sum_{j=1}^{3} u_{i_1 j_1} u_{i_2 j_2} \ldots u_{i_n j_n} T_{j_1 j_2 \ldots j_n} .$$

(a) For $n = 1$, we have a 1st-rank tensor, with $3^1 = 3$ components, T_i $(i = 1, 2, 3)$.

$$T_i' = \sum_{j=1}^{3} u_{ij} T_j, \text{ identical with a vector.}$$

(b) For $n = 0$, $S' = S$, and S is a scalar.

(c) For $n = 2$, we have a 2nd-rank tensor, with $3^2 = 9$ components, T_{ij} $(i, j = 1, 2, 3)$.

$$T_{ij}' = \sum_{k,l=1}^{3} u_{ik} u_{jl} T_{kl} .$$

Example: Consider $x_i\, x_j$. We know this to transform as:

$$(x_i\, x_j)' = x_i' x_j' = \sum_{kl=1}^{3} u_{ik} x_k\, u_{jl} x_l = \sum_{kl=1}^{3} u_{ik} u_{jl}\, (x_k\, x_l) .$$

Hence, $x_i\, x_j$ is a 2nd-rank tensor.

2.2 2nd-Rank Isotropic (Invariant) Tensor

Define $\delta_{ij} \equiv \begin{cases} 1 & \text{for } i = j \\ 0 & \text{for } i \neq j \end{cases}$.

$$\sum_{kl=1}^{3} u_{ik} u_{jl} \, \delta_{kl} = \sum_{k=1}^{3} u_{ik} u_{jk} = \delta_{ij},$$

which follows from the properties of the rotation transformation.

Hence, not only is δ_{ij} a good 2nd-rank tensor (by virtue of its transformation properties), but it is an *invariant* tensor, i.e. its components are the same in any coordinate system.

2.3 Contraction

Theorem: If $T_{i_1 i_2 \dots i_n}$ is an nth rank tensor, then

$$\sum_{i_1 i_2=1}^{3} \delta_{i_1 i_2} T_{i_1 i_2 i_3 \dots i_n} = O_{i_3 i_4 \dots i_n}$$

yields a tensor of rank $(n-2)$.

Proof: Define $O'_{i_3 i_4 \dots i_n} \equiv \sum_{i_1 i_2=1}^{3} \delta'_{i_1 i_2} T'_{i_1 i_2 \dots i_n}$

$$= \sum_{i_1 i_2=1}^{3} \delta_{i_1 i_2} \sum_{j_i=1}^{3} u_{i_1 j_1} u_{i_2 j_2} u_{i_3 j_3} \dots u_{i_n j_n} T_{j_1 j_2 j_3 \dots j_n}$$

$$= \sum_{i_1=1}^{3} \sum_{j_i=1}^{3} u_{i_1 j_1} u_{i_1 j_2} u_{i_3 j_3} \dots u_{i_n j_n} T_{j_1 j_2 j_3 \dots j_n}$$

$$= \sum_{j_i=1}^{3} \delta_{j_1 j_2} u_{i_3 j_3} \dots u_{i_n j_n} T_{j_1 j_2 j_3 \dots j_n} = \sum_{j_i=1}^{3} u_{i_3 j_3} \dots u_{i_n j_n} \delta_{j_1 j_2} T_{j_1 j_2 \dots j_n}$$

$$= \sum_{j_i=1}^{3} u_{i_3 j_3} u_{i_4 j_4} \dots u_{i_n j_n} O_{j_3 j_4 \dots j_n}.$$

Hence, $O_{i_3 i_4 \dots i_n}$ transforms like a tensor of rank $(n-2)$.

2.4 Outer Product Theorem

Theorem: If $S_{i_1 i_2 \dots i_n}$ is a tensor of rank n and $T_{j_1 j_2 \dots j_m}$ is a tensor of rank m, then $S_{i_1 i_2 \dots i_n} T_{j_1 j_2 \dots j_m}$ is a tensor of rank $(m + n)$ with 3^{m+n} components.

Proof: Define $(S_{i_1 i_2 \dots i_n} T_{j_1 j_2 \dots j_m})' \equiv S'_{i_1 i_2 \dots i_n} T'_{j_1 j_2 \dots j_m}$

$$= \sum_{k_i=1}^{3} u_{i_1 k_1} u_{i_2 k_2} \dots u_{i_n k_n} S_{k_1 k_2 \dots k_n} \sum_{l_i=1}^{3} u_{j_1 l_1} u_{j_2 l_2} \dots u_{j_m l_m} T_{l_1 l_2 \dots l_m}$$

$$= \sum_{k_i l_i=1}^{3} \left[u_{i_1 k_1} u_{i_2 k_2} \dots u_{i_n k_n} u_{j_1 l_1} u_{j_2 l_2} \dots u_{j_m l_m} \right] \left[S_{k_1 k_2 \dots k_n} T_{l_1 l_2 \dots l_m} \right]$$

Hence, $S_{i_1 i_2 \dots i_n} T_{j_1 j_2 \dots j_m}$ transforms like a tensor of rank $(m + n)$.

By virtue of the above proof: If A_i and B_i are components of a vector, then $A_i B_j$ is a tensor of the 2nd rank and can be written as $A_i B_j = T_{ij}$. It is not necessarily true, however, that any T_{ij} can be written as the outer product of two vectors.

Consider, for example, the expression involving two good 1st-rank tensors A_i and B_i:

$$\sum_{ij=1}^{3} \delta_{ij} A_i B_j = \sum_{i=1}^{3} A_i B_i ,$$

a scalar (tensor of rank 2 - 2 = 0).

If A_i, B_i and C_i are good 1st-rank tensors, then $A_{i_1} B_{i_2} C_{i_3} = T_{i_1 i_2 i_3}$, a 3rd-rank tensor. We now perform a contraction on i_2 and i_3 and on i_1 and i_2 and compare.

$$\sum_{i_2=1}^{3} T_{i_1 i_2 i_2} = \sum_{i_2=1}^{3} A_{i_1} B_{i_2} C_{i_2} = A_{i_1} (\mathbf{B} \cdot \mathbf{C}) .$$

$$\sum_{i_1=1}^{3} T_{i_1 i_1 i_3} = \sum_{i_1=1}^{3} A_{i_1} B_{i_1} C_{i_3} = C_{i_3} (\mathbf{A} \cdot \mathbf{B})$$

and we note that the two resulting expressions are quite different.

2.5 3rd-Rank Isotropic (Invariant) Tensor

Define

$$\varepsilon_{ijk} \equiv \begin{cases} +1 & \text{for } ijk \text{ an even permutation of } 1, 2, 3 \\ -1 & \text{for } ijk \text{ an odd permutation of } 1, 2, 3 \\ 0 & \text{for all other cases} \end{cases}$$

Theorem: ε_{ijk} is an invariant tensor of the 3rd rank.

Proof: Form

$$\sum_{l_i=1}^{3} u_{i_1 l_1} u_{i_2 l_2} u_{i_3 l_3} \varepsilon_{l_1 l_2 l_3} = \begin{vmatrix} u_{i_1 1} & u_{i_1 2} & u_{i_1 3} \\ u_{i_2 1} & u_{i_2 2} & u_{i_2 3} \\ u_{i_3 1} & u_{i_3 2} & u_{i_3 3} \end{vmatrix}$$

$$= i' \cdot (j' \times k') \times \begin{cases} +1 & \text{for even permutation of } i_1\, i_2\, i_3 \\ -1 & \text{for odd permutation of } i_1\, i_2\, i_3 \\ 0 & \text{in all other cases, i.e. 2 identical indices} \end{cases}$$

$$= \varepsilon_{i_1 i_2 i_3},$$

thus showing both transformation property and invariance of ε_{ijk}.

2.6 Examples and Applications

Vector (Cross) Product: If A_i and B_i are 1st-rank tensors, then $A_i B_j$ is a 2nd-rank tensor and $\varepsilon_{ijk} A_l B_m$ is a 5th-rank tensor $\equiv T_{ijklm}$.

Consider the 1st-rank tensor:

$$\sum_{jk=1}^{3} \varepsilon_{ijk} A_j B_k = O_i .$$

From the properties of ε_{ijk}, $O_1 = A_2 B_3 - A_3 B_2$, which is identical to the expression for the components of a cross product. Hence, $O_i = (\mathbf{A} \times \mathbf{B})_i$.

The vector (cross) product can therefore be written as

$$(\mathbf{A} \times \mathbf{B})_i \equiv \sum_{jk=1}^{3} \varepsilon_{ijk} A_j B_k .$$

Scalar Product:

$$\mathbf{A}\cdot\mathbf{B} = \sum_{i=1}^{3} A_i B_i = \sum_{ij=1}^{3} \delta_{ij} A_i B_j .$$

0th Rank Tensor:

$$S = \sum_{ijk=1}^{3} T_{ijk}\,\varepsilon_{ijk},$$

a 0th-rank tensor with 1 component,

$$= T_{123} + T_{231} + T_{312} - T_{213} - T_{132} - T_{321} .$$

2.7 Geometrical Representation of Tensors

Problem: Given a tensor of 2nd rank. Construct a vector from it, i.e. represent a tensor geometrically.

Solution:

(a) Consider first an anti-symmetric tensor of 2nd rank, T_{ij}, having $\frac{(n^2 - n)}{2}$ components (in 3 dimensions, only 3 independent components).

First form a 4th-rank tensor and then contract:

$$\sum_{ijk=1}^{3} T_{ij}\,\delta_{kl}\,\varepsilon_{ijk} = \sum_{ij=1}^{3} T_{ij}\,\varepsilon_{ijl} = S_l ,$$

thereby showing that there exists a direct correspondence between vectors and 2nd-rank anti-symmetric tensors.

Problem: Given a vector A_i. Construct a 2nd-rank tensor.

Solution: $\sum_{i=1}^{3} \varepsilon_{ijk} A_i = S_{jk}\,(= -S_{kj})$.

Hence, from a vector one can construct an anti-symmetric tensor of the 2nd rank. Now we wish to show the symmetry (anti-symmetry) property to be invariant, i.e. if $S_{ij} = -(+)S_{ji}$, $S'_{ij} = -(+)S'_{ji}$.

$$\left\{\begin{aligned} S'_{ij} &= \sum_{kl=1}^{3} u_{ik}u_{jl}S_{kl} \\ S'_{ji} &= \sum_{kl=1}^{3} u_{jl}u_{ik}S_{lk} \end{aligned}\right\} \qquad \begin{aligned} S_{kl} &= +S_{lk} \Rightarrow S'_{ij} = +S'_{ji}. \\ S_{kl} &= -S_{lk} \Rightarrow S'_{ij} = -S'_{ji}. \end{aligned}$$

$$\begin{aligned} S_{11} &= 0 & S_{12} &= +A_3 & S_{13} &= -A_2 \\ S_{21} &= -A_3 & S_{22} &= 0 & S_{23} &= +A_1 \\ S_{31} &= +A_2 & S_{32} &= -A_1 & S_{33} &= 0 \end{aligned}$$

$$S_{ij} = \begin{pmatrix} 0 & A_3 & -A_2 \\ -A_3 & 0 & A_1 \\ A_2 & -A_1 & 0 \end{pmatrix}.$$

Given T_{ij}: $S_l = \sum_{ij=1}^{3} \varepsilon_{ijl} T_{ij}$.

$$\begin{aligned} S_1 &= T_{23} - T_{32} = 2\,T_{23}. \\ S_2 &= T_{31} - T_{13} = 2\,T_{31}. \\ S_3 &= T_{12} - T_{21} = 2\,T_{12}. \end{aligned}$$

$$S_i = \begin{pmatrix} 2\,T_{23} \\ 2\,T_{31} \\ 2\,T_{12} \end{pmatrix}.$$

(b) Consider a symmetric tensor of 2nd-rank, $T_{ij} = T_{ji}$. Can we represent it geometrically?

T_{ij} will have $n + (n^2 - n)/2$ independent components (6 in 3 dimensions). If we try to form

$$\sum_{ij=1}^{3} \varepsilon_{ijl} T_{ij} = S_l,$$

then $S_l = 0$ [e.g. $S_2 = (\varepsilon_{312}T_{31} + \varepsilon_{132}T_{13}) = \varepsilon_{312}(T_{31} - T_{13}) = 0$], so that we cannot form a vector from a 2nd-rank symmetric tensor.

Consider instead:

$$Ax^2 + By^2 + Cz^2 + Dxy + Eyz + Fzx = 1,$$

a general quadratic surface with 6 constants. Hence, we can represent a symmetric 2nd-rank tensor by a quadratic surface.

Theorem: Any symmetric 2nd-rank tensor T_{ij} can be represented uniquely by a surface

$$\sum_{ij=1}^{3} T_{ij} x_i x_j = \pm 1,$$

where the sign is determined by the sign of the determinant det $|T_{ij}|$. (The sign convention is needed; otherwise, if, for example, $T_{ij} = -\delta_{ij}$, then $\Sigma T_{ij} x_i x_j = -x^2 - y^2 - z^2 = 1$, which does not exist in real space.)

Under a rotation, $T_{ij} \rightarrow T_{ij}$ and $\Sigma T_{ij} x_i x_j \rightarrow \Sigma T_{ij} x_i' x_j'$. The sign, since it is determined by the determinant, itself a scalar, is invariant to a rotation, and the sign convention holds under the transformation.

Proof: We must show that the quantity $\Sigma T_{ij} x_i x_j$ has proper transformation properties. We note that it is a scalar, hence if $= \pm 1$ in one system, it remains $= \pm 1$ in every system. (We have already shown the sign convention to be invariant.) Furthermore, the physical surface itself is invariant to a rotation.

For the case $\sum_{i=1}^{3} x_i^2 = 1$, what is the symmetric tensor corresponding to this surface?

Evidently, $T_{ij} = \delta_{ij}$, $|T_{ij}|$ is $+$, and $T_{ij} = +\delta_{ij}$. For the case $\sum_{i=1}^{3} x_i^2 = -1$, $|T_{ij}|$ is $-$, and

$$T_{ij} = -\delta_{ij}.$$

Hence:

Given a 2nd-rank symmetric tensor, one can find a quadratic surface.

Given a quadratic surface, one can find a symmetric tensor of 2nd rank.

Principal Axes: For every surface $\sum_{ij=1}^{3} T_{ij}x_i x_j = \pm 1$, there exists a rotation such that $T_{ij} \rightarrow \lambda_i \delta_{ij}$, where

$$\sum_{ij=1}^{3} \lambda_i \delta_{ij} x_i' x_j' = \sum_{ij=1}^{3} \lambda_i x_i'^2 = \pm 1,$$

where the λ_i are the eigenvalues of T_{ij}. The new axes are called the principal axes.

Arbitrary 2nd-Rank Tensors: What is the significance of an arbitrary tensor of the 2nd rank, R_{ij}?

Define

$$S_{ij} \equiv 1/2\,(R_{ij} - R_{ji}) \quad \text{so that} \quad S_{ij} = -S_{ji}.$$

$$S_{ij}' \equiv 1/2\,(R_{ij} + R_{ji}) \quad \text{so that} \quad S_{ij} = +S_{ji}.$$

Then $R_{ij} = S_{ij} + S_{ij}'$, the sum of a symmetric and an anti-symmetric 2nd-rank tensor.

Hence: Any arbitrary 2nd-rank tensor can be represented by a vector (anti-symmetric part) and a quadratic surface (symmetric part).

2.8 Moment of Inertia Tensor

For the body ilustrated in Fig. 2-1, angular momentum **L** is defined as:

$$\mathbf{L} \equiv \int \mathbf{r}\times(\rho\mathbf{v}\,d\tau).$$

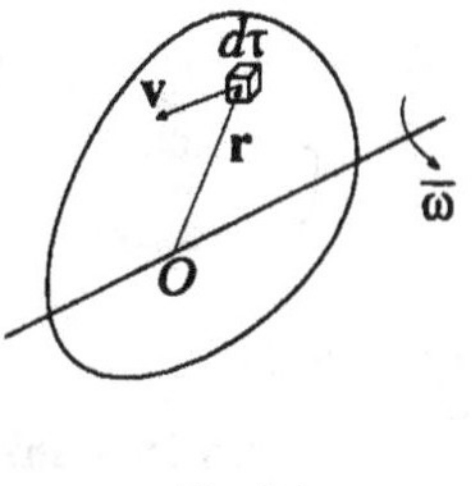

Fig. 2-1

Substituting $\mathbf{v} = \overline{\omega}\times\mathbf{r}$,

$$\mathbf{L} = \int \rho(\mathbf{r}\times\overline{\omega}\times\mathbf{r})\,d\tau = \int \rho\left[r^2\overline{\omega} - \mathbf{r}(\overline{\omega}\cdot\mathbf{r})\right]d\tau.$$

In tensor notation,

$$L_i = \int \rho\left[r^2\omega_i - \omega_j r_j r_i\right]d\tau.$$

Define the symmetric, 2nd rank tensor I_{ij}:

$$I_{ij} \equiv \int \rho\left[r^2\delta_{ij} - r_i r_j\right]d\tau,$$

so that

$$\sum_{j=1}^{3} I_{ij}\omega_j = \int \rho\left[r^2\omega_i - r_i\omega_j r_j\right]d\tau = L_i .$$

Rewriting,

$$L_i = \sum_{j=1}^{3} I_{ij}\omega_j$$

Note that, in general, $\mathbf{L}$ is not in the same direction as $\overline{\omega}$. For principal axes, however, $L_i = \lambda_i\omega_i$. If no external torques act, then $\mathbf{L} = 0$ and $\mathbf{L}$ is fixed in space while $\overline{\omega}$ precesses about it.

Physical Significance of I_{ij} : If we define the moment of inertia of a body about an axis (Fig. 2-2) as

$$I \equiv \int \rho d^2 d\tau,$$

where d is the perpendicular distance to the axis, and if e_i (i =1, 2, 3) be unit vectors defined by their direction cosines, then

$$I = \sum_{ij} e_i I_{ij} e_j,$$

which is a scalar invariant. Hence we can rotate to any desired coordinate system. To verify the result,

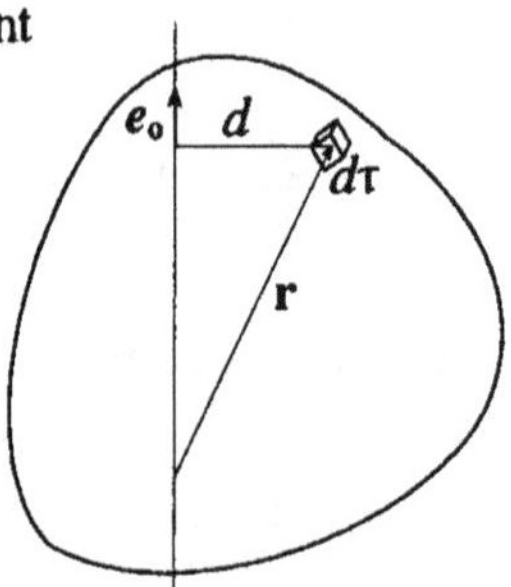

Fig. 2-2

$$\sum_{ij} e_i e_j I_{ij} = \sum_{ij} e_i e_j \int \rho\left[r^2\delta_{ij} - r_i r_j\right]d\tau .$$

Choosing the z-axis along e: (0, 0, 1), we have

$$I = e_3 e_3 I_{33} = \int \rho d\tau\left[r^2 - z^2\right] = \int \rho d^2 d\tau .$$

Chapter 3

Fields

3.1. Tensor Field

3.1.1 Definition of Tensor Field

A tensor field of the nth rank, $T_{i_1 i_2 \ldots i_n}$ is the totality of 3^n functions of space ($i_1 i_2 \ldots i_n = 1,2,3$), which for any given point in space ($x_1 x_2 x_3$) constitutes an nth-rank tensor.

Examples:

$n = 0$ → scalar field, one component, e.g., $\varphi(\mathbf{r})$.
$n = 1$ → vector field, $3^1 = 3$ components, e.g., $\mathbf{E}(\mathbf{r})$.

Theorem: (Fundamental Theorem in Tensor Field Analysis)
If $T_{i_1 i_2 \ldots i_n}$ is a tensor field of the nth rank, then $\frac{\partial}{\partial x_i} T_{i_1 i_2 \ldots i_n}$ defines a tensor field of rank $(n + 1)$.

Proof:

$$\left(\frac{\partial}{\partial x_i} T_{i_1 i_2 \ldots i_n}\right)' = \frac{\partial}{\partial x_i'} T'_{i_1 i_2 \ldots i_n}.$$

Under a rotation of coordinates,

$$x_i' = \sum_{j=1}^{3} u_{ij} x_j \quad \text{or} \quad x_j = \sum_{j=1}^{3} u_{ij} x_i' \quad \rightarrow \quad \frac{\partial x_j}{\partial x_i'} = u_{ij}.$$

$$\frac{\partial}{\partial x_i'} = \sum_{j=1}^{3}\frac{\partial}{\partial x_j}\frac{\partial x_j}{\partial x_i'} = \sum_{j=1}^{3} u_{ij}\frac{\partial}{\partial x_j},$$

i.e. the quantity $\frac{\partial}{\partial x_j}$ transforms like a tensor of the 1st rank.

$$\left(\frac{\partial}{\partial x_i}T_{i_1 i_2 \ldots i_n}\right)' = \sum_{k=1}^{3} u_{ik}\frac{\partial}{\partial x_k}\sum_{j_i=1}^{3} u_{i_1 j_1}u_{i_2 j_2}\ldots u_{i_n j_n}T_{j_1 j_2 \ldots j_n}$$

$$= \sum_{k j_i=1}^{3} u_{ik}u_{i_1 j_1}u_{i_2 j_2}\ldots u_{i_n j_n}\frac{\partial}{\partial x_k}T_{j_1 j_2 \ldots j_n}$$

which is the law for the transformation of a tensor of rank $(n+1)$.

3.1.2 Operators of a Tensor Field

(a) $\nabla\varphi$ = gradient φ → $\nabla\varphi = \frac{\partial\varphi}{\partial x_i}$, a 1st-rank tensor, hence a vector.

(b) $\nabla\cdot\mathbf{A}$ = divergence of $\mathbf{A}$ → $\nabla\cdot\mathbf{A} = \sum_i \frac{\partial A_i}{\partial x_i}$, a scalar.

(c) $\nabla\times\mathbf{A}$ = curl of $\mathbf{A}$ → $(\nabla\times\mathbf{A})_i = \sum_{jk}\varepsilon_{ijk}\frac{\partial A_k}{\partial x_j}$, a 1st-rank tensor, hence a vector.

(d) $\Delta\varphi = \nabla^2\varphi$ = Laplacian of φ → $\nabla^2\varphi = \sum_{i=1}^{3}\frac{\partial^2\varphi}{\partial x_i^2}$, a scalar.

(e) $\Delta\mathbf{A} = \nabla^2\mathbf{A}$ = Laplacian of $\mathbf{A}$ → $\left[\nabla^2\mathbf{A}\right]_j = \sum_{i=1}^{3}\frac{\partial^2 A_j}{\partial x_i^2}$, a 1st-rank tensor, hence a vector.

Vector Identities

(1) div curl = 0 → $\nabla\cdot(\nabla\times\mathbf{A}) = 0$ for any $\mathbf{A}$.

$$\sum_i \frac{\partial}{\partial x_i}\left\{\sum_{ijk}\varepsilon_{ijk}\frac{\partial A_k}{\partial x_j}\right\} = \sum_i \varepsilon_{ijk}\frac{\partial^2 A_k}{\partial x_i\partial x_j} = -\sum_i \varepsilon_{jik}\frac{\partial^2 A_k}{\partial x_j \partial x_i} = 0.$$

(2) curl grad = 0 $\rightarrow$ $\nabla\times(\nabla\varphi) = 0$ for any φ.

$$[\nabla\times(\nabla\varphi)]_i = \sum_{jk}\varepsilon_{ijk}\frac{\partial}{\partial x_j}\frac{\partial}{\partial x_k}\varphi = -\sum_{jk}\varepsilon_{ikj}\frac{\partial}{\partial x_k}\frac{\partial}{\partial x_j}\varphi = 0.$$

(3) $\nabla\times(\nabla\times\mathbf{A}) = \nabla(\nabla\cdot\mathbf{A}) - \nabla^2\mathbf{A}$.

$$[\nabla\times(\nabla\times\mathbf{A})]_i = \sum_{jk}\varepsilon_{ijk}\frac{\partial}{\partial x_j}\left\{\sum_{lm}\varepsilon_{klm}\frac{\partial A_m}{\partial x_l}\right\} = \sum_{jklm}\varepsilon_{ijk}\varepsilon_{lmk}\frac{\partial}{\partial x_j}\frac{\partial A_m}{\partial x_l}$$

$$= \sum_{jlm}\left[\delta_{il}\delta_{jm} - \delta_{im}\delta_{lj}\right]\frac{\partial}{\partial x_j}\frac{\partial A_m}{\partial x_l}$$

$$= \sum_j\frac{\partial}{\partial x_i}\frac{\partial A_j}{\partial x_j} - \sum_l\frac{\partial^2 A_i}{\partial x_l^2} = \left[\nabla(\nabla\cdot\mathbf{A}) - \nabla^2\mathbf{A}\right]_i .$$

(See Problems for the last step.)

Tensor Fields in Electrostatics

Scalar Field (Electrostatic potential φ):
In 3-dimensional space, φ = const. defines lines of constant electrostatic potential, i.e. equipotentials.

For a point charge, $\varphi = e/r$, φ = const. gives equipotential spherical surfaces about the source e (Fig. 3-1).

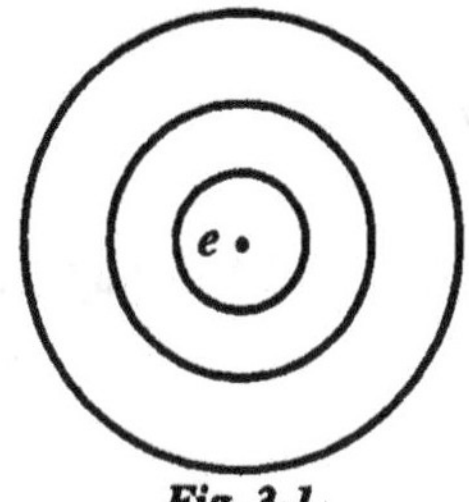

Fig. 3-1

Vector Field (Electrostatic field $\mathbf{E}$):
A line of force is that line such that $\mathbf{E}$ is tangent to the line at that point (Fig. 3-2). This means that $\mathbf{E} = \alpha ds$, or

$$E_{x_1} = \alpha dx_1 .$$
$$E_{x_2} = \alpha dx_2 .$$
$$E_{x_3} = \alpha dx_3 .$$

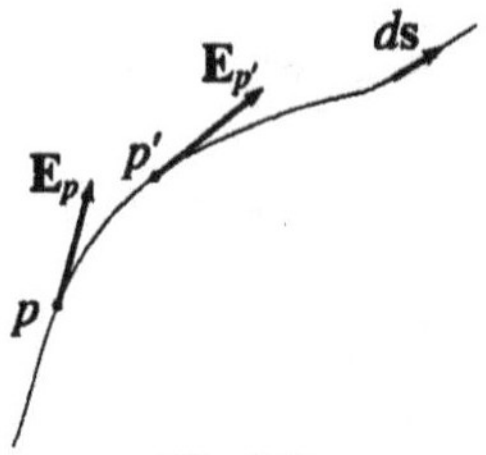

Fig. 3-2

Thus the analytical expression of a line of force is given by

$$\frac{dx_1}{E_1} = \frac{dx_2}{E_2} = \frac{dx_3}{E_3},$$

where ds is a line element:(dx_1, dx_2, dx_3). Lines of force are arbitrarily drawn such that the magnitude of **E** equals the number of lines per unit area perpendicular to **E** (Fig. 3-3).

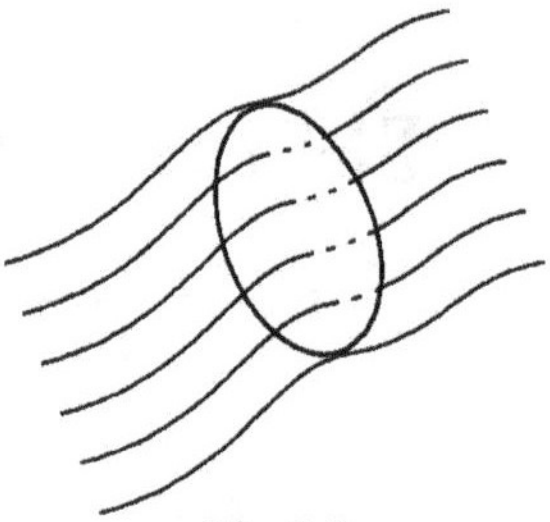

Fig. 3-3

If $\mathbf{E} = -\nabla\varphi$, what is the relationship between the equipotential surfaces and the lines of force at the surface φ = constant?
If $d\mathbf{r}$ is on the equipotential surface, then

$$d\varphi = \nabla\varphi\cdot d\mathbf{r} = \sum_{i=1}^{3}\frac{\partial\varphi}{\partial x_i}dx_i$$

$$= -\mathbf{E}\cdot d\mathbf{r} = 0,$$

so that **E** is perpendicular to $d\mathbf{r}$. Hence, at the surface φ = const., the lines of force are perpendicular to the equipotential surface.

Fig. 3-4 maps the field lines and the equipotential surfaces for the case of two equal and oppositely charged point charges.

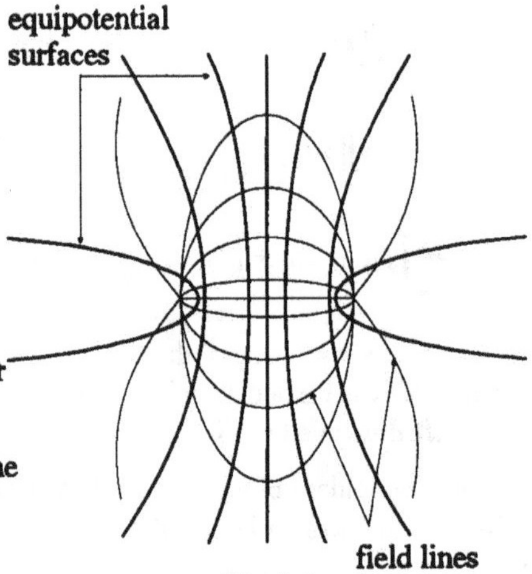

Fig. 3-4

3.2 Gauss' Theorem

Let V in Fig. 3-5 include only the shaded volume.

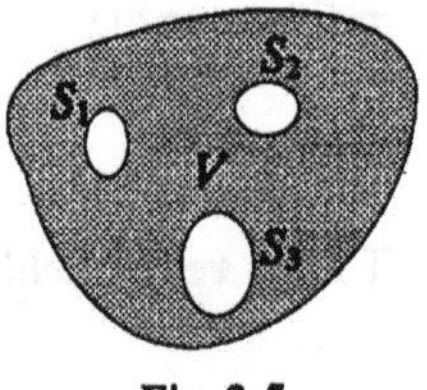

Fig. 3-5

Theorem:

$$\sum_{i_1}\int_V \frac{\partial}{\partial x_{i_1}} T_{i_1 i_2 \dots i_n}\, d\tau = \int_{\substack{\text{all}\\ \text{surfaces}}} \sum_{i_1} T_{i_1 i_2 \dots i_n}\, dS_{i_1},$$

where $d\mathbf{S}:(dS_1, dS_2, dS_3)$ is parallel to the outward drawn normal and has magnitude equal to $|d\mathbf{S}|$.

Proof: Consider the infinitesimal cubical volume element $dx_1dx_2dx_3$ shown in Fig. 3-6.

$$\sum_{i_1}\int\frac{\partial}{\partial x_{i_1}}T_{i_1i_2\cdots i_n}\,d\tau = \int\frac{\partial}{\partial x_1}T_{1i_2\cdots i_n}\,d\tau + \int\frac{\partial}{\partial x_2}T_{2i_2\cdots i_n}\,d\tau + \int\frac{\partial}{\partial x_3}T_{3i_2\cdots i_n}\,d\tau$$

$$= \int\left|T_1i_2\ldots i_n\right|_{x_1}^{x_1+dx_1}dx_2dx_3 + \int\left|T_2i_2\ldots i_n\right|_{x_2}^{x_2+dx_2}dx_1dx_3$$

$$+ \int\left|T_3i_2\ldots i_n\right|_{x_3}^{x_3+dx_3}dx_2dx_1$$

$$= \int\left(T_{1i_2\cdots i_n}\Big|_{ABCD} - T_{1i_2\cdots i_n}\Big|_{EFGH}\right)dx_2dx_3 + \int\left(-T_{2i_2\cdots i_n}\Big|_{ABFE} + T_{2i_2\cdots i_n}\Big|_{CDHG}\right)dx_1dx_3$$

$$+ \int\left(-T_{3i_2\cdots i_n}\Big|_{AEHD} + T_{3i_2\cdots i_n}\Big|_{BCGF}\right)dx_1dx_2$$

$$= \int T_{1i_2\cdots i_n}\,dS_1 + \int T_{2i_2\cdots i_n}\,dS_2 + \int T_{3i_2\cdots i_n}\,dS_3 = \sum_{i=1}^{3}\int T_{ii_2\cdots i_n}\,dS_i\,.$$

We have thus shown Gauss' theorem to be true for the particular volume element $d\tau$ illustrated in Fig. 3-6.

If we now condider volume V to be divided up into small cubical volume elements, the surface integrations of neighboring cubes will cancel each other out, thus leaving us with a surface integration over the outer surface.

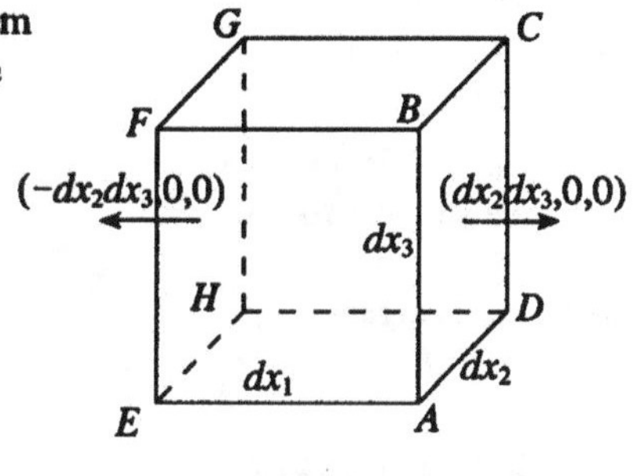

Fig. 3-6

Special Cases:

(a) If T is a vector field T_i: $\displaystyle\sum_i\int_V\frac{\partial T_i}{\partial x_i}\,d\tau = \int_V\nabla\cdot\mathbf{T}\,d\tau = \int_S\sum_i T_i dS_i = \int_S\mathbf{T}\cdot d\mathbf{S}.$

Hence,

$$\int_V\nabla\cdot\mathbf{F}\,d\tau = \int_S\mathbf{F}\cdot d\mathbf{S}.$$

(b) If T_{ij} is a 2nd-rank tensor, $\sum_i \int_V \frac{\partial T_{ij}}{\partial x_i} d\tau = \int_S T_{ij} dS_i$.

How do we use Gauss' theorem for the case $\int_V \frac{\partial}{\partial x_j} T_{i_1 i_2 \cdots i_n} d\tau$?

If we define $R_{iji_1 i_2 \cdots i_n} \equiv \delta_{ij} T_{i_1 i_2 \cdots i_n}$,

$$\sum_i \frac{\partial}{\partial x_i} R_{iji_1 i_2 \cdots i_n} = \sum_i \frac{\partial}{\partial x_i} \delta_{ij} T_{i_1 i_2 \cdots i_n} = \frac{\partial}{\partial x_j} T_{i_1 i_2 \cdots i_n}.$$

Then, applying Gauss' theorem:

$$\int_V \frac{\partial}{\partial x_j} T_{i_1 i_2 \cdots i_n} d\tau = \int_V \sum_i \frac{\partial}{\partial x_i} R_{iji_1 i_2 \cdots i_n} d\tau = \sum_i \int R_{iji_1 i_2 \cdots i_n} dS_i$$

$$= \sum_i \int \delta_{ij} T_{i_1 i_2 \cdots i_n} dS_i = \int T_{i_1 i_2 \cdots i_n} dS_j.$$

Hence, for $i \neq i_1 i_2 \ldots i_n$,

$$\int_V \frac{\partial}{\partial x_i} T_{i_1 i_2 \cdots i_n} d\tau = \int_S T_{i_1 i_2 \cdots i_n} dS_i.$$

3.2.1 Applications in Physics

(a) Equation of Continuity: Consider a fluid of density ρ and velocity **v**, and consider a volume V, fixed in space, not moving with the fluid. (ρ and **v** are evaluated at a particular point in space.)

At any instant the flux of fluid passing outward through an element of surface $d\mathbf{S}$ (Fig. 3-7) is that contained in an oblique cylinder of height υ and base dS. The mass then will be given by $\rho d\tau = \rho \mathbf{v} \cdot d\mathbf{S}$. Hence, the mass of fluid flowing outward per unit time through the entire closed surface is given by

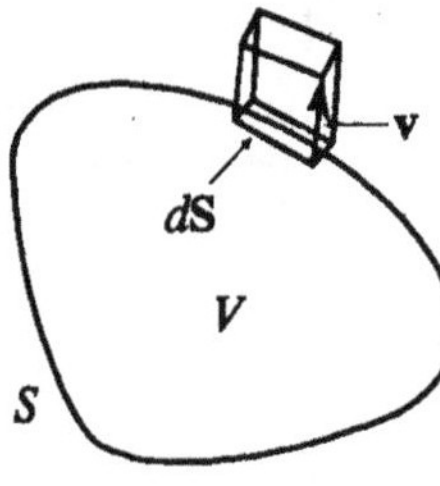

Fig. 3-7

$$\int_S \rho \mathbf{v} \cdot d\mathbf{S},$$

which, by Gauss' theorem, must equal $\int_V \nabla \cdot \rho \mathbf{v} d\tau$.

But by the law of conservation of mass, the time rate of change of mass within V, which is expressed by

$$\frac{\partial}{\partial t}\int_V \rho d\tau = \int_V \frac{\partial \rho}{\partial t} d\tau,$$

must equal the negative of the outgoing flux, i.e.

$$\int_V \frac{\partial \rho}{\partial t} d\tau = -\int_V \nabla\cdot\rho\mathbf{v} d\tau.$$

[For a positive outward flux, there must be a decrease in mass.]

Since the above holds for all volume elements $d\tau$,

$$\frac{\partial \rho}{\partial t} + \nabla\cdot\rho\mathbf{v} = 0$$

at each point of the fluid. This is the equation of continuity, and is seen to be a consequence of the law of conservation of mass, and holds whether or not the fluid is compressible.

Fluid incompressibility is expressed by $\frac{d\rho}{dt} = \frac{\partial \rho}{\partial t} + \nabla\rho\cdot\mathbf{v} = 0$. But, since

$$\nabla\cdot\rho\mathbf{v} = \rho\nabla\cdot\mathbf{v} + \nabla\rho\cdot\mathbf{v},$$

$$\frac{\partial \rho}{\partial t} + \nabla\cdot\rho\mathbf{v} = \frac{\partial \rho}{\partial t} + \nabla\rho\cdot\mathbf{v} + \rho\nabla\cdot\mathbf{v} = 0 \quad \rightarrow \quad \nabla\cdot\mathbf{v} = 0,$$

the condition of incompressibility.

If the fluid density is uniform throughout, ρ is independent of position, i.e. $\nabla\rho = 0$, while if it is incompressible, $\nabla\cdot\mathbf{v} = 0$, so that

$$\int_S \rho\mathbf{v}\cdot d\mathbf{S} = \int_V \nabla\cdot\rho\mathbf{v} d\tau = \int_V [\rho\nabla\cdot\mathbf{v} + \nabla\rho\cdot\mathbf{v}] d\tau = 0,$$

Thus,

$$\int_S \rho\mathbf{v}\cdot d\mathbf{S} = 0,$$

which says that the net outward flux is zero, or the mass leaving S per unit time equals the mass entering per unit time.

(b) Magnetostatics:

Theorem: If $\nabla\cdot\mathbf{H} = 0$ in a region R, then lines of force of $\mathbf{H}$ are continuous in R.

Proof: Consider a tube formed of lines of force parallel to $\mathbf{H}$ with end surfaces S_1 and S_2 perpendicular to $\mathbf{H}$ (Fig 3-8).

$$\oint_{\text{tube}}\mathbf{H}\cdot d\mathbf{S} = \int_{S_1S_2}\mathbf{H}\cdot d\mathbf{S} = \int_{S_1}\mathbf{H}\cdot d\mathbf{S} + \int_{S_2}\mathbf{H}\cdot d\mathbf{S} = \int_V \nabla\cdot\mathbf{H}d\tau = 0,$$

by hypothesis.

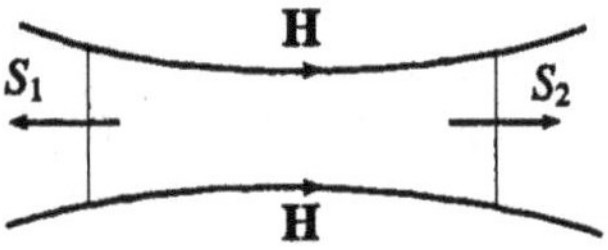

Fig. 3-8

Hence, the number of lines of force through S_1 (negative) plus the number of lines of force through S_2 (positive) $= -N_1 + N_2 = 0$. Since the volume can be chosen as small as desired, still remaining in region R, the lines of force must be continuous, i.e. lines of force must either form closed curves or extend from $-\infty$ to $+\infty$. Thus, if a line of force is created or annihilated at a point, a source exists at that point where lines of field fail to remain continuous, i.e. $\nabla\cdot\mathbf{A} = 0$ there. As an example: $\nabla\cdot\mathbf{E} = 4\pi\rho$, and a charge exists at the discontinuity.

(c) Stress Tensor:

Consider a deformable body immersed in a medium such as a gas, liquid or solid. What are the forces exerted on the body? In general, there will be two kinds of force acting on the body in V (Fig. 3-9).

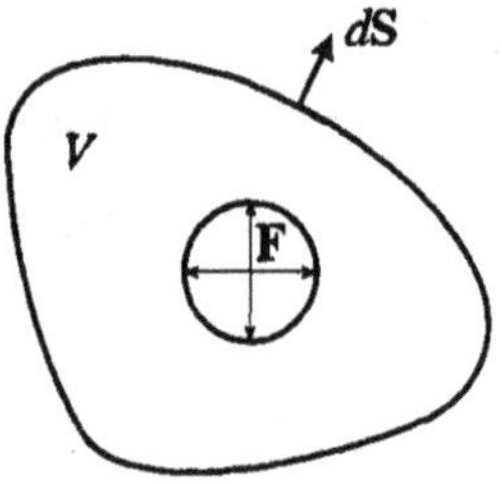

Fig. 3-9

(a) Body force: $\int_V \mathbf{F}d\tau$ — electrical, gravitational — some function of charge density, mass density, etc., expressed by action at a distance.

(b) Surface force: that force exerted across the surface of the medium, i.e. some function of the particular medium expressed by an action through the medium. We seek the form that such a force takes.

Let f_i = the force exerted on $d\mathbf{S}$ by the exterior of V upon the interior of V. For small areas, most generally,

$$f_i = \sum_{j=1}^{3} T_{ij} dS_j ,$$

where we neglect terms of order dS_j^2, and where $f = 0$ if $dS = 0$, i.e. a linear combination. Since f_i and dS_j are vectors, T_{ij} is a 2nd-rank tensor, called the *stress tensor*.

The total force is given by

$$\int_V \rho a_i d\tau = \int_V F_i d\tau + \int_S \sum_j T_{ij} dS_j = \int_V F_i d\tau + \sum_j \frac{\partial}{\partial x_j} \int_V T_{ij} d\tau ,$$

where the last term follows from Gauss's theorem. Since the above must be true for any volume element $d\tau$,

$$F_i + \sum_{j=1}^{3} \frac{\partial}{\partial x_j} T_{ij} = \rho a_i ,$$

yielding the equation of motion of deformable bodies.

Physical Interpretation of T_{ij}:

$$f_1 = T_{11} dS_1 + T_{12} dS_2 + T_{13} dS_3 .$$

$$f_2 = T_{21} dS_1 + T_{22} dS_2 + T_{23} dS_3 .$$

$$f_3 = T_{31} dS_1 + T_{32} dS_2 + T_{33} dS_3 .$$

If we choose $d\mathbf{S}$ parallel to the x-axis ($d\mathbf{S} = \mathbf{i}_0 dS$), then

$$f_x = T_{11} dS . \quad f_y = T_{21} dS . \quad f_z = T_{31} dS .$$

In this case, T_{11} represents tension or compression (force in the direction of $d\mathbf{S}$) while T_{21} and T_{31} represent shearing forces.

From considerations of rotational stability of the medium, it can be shown that $T_{ij} = T_{ji}$ if we wish to avoid angular accelerations. Hence, T_{ij} is a 2nd-rank symmetric tensor, and one can therefore diagonalize it to $T_{ij}' = \lambda_i \delta_{ij}$.

For an element of area dS parallel to the x-axis,

$$\left.\begin{cases} f_x = \lambda_1 dS. \\ f_y = 0 \\ f_z = 0 \end{cases}\right\} \quad \begin{array}{lll} \lambda_1 > 0 & \rightarrow & \text{tension} \\ \lambda_1 < 0 & \rightarrow & \text{compression} \end{array}$$

Once the principal axes are found, the stress tensor represents a tension or compression along these principal axes. For any other axes, T_{ij} represents shearing forces.

Example: For an inviscid fluid (no viscosity), $T_{ij} = -p\delta_{ij}$, which represents a pressure regardless of choice of axes.

$$\sum_{j=1}^{3} \frac{\partial}{\partial x_j} T_{ij} = \sum_{j=1}^{3} -\frac{\partial p}{\partial x_j}\delta_{ij} = -\frac{\partial p}{\partial x_i} = -[\nabla p]_i ,$$

so that the equation of motion becomes

$$\rho a_i = F_i - \frac{\partial p}{\partial x_i}. \quad \rightarrow \quad \rho \mathbf{a} = \mathbf{F} - \nabla p.$$

Let $\mathbf{v}(\mathbf{r}, t)$ represent the velocity of a given particle. What is the relation between $\mathbf{a}$ and $\mathbf{v}$? Is it $\mathbf{a} = \dot{\mathbf{v}}$? The answer is no, since $\mathbf{a} = \dot{\mathbf{v}}$ applies to a fixed point in space. Here, we consider a point in the medium and seek the velocity at that point at a particular instant in time.

We have $\mathbf{r}$ at time $t = t_o$ and $\mathbf{r} + \mathbf{v}\delta t$ at time $t = t_o + \delta t$. Then

$$\begin{aligned}
\mathbf{a}\delta t &= \mathbf{v}(\mathbf{r}+\mathbf{v}\delta t, t_o+\delta t) - \mathbf{v}(\mathbf{r}, t_o) \\
&= \mathbf{v}(\mathbf{r}, t_o) + \left[\frac{\partial \mathbf{v}}{\partial t}\right]_{t=t_o} \delta t + \frac{1}{2}\left[\frac{\partial^2 \mathbf{v}}{\partial t^2}\right]_{t=t_o} \delta t^2 + \ldots . \\
&\quad + \frac{\partial \mathbf{v}}{\partial x_1}\delta x_1 + \frac{\partial \mathbf{v}}{\partial x_2}\delta x_2 + \frac{\partial \mathbf{v}}{\partial x_3}\delta x_3 + \ldots . - \mathbf{v}(\mathbf{r}, t_o) \\
&= \frac{\partial \mathbf{v}}{\partial t}\delta t + \sum_{i=1}^{3} \delta x_i \frac{\partial \mathbf{v}}{\partial x_i} = \frac{\partial \mathbf{v}}{\partial t}\delta t + \sum_{i=1}^{3} \upsilon_i \delta t \frac{\partial \mathbf{v}}{\partial x_i} = \frac{\partial \mathbf{v}}{\partial t}\delta t + (\mathbf{v}\cdot\nabla)\mathbf{v}\delta t .
\end{aligned}$$

Thus, Euler's equation, to second order in δt and δx_i, becomes

$$\rho\left[\frac{\partial \mathbf{v}}{\partial t} + (\mathbf{v}\cdot\nabla)\mathbf{v}\right] = \mathbf{F} - \nabla p .$$

(d) Electrostatics:

Let σ = charge density. For a point charge at the origin, what is σ over all space? For $\mathbf{r} \neq 0$, $\sigma(\mathbf{r}) = 0$, where $\int\limits_{\substack{\text{all}\\\text{space}}} \sigma(\mathbf{r}) d\tau = e$.

We see that $\sigma(\mathbf{r})$ is a singular function, becoming infinite at the origin, but such that $\int\limits_{\substack{\text{all}\\\text{space}}} \sigma(\mathbf{r}) d\tau = e$. We introduce the Dirac δ-function, such that

$$\sigma(\mathbf{r}) = e\,\delta^3(\mathbf{r}) \begin{cases} \delta^3(\mathbf{r}) = \delta(x)\delta(y)\delta(z) \\ \delta(x) = 0 \text{ for } x \neq 0 \\ \int_{-a}^{b} \delta(x)dx = 1 \text{ for } a,b > 0 \end{cases}$$

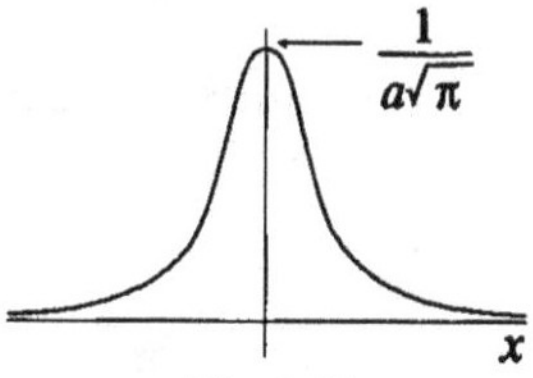

Fig. 3-10

Consider a function like $(1/a\sqrt{\pi})e^{-x^2/a^2}$, the Gaussian shown in Fig. 3-10. If we let the width approach zero, while the height approaches infinity in such a manner that the enclosed area remains equal to 1, we approach a delta-function. Such a delta-function is a continuous function whose width we let approach zero while the area is kept constant and equal to 1.

Returning to the problem above: What is the potential and field due to this point charge at points away from the singularity? From Coulomb's law,

$$\varphi = \frac{e}{r} \quad \rightarrow \quad \mathbf{E} = -\nabla\varphi = \frac{e}{r^3} r_o .$$

$$\nabla\cdot\mathbf{E} = \sum_{i=1}^{3} \frac{\partial}{\partial x_i}\left(\frac{eX_i}{r^3}\right) = \frac{e}{r^3}\sum_{i=1}^{3}\frac{\partial X_i}{\partial x_i} + \sum_{i=1}^{3}\frac{-3ex_i}{r^4}\frac{\partial r}{\partial x_i}$$

$$= \frac{3e}{r^3} - \frac{3e}{r^4}\sum_{i=1}^{3}\frac{x_i^2}{r} = 0.$$

Hence, at $r \neq 0$,

$$\nabla\cdot\mathbf{E} = 0 \quad \rightarrow \quad \Delta\varphi = \nabla^2\varphi = \nabla\cdot\nabla\varphi = -\nabla\cdot\mathbf{E} = 0.$$

To evaluate $\nabla\cdot\mathbf{E}$ at the origin, where $\mathbf{E}$ is not defined, we consider a volume of radius ε enclosing the charge, and imagine that $\nabla\cdot\mathbf{E}$ is defined over the volume. From Gauss' law,

$$\int_V \nabla\cdot\mathbf{E}d\tau = \oint_S \mathbf{E}\cdot d\mathbf{S} = \oint_S \frac{e\mathbf{r}_o}{r^3}\cdot d\mathbf{S} = \oint_S \frac{e\mathbf{r}_o}{\varepsilon^3}\cdot d\mathbf{S} = \oint_S \frac{edS}{\varepsilon^2}$$

$$= \oint_S \frac{e\varepsilon^2 d\Omega}{\varepsilon^2} = 4\pi e.$$

Remembering that $\sigma(\mathbf{r}) = e\delta^3(\mathbf{r})$ and that $\int_{\substack{\text{all}\\\text{space}}} \sigma(\mathbf{r})d\tau = e$, it follows, by comparison:

$$\nabla\cdot\mathbf{E} = 4\pi e\delta^3(\mathbf{r}). \quad \rightarrow \quad \Delta\varphi = -4\pi e\delta^3(\mathbf{r}).$$

If we have some Gaussian distribution of charge, and then let this distribution shrink to zero in a manner described above, then, since the charge distribution becomes a delta-function, it will be found that $\nabla\cdot\mathbf{E} = 4\pi e\delta^3(\mathbf{r})$.

Generalizing:

(a) For a discrete distribution (Fig. 3-11), the potential at point P is given by

$$V_P = \sum_i \frac{q_i}{R_i}, \quad R_i = |\mathbf{r} - \mathbf{r}'|.$$

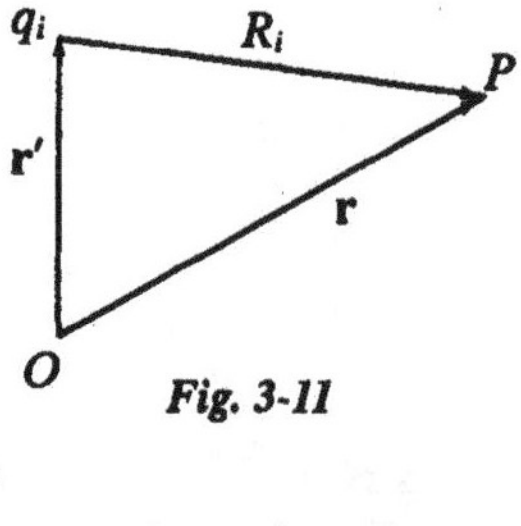

Fig. 3-11

(b) For a continuous distribution $\rho(\mathbf{r}')$, the potential at point P is given by

$$V(\mathbf{r}) = \iiint \frac{\rho(\mathbf{r}')}{|\mathbf{r} - \mathbf{r}'|} dV',$$

where the limits are such as to include all of $\rho(\mathbf{r}')$.

If $\rho(\mathbf{r}')$ should include point charges at point a, for example, we introduce the Dirac delta-function, namely, the limiting form of Fig. 3-12, given by

$$\delta(x - a) = \begin{cases} 0 \text{ if } x \neq a \\ \infty \text{ if } x = a \end{cases}$$

where

$$\int_\alpha^\beta \delta(x - a)dx = \begin{cases} 1 \text{ if } \alpha < a < \beta \\ 0 \text{ otherwise} \end{cases}$$

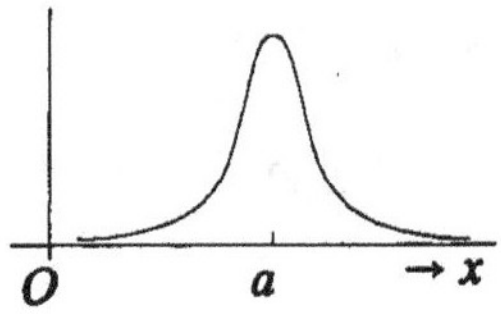

Fig. 3-12

Properties of delta-functions

(a) $\delta(x) = \delta(-x)$

(b) $\int \delta(x-a) f(x) dx = f(a)$ for $f(x)$ continuous

(c) $\delta^3(\mathbf{r}) = \delta(x)\delta(y)\delta(z)$

(d) $\iiint \delta^3(\mathbf{r}) = \begin{cases} 1 \text{ if range of integration includes } (000) \\ 0 \text{ otherwise} \end{cases}$

(e) $\iiint_V \delta^3(\mathbf{r}) f(xyz) d^3\mathbf{r} = \begin{cases} f(000) \text{ if } (000) \text{ is in } V \\ 0 \text{ otherwise} \end{cases}$

Considering some arbitrary distribution of charge (Fig. 3-13),

$$\varphi(\mathbf{r}) = \int_V \frac{\rho(\mathbf{r}')}{|\mathbf{r}' - \mathbf{r}|} d\tau'.$$

$$\Delta_r \varphi(\mathbf{r}) = \int_V \rho(\mathbf{r}') \Delta_r \left[\frac{1}{|\mathbf{r}' - \mathbf{r}|}\right] d\tau'.$$

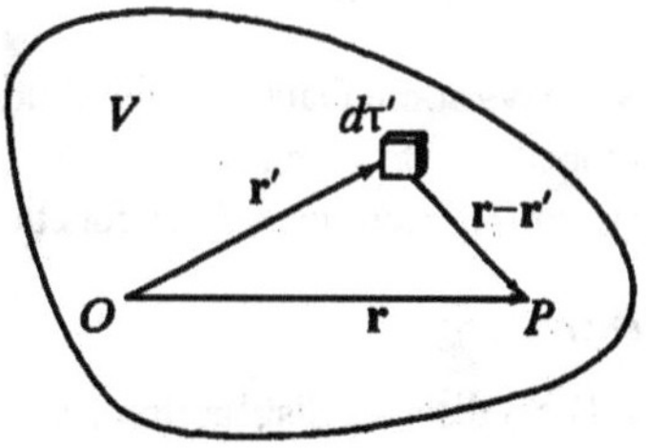

Fig. 3-13

A) Assuming Coulomb's law is true, find an expression for $\Delta_r\varphi$, the distribution of potential.

We see that $\Delta_r \frac{1}{r} = -4\pi\delta^3(\mathbf{r}) \rightarrow \Delta_r \frac{1}{|\mathbf{r}' - \mathbf{r}|} = -4\pi\delta^3(\mathbf{r}' - \mathbf{r})$. Thus,

$$\Delta_r \varphi(\mathbf{r}) = \int -\rho(\mathbf{r}') 4\pi\delta^3(\mathbf{r}' - \mathbf{r}) = -4\pi\rho(\mathbf{r}).$$

$$\Delta_r \varphi(\mathbf{r}) = -4\pi\rho(\mathbf{r}) \quad \text{Poisson's equation,}$$

or the equivalent,

$$\nabla \cdot \mathbf{E}(\mathbf{r}) = 4\pi\rho(\mathbf{r}) \quad \text{Maxwell's 1st source equation.}$$

B) Given a distribution of charge $\rho(\mathbf{r})$; what will be the most general solution for $\varphi(\mathbf{r})$, a solution to Poisson's equation?

Theorem:

$$\varphi(\mathbf{r}) = \varphi_o(\mathbf{r}) + \int_V \frac{\rho(\mathbf{r}')}{|\mathbf{r}' - \mathbf{r}|} d\tau' \text{ with } \Delta\varphi_o = 0.$$

Proof: Define

$$\varphi_o(\mathbf{r}) = \varphi(\mathbf{r}) - \int_V \frac{\rho(\mathbf{r}')}{|\mathbf{r}' - \mathbf{r}|} d\tau',$$

so that

$$\Delta\varphi_o(\mathbf{r}) = \Delta\varphi(\mathbf{r}) - \int_V \rho(\mathbf{r}')\,\Delta\frac{1}{|\mathbf{r}' - \mathbf{r}|} d\tau', = \Delta\varphi(\mathbf{r}) + 4\pi\rho(\mathbf{r}) = 0.$$

In general, $O\varphi = f(\mathbf{r})$ and we seek the general solution for φ.

Theorem: If $O_r G(\mathbf{r}) = \delta^3(\mathbf{r})$, then $\varphi(\mathbf{r}) = \int G(\mathbf{r}' - \mathbf{r}) f(\mathbf{r}')d\tau' + \varphi_o(\mathbf{r})$, where $O\varphi_o(\mathbf{r}) = 0$, yields the most general solution.

Proof:

$$O_r\varphi(\mathbf{r}) = \int O_r G(\mathbf{r}' - \mathbf{r}) f(\mathbf{r}')d\tau' + O_r\varphi_o(\mathbf{r}) = \int \delta^3(\mathbf{r}' - \mathbf{r}) f(\mathbf{r}')d\tau' = f(\mathbf{r}).$$

The problem is reduced to finding a Green's function $G(\mathbf{r})$ such that $O_r G(\mathbf{r}) = \delta^3(\mathbf{r})$.

(e) Heat Flow:

If κ is the heat conductivity, T the temperature, and $\mathbf{Q}$ the heat flux, or the amount of heat energy crossing unit area per unit time, then $\mathbf{Q} = -\kappa\nabla T$. $\mathbf{Q}$ is opposite in direction to the greatest rate of change of temperature, by virtue of the 2nd law of thermodynamics.

If E represents the energy content, then

$$\frac{\partial}{\partial t}\int_V E d\tau + \oint_S \mathbf{Q}\cdot d\mathbf{S} = 0,$$

by virtue of the law of conservation of energy.

Now $dE = c_p dT$, where c_p is the specific heat per unit volume at constant pressure, and is assumed constant throughout volume V. Then $E = c_p T$ and

$$\int_V c_p \frac{\partial T}{\partial t} d\tau + \oint_S \mathbf{Q}\cdot d\mathbf{S} = 0 \quad \rightarrow \quad \int_V \left\{ c_p \frac{\partial T}{\partial t} + \nabla\cdot\mathbf{Q} \right\} d\tau = 0.$$

But since the above holds for all $d\tau$, and since $\nabla\cdot\mathbf{Q} = -\kappa\nabla^2 T$, it follows that

$$\frac{\partial T}{\partial t} = \frac{\kappa}{c_p}\Delta T,$$

which is the heat flow equation. Thus, knowing the spatial distribution of temperature, one knows how it varies with time, and vice versa.

Singularity problem in the application of Gauss' theorem:

Let $\mathbf{V} = \nabla\left(\frac{e^{-\lambda r}}{r}\right)$. (Note that a singularity exists at $r = 0$). Then by Gauss' theorem, $\int_V \nabla\cdot\mathbf{V} d\tau = \oint_S \mathbf{V}\cdot\mathbf{n}_o dS$.

$$\mathbf{V} = \mathbf{r}_o \frac{\partial}{\partial r}\left(\frac{e^{-\lambda r}}{r}\right) = \frac{\mathbf{r}}{r}\left(\frac{-e^{-\lambda r}}{r^2} - \frac{\lambda e^{-\lambda r}}{r}\right)$$

$$= -\frac{\mathbf{r}e^{-\lambda r}}{r^3}(1 + \lambda r).$$

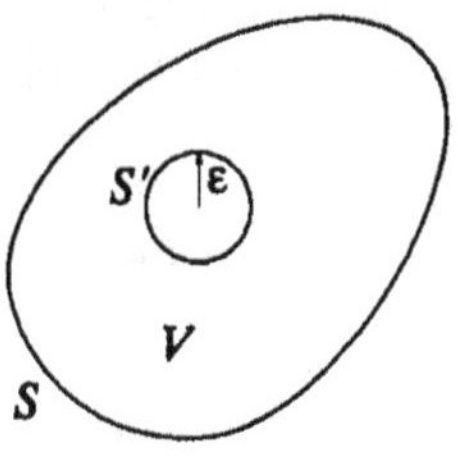

Fig. 3-14

Consider $\int_V \nabla\cdot\mathbf{V} d\tau = \int_S \mathbf{V}\cdot\mathbf{n}_o dS + \int_{S'} \mathbf{V}\cdot\mathbf{n}_o dS'$. Letting the outer sphere $S \rightarrow \infty$,

$\int_{S\rightarrow\infty} -\frac{\mathbf{r}e^{-\lambda r}}{r^3}(1 + \lambda r)\cdot\mathbf{r}_o dS \rightarrow 0$ as $r \rightarrow 0$ by virtue of $e^{-\lambda r}$ (Fig. 3-14).

$$\int_{S'} \mathbf{V}\cdot\mathbf{n}_o dS' = -\int_{S'} \frac{\mathbf{r}_o e^{-\lambda r}}{r^2}(1 + \lambda r)\cdot\mathbf{r}_o \varepsilon^2 d\Omega = \int e^{-\lambda\varepsilon}(1 + \lambda\varepsilon) d\Omega$$

$$= 4\pi e^{-\lambda\varepsilon}(1 + \lambda\varepsilon) \quad \rightarrow \quad 4\pi \text{ as } \varepsilon \rightarrow 0.$$

Hence we see that $\int_V \nabla\cdot\mathbf{V} d\tau = 4\pi$ over the region excluding the singularity, whereas had we applied Gauss' theorem blindly, the result would be zero.

3.3 Stokes' Theorem

Let S be a two-dimensional surface bounded by a closed curve s. Then $d\mathbf{S}$ is a surface element of S parallel to the normal of S, and $d\mathbf{s}$ is a line element of s parallel to the tangent of curve s. Choose directions such that the right-hand screw rule applies (Fig. 3-15).

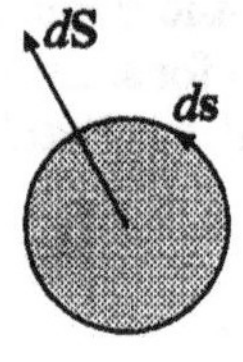

Fig. 3-15

Theorem: If $\mathbf{A}$ is any vector point function, then

$$\oint_s \mathbf{A}\cdot d\mathbf{s} = \int_S (\nabla\times\mathbf{A})\cdot d\mathbf{S}.$$

(See problems for the case of arbitrary nth rank tensor.)

Proof: Consider an infinitesimal rectangle. Since the equations are scalar, we can choose a convenient system of coordinates (Fig. 3-16). In this coordinate system, $d\mathbf{S}$: $(0,0,dx_1dx_2)$.

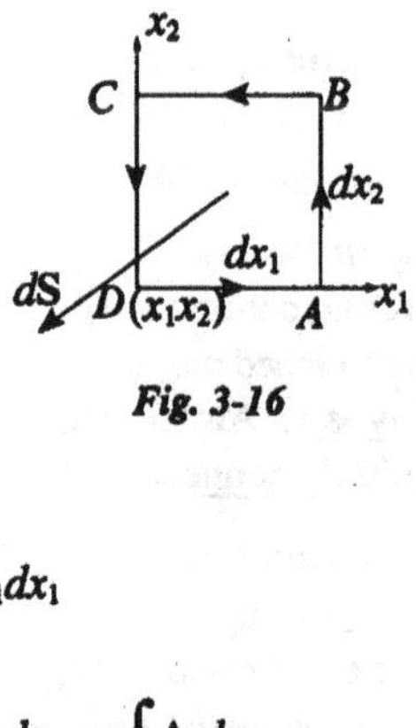

Fig. 3-16

$$\begin{aligned}\int_S (\nabla\times\mathbf{A})\cdot d\mathbf{S} &= \int_S \left(\frac{\partial A_2}{\partial x_1} - \frac{\partial A_1}{\partial x_2}\right) dx_1 dx_2 \\ &= \int A_2\Big|_{x_1}^{x_1+\delta x_1} dx_2 - \int A_1\Big|_{x_2}^{x_2+\delta x_2} dx_1 \\ &= \int_A^B A_2 dx_2 - \int_D^C A_2 dx_2 - \int_C^B A_1 dx_1 + \int_D^A A_1 dx_1 \\ &= \int_A^B A_2 dx_2 + \int_B^C A_1 dx_1 + \int_C^D A_2 dx_2 + \int_D^A A_1 dx_1 = \oint_s \mathbf{A}\cdot d\mathbf{s}\end{aligned}$$

over the complete rectangle. For any arbitrary surface (not necessarily 2-dimensional), the surface can always be divided into small rectangles (Fig. 3-17). We have proven Stokes' theorem for any arbitrary rectangle; hence it will be true for all the rectangles of S. For neighboring rectangles, the line integrals over the common paths cancel and we are left with the line integral of $\mathbf{A}\cdot d\mathbf{s}'$ over the zig-zag path shown.

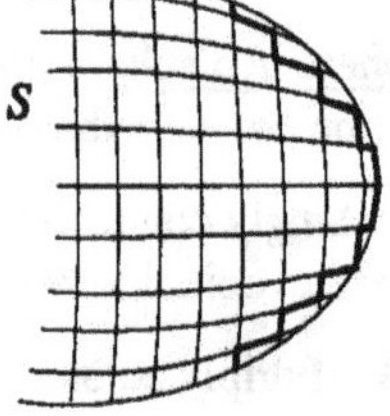

Fig. 3-17

Consider any small section as shown in Fig. 3-18. We would like to show that as the section gets smaller,

$$\int_{s'} \to \int_{s} \text{ i.e. } \int_{s'} \mathbf{A}\cdot d\mathbf{s}' \to \int_{s} \mathbf{A}\cdot d\mathbf{s}.$$

Consider $F_1 dx_1' + F_2 dx_2'$, where $\mathbf{F}:(F_1, F_2)$. Note that for small sections, $d\mathbf{s}:(dx_1', dx_2')$. Hence, $F_1 dx_1' + F_2 dx_2' \to \mathbf{F}\cdot d\mathbf{s}$, so that

$$\int_{s'} \mathbf{F}\cdot d\mathbf{s}' \to \int_{s} \mathbf{F}\cdot d\mathbf{s}.$$

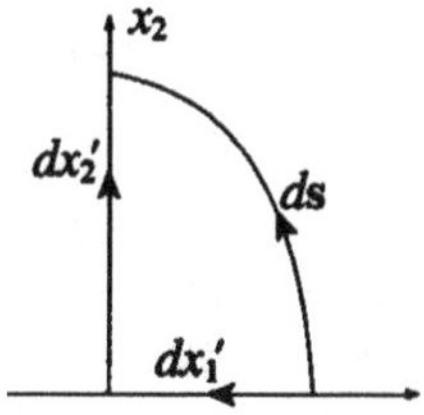

Fig. 3-18

3.4 Connectivity of Space

Definition (1): If any point in a region R can be reached from any other point in R through a continuous motion in R, then R is defined to be a connected region. A connected region has a single closed outer boundary S.

Definition (2): If any closed curve in a connected region R_1 can be made, through continuous motion and deformation in R_1, to coincide with any other closed curve in R_1, then R_1 is defined to be a singly connected region of type 1. Any region of type 1 that is not single connected is defined to be multiply connected of type 1.

Definition (3): If the outer boundary S of a connected region R_2 can be made, through continuous deformation, to shrink to zero, then R_2 is a singly connected region of type 2. Any region of type 2 that is not singly connected is defined to multiply connected of type 2.

Examples:

Sphere in a sphere (Fig. 3-19): We have a connected region between the two concentric spheres.

(a) Singly connected of type 1, since curve 1 can be made to coincide with curve 2.

(b) Multiply connected of type 2, since the outer boundary S cannot be made to shrink to zero.

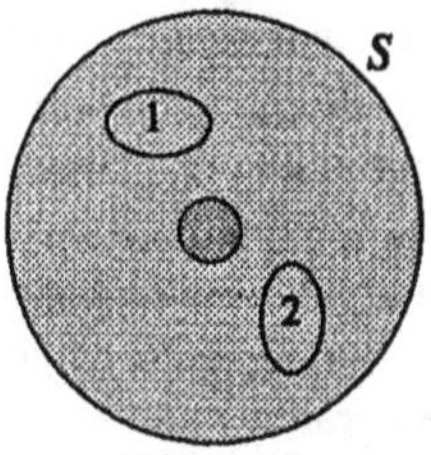

Fig. 3-19

Donut (Fig. 3-20):

(a) Singly connected of type 2, since curve 1 cannot be made to coincide with curve 2.

(b) Multiply connected of type 1, since the outer boundary S can be made to shrink to S', and S' can be made to shrink to zero.

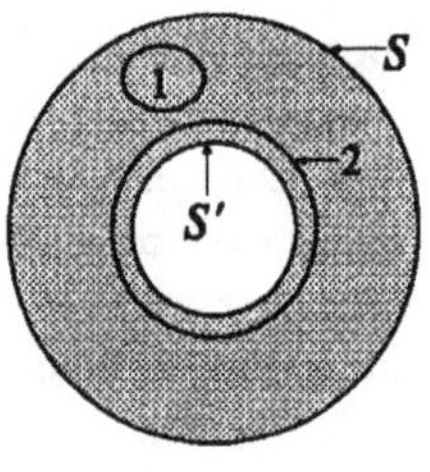

Fig. 3-20

Theorem: If $\nabla\times\mathbf{E} = 0$ in a singly connected region (type 1 and 2), as in a sphere, then there exists a scalar function φ in R such that $\mathbf{E} = \nabla\varphi$.

Proof: Define $\varphi(\mathbf{r}) \equiv \int_o^{\mathbf{r}} \mathbf{E}\cdot d\mathbf{s}$, where o is arbitrarily chosen (Fig. 3-21). Using Stokes' theorem, we show that $\varphi(\mathbf{r})$ does not depend on the path but only on $\mathbf{r}$.

$\oint_s \mathbf{E}\cdot d\mathbf{s} = \int_S (\nabla\times\mathbf{E})\cdot d\mathbf{S} = 0$ by hypothesis, hence $\int_{(1)}^{\mathbf{r}} \mathbf{E}\cdot d\mathbf{s} = \int_{(2)}^{\mathbf{r}} \mathbf{E}\cdot d\mathbf{s}$ and $\varphi(\mathbf{r}) \equiv \int_o^{\mathbf{r}} \mathbf{E}\cdot d\mathbf{s}$ is a function of r only.

But $\varphi(\mathbf{r}) \equiv \int_o^{\mathbf{r}} \mathbf{E}\cdot d\mathbf{s} \rightarrow d\varphi(\mathbf{r}) = \mathbf{E}\cdot d\mathbf{s}$ and $\mathbf{E}(\mathbf{r}) = \nabla\varphi(\mathbf{r})$.

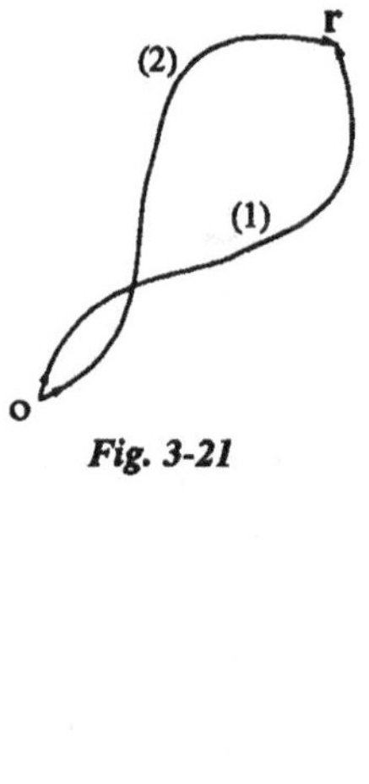

Fig. 3-21

We could also show this by considering the following:

If $\int f(x)dx = g(x)$, then

$$\frac{\partial g(x)}{\partial x} = f(x), \quad \int_o^x f(x)dx = g(x) - g(0).$$

$$\frac{\partial}{\partial x}\int_o^x f(x)dx = \frac{\partial}{\partial x}[g(x) - g(o)] = f(x).$$

Since $d\mathbf{s}:(dx_1,dx_2,dx_3)$ and $\mathbf{E}:(E_1,E_2,E_3)$,

$$\varphi(\mathbf{r}) = \int_o^{\mathbf{r}} \mathbf{E}(\mathbf{r})\cdot d\mathbf{s} = \int_o^{\mathbf{r}} (E_1dx_1 + E_2dx_2 + E_3dx_3).$$

$$\frac{\partial\varphi}{\partial x_1} = E_1. \quad \frac{\partial\varphi}{\partial x_2} = E_2. \quad \frac{\partial\varphi}{\partial x_3} = E_3. \quad \rightarrow \quad \mathbf{E} = \nabla\varphi.$$

Suppose now that the region is singly connected type 1, multiply connected type 2, as a sphere in a sphere. We can make this region all singly connected by a proper cut, as shown in Fig. 3-22. Then φ exists within the region R', which excludes all shaded regions. The question now becomes: what is the behavior of φ across the shaded boundary? If φ is continuous across the boundary, then it can be shrunk to zero.

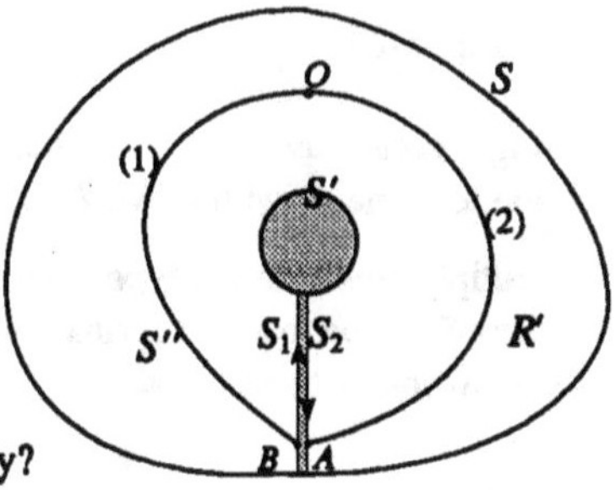

Fig. 3-22

Theorem: If S' is completely enclosed in S, then a function φ exists everywhere in $R = R' + \|$, where R is a singly connected type 1, multiply connected type 2 region.

Proof: Consider

$$\oint_{(2)+(1)+\|} \mathbf{E}\cdot d\boldsymbol{s} = \int_{S''+S_1+S_2} \nabla\times\mathbf{E}\cdot \boldsymbol{n}_o dS = \int_{AOB} \mathbf{E}\cdot d\boldsymbol{s} + \int_{B\uparrow A\downarrow} \mathbf{E}\cdot d\boldsymbol{s}.$$

Since $\mathbf{E}$ is finite and continuous, with continuous derivatives in R, as we shrink the shaded region to zero, $\int_{B\uparrow A\downarrow} \mathbf{E}\cdot d\boldsymbol{s} \to 0$. Hence,

$$\oint_{(2)+(1)+\|} \mathbf{E}\cdot d\boldsymbol{s} = \int_{S''} \nabla\times\mathbf{E}\cdot \boldsymbol{n}_o dS = \varphi_+ - \varphi_-,$$

where

$$\varphi_+ = \int_0^B \mathbf{E}\cdot d\boldsymbol{s} \text{ and } \varphi_- = \int_0^A \mathbf{E}\cdot d\boldsymbol{s}.$$

If S' is completely enclosed in S, we set S'' to be a cap S_3 (Fig. 3-23), on which $\nabla\times\mathbf{E} = 0$ everywhere. Hence,

$$\int_{S''-S_3} \nabla\times\mathbf{E}\cdot d\mathbf{S} = 0 = \varphi_+ - \varphi_-.$$

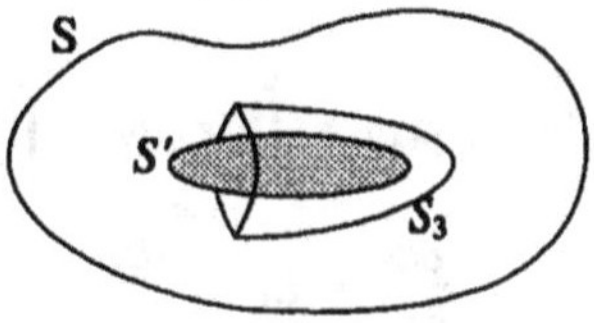

Fig. 3-23

Therefore, $\varphi_+ = \varphi_-$ and φ is continuous across the cut.

It is of interest to consider what happens if we choose S_3 to cut through S', within which $\nabla\times\mathbf{E} \neq 0$ (Fig. 3-24). Consider

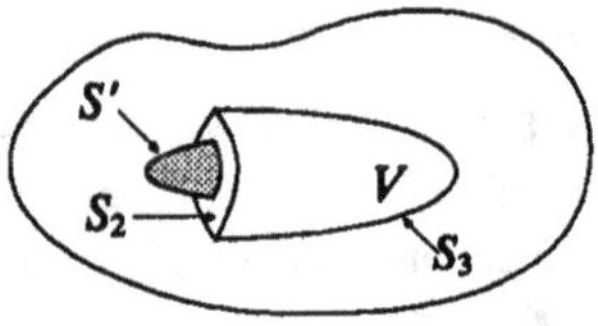

Fig. 3-24

$$\int_{S_2+S_3} \mathbf{E}\cdot d\mathbf{S} = \oint \mathbf{E}\cdot d\mathbf{S} = \int_V (\nabla\cdot\mathbf{E})d\tau .$$

From Gauss' theorem,

$$\int_{S_2+S_3} \nabla\times\mathbf{E}\cdot d\mathbf{S} = \oint \nabla\times\mathbf{E}\cdot d\mathbf{S} = \int_V \nabla\cdot(\nabla\times\mathbf{E})d\tau = 0 .$$

Hence,

$$\int_{S_2} \nabla\times\mathbf{E}\cdot d\mathbf{S} + \int_{S_3} \nabla\times\mathbf{E}\cdot d\mathbf{S} = 0 .$$

But

$$\int_{S_3} \nabla\times\mathbf{E}\cdot d\mathbf{S} = 0 \quad \rightarrow \quad \int_{S_2} \nabla\times\mathbf{E}\cdot d\mathbf{S} = 0,$$

in which case $\varphi_+ = \varphi_-$ and again φ is continuous across the cut, and therefore defined in the entire region surrounding S'. Physically, we can see the above result by considering a region S_1, within which $\nabla\times\mathbf{E} \neq 0$ and outside of which $\nabla\times\mathbf{E} = 0$. Defining $\mathbf{j} = \nabla\times\mathbf{E}$,

$$\begin{cases} \mathbf{j} = 0 \ \text{ outside } S_1 \\ \mathbf{j} \neq 0 \ \text{ within } S_1 \end{cases}$$

Within S_1, $\nabla\cdot\mathbf{j} = 0$, and we have already shown that if $\nabla\cdot\mathbf{j} = 0$ in a region, then lines of $\mathbf{j}$ are continuous (Fig. 3-25). Then we are considering

$$\int_{S_1} \nabla\times\mathbf{E}\cdot d\mathbf{S} = \int_{S_1} \mathbf{j}\cdot d\mathbf{S} .$$

Although $\mathbf{j} \neq 0$ within S_1, we get negative and positive contributions to $\mathbf{j}\cdot d\mathbf{S}$ and the surface integral vanishes. We showed this result above using Gauss' theorem.

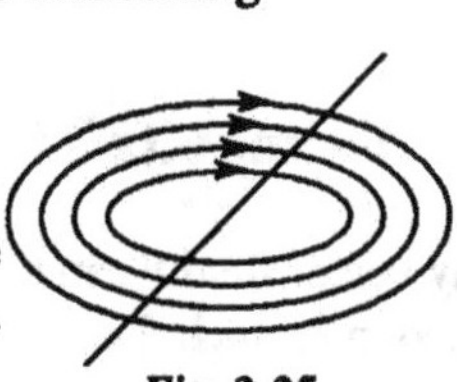

Fig. 3-25

We now wish to restate the above important theorem, together with some of its uses.

Theorem:

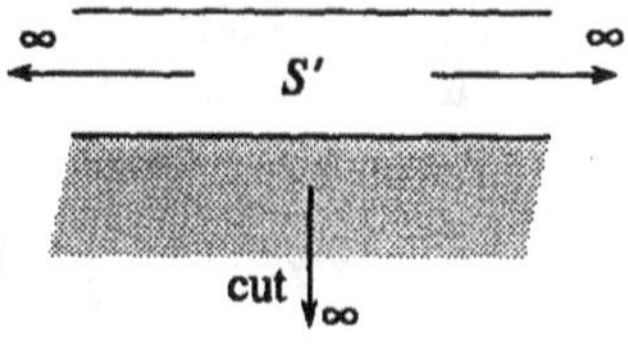

Fig. 3-26

(1) If $\nabla\times\mathbf{E} \neq 0$ only within a finite region S' which is singly connected, types 1 and 2, then outside the region, you can represent $\mathbf{E} = \nabla\varphi$.

(2) If the region S' extends to infinity, we can introduce a cut, extending to infinity, as shown in Figs. 3-26 and 3-27, so that φ is defined everywhere outside of S', but is discontinuous across the cut.

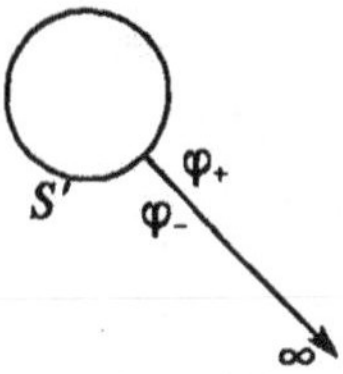

Fig. 3-27

Magnetostatics:

If $\mathbf{H}$ is the magnetic field and $\mathbf{j}$ the current density, then

$$\nabla\times\mathbf{H} = (4\pi\mathbf{j})/c \Rightarrow \nabla\cdot(\nabla\times\mathbf{H}) = (4\pi/c)\nabla\cdot\mathbf{J} = 0.$$

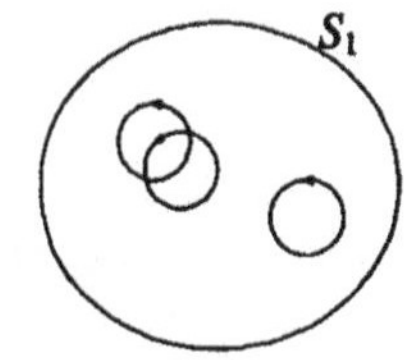

Fig. 3-28

$\nabla\cdot\mathbf{j} = 0$ implies continuous lines of $\mathbf{j}$, or no monopoles, i.e. current loops. For a region S_1, within which $\mathbf{j} \neq 0$ (i.e. $\nabla\times\mathbf{H} \neq 0$), we have current loops inside (Fig. 3-28). Outside S_1 we can represent $\mathbf{H} = \nabla\varphi$.

Electrostatics:

Within S_1, $\nabla\times\mathbf{E} \neq 0$, so that outside of S_1, $\mathbf{E} = -\nabla\varphi$.

Hydrodynamics:

If $\mathbf{v}$ is the velocity at a point, and $\nabla\times\mathbf{v} \neq 0$ only in a finite region, then outside this region, $\mathbf{v} = \nabla\varphi$ ("potential flow").

3.5 Helmholtz Theorem

Any vector field F can be represented by a scalar function φ and a vector function $\mathbf{A}$, such that

$$\mathbf{F} = -\nabla\varphi + \nabla\times\mathbf{A}, \quad \text{where} \quad \nabla\cdot\mathbf{A} = 0.$$

Proof: Let $\mathbf{F}$ approach zero faster than r^{-2}, i.e.

$$\mathbf{F} < \frac{1}{r^{2+\eta}} \text{ as } r \to \infty, \text{ where } \eta > 0.$$

Then

$$\mathbf{G} = \int_{V'} \frac{\mathbf{F}(\mathbf{r}')}{|\mathbf{r}-\mathbf{r}'|} d\tau' \text{ exists and } \to 0 \text{ as } r \to \infty.$$

Using Dirac delta-function properties, where ∇_r denotes derivatives with respect to variable r,

$$\nabla_r^2 \mathbf{G} = \int \nabla_r^2 \left[\frac{1}{|\mathbf{r}-\mathbf{r}'|}\right] \mathbf{F}(\mathbf{r}') d\tau' = -\int 4\pi\delta^3(\mathbf{r}-\mathbf{r}')\mathbf{F}(\mathbf{r}') d\tau' = -4\pi\mathbf{F}(\mathbf{r}).$$

But $\nabla\times\nabla\times\mathbf{G} = \nabla(\nabla\cdot\mathbf{G}) - \nabla^2\mathbf{G}$, so that

$$\mathbf{F}(\mathbf{r}) = -\frac{1}{4\pi}\left[\nabla(\nabla\cdot\mathbf{G}) - \nabla\times\nabla\times\mathbf{G}\right].$$

Define

$$\varphi = \frac{1}{4\pi}\nabla\cdot\mathbf{G} = \frac{1}{4\pi}\nabla\cdot\int\frac{\mathbf{F}(\mathbf{r}')}{|\mathbf{r}-\mathbf{r}'|} d\tau'.$$

$$\mathbf{A} = \frac{1}{4\pi}\nabla\times\mathbf{G} = \frac{1}{4\pi}\nabla\times\int\frac{\mathbf{F}(\mathbf{r}')}{|\mathbf{r}-\mathbf{r}'|} d\tau'.$$

Then $\mathbf{F}(\mathbf{r}) = -\nabla\varphi + \nabla\times\mathbf{A}$ and $\nabla\cdot\mathbf{A} = \frac{1}{4\pi}\nabla\cdot(\nabla\times\mathbf{G}) = 0$.

Let us now evaluate $\nabla_r\cdot\mathbf{G}(\mathbf{r})$.

$$\nabla_r\cdot\mathbf{G}(\mathbf{r}) = \nabla_r\cdot\int_V \frac{\mathbf{F}(\mathbf{r}')}{|\mathbf{r}-\mathbf{r}'|} d\tau' = \int_V \nabla_r \frac{1}{|\mathbf{r}-\mathbf{r}'|}\cdot\mathbf{F}(\mathbf{r}') d\tau'$$

$$= -\int_V \nabla_{r'} \frac{1}{|\mathbf{r}-\mathbf{r}'|}\cdot\mathbf{F}(\mathbf{r}') d\tau'$$

$$= -\int_V \nabla_{r'}\cdot\frac{\mathbf{F}(\mathbf{r}')}{|\mathbf{r}-\mathbf{r}'|} d\tau' + \int_V \frac{\nabla_{r'}\cdot\mathbf{F}(\mathbf{r}')}{|\mathbf{r}-\mathbf{r}'|} d\tau'$$

$$= -\int_S \frac{\mathbf{F}(\mathbf{r}')}{|\mathbf{r}-\mathbf{r}'|}\cdot d\mathbf{S} + \int_V \frac{\nabla_{r'}\cdot\mathbf{F}(\mathbf{r}')}{|\mathbf{r}-\mathbf{r}'|} d\tau' \to \int_V \frac{\nabla_{r'}\cdot\mathbf{F}(\mathbf{r}')}{|\mathbf{r}-\mathbf{r}'|} d\tau'.$$

since the first term on the right side of the equal sign vanishes, by virtue of the behavior of **F** as $r \to \infty$.

Hence,

$$4\pi\varphi = \nabla_r\cdot\mathbf{G}(\mathbf{r}) = \int_V \frac{\nabla_{r'}\cdot\mathbf{F}(\mathbf{r}')}{|\mathbf{r}-\mathbf{r}'|}\,d\tau'.$$

Let us now evaluate $\nabla_r\times\mathbf{G}(\mathbf{r})$.

$$\nabla_r\times\mathbf{G}(\mathbf{r}) = \int_V \nabla_r\times\frac{\mathbf{F}(\mathbf{r}')}{|\mathbf{r}-\mathbf{r}'|}\,d\tau' = -\int_V \nabla_{r'}\frac{1}{|\mathbf{r}-\mathbf{r}'|}\times\mathbf{F}(\mathbf{r}')d\tau'$$

$$= -\int_V \nabla_{r'}\times\frac{\mathbf{F}(\mathbf{r}')}{|\mathbf{r}-\mathbf{r}'|} + \int_V \frac{\nabla_{r'}\times\mathbf{F}(\mathbf{r}')}{|\mathbf{r}-\mathbf{r}'|}\,d\tau'.$$

But

$$\int_V (\nabla\times u\mathbf{A})_i d\tau = \int_V \sum_{jk}\varepsilon_{ijk}\frac{\partial}{\partial x_j}uA_k d\tau = \sum_j\int_V \frac{\partial}{\partial x_j}\left(\sum_k \varepsilon_{ijk}uA_k\right)d\tau$$

$$= \sum_j\int_S\sum_k \varepsilon_{ijk}uA_k dS_j = \int_S d\mathbf{S}\times u\mathbf{A} = -\int_S u\mathbf{A}\times d\mathbf{S}.$$

Hence

$$-\int_V \nabla_{r'}\times\frac{\mathbf{F}(\mathbf{r}')}{|\mathbf{r}-\mathbf{r}'|}\,d\tau' = \int_S \frac{\mathbf{F}(\mathbf{r}')}{|\mathbf{r}-\mathbf{r}'|}\times d\mathbf{S}' \Rightarrow 0 \text{ as } r\to\infty,$$

and

$$4\pi\mathbf{A} = \nabla_r\times\mathbf{G}(\mathbf{r}) = \int_V \frac{\nabla_{r'}\times\mathbf{F}(\mathbf{r}')}{|\mathbf{r}-\mathbf{r}'|}\,d\tau'.$$

Substituting the expressions for φ and $\mathbf{A}$ into $\mathbf{F} = -\nabla\varphi + \nabla\times\mathbf{A}$,

$$\mathbf{F}(\mathbf{r}) = -\frac{1}{4\pi}\nabla\int_V \frac{\nabla'\cdot\mathbf{F}(\mathbf{r}')}{|\mathbf{r}-\mathbf{r}'|}\,d\tau' + \frac{1}{4\pi}\nabla\times\int_V \frac{\nabla'\times\mathbf{F}(\mathbf{r}')}{|\mathbf{r}-\mathbf{r}'|}\,d\tau'.$$

If we let $\underset{\text{everywhere}}{(\nabla\cdot\mathbf{F})} = -\nabla^2\varphi = 4\pi\rho$ and $\underset{\text{everywhere}}{(\nabla\times\mathbf{F})} = \nabla\times\nabla\times\mathbf{A} = 4\pi\mathbf{j}/c$, then

$\varphi = \int \frac{\rho(\mathbf{r}')}{|\mathbf{r}-\mathbf{r}'|}\,d\tau'$ represents the solution of the scalar form of Poisson's equation and $\mathbf{A} = \int \frac{\mathbf{j}(\mathbf{r}')}{c|\mathbf{r}-\mathbf{r}'|}\,d\tau'$ represents the solution of the vector form of Poisson's equation.

Corollary (i):

If $\nabla^2\varphi = 0$ and $\varphi \to < \frac{1}{r^{1+\eta}}$ as $r \to \infty$, then $\varphi = 0$.

Proof: If $\mathbf{F} = -\nabla\varphi$, then $\nabla\cdot\mathbf{F} = -\nabla^2\varphi = 0$ by hypothesis, so that

$$\varphi = -\frac{1}{4\pi}\int\frac{\nabla'\cdot\mathbf{F}(\mathbf{r}')}{|\mathbf{r}-\mathbf{r}'|}d\tau' = 0.$$

Note that $\nabla\times\mathbf{F} = \nabla\times\nabla\varphi = 0$, so that $\mathbf{A} = \frac{1}{4\pi}\int_V\frac{\nabla'\times\mathbf{F}(\mathbf{r}')}{|\mathbf{r}-\mathbf{r}'|}d\tau' = 0$ as well.

The restriction placed on $\mathbf{F}$, enabling us to use Helmholtz' theorem, is here satisfied, since if $\varphi < \frac{1}{r^{1+\eta}}$, then $\mathbf{F} < \frac{1}{r^{2+\eta}}$ as $r \to \infty$.

Corollary (ii):

If $\mathbf{F}$ does not approach 0 as $r \to \infty$, but $\underset{r\to\infty}{\nabla\times\mathbf{F}} = \frac{4\pi\mathbf{j}}{c} \to < \frac{1}{r^{2+\eta}}$, then Helmholtz' theorem still holds.

Proof: Define $\mathbf{A}(\mathbf{r}) = \frac{1}{c}\int_V\frac{\mathbf{j}(\mathbf{r}')}{|\mathbf{r}-\mathbf{r}'|}d\tau'$.

$$\nabla_r\cdot\mathbf{A}(\mathbf{r}) = -\frac{1}{c}\int_V\nabla'\frac{1}{|\mathbf{r}-\mathbf{r}'|}\cdot\mathbf{j}(\mathbf{r}')d\tau'$$

$$= -\frac{1}{c}\int_V\nabla'\cdot\frac{\mathbf{j}(\mathbf{r}')}{|\mathbf{r}-\mathbf{r}'|}d\tau' + \frac{1}{c}\int_V\frac{\nabla'\cdot\mathbf{j}(\mathbf{r}')}{|\mathbf{r}-\mathbf{r}'|}d\tau'$$

$$= -\frac{1}{c}\int_S\frac{\mathbf{j}(\mathbf{r}')}{|\mathbf{r}-\mathbf{r}'|}\cdot d\mathbf{S}' + \frac{1}{c}\int_V\frac{\nabla'\cdot\mathbf{j}(\mathbf{r}')}{|\mathbf{r}-\mathbf{r}'|}d\tau'.$$

But

$$\frac{\nabla'\cdot\mathbf{j}'}{c} = \frac{\nabla'\cdot[\nabla'\times\mathbf{F}(\mathbf{r}')]}{4\pi} = 0 \text{ and } \int_S \to 0 \text{ as } r \to \infty.$$

Hence, $\nabla_r\cdot\mathbf{A}(\mathbf{r}) = 0$, so that $\nabla\times\nabla\times\mathbf{A} = \nabla(\nabla\cdot\mathbf{A}) - \nabla^2\mathbf{A} = -\nabla^2\mathbf{A}$. From the definition of $\mathbf{A}$ above, $\nabla^2\mathbf{A} = (4\pi)\mathbf{j}/c = \nabla\times\mathbf{F} \Rightarrow \nabla\times[\mathbf{F} - \nabla\times\mathbf{A}] = 0$.

Hence, $\mathbf{F} - \nabla\times\mathbf{A} = -\nabla\varphi$ and $\mathbf{F} = -\nabla\varphi + \nabla\times\mathbf{A}$.

Corollary (iii): Consider a sphere of radius R.

Define

$$\mathbf{F}'(\mathbf{r}) = \mathbf{F}(\mathbf{r}) \text{ for } |\mathbf{r}| < R.$$

$$\mathbf{F}'(\mathbf{r}) = 0, \text{ for } |\mathbf{r}| \geq R.$$

Now define $\mathbf{j}'(\mathbf{r}) = c\dfrac{\nabla\times\mathbf{F}'(\mathbf{r})}{4\pi}$ everywhere, so that $\mathbf{j}' = 0$ outside R and $\mathbf{j}' = c\dfrac{\nabla\times\mathbf{F}(\mathbf{r})}{4\pi}$ inside R, so that $\nabla\cdot\mathbf{j}'(\mathbf{r}) = 0$ everywhere. Now define $\mathbf{A}(\mathbf{r}) = \dfrac{1}{c}\displaystyle\int_V \frac{\mathbf{j}'(\mathbf{r}')}{|\mathbf{r}-\mathbf{r}'|}\,d\tau'$ and investigate $\nabla_r\cdot\mathbf{A}(\mathbf{r})$.

$$c\nabla_r\cdot\mathbf{A}(\mathbf{r}) = -\int_V \nabla'\frac{1}{|\mathbf{r}-\mathbf{r}'|}\cdot\mathbf{j}'(\mathbf{r}')\,d\tau'$$

$$= -\int_V \nabla'\cdot\frac{\mathbf{j}'(\mathbf{r}')}{|\mathbf{r}-\mathbf{r}'|}\,d\tau' + \int_V \frac{\nabla'\cdot\mathbf{j}'(\mathbf{r}')}{|\mathbf{r}-\mathbf{r}'|}\,d\tau'$$

$$= -\int_S \frac{\mathbf{j}'(\mathbf{r}')}{|\mathbf{r}-\mathbf{r}'|}\cdot d\mathbf{S}' + 0 \;\rightarrow\; 0 \text{ for } S > R.$$

Hence, $\nabla_r\cdot\mathbf{A}(\mathbf{r}) = 0$.

We have the vector identity, $\nabla\times\nabla\times\mathbf{A} = \nabla(\nabla\cdot\mathbf{A}) - \nabla^2\mathbf{A}$. But since $\nabla\cdot\mathbf{A} = 0$, $\nabla\times\nabla\times\mathbf{A} = -\nabla^2\mathbf{A} = 4\pi\mathbf{j}/c$ from the definition of $\mathbf{A}$ and from Poisson's equation.

Hence,

$$\nabla\times[\mathbf{F}' - \nabla\times\mathbf{A}] = \frac{4\pi}{c}[\mathbf{j}' - \mathbf{j}] \quad \begin{array}{l} = 0 \text{ inside } R \\ \neq 0 \text{ outside } R \end{array}$$

Then within a singly connected region R, such as a sphere, $\mathbf{F} - \nabla\times\mathbf{A} = -\nabla\varphi$, and since the region R is arbitrary, we may make it as large as we please, thus proving the theorem for every point except at infinity, which can be included if $|\mathbf{j}| \rightarrow (1/r^{2+\delta})$, which is usually the case.

Corollary (iv):

$$\mathbf{F} = -\nabla\varphi + \nabla\times\mathbf{A}'$$

where $\nabla\cdot\mathbf{A}'$ equals any given function (not necessarily 0).

Proof: We know that **F** can be written as $\mathbf{F} = -\nabla\varphi + \nabla\times\mathbf{A}$, where $\nabla\cdot\mathbf{A} = 0$. Let us define χ such that $\nabla^2\chi = f$ and construct the vector $\mathbf{A}'$ from the vectors **A** and $\nabla\chi$.

Define $\mathbf{A}' \equiv \mathbf{A} + \nabla\chi \Rightarrow \nabla\cdot\mathbf{A}' = \nabla\cdot\mathbf{A} + \nabla^2\chi = f$ and $\nabla\times\mathbf{A}' = \nabla\times\mathbf{A}$. If we choose a χ_o such that $\nabla^2\chi_o = 0$, and define a new vector **B** such that $\mathbf{B} \equiv \mathbf{A} + \nabla\chi_o$, then

$$\nabla\cdot\mathbf{B} = \nabla\cdot\mathbf{A} + \nabla^2\chi_o = \nabla\cdot\mathbf{A}.$$

$$\nabla\times\mathbf{B} = \nabla\times\mathbf{A} + \nabla\times(\nabla\chi_o) = \nabla\times\mathbf{A}.$$

Thus we see that the vector potentials **A** and **B** are not unique, i.e.

$$\mathbf{F} = -\nabla\varphi + \nabla\times\mathbf{B} = -\nabla\varphi + \nabla\times\mathbf{A}.$$

3.6 Equivalent Forms of Gauss' and Stokes' Theorems

3.6.1 Gauss' Theorem (refer to problems)

In its general form,

$$\int_V \frac{\partial}{\partial x_i} T_{j_1 j_2 \ldots j_n} d\tau = \int_S T_{j_1 j_2 \ldots j_n} dS_i .$$

$$\sum_{jk} \int_V \varepsilon_{ijk} \frac{\partial}{\partial x_j} A_{kl}\, d\tau = \sum_j \int_V \sum_k \varepsilon_{ijk} A_{kl}\, dS_j .$$

In its special form,

$$\int_V \frac{\partial}{\partial x_i} T_{ijk\ldots} d\tau = \sum_i \int_S T_{ijk\ldots} dS_i .$$

The above reduce to

$$\int_V \nabla\varphi d\tau = \int_S \varphi d\mathbf{S},$$

for T = tensor of 0th rank, i.e. a scalar.

$$\int_V \nabla\times\mathbf{A}d\tau = -\int_S \mathbf{A}\times d\mathbf{S} \text{ and } \int_V \nabla\cdot\mathbf{A}d\tau = -\int_S \mathbf{A}\cdot d\mathbf{S},$$

for T = tensor of 1st rank, i.e. a vector.

3.6.2 Stokes' Theorem (refer to problems)

In its general form,

$$\sum_{ijk}\int_S \varepsilon_{ijk}\frac{\partial}{\partial x_j} T_{ki_1i_2\ldots}dS_i = \sum_i\int_s T_{ii_1i_2\ldots}ds_i.$$

$$\sum_{ij}\int_S \varepsilon_{ijk}\frac{\partial}{\partial x_j} T_{i_1i_2\ldots}dS_i = \sum_i\int_s T_{i_1i_2\ldots}ds_k.$$

These reduce to

$$\int_S d\mathbf{S}\times\nabla\varphi = \int_s \varphi d\mathbf{s},$$

for T = tensor of 0th rank, i.e. a scalar.

$$\int_S \nabla\times\mathbf{A}\cdot d\mathbf{S} = \int_s \mathbf{A}\cdot d\mathbf{s},$$

for T = tensor of 1st rank, i.e. a vector.

3.7 Maxwell's Equations

3.7.1 Magnetostatics

Define a vector potential

$$\mathbf{A}(\mathbf{r}) = \frac{1}{c}\int_V \frac{\mathbf{j}(\mathbf{r}')}{|\mathbf{r}-\mathbf{r}'|}d\tau'.$$

$$\nabla\times\mathbf{A} = \frac{1}{c}\int \mathbf{j}(\mathbf{r}')\times\left(\nabla\frac{-1}{|\mathbf{r}-\mathbf{r}'|}\right)d\tau'.$$

Since

$$\nabla\frac{-1}{|\mathbf{r}-\mathbf{r}'|} = -\frac{1}{|\mathbf{r}-\mathbf{r}'|^2}\nabla|\mathbf{r}-\mathbf{r}'| = -\frac{(\mathbf{r}-\mathbf{r}')}{|\mathbf{r}-\mathbf{r}'|^3},$$

$$\nabla\times\mathbf{A} = \frac{1}{c}\int\frac{\mathbf{j}(\mathbf{r}')\times(\mathbf{r}-\mathbf{r}')}{|\mathbf{r}-\mathbf{r}'|^3}\,d\tau'.$$

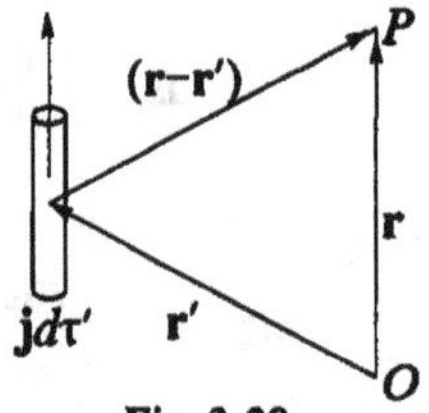

Fig. 3-29

Ampere showed experimentally that the magnetic field $\mathbf{H}_P$ (Fig. 3-29) due to current $\mathbf{j}d\tau'$ is given by

$$\mathbf{H}_P = \int\frac{\mathbf{j}(\mathbf{r}')\times(\mathbf{r}-\mathbf{r}')}{c|\mathbf{r}-\mathbf{r}'|^3}\,d\tau'.$$

Hence, $\mathbf{H} = \nabla\times\mathbf{A} \Rightarrow \nabla\cdot\mathbf{H} = 0$.

Note that $\nabla\times\mathbf{H} = \nabla\times(\nabla\times\mathbf{A}) = \nabla(\nabla\cdot\mathbf{A}) - \nabla^2\mathbf{A}$, and if $\nabla\cdot\mathbf{j} = 0$, then $\nabla\cdot\mathbf{A} = 0$, since

$$\nabla\cdot\mathbf{A} = \frac{1}{c}\int\nabla\cdot\frac{\mathbf{j}'}{|\mathbf{r}-\mathbf{r}'|}d\tau' = \frac{1}{c}\int\nabla\frac{1}{|\mathbf{r}-\mathbf{r}'|}\cdot\mathbf{j}'d\tau' = -\frac{1}{c}\int\nabla'\cdot\frac{\mathbf{j}'}{|\mathbf{r}-\mathbf{r}'|}d\tau'$$

$$= -\frac{1}{c}\int_\infty\frac{\mathbf{j}'\cdot d\mathbf{S}'}{|\mathbf{r}-\mathbf{r}'|} \to 0 \text{ as } r\to\infty.$$

Thus, $\nabla\times\mathbf{H} = -\nabla^2\mathbf{A} = 4\pi\mathbf{j}/c$, which defines magnetostatics.

What is the significance of $\nabla\cdot\mathbf{j} = 0$? In general, $\nabla\cdot\mathbf{j} \neq 0$. Consider the continuity equation $\nabla\cdot\rho\mathbf{v} = -\partial\rho/\partial t$ or $\nabla\cdot\mathbf{j} = -\partial\rho/\partial t$. Thus, the static case implies $\partial\rho/\partial t = 0$, which in turn implies $\nabla\cdot\mathbf{j} = 0$.

Let us therefore write $\nabla\times\mathbf{H} = 4\pi\mathbf{j}/c + \vec{\chi}$, where $\vec{\chi} = 0$ when things do not change with time. Thus, $\nabla\cdot(\nabla\times\mathbf{H}) = 0 = (4\pi/c)\nabla\cdot\mathbf{j} + \nabla\cdot\vec{\chi}$; in other words,

$$\nabla\cdot\vec{\chi} = -\frac{4\pi}{c}\nabla\cdot\mathbf{j} = \frac{4\pi}{c}\left(\frac{\partial\rho}{\partial t}\right) = \frac{1}{c}\frac{\partial(4\pi\rho)}{\partial t} = \frac{1}{c}\frac{\partial(\nabla\cdot\mathbf{E})}{\partial t} = \frac{\nabla\cdot\dot{\mathbf{E}}}{c}.$$

Hence, we choose $\vec{\chi} = \dot{\mathbf{E}}/c$, so that $\nabla\times\mathbf{H} = 4\pi\mathbf{j}/c + \dot{\mathbf{E}}/c$, where $\dot{\mathbf{E}}/c$ is denoted the displacement current. For constant electric fields, i.e. no time variations, $\dot{\mathbf{E}} = 0$.

3.7.2 Faraday's Law

Faraday's law states (Fig. 3-30) that

$$\oint_s \mathbf{E}\cdot d\mathbf{s} = -\frac{1}{c}\frac{\partial}{\partial t}\int_S \mathbf{H}\cdot d\mathbf{S}.$$

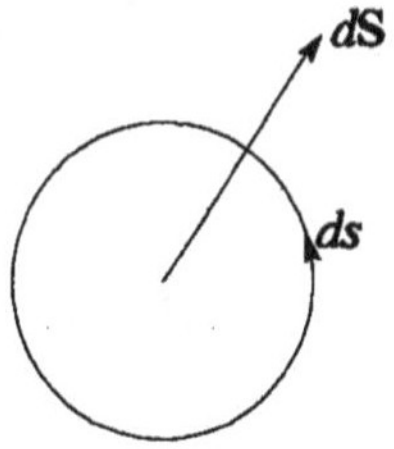

Fig.3-30

Applying Stokes' theorem to the above,

$$\int_S \nabla\times\mathbf{E}\cdot d\mathbf{S} + \frac{1}{c}\frac{\partial}{\partial t}\int_S \mathbf{H}\cdot d\mathbf{S} = 0.$$

Hence,

$$\nabla\times\mathbf{E} = -\frac{1}{c}\frac{\partial \mathbf{H}}{\partial t}.$$

Summarizing Maxwell's equations for electrodynamics:

(1) $\nabla\cdot\mathbf{E} = 4\pi\rho$, the electric field arising from a charge distribution ρ.

(2) $\nabla\cdot\mathbf{H} = 0$, implying the non-existence of magnetic monopoles.

(3) $\nabla\times\mathbf{E} = -\dot{\mathbf{H}}/c$, arising from Faraday's law of electromagnetic induction.

(4) $\nabla\times\mathbf{H} = 4\pi\mathbf{j}/c + \dot{\mathbf{E}}/c$, arising from Ampere's law and the equation of continuity.

3.7.3 Wave Equations

If $\nabla\cdot\mathbf{H} = 0$ everywhere, then we know (see Problems) that

(2a) $\mathbf{H} = \nabla\times\mathbf{A}$,

where $\nabla\cdot\mathbf{A}$ can equal any function we please. Substituting (2a) into (3),

$$\nabla\times\mathbf{E} = -\frac{1}{c}\nabla\times\dot{\mathbf{A}} \quad\rightarrow\quad \nabla\times\left[\mathbf{E} + \frac{1}{c}\dot{\mathbf{A}}\right] = 0.$$

But we have shown that if the curl of a vector vanishes, then the vector can be expressed as the gradient of a scalar. Hence,

(3a) $\mathbf{E} = -\nabla\varphi - \dot{\mathbf{A}}/c.$

Substituting (3a) into (1), we get

(5) $-\Delta\varphi - (\nabla\cdot\dot{\mathbf{A}})/c = 4\pi\rho.$

[If we choose the "Lorentz gauge," such that $\nabla\cdot\mathbf{A} = -\dot{\varphi}/c$, then we have $\nabla^2\varphi - \ddot{\varphi}/c^2 = -4\pi\rho$.]

Substituting (2a) and (3a) into (4) we get

$$\nabla\times(\nabla\times\mathbf{A}) - \frac{1}{c}\left(-\nabla\dot{\varphi} - \frac{1}{c}\ddot{\mathbf{A}}\right) = -\Delta\mathbf{A} + \frac{1}{c^2}\ddot{\mathbf{A}} + \nabla(\nabla\cdot\mathbf{A} + \dot{\varphi}/c) = \frac{4\pi\mathbf{j}}{c},$$

and, using the Lorentz gauge again:

(6) $\Delta\mathbf{A} - \frac{1}{c^2}\ddot{\mathbf{A}} = -\frac{4\pi\mathbf{j}}{c}.$

Defining the D'Alembertian operator $\Box^2 \equiv \Delta - \frac{1}{c^2}\frac{\partial^2}{\partial t^2}$, we can rewrite the above wave equations as

$$\Box^2\mathbf{A} = -\frac{4\pi\mathbf{j}}{c}. \qquad \Box^2\varphi = -4\pi\rho.$$

Consider the homogeneous case (i.e. within a vacuum), for which $\rho = \mathbf{j} = 0$. The wave equations then become $\Box^2\mathbf{A} = \Box^2\varphi = 0$, and Maxwell's equations become

$$\nabla\cdot\mathbf{H} = 0. \qquad \nabla\cdot\mathbf{E} = 0.$$

$$\nabla\times\mathbf{E} = -\dot{\mathbf{H}}/c \quad\Rightarrow\quad \nabla\times(\nabla\times\mathbf{E}) = -\nabla\times\dot{\mathbf{H}}/c = -\ddot{\mathbf{E}}/c^2 = -\nabla^2\mathbf{E}.$$

$$\nabla\times\mathbf{H} = \dot{\mathbf{E}}/c \quad\Rightarrow\quad \nabla\times(\nabla\times\mathbf{H}) = \nabla\times\dot{\mathbf{E}}/c = -\ddot{\mathbf{H}}/c^2 = -\nabla^2\mathbf{H}.$$

Thus **H** and **E** satisfy the same equations as do **A** and φ, namely:

$$\Box^2\mathbf{H} = 0. \qquad \Box^2\mathbf{E} = 0.$$

Thus, in a vacuum the function $\chi = \varphi, A_i, H_i, E_i$ all satisfy $\Box^2\chi = 0$.

Let $\chi = f(z)$, where $z = t - (\mathbf{u}_o\cdot\mathbf{r})/c$, and where f is an arbitrary function of z and $\mathbf{u}_o$ is a unit vector in any direction.

$$\frac{\partial\chi}{\partial x_i} = \frac{df}{dz}\frac{\partial z}{\partial x_i} = \frac{df}{dz}\left(-\frac{u_i}{c}\right).$$

$$\frac{\partial^2\chi}{\partial x_i^2} = \frac{d^2f}{dz^2}\left(-\frac{u_i}{c}\right)\frac{\partial z}{\partial x_i} = \frac{u_i^2}{c^2}\frac{d^2f}{dz^2}.$$

$$\frac{\partial\chi}{\partial t} = \frac{df}{dz}\frac{\partial z}{\partial t} = \frac{df}{dz}. \qquad \frac{\partial^2\chi}{\partial t^2} = \frac{d^2f}{dz^2}.$$

Since $\sum_i u_i^2 = 1$, we have $\Box^2\chi = \frac{d^2f}{dz^2}\left\{\sum_i\frac{u_i^2}{c^2} - \frac{1}{c^2}\right\} = 0$, thereby showing that $\chi = f(z)$ satisfies $\Box^2\chi = 0$. If χ_1 and χ_2 are solutions, then by virtue of the linear operations ∇ and $\partial/\partial t$, the sum of χ_1 and χ_2 is also a solution. Hence, the most general solution of $\Box^2\chi = 0$ is given by

$$\chi = \sum_{\substack{\text{set}\\ \text{of } u_j}} f_j\left(t - \frac{\mathbf{u}_j\cdot\mathbf{r}}{c}\right).$$

3.7.4 Physical Significance of Solution to Wave Equation

Consider a particular solution, namely ψ = const. at $t = t_o$. (Fig. 3-31).

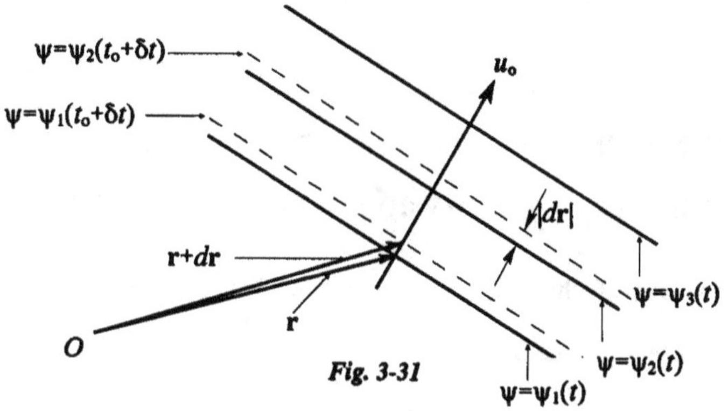

Fig. 3-31

We are implying that $z = t_0 - (u_i \cdot \mathbf{r})/c$ = constant, which are clearly seen to be plane surfaces. Note that along z = const., $dz = 0$, and $dt = 0$, so that $u_0 \cdot d\mathbf{r} = 0$ and u_0 is perpendicular to $d\mathbf{r}$. Let a later time $t = t_0 + \delta t$ be represented by the dotted lines, as shown in Fig. 3-31. Then at this later time (fixed at $t = t_0 + \delta t$), $dz = 0$, i.e. the same surface is under consideration at a later time.

$$dz = 0 = dt - \frac{d\mathbf{r} \cdot u_0}{c} \quad \rightarrow \quad |d\mathbf{r}| = c\,dt,$$

where c = phase velocity, i.e. that with which the planes move.

For a sinusoidal function,

$$\psi = \psi_0 \begin{matrix}\sin \\ \cos\end{matrix} (\mathbf{k} \cdot \mathbf{r} - \omega t), \text{ where } \omega = c|\mathbf{k}|$$

(Fig. 3-32). For $\mathbf{r}$ parallel to $\mathbf{k}$,

$$\psi = \psi_0 \begin{matrix}\sin \\ \cos\end{matrix} (kr - \omega t).$$

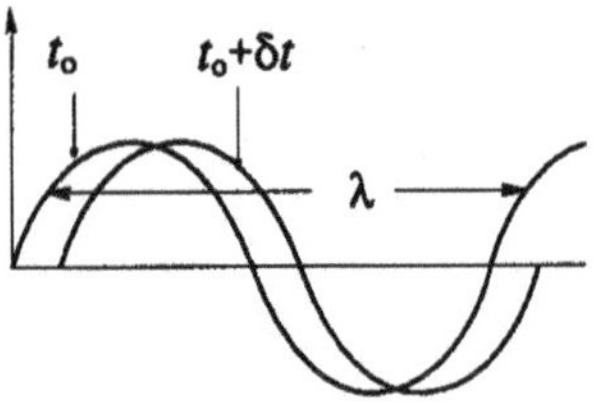

Fig. 3-32

2-dimensional analysis (Fig. 3-33):

At a fixed point, things repeat after a time $T \equiv (2\pi)/\omega$, and at a given time, things repeat after a distance $\lambda = (2\pi)/k$, the wavelength.

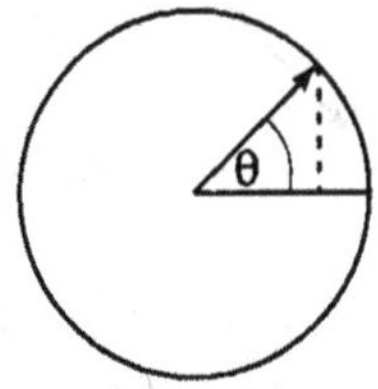

Fig. 3-33

ω ≡ angular velocity of phase.

k ≡ wave number (the number of waves in a distance 2π).

3.8 Curvilinear Orthogonal Coordinate Systems

A. Spherical Coordinates

A point is determined by (r, θ, φ) (Fig. 3-34). At a point P we have the orthogonal system formed by the triad shown in Fig. 3-35. This system is called orthogonal because the triad formed at every point by unit vectors in the direction of increasing r, θ and φ is orthogonal. This is reflected in the fact that ds^2 contains no mixed coordinate differentials.

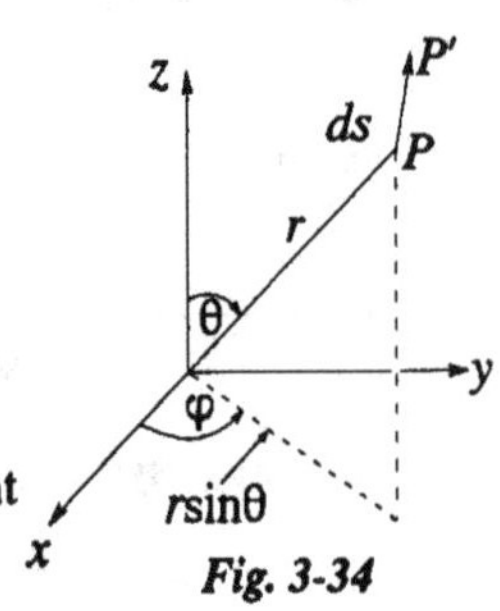

Fig. 3-34

$$ds^2 = dr^2 + r^2 d\theta^2 + r^2 \sin^2\theta d\varphi^2.$$

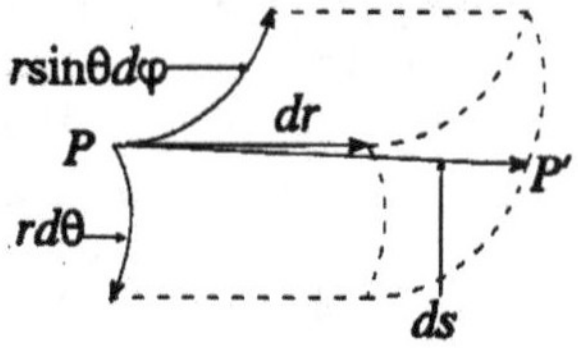

Fig. 3-35

B. Cylindrical Coordinates

A point is determined by (ρ, z, φ) (Fig. 3-36). At a point P we have the orthogonal system formed by the triad shown in Fig. 3-37.

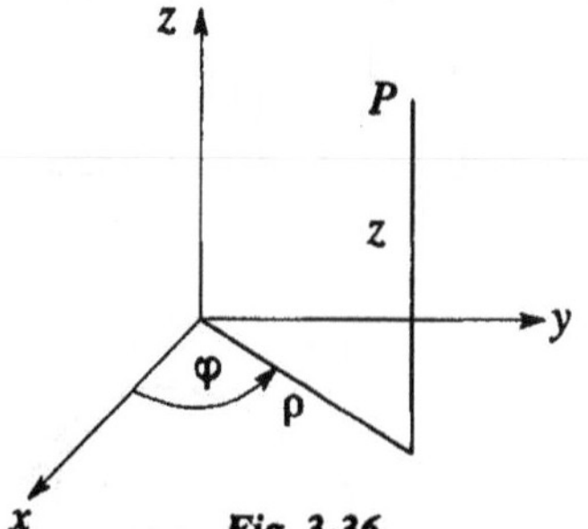

Fig. 3-36

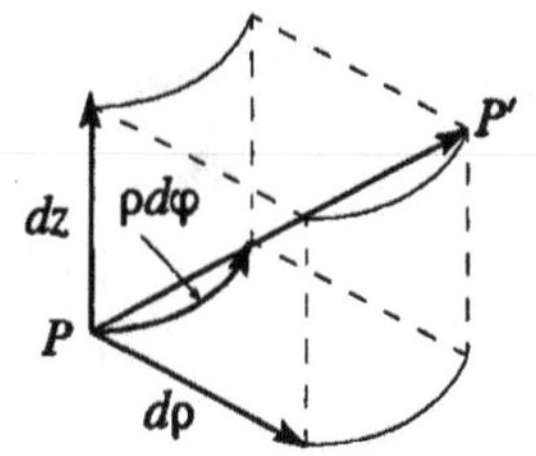

Fig. 3-37

In this case, $ds^2 = d\rho^2 + \rho^2 d\varphi^2 + dz^2$.

In general the transformation from cartesian rectangular to orthogonal curvilinear coordinates is given by $x_1\, x_2\, x_3 \rightarrow \xi_1\, \xi_2\, \xi_3$, where

$$x_1 = x_1(\xi_1\, \xi_2\, \xi_3); \quad x_2 = x_2(\xi_1\, \xi_2\, \xi_3); \quad x_3 = x_3(\xi_1\, \xi_2\, \xi_3).$$

$$dx_1 = \frac{\partial x_1}{\partial \xi_1} d\xi_1 + \frac{\partial x_1}{\partial \xi_2} d\xi_2 + \frac{\partial x_1}{\partial \xi_3} d\xi_3.$$

$$dx_2 = \frac{\partial x_2}{\partial \xi_1} d\xi_1 + \frac{\partial x_2}{\partial \xi_2} d\xi_2 + \frac{\partial x_2}{\partial \xi_3} d\xi_3.$$

$$dx_3 = \frac{\partial x_1}{\partial \xi_1} d\xi_1 + \frac{\partial x_3}{\partial \xi_2} d\xi_2 + \frac{\partial x_3}{\partial \xi_3} d\xi_3.$$

In the x-coordinate system,

$$ds^2 = \sum_i dx_i^2 = \sum_k dx_k dx_k.$$

Since we can write $dx_i = \sum_j \frac{\partial x_i}{\partial \xi_j} d\xi_j$, the transformed ds^2 becomes:

$$ds^2 = \sum_{ijk} \frac{\partial x_i}{\partial \xi_j}\frac{\partial x_i}{\partial \xi_k} d\xi_j d\xi_k = \sum_{jk} g_{jk} d\xi_j d\xi_k ,$$

where

$$g_{jk} = g_{kj} = \sum_i \frac{\partial x_i}{\partial \xi_j}\frac{\partial x_i}{\partial \xi_k}.$$

In the expression $ds^2 = \sum_{jk} g_{jk} d\xi_j d\xi_k$,one would naturally expect to get cross terms such as $d\xi_1 d\xi_3$. However, if $ds^2 = \sum_i h_i^2 d\xi_i^2$, then the ξ-system defines an orthogonal system, and if $dx_i dxi_j \neq 0$, then we do not have an orthogonal system.

For spherical coordinates:

$$\xi_1 = r \qquad h_1 = 1 = \sqrt{g_{11}}.$$

$$\xi_2 = \theta \qquad h_1 = r = \sqrt{g_{22}}.$$

$$\xi_3 = \varphi \qquad h_3 = r\sin\theta = \sqrt{g_{33}}.$$

For cylindrical coordinates:

$$\xi_1 = \rho \qquad h_1 = 1 = \sqrt{g_{11}}.$$

$$\xi_2 = \varphi \qquad h_1 = \rho = \sqrt{g_{22}}.$$

$$\xi_3 = z \qquad h_3 = 1 = \sqrt{g_{33}}.$$

For any set of orthogonal curvilinear coordinates (Fig. 3-38), we have then

$$ds^2 = \sum_i h_i^2 d\xi_i^2 = \sum_i g_{ii} d\xi_i^2 ,$$

where

$$g_{ii} = h_i^2 = \sum_j \left(\frac{\partial x_j}{\partial \xi_i}\right)^2.$$

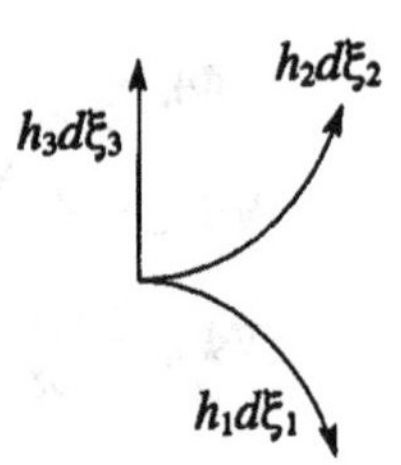

Fig. 3-38

3.8.1 Vector Components

To enable us to obtain vector components in orthogonal curvilinear coordinate systems, we define

$$(\mathbf{A})_{i_{\text{at }P}} \equiv (\mathbf{A}\cdot\boldsymbol{e}_i)_{\text{at }P},$$

where $\boldsymbol{e}_i$ is a unit vector in the ith direction, i.e. tangent to the ith coordinate. Direction i is such that $d\xi_j = d\xi_k = 0$.

Gradient (∇)

$$(\nabla\Phi)_i = \left(\frac{\delta\Phi}{\delta s}\right)_i,$$

where s is a distance. But along direction i, $d\xi_j = d\xi_k = 0$ and $\delta s = h_i\delta\xi_i$. Hence,

$$(\nabla\Phi)_i = \frac{1}{h_i}\frac{\partial\Phi}{\partial\xi_i}.$$

Spherical Coordinates

$$(\nabla\Phi)_r = \frac{\partial\Phi}{\partial r}$$

$$(\nabla\Phi)_\theta = \frac{1}{r}\frac{\partial\Phi}{\partial\theta} \quad\Rightarrow\quad \nabla\Phi = \frac{\partial\Phi}{\partial r}\boldsymbol{e}_r + \frac{1}{r}\frac{\partial\Phi}{\partial\theta}\boldsymbol{e}_\theta + \frac{1}{r\sin\theta}\frac{\partial\Phi}{\partial\varphi}\boldsymbol{e}_\varphi .$$

$$(\nabla\Phi)_\varphi = \frac{1}{r\sin\theta}\frac{\partial\Phi}{\partial\varphi}$$

Cylindrical Coordinates

$$(\nabla\Phi)_\rho = \frac{\partial\Phi}{\partial\rho}$$

$$(\nabla\Phi)_\varphi = \frac{1}{\rho}\frac{\partial\Phi}{\partial\varphi} \quad\Rightarrow\quad \nabla\Phi = \frac{\partial\Phi}{\partial\rho}\boldsymbol{e}_\rho + \frac{1}{\rho}\frac{\partial\Phi}{\partial\varphi}\boldsymbol{e}_\varphi + \frac{\partial\Phi}{\partial z}\boldsymbol{e}_z .$$

$$(\nabla\Phi)_z = \frac{\partial\Phi}{\partial z}$$

Divergence ($\nabla\cdot\mathbf{A}$)

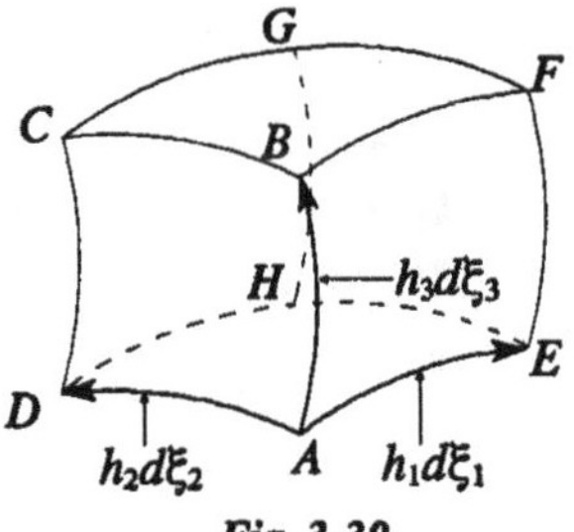

Fig. 3-39

One could plug in **A** and $\nabla\cdot$ in the $i_o j_o k_o$ system and write out $\nabla\cdot\mathbf{A}$ involving transformations from x to ξ systems. Instead, we derive $\nabla\cdot\mathbf{A}$ applying Gauss' divergence theorem as follows:

In general, $\int_V \nabla\cdot\mathbf{A}d\tau = \oint_S \mathbf{A}\cdot d\mathbf{S}$, and let us consider a small volume $\delta\tau$ (Fig. 3-39). Then,

$$\nabla\cdot\mathbf{A} = \lim_{\delta\tau\to 0}\frac{1}{\delta\tau}\oint_S \mathbf{A}\cdot d\mathbf{S}.$$

$$\delta\tau = \prod_{i=1}^{3} h_i d\xi_i = h_1h_2h_3 d\xi_1 d\xi_2 d\xi_3.$$

$$\int(\mathbf{A}\cdot d\mathbf{S})_{\text{at } ABCD} = -A_1h_2h_3d\xi_2d\xi_3 \text{ (at } \xi_1).$$

$$\int(\mathbf{A}\cdot d\mathbf{S})_{\text{at } EFGH} = +A_1h_2h_3d\xi_2d\xi_3 \text{ (at } \xi_1 + d\xi_1).$$

The sum of these terms yields

$$\frac{\partial}{\partial\xi_1}(A_1h_2h_3)d\xi_2d\xi_3d\xi_1.$$

Doing the same for $d\mathbf{S}_2$ and $d\mathbf{S}_3$ merely involves cyclical permutation of indices. Hence,

$$\nabla\cdot\mathbf{A} = \frac{1}{\prod_i^3 h_i d\xi_i}\left[\frac{\partial(A_1h_2h_3)}{\partial\xi_1} + \frac{\partial(A_2h_3h_1)}{\partial\xi_2} + \frac{\partial(A_3h_1h_2)}{\partial\xi_3}\right]d\xi_1d\xi_2d\xi_3,$$

or

$$\nabla\cdot\mathbf{A} = \frac{1}{h_1h_2h_3}\left[\frac{\partial(A_1h_2h_3)}{\partial\xi_1} + \frac{\partial(A_2h_3h_1)}{\partial\xi_2} + \frac{\partial(A_3h_1h_2)}{\partial\xi_3}\right].$$

Spherical Coordinates

$$\begin{aligned} h_1 &= 1 \\ h_2 &= r \qquad \Rightarrow \quad h_1h_2h_3 = r^2\sin\theta. \\ h_3 &= r\sin\theta \end{aligned}$$

$$\nabla\cdot\mathbf{A} = \frac{1}{r^2\sin\theta}\left[\frac{\partial(A_1r^2\sin\theta)}{\partial r} + \frac{\partial(A_2r\sin\theta)}{\partial\theta} + \frac{\partial(A_3r)}{\partial\varphi}\right]$$

$$= \frac{1}{r^2}\frac{\partial(A_rr^2)}{\partial r} + \frac{1}{r\sin\theta}\frac{\partial(A_\theta\sin\theta)}{\partial\theta} + \frac{1}{r\sin\theta}\frac{\partial(A_\varphi)}{\partial\varphi}.$$

Cylindrical Coordinates

$$\begin{aligned} h_1 &= 1 \\ h_2 &= \rho \quad \Rightarrow \quad h_1h_2h_3 = \rho. \\ h_3 &= 1 \end{aligned}$$

$$\nabla\cdot\mathbf{A} = \frac{1}{\rho}\left[\frac{\partial(A_\rho\rho)}{\partial\rho} + \frac{\partial(A_\varphi)}{\partial\varphi} + \frac{\partial(A_z\rho)}{\partial z}\right]$$

$$= \frac{1}{\rho}\frac{\partial(A_\rho\rho)}{\partial\rho} + \frac{1}{\rho}\frac{\partial(A_\varphi)}{\partial\varphi} + \frac{\partial(A_z\rho)}{\partial z}.$$

Laplacian **($\nabla^2\Phi$)**

$$\nabla^2\Phi = \nabla\cdot\nabla\Phi$$

$$= \frac{1}{h_1h_2h_3}\left\{\frac{\partial}{\partial\xi_1}\left(\nabla\Phi_1h_2h_3\right) + \frac{\partial}{\partial\xi_2}\left(\nabla\Phi_2h_3h_1\right) + \frac{\partial}{\partial\xi_3}\left(\nabla\Phi_3h_1h_2\right)\right\}.$$

Using expressions previously developed for $\nabla\Phi$, we have:

Spherical Coordinates

$$\Delta\Phi = \frac{1}{r^2}\frac{\partial}{\partial r}\left(r^2\frac{\partial\Phi}{\partial r}\right) + \frac{1}{r^2\sin\theta}\frac{\partial}{\partial\theta}\left(\sin\theta\frac{\partial\Phi}{\partial\theta}\right) + \frac{1}{r^2\sin^2\theta}\frac{\partial^2\Phi}{\partial\varphi^2}.$$

<u>*Cylindrical Coordinates*</u>

$$\Delta\Phi = \frac{1}{\rho}\frac{\partial}{\partial\rho}\left(\rho\frac{\partial\Phi}{\partial\rho}\right) + \frac{1}{\rho^2}\frac{\partial^2\Phi}{\partial^2\varphi} + \frac{\partial^2\Phi}{\partial z^2}.$$

Curl **($\nabla\times\mathbf{A}$)**

As we did in the case of the divergence, we derive $\nabla\times\mathbf{A}$ applying Stokes' theorem,

$$\int_S \nabla\times\mathbf{A}\cdot d\mathbf{S} = \oint_s \mathbf{A}\cdot d\mathbf{s},$$

to an infinitesimal surface area S that is parallel to the ith axis. Setting $|\mathbf{S}| = S^i$,

$$(\nabla\times\mathbf{A})_i = \lim_{S^i\to 0}\frac{1}{S^i}\oint \mathbf{A}\cdot d\mathbf{s}.$$

Construct a surface so that $d\mathbf{S}$ is parallel to axis 1 (Fig. 3-40), and consider $\oint_s \mathbf{A}\cdot d\mathbf{s}$.

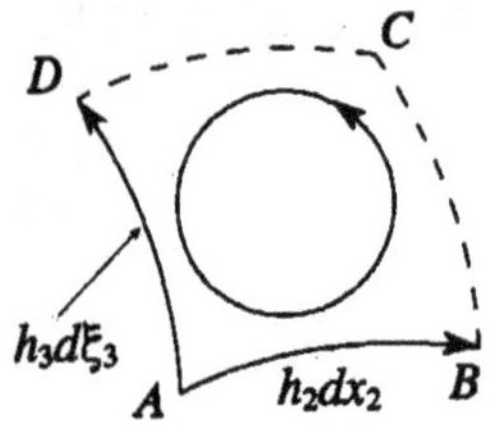

Fig. 3-40

Contribution from AB: $A_2h_2d\xi_2$ (at ξ_3).

Contribution from BC: $A_3h_3d\xi_3$ (at $\xi_2 + d\xi_2$).

Contribution from CD: $-A_2h_2d\xi_2$ (at $\xi_3 + d\xi_3$).

Contribution from DA: $-A_3h_3d\xi_3$ (at ξ_2).

Hence,

$$\oint_s \mathbf{A}\cdot d\mathbf{s} = \left[-\frac{\partial}{\partial\xi_3}(A_2h_2) + \frac{\partial}{\partial\xi_2}(A_3h_3)\right]d\xi_2 d\xi_3.$$

Substituting $S^1 = h_2h_3d\xi_2d\xi_3$, $(\nabla\times\mathbf{A})_1 = \frac{1}{h_2h_3}\left[\frac{\partial}{\partial\xi_2}(A_3h_3) - \frac{\partial}{\partial\xi_3}(A_2h_2)\right]$.

Employing cyclic interchange,

$$(\nabla\times\mathbf{A})_i = \frac{1}{h_jh_k}\left[\frac{\partial}{\partial\xi_j}(A_kh_k) - \frac{\partial}{\partial\xi_k}(A_jh_j)\right].$$

Spherical Coordinates

$$(\nabla\times\mathbf{A})_r = \frac{1}{r^2\sin\theta}\left[\frac{\partial}{\partial\theta}(r\sin\theta A_\varphi) - \frac{\partial}{\partial\varphi}(rA_\theta)\right].$$

$$(\nabla\times\mathbf{A})_\theta = \frac{1}{r\sin\theta}\left[\frac{\partial}{\partial\varphi}(A_r) - \frac{\partial}{\partial r}(r\sin\theta A_\varphi)\right].$$

$$(\nabla\times\mathbf{A})_\varphi = \frac{1}{r}\left[\frac{\partial}{\partial r}(rA_\theta) - \frac{\partial}{\partial\theta}(A_r)\right].$$

Cylindrical Coordinates

$$(\nabla\times\mathbf{A})_\rho = \frac{1}{\rho}\left[\frac{\partial}{\partial\varphi}(A_z) - \frac{\partial}{\partial z}(\rho A_\varphi)\right].$$

$$(\nabla\times\mathbf{A})_\varphi = \left[\frac{\partial}{\partial z}(A_\rho) - \frac{\partial}{\partial\rho}(A_z)\right].$$

$$(\nabla\times\mathbf{A})_z = \frac{1}{\rho}\left[\frac{\partial}{\partial\rho}(\rho A_\varphi) - \frac{\partial}{\partial\varphi}(A_\rho)\right].$$

Chapter 4

Matrix and Vector Algebra in N-Dimensional Space

4.1 Algebra of N-Dimensional Complex Space

The generalization of a radial vector, $\mathbf{r}_{PQ}$, in 3-dimensional space is given by $\mathbf{r}_{PQ}:(x_1-y_1, x_2-y_2 \ldots . x_n-y_n)$ (Fig. 4-1), where x_i and y_i are complex.

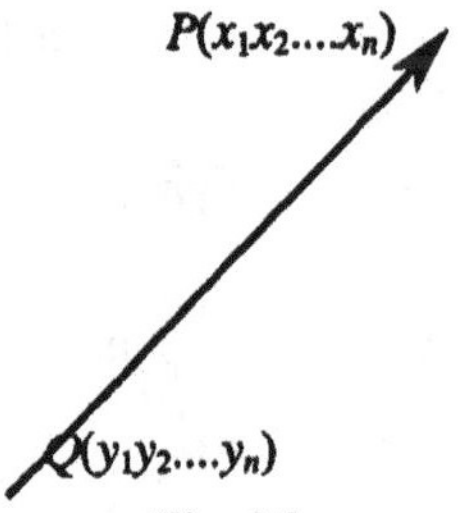

Fig. 4-1

[A complex number z can be written as $z = a + bi$ with a and b real and $i^2 = -1$. a is called the real part and b the imaginary part of z. The complex conjugate of z is $z^* \equiv a - bi$, and when multiplied by z yields $z^*z = (a - bi)(a + bi) = a^2 + b^2 \geq 0$.]

We define $|\mathbf{r}_{PQ}|^2$, the square of the distance between P and Q, as

$$|\mathbf{r}_{PQ}|^2 = \sum_{i=1}^{N}(x_i - y_i)^*(x_i - y_i) \geq 0.$$

Consider the 1-dimensional complex space (Fig. 4-2), where $z = x + iy$. Then $\mathbf{r}_{PQ}:(z - z')$ and

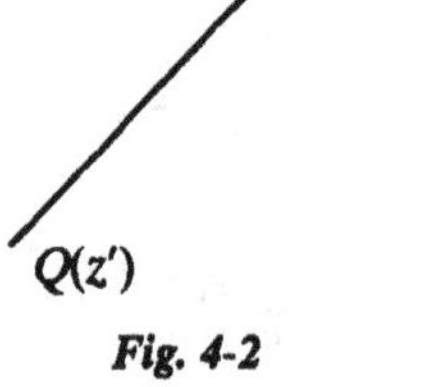

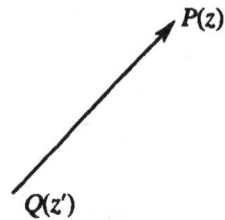

Fig. 4-2

$$|\mathbf{r}_{PQ}|^2 = (z - z')^*(z - z') = (x - x')^2 + (y - y')^2.$$

But, since this is ds^2 for a 2-dimensional real space, we see that there is an analogy between a 1-dimensional complex space and a 2-dimensional real space.

4.1.1 Rotations in N-Dimensional Space

For a rotation of coordinate axes, wherein a point P remains fixed but its components are referred to a new set of coordinate axes, we have the case $P(x_1x_2...x_N) \rightarrow P'(x_1'x_2'....x_N')$. As discussed earlier, such a rotation is defined by

$$x_i' = \sum_{j=1}^{N} u_{ij}x_j.$$

Theorem: If under a rotation $x_i' = \sum_{j=1}^{N} u_{ij}x_j$, the quantity $dq^2 = |\mathbf{r}_{PQ}|^2 = \sum_{i=1}^{N}(x_i - y_i)^*(x_i - y_i)$ is an invariant, then the transformation matrix u_{ij} satisfies $\sum_{i=1}^{N} u_{ij}^* u_{ik} = \delta_{jk}$, where δ_{jk} is the familiar Kronecker delta. U is said to be a unitary matrix and the transformation is said to be unitary.

Proof: Applying the rotation definition to any points P and Q:

$$Q: \quad y_i' = \sum_{j=1}^{N} u_{ij}y_j \quad \Rightarrow \quad y_i'^* = \sum_{j=1}^{N} u_{ij}^* y_j^*.$$

$$P: \quad x_i' = \sum_{j=1}^{N} u_{ij}x_j \quad \Rightarrow \quad x_i'^* = \sum_{j=1}^{N} u_{ij}^* x_j^*.$$

By hypothesis,

$$\sum_{i=1}^{N}(x_i' - y_i')^*(x_i' - y_i') = \sum_{j=1}^{N}(x_j - y_j)^*(x_j - y_j).$$

$$\sum_i [x_i'^* x_i' + y_i'^* y_i' - (y_i'^* x_i' + x_i'^* y_i')]$$

$$= \sum_j [x_j^* x_j + y_j^* y_j - (y_j^* x_j + x_j^* y_j)].$$

But $\sum_i x_i'^* x_i' = \sum_i x_i^* x_i$ and $\sum_i y_i'^* y_i' = \sum_i y_i^* y_i$, since these represent the squares of distances from the origin to points P and Q in the primed and unprimed systems and, by hypothesis, are equal.

Our expression therefore reduces to:

$$\sum_{i=1}^{N}[y_i'^*x_i' + x_i'^*y_i'] = \sum_{j=1}^{N}[y_j^*x_j + x_j^*y_j].$$

Evaluating the left side:

$$\sum_{i=1}^{N}[y_i'^*x_i' + x_i'^*y_i'] = \sum_{i=1}^{N}\left\{\sum_{jk=1}^{N}u_{ij}^*u_{ik}\,y_j^*x_k + \sum_{jk=1}^{N}u_{ij}^*u_{ik}\,x_j^*y_k\right\}$$

$$= \sum_{ijk=1}^{N}u_{ij}^*u_{ik}\,[y_j^* + x_j^*y_k].$$

Evaluating the right side:

$$\sum_{j=1}^{N}[y_j^*x_j + x_j^*y_j] = \sum_{jk=1}^{N}\delta_{kj}\,[y_j^*x_k + x_j^*y_k].$$

Combining:

$$(1)\qquad \sum_{jk=1}^{N}\left\{\left(\sum_{i=1}^{N}[u_{ij}^*u_{ik}] - \delta_{kj}\right)[y_j^*x_k + x_j^*y_k]\right\} = 0.$$

It is from the above expression that we wish to derive our result, namely,
$\sum_{i=1}^{N}u_{ij}^*u_{ik} = \delta_{jk}$.

Choosing $x_k = \delta_{k_o k}$ and $y_j = \delta_{j_o j}$, so that $x_k^* = x_k$ and $y_j^* = y_j$, (1) becomes:

$$(2)\qquad \left(\sum_{i=1}^{N}[u_{ij_o}^*u_{ik_o}] - \delta_{j_o k_o}\right) + \left(\sum_{i=1}^{N}[u_{ik_o}^*u_{ij_o}] - \delta_{j_o k_o}\right) = 0.$$

Choosing $x_k = i\delta_{k_o k}$ and $y_j = \delta_{j_o j}$, so that $x_k^* = -x_k$ and $y_j^* = y_j$, (1) becomes:

$$(3)\qquad i\left(\sum_{i=1}^{N}[u_{ij_o}^*u_{ik_o}] - \delta_{j_o k_o}\right) - i\left(\sum_{i=1}^{N}[u_{ik_o}^*u_{ij_o}] - \delta_{j_o k_o}\right) = 0.$$

We note that equations (2) and (3) are of the form $A + B = 0$ and $A - B = 0$, so that $A = B = 0$. We have, therefore,

$$\sum_{i=1}^{N} u_{ij_o}{}^* u_{ik_o} = \delta_{j_o k_o},$$

where j_o and k_o are arbitrary.

4.2 Matrix Algebra

Definition: M is defined to be a $m \times n$ (rows×columns) matrix if M has $m \times n$ elements and if these elements are arranged in the form

$$\begin{pmatrix} M_{11} & M_{12} & M_{13} & \cdot & \cdot & \cdot & M_{1n} \\ M_{21} & M_{22} & \cdot & \cdot & \cdot & \cdot & M_{2n} \\ \cdot & \cdot & \cdot & \cdot & \cdot & \cdot & \cdot \\ \cdot & \cdot & \cdot & \cdot & \cdot & \cdot & \cdot \\ \cdot & \cdot & \cdot & \cdot & \cdot & \cdot & \cdot \\ M_{m1} & M_{m2} & \cdot & \cdot & \cdot & \cdot & M_{mn} \end{pmatrix}$$

Definition: If A is an $m \times n$ matrix and B is an $n \times l$ matrix, then the product $A \cdot B$ is defined to be an $m \times l$ matrix, and

$$(A \cdot B)_{ij} = \sum_{k=1}^{n} A_{ik} B_{kj}, \qquad i = 1, 2, \ldots . m, \quad j = 1, 2, \ldots . l.$$

Since $B \cdot A$ has no meaning unless $l = m$, it is not always the case that $B \cdot A = A \cdot B$. The usefulness of this matrix notation is easily seen if one considers any linear transformation given by $x_i' = \sum_{j=1}^{n} M_{ij} x_j$. Denote the $(n \times 1)$ matrix x', the $(n \times 1)$ matrix x, and the square matrix M_{ij}, respectively, by:

$$(x') = \begin{pmatrix} x_1' \\ x_2' \\ \cdot \\ \cdot \\ \cdot \\ x_n' \end{pmatrix} \quad (x) = \begin{pmatrix} x_1 \\ x_2 \\ \cdot \\ \cdot \\ \cdot \\ x_n \end{pmatrix} \quad (M)_{ij} = \begin{pmatrix} M_{11} & M_{12} & M_{13} & \cdot & \cdot & \cdot & M_{1n} \\ M_{21} & M_{22} & \cdot & \cdot & \cdot & \cdot & M_{2n} \\ \cdot & \cdot & \cdot & \cdot & \cdot & \cdot & \cdot \\ \cdot & \cdot & \cdot & \cdot & \cdot & \cdot & \cdot \\ \cdot & \cdot & \cdot & \cdot & \cdot & \cdot & \cdot \\ M_{n1} & M_{n2} & \cdot & \cdot & \cdot & \cdot & M_{nn} \end{pmatrix}$$

We can then write the transformation as $x' = Mx$, a matrix equation which embodies all n equations of the given linear transformation.

As in all of algebra, we will have need for certain convenient notations. Therefore, let us introduce some new definitions.

Transpose: $(\tilde{A})_{ij} \equiv A_{ji}$. If A is a $m\times n$ matrix, $\tilde{A}$ is a $n\times m$ matrix.

Complex Conjugate: $(A^*)_{ij} \equiv (A_{ij})^*$. If A is $m\times n$, A^* is $m\times n$ also.

Adjoint: $(A^+)_{ij} \equiv (A_{ji})^* = (\tilde{A})^*_{ij} = (\tilde{A}^*)_{ij}$. If A is $m\times n$, A^+ is $n\times m$.

Matrices that are well defined, though not necessarily commutable, obey the following rules:

[1] *Associative Law:* $(A\cdot B)\cdot C = A\cdot(B\cdot C)$.

Proof:

$$\begin{aligned}[(A\cdot B)\cdot C]_{ij} &= \sum_k (A\cdot B)_{ik}C_{kj} = \sum_{kl} A_{il}B_{lk}C_{kj} = \sum_{kl} A_{il}(B_{lk}C_{kj}) \\ &= \sum_l A_{il}(B\cdot C)_{lj} = [A\cdot(B\cdot C)]_{ij}.\end{aligned}$$

[2] *Transpose of Product:* $(\widetilde{A\cdot B}) = \tilde{B}\cdot\tilde{A}$.

Proof:

$$(\widetilde{A\cdot B})_{ij} = (A\cdot B)_{ji} = \sum_k A_{jk}B_{ki} = \sum_k (\tilde{B})_{ik}(\tilde{A})_{kj} = (\tilde{B}\cdot\tilde{A})_{ij}.$$

[3] *Complex Conjugate of Product:* $(A\cdot B)^* = A^*\cdot B^*$.

Proof:

$$[(A\cdot B)^*]_{ij} = [(A\cdot B)_{ij}]^* = \sum_k A^*_{ik}B^*_{kj} = \sum_k (A^*)_{ik}(B^*)_{kj} = (A^*\cdot B^*)_{ij}.$$

[4] *Adjoint of Product:* $(A\cdot B)^+ = B^+\cdot A^+$.

Proof:

$$[(A\cdot B)^+]_{ij} = [(\widetilde{A\cdot B})^*]_{ij} = (\widetilde{A^*\cdot B^*})_{ij} = (\tilde{B}^*\cdot\tilde{A}^*)_{ij} = (B^+\cdot A^+)_{ij}.$$

We saw before that $\sum_{k=1}^{n} M_{ik}x_k$ could be written as $x' = Mx$. If we take the transpose of both sides, we have $\tilde{x}' = \tilde{x}\tilde{M}$, where

$$\tilde{x}' = (x_1' \, x_2' \ldots x_n').$$

$$\tilde{x} = (x_1 \, x_2 \ldots x_n).$$

$$\tilde{M} = \begin{pmatrix} M_{11} & M_{21} & M_{31} & \cdot & \cdot & \cdot & M_{n1} \\ M_{12} & M_{22} & \cdot & \cdot & \cdot & \cdot & M_{n2} \\ \cdot & \cdot & \cdot & \cdot & \cdot & \cdot & \cdot \\ \cdot & \cdot & \cdot & \cdot & \cdot & \cdot & \cdot \\ \cdot & \cdot & \cdot & \cdot & \cdot & \cdot & \cdot \\ M_{1n} & M_{2n} & \cdot & \cdot & \cdot & \cdot & M_{nn} \end{pmatrix}.$$

In the future we shall be dealing only with $n \times n$, $n \times 1$ and $1 \times n$ matrices. Hence, let us adopt the following notations:

$X\ Y\ A\ B\ C\ M$ will denote arbitrary $n \times n$ matrices.

$\sigma\ \tau$ will denote particular 2×2 matrices.

$x\ y\ z\ \psi\ \varphi$ will denote $n \times 1$ (column) matrices.

$\tilde{x}\ \tilde{y}\ \tilde{z}\ \tilde{\psi}\ \tilde{\varphi}$ will denote $1 \times n$ (row) matrices.

An important additional matrix is the unit matrix I, defined as $I_{ij} \equiv \delta_{ij}$. Then for any matrix A, $I{\cdot}A = A{\cdot}I$, since

$$(I{\cdot}A)_{ik} = \sum_l \delta_{il} A_{lk} = A_{ik}.$$

$$(A{\cdot}I)_{ik} = \sum_l A_{il}\, \delta_{lk} = A_{ik}.$$

Furthermore, $\tilde{I} = I = I^* = I^\dagger$.

4.2.1 Commutation of Matrices

Definition: If $A{\cdot}B = B{\cdot}A$, then A commutes with B.

Definition: If α is a pure number, $(\alpha A)_{ij} \equiv \alpha A_{ij}$.

As an example of non-commutation, consider the matrices describing electron spin, the so-called Pauli spin matrices, defined as:

$$\sigma_x = \begin{pmatrix} 0 & 1 \\ 1 & 0 \end{pmatrix}. \quad \sigma_x = \begin{pmatrix} 0 & -i \\ i & 0 \end{pmatrix}. \quad \sigma_x = \begin{pmatrix} 1 & 0 \\ 0 & -1 \end{pmatrix}.$$

Employing matrix multiplication,

$$\sigma_x\sigma_z = \begin{pmatrix} 0 & -1 \\ 1 & 0 \end{pmatrix} = -i\sigma_y . \quad \sigma_z\sigma_x = \begin{pmatrix} 0 & 1 \\ -1 & 0 \end{pmatrix} = i\sigma_y .$$

$$\sigma_x\sigma_y = \begin{pmatrix} i & 0 \\ 0 & -i \end{pmatrix} = i\sigma_z . \quad \sigma_y\sigma_x = \begin{pmatrix} -i & 0 \\ 0 & i \end{pmatrix} = -i\,\sigma_z .$$

$$\sigma_y\sigma_z = \begin{pmatrix} 0 & i \\ i & 0 \end{pmatrix} = i\sigma_x . \quad \sigma_z\sigma_y = \begin{pmatrix} 0 & -i \\ -i & 0 \end{pmatrix} = -i\sigma_x .$$

It now becomes convenient to define:

$$(M+N)_{ij} = M_{ij} + N_{ij} . \quad (\alpha M)_{ij} = \alpha M_{ij} . \quad (x+y)_{i1} = x_{i1} + y_{i1} ,$$

where α is a pure number. Then, from the 6 relations above, we have

$$\sigma_x\sigma_z + \sigma_z\sigma_x = 0 \quad \Rightarrow \quad \sigma_i\sigma_j + \sigma_j\sigma_i = 2\delta_{ij} ,$$

where $i, j \rightarrow x, y, z$.

Note:

$$\text{For } i = j = x, \quad 2\sigma_x\sigma_x = \begin{pmatrix} 0 & 1 \\ 1 & 0 \end{pmatrix}\begin{pmatrix} 0 & 1 \\ 1 & 0 \end{pmatrix} = 2\begin{pmatrix} 1 & 0 \\ 0 & 1 \end{pmatrix} = 2\delta_{ij} .$$

$$\text{For } i = j = y, \quad 2\sigma_y\sigma_y = \begin{pmatrix} 0 & -i \\ i & 0 \end{pmatrix}\begin{pmatrix} 0 & -i \\ i & 0 \end{pmatrix} = 2\begin{pmatrix} 1 & 0 \\ 0 & 1 \end{pmatrix} = 2\delta_{ij} .$$

$$\text{For } i = j = z, \quad 2\sigma_z\sigma_z = \begin{pmatrix} 1 & 0 \\ 0 & -1 \end{pmatrix}\begin{pmatrix} 1 & 0 \\ 0 & -1 \end{pmatrix} = 2\begin{pmatrix} 1 & 0 \\ 0 & 1 \end{pmatrix} = 2\delta_{ij} .$$

4.2.2 Matrix Inverse

Does such a thing as a matrix inverse exist? Since this is very important, we shall need to discuss some major theorems in algebra before we can answer the above question.

Theorem (1): If $\sum_{j=1}^{n} A_{ij}x_j = 0$, and if $(x) = \begin{pmatrix} x_1 \\ x_2 \\ \cdot \\ \cdot \\ x_n \end{pmatrix} \neq 0$, $\det|A| = 0$, i.e. the only solution for x_j is $\det|A| = 0$.

Theorem (2): If $\det|A| \neq 0$, and if $Ax = 0$, then $x = 0$.

Theorem (3): If $\det|A| \neq 0$, and if $AB = 0$, then $B = 0$.

Proof: Consider $\sum_j A_{ij}B_{jk} = 0$, and fix a particular (k). Then $B_{j(k)}$ is identical with a column matrix x, i.e.

$$B_{j(k)} = \begin{pmatrix} B_{1k} \\ B_{2k} \\ \cdot \\ \cdot \\ B_{nk} \end{pmatrix} = x_k .$$

But, from Theorem (2), $x_k = 0$, and since this holds for all k, it holds for all $B_{j(k)}$. Hence, $B = 0$.

Theorem (4): If $\det|A| \neq 0$, and if $y \neq 0$, then there exists an $x \neq 0$ such that $Ax = y$.

Theorem (5): If $\det|A| \neq 0$, there exists an A^{-1} such that $A^{-1}A = AA^{-1} = I$.

Proof: From Theorem (4), there exists an x^j ($n\times1$ column matrix), such that

$$Ax^j = \begin{pmatrix} 0 \\ 0 \\ 1 \\ 0 \\ \cdot \\ 0 \end{pmatrix}, \text{ where the } j\text{th element is 1.}$$

In other words,

$$Ax^1 = \begin{pmatrix} 1 \\ 0 \\ 0 \\ 0 \\ \cdot \\ \cdot \\ 0 \end{pmatrix}, \quad Ax^2 = \begin{pmatrix} 0 \\ 1 \\ 0 \\ 0 \\ \cdot \\ \cdot \\ 0 \end{pmatrix}, \quad \cdot \quad Ax^5 = \begin{pmatrix} 0 \\ 0 \\ 0 \\ 0 \\ 1 \\ \cdot \\ 0 \end{pmatrix}.$$

Define

$$A^{-1} = \begin{pmatrix} x_1^1 & x_1^2 & x_1^3 & \cdot & \cdot & \cdot & x_1^n \\ x_2^1 & x_2^2 & \cdot & \cdot & \cdot & \cdot & x_2^n \\ \cdot & \cdot & \cdot & \cdot & \cdot & \cdot & \cdot \\ \cdot & \cdot & \cdot & \cdot & \cdot & \cdot & \cdot \\ \cdot & \cdot & \cdot & \cdot & \cdot & \cdot & \cdot \\ x_n^1 & x_n^2 & \cdot & \cdot & \cdot & \cdot & x_n^n \end{pmatrix} \Rightarrow AA^{-1} = \begin{pmatrix} 1 & 0 & 0 & \cdot & \cdot & \cdot & 0 \\ 0 & 1 & 0 & \cdot & \cdot & \cdot & 0 \\ 0 & 0 & 1 & \cdot & \cdot & \cdot & 0 \\ \cdot & \cdot & \cdot & \cdot & \cdot & \cdot & \cdot \\ \cdot & \cdot & \cdot & \cdot & \cdot & \cdot & \cdot \\ 0 & 0 & 0 & \cdot & \cdot & \cdot & 1 \end{pmatrix} = I.$$

Consider the expression $AA^{-1} = I$. Multiply both sides of the equation on the right by A, obtaining

$$AA^{-1}A = IA = AI \quad \Rightarrow \quad A(A^{-1} - I) = 0.$$

But $\det|A| \neq 0 \Rightarrow A^{-1}A = I$. We have thereby shown that if $\det|A| \neq 0$, $A^{-1}A = AA^{-1} = I$. If on the other hand, $\det|A| = 0$, we cannot, in general, find the inverse of A.

4.2.3 Special Matrices

When dealing with n-dimensional space, we considered the transformation (rotation) $x_i' = \sum_{j=1}^{n} u_{ij}x_j$, where $\sum_i x_i'^* x_i' = \sum_i x_i^* x_i$, i.e. ds^2 is an invariant. From this we proceeded to show that

$$\sum_j u_{ji}^* u_{jk} = \delta_{ik} \quad \Rightarrow \quad U^+U = I.$$

We ask the following: Does it follow from the above that $UU^+ = I$, i.e. is it true that $U^+ = U^{-1}$? We must then ask ourselves whether or not $\det|U| = 0$.

Theorem: If $U^+U = I$, then $\det|U| = e^{i\theta}$, where θ is real, and $UU^+ = U^+U = I$.

Proof: We first employ some properties of determinants.

Lemma (1) $\det|\tilde{A}| = \det|A|$.

Lemma (2) $\det|A^+| = \det|\tilde{A}^*|. = \det|A^*| = [\det|A|]^*$.

Lemma (3) $\det|AB| = \det|A|\cdot\det|B|$.

We proceed to prove Lemma (3), (1) and (2) being fairly evident.

$$\det|A| = \sum_i (-1)^{P_i} A_{1i_1} A_{2i_2} \ldots A_{ni_n},$$

where $i_1 i_2,\ldots,i_n$ is some permutation of $1,2,\ldots,n$ and $(-1)^{P_i}$ is $+1/-1$ if $i_1 i_2,\ldots,i_n$ is an even/odd permutation of $1,2,\ldots,n$. Then

$$\det|A|\cdot\det|B| = \sum_{ij} (-1)^{P_i+P_j} A_{1i_1} A_{2i_2} \ldots . A_{ni_n} B_{1i_1} B_{2i_2} \ldots . B_{ni_n}.$$

Let us introduce the following symbolic permutation notation:

$$\underset{\text{I}}{\begin{pmatrix} 1\ 2\ 3 \cdots n \\ i_1\ i_2\ i_3 \cdots i_n \end{pmatrix}\begin{pmatrix} 1\ 2\ 3 \cdots n \\ j_1\ j_2\ i_3 \cdots j_n \end{pmatrix}} = \underset{\text{II}}{\begin{pmatrix} 1\ 2\ 3 \cdots n \\ i_1\ i_2\ i_3 \cdots i_n \end{pmatrix}\begin{pmatrix} i_1\ i_2\ i_3 \cdots i_n \\ k_1\ k_2\ k_3 \cdots k_n \end{pmatrix}},$$

where the k_i are defined from the j_i, i.e. there is a direct one-to-one correspondence between forms I and II.

Example:

$$\begin{pmatrix} 1\,2\,3 \cdots n \\ 3\ 1\ \cdots\cdots \end{pmatrix}\begin{pmatrix} 1 & 2 & 3 & \cdots n \\ 10 & 101 & 76 & \cdots\cdots \end{pmatrix} = \begin{pmatrix} 1\,2\,3 \cdots n \\ 3\ 1\ \cdots\cdots \end{pmatrix}\begin{pmatrix} 3 & 1 & \cdots\cdots \\ 76 & 10 & \cdots\cdots \end{pmatrix},$$

so that

$$\det|A|\cdot\det|B| = \sum_k (-1)^{P_k} A_{1i_1} B_{i_1k_1} A_{2i_2} B_{i_2k_2} \ldots . A_{ni_n} B_{i_nk_n},$$

where $(-1)^{P_k} = (-1)^{P_i+P_j}$, so that

$P_i(+)$ and $P_j(-) \rightarrow P_k(-)$. $P_i(+)$ and $P_j(+) \rightarrow P_k(+)$.
$P_i(+)$ and $P_j(-) \rightarrow P_k(-)$. $P_i(-)$ and $P_j(-) \rightarrow P_k(+)$.

Thus, $\det|AB| = \det|A|\cdot\det|B|$.

Hence, from $\det|U^+U| = \det I = 1$, it follows that

$$\det|U^+U| = \det|U^+|\det|U| = (\det|U|)^* \det|U| = (\det|U|)^2 = 1.$$

Hence, the most general form of $\det|U|$ is $\det|U| = e^{i\theta} \neq 0$.

Definition: If $UU^+ = U^+U = I$, then U is a unitary matrix.

Theorem: If $x' = Ux$ and U is unitary, then $x'^+x' = x^+x$.

Proof: We have already shown that if U is unitary (i.e. $\sum_i u_{ij}^* u_{ik} = \delta_{jk}$), then length^2 is invariant, which is to say that $\sum_i x_i'^* x_i' = \sum_i x_i^* x_i$. Employing matrix algebra,

$$x' = Ux \Rightarrow x'^+x' = x^+U^+Ux = x^+(U^+U)x = x^+x.$$

When you insist that length^2 remain unchanged, i.e. $\sum_i u_{ij}^* u_{ik} = \delta_{jk}$, you are in essence specifying a unitary transformation, $x_i' = \sum_i u_{ij}x_j$. This is what is done in Quantum Mechanics, where length is defined by $(x^+x)^{1/2}$. In the study of Special Relativity, however, we define length by $(\tilde{x}x)^{1/2}$, and we demand that $\tilde{x}x$ remain unchanged, i.e. $\sum_i x_i'x_i' = \sum_i x_i x_i$.

Theorem: If $x' = Vx$ and if $\tilde{x}'x' = \tilde{x}x$, then $\tilde{V}V = V\tilde{V} = I$ (i.e. $\tilde{V} = V^{-1}$).

Proof of Converse: $x' = Vx$ and $\tilde{V}V = V\tilde{V} = I \rightarrow \tilde{x}'x' = \tilde{x}\tilde{V}Vx = \tilde{x}x$.

Proof of Theorem: If $\tilde{x}'x' = \tilde{x}x$, that is the length of a vector, defined by $d_{pq}{}^2 = (x \tilde{-} y)(x - y)$, remains unchanged, and $x' = Vx$, then

$$\tilde{x}'x' - \tilde{x}x = 0 \Rightarrow \tilde{x}\tilde{V}Vx - \tilde{x}x = 0 \Rightarrow \tilde{x}(\tilde{V}V - I)x = 0.$$

We would like to show that $\tilde{V}V - I = 0$. We will show it to be the case for a specific x, although the relation actually holds for all x. (We are using a specific x to find the value of $\tilde{V}V - I$.)

Let x have as its only element the ith element, equal to unity, i.e. let

$$x = \begin{pmatrix} 0 \\ 0 \\ 1 \\ 0 \\ \cdot \\ \cdot \\ 0 \end{pmatrix} \Rightarrow \tilde{x}x = (0\,0\,1\,0 \cdots 0)\begin{pmatrix} 0 \\ 0 \\ 1 \\ 0 \\ \cdot \\ \cdot \\ 0 \end{pmatrix} = 1.$$

It follows then that

$$\tilde{x}(\tilde{V}V - I)x = (0\,0\,1\,0 \cdots 0)\begin{pmatrix} (\tilde{V}V - I)_{1i} \\ (\tilde{V}V - I)_{2i} \\ (\tilde{V}V - I)_{3i} \\ \cdot \\ \cdot \\ \cdot \\ (\tilde{V}V - I)_{ni} \end{pmatrix} = (\tilde{V}V - I)_{ii} = 0.$$

Hence, $\tilde{x}(\tilde{V}V - I)x = (\tilde{V}V - I)_{ii} = 0$ and all diagonal elements of $(\tilde{V}V - I) = 0$.

If we let x have as its only elements the ith and jth elements, both equal to unity, i.e. let

$$x = \begin{pmatrix} 0 \\ 0 \\ 1 \\ 0 \\ 1 \\ \cdot \\ 0 \end{pmatrix} \Rightarrow \tilde{x}(\tilde{V}V - I)x = (0\,0\,1\,0\,1 \cdot 0)\begin{pmatrix} (\tilde{V}V - I)_{1i} + (\tilde{V}V - I)_{1j} \\ (\tilde{V}V - I)_{2i} + (\tilde{V}V - I)_{2j} \\ (\tilde{V}V - I)_{3i} + (\tilde{V}V - I)_{3j} \\ \cdot \\ \cdot \\ \cdot \\ (\tilde{V}V - I)_{ni} + (\tilde{V}V - I)_{nj} \end{pmatrix}$$

$$= \left[(\tilde{V}V - I)_{ii} + (\tilde{V}V - I)_{ij} + (\tilde{V}V - I)_{ji} + (\tilde{V}V - I)_{jj}\right].$$

Hence,

$$\tilde{x}(\tilde{V}V - I)x = (\tilde{V}V - I)_{ii} + (\tilde{V}V - I)_{ji} + (\tilde{V}V - I)_{ij} + (\tilde{V}V - I)_{jj} = 0.$$

But $(\tilde{V}V - I)_{ii} = (\tilde{V}V - I)_{jj} = 0$, so that

$$(\tilde{V}V - I)_{ij} + (\tilde{V}V - I)_{ji} = 0 \quad \rightarrow \quad (\tilde{V}V - I) + (\widetilde{\tilde{V}V} - I) = 0,$$

and since $(\widetilde{\tilde{V}V} - I) = (\tilde{V}V - I)$, it follows that $(\tilde{V}V - I) = 0$ or $\tilde{V}V = I$.

Since $\tilde{V}V = I$, we wish to know whether it is true that $\tilde{V}V = V\tilde{V} = I$, i.e. is it true that $\tilde{V} = V^{-1}$?

Multiply by V to the left of both sides of the expression $\tilde{V}V = I$ to get $V\tilde{V}V = VI$ or $(I - V\tilde{V})V = 0$. We see that the only solution for $I = V\tilde{V}$ is for $\det|V| \neq 0$. Thus we investigate $\det|V|$.

Since $\tilde{V}V = I$, $\det|\tilde{V}V| = \det|\tilde{V}|\det|V| = (\det|V|)^2 = 1$. Thus, $\det|V| = \pm 1$ and $\tilde{V}V = V\tilde{V} = I$, where $\tilde{V} = V^{-1}$.

Definition: A matrix V having the property that $\tilde{V}V = V\tilde{V} = I$ is called an orthogonal matrix.

In real space, a unitary transformation becomes identical to an orthogonal transformation. Why? A unitary transformation retains length^2, where $d_{pq}^2 = \sum_i (x_i - y_i)^*(x_i - y_i) = (x - y)^+(x - y)$. But if space is real, $(x - y)^+ = (\widetilde{x - y})^* = (\widetilde{x - y})$ and $d_{pq}^2 = (\widetilde{x - y})(x - y)$, which is the length^2 definition for an orthogonal transformation.

Furthermore, if V is real, then $V^* = V \rightarrow V^+ = \tilde{V}^* = \tilde{V}$. Thus, $\tilde{V}V = V^+V = VV^+ = I$, which defines a unitary matrix. Hence we have that a real, orthogonal matrix is a unitary matrix.

Dual Role of a Matrix:

Consider the transformation $y = Ax$ (Fig. 4-3).

(1) The transformation can be considered as a point transformation within a given coordinate system.

(2) The transformation can also be considered as a coordinate transformation from the x coordinate system to the y coordinate system.

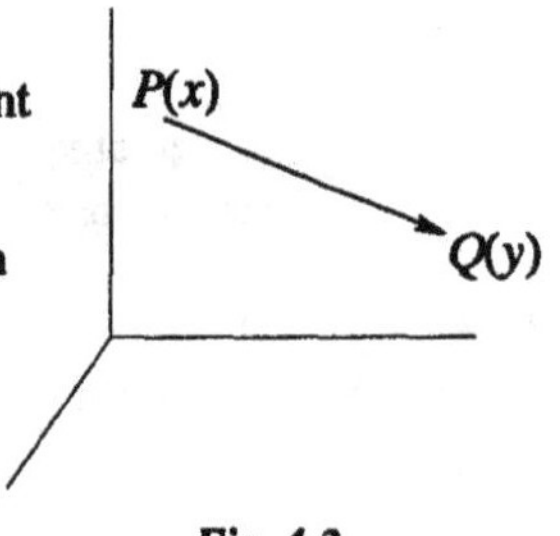

Fig. 4-3

In the physical applications we shall be dealing with, we usually require that the length of a vector be kept the same. Therefore, we have to know what we mean by length. We have fundamentally two definitions, one applying to Quantum Mechanics (QM) and the other to Special Relativity (SR).

Definition I (QM): $d_{pq}^2 = \sum_i (x_i - y_i)^*(x_i - y_i) \quad \Rightarrow \quad d_{pq}^2 = (x-y)^+(x-y).$

Definition II (SR): $d_{pq}^2 = (x \tilde{-} y)(x-y).$

<u>Summary</u>

[I] If $x' = Ux$, then $U^+U = UU^+ = I$ if and only if $x'^+x' = x^+x$.

Necessity: $x'^+x' = x^+U^+Ux = x^+x.$

Sufficiency: See pg. 64, followed by showing that $\det|U| = e^{i\theta}$.

[II] If $x' = Vx$, then $\tilde{V}V = V\tilde{V} = I$ if and only if $\tilde{x}'x' = \tilde{x}x$.

Necessity: $\tilde{x}'x' = \tilde{x}\tilde{V}Vx = \tilde{x}x.$

Sufficiency: See pg. 73, followed by showing that $\det|V| = \pm 1$.

4.3 Examples of Matrices

4.3.1 Orthogonal Matrix

Theorem: The most general form of an orthogonal matrix in 2-dimensional space is given by

$$V = \begin{pmatrix} \cos\varphi & \sin\varphi \\ -\sin\varphi & \cos\varphi \end{pmatrix} \quad \text{or} \quad V' = \begin{pmatrix} \cos\varphi & \sin\varphi \\ \sin\varphi & -\cos\varphi \end{pmatrix},$$

where $\det|V| = +1$ (or $\det|V'| = -1$), and where $\varphi = \theta + i\chi$, where φ is complex and θ and χ are real.

Proof:

$$\tilde{V}V = 1 \quad \Rightarrow \quad \begin{pmatrix} V_{11} & V_{12} \\ V_{21} & V_{22} \end{pmatrix}\begin{pmatrix} V_{11} & V_{21} \\ V_{12} & V_{22} \end{pmatrix} = I \quad \Rightarrow \quad \begin{array}{l} V_{11}^2 + V_{12}^2 = 1. \\ V_{11}V_{21} + V_{12}V_{22} = 0. \\ V_{21}^2 + V_{22}^2 = 1. \end{array}$$

If $\varphi = \cos^{-1}V_{11}$, so that $V_{11} = \cos\varphi$ and $V_{12} = \sin\varphi$, then

$$\frac{V_{22}}{V_{21}} = -\frac{V_{11}}{V_{12}} = -\frac{\cos\varphi}{\sin\varphi}.$$

The negative sign can be absorbed by having either $V_{21} = \sin\varphi$, $V_{22} = -\cos\varphi$ or $V_{22} = \cos\varphi$, $V_{21} = -\sin\varphi$. Thus the reason for the two forms, V and V'.

Observe that

$$V = \begin{pmatrix} \cos\varphi & \sin\varphi \\ -\sin\varphi & \cos\varphi \end{pmatrix}$$

represents a 2-dimensional real coordinate rotation if φ is real (Fig. 4-4).

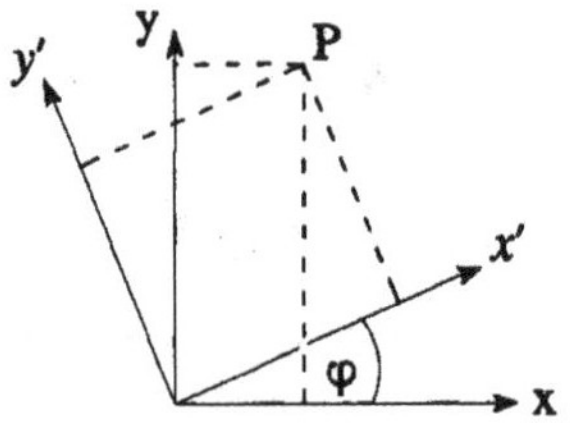

Fig. 4-4

$$x' = Vx \quad \Rightarrow \quad \begin{cases} x_1' = x_1\cos\varphi + x_2\sin\varphi. \\ x_2' = -x_1\sin\varphi + x_2\cos\varphi. \end{cases}$$

Since this represents a rotation,

$$\sum_i V_{ij}V_{ik} = \delta_{ik} \quad \Rightarrow \quad \tilde{V}V = I.$$

Now V' can be obtained from V by a reflection, i.e. $V' = RV = \begin{pmatrix} 1 & 0 \\ 0 & -1 \end{pmatrix}V$, where we denote R as the reflection matrix. Note then that $x' = Rx$ yields $x_1' = x_1$ and $x_2' = -x_2$, i.e. a mirror reflection about the x_1-axis (Fig. 4-5).

We speak of $\det|V| = +1$ for the "proper" orthogonal matrix, and of $\det|V'| = -1$ for the "improper" orthogonal matrix.

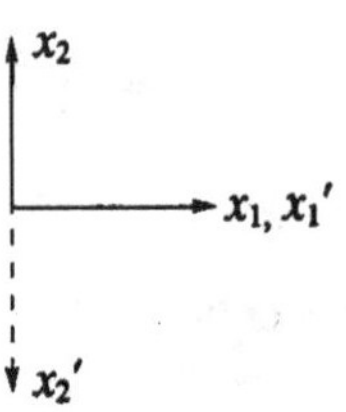

Fig. 4-5

Theorem:

(1) If V_1 and V_2 are orthogonal, then $V_1 \cdot V_2$ is orthogonal.

(2) If U_1 and U_2 are unitary, then $U_1 \cdot U_2$ is unitary.

Proof:

(1)
$$(\widetilde{V_1 V_2})(V_1 V_2) = \tilde{V}_2 \tilde{V}_1 V_1 V_2 = I.$$
$$(V_1 V_2)(\widetilde{V_1 V_2}) = V_1 V_2 \tilde{V}_2 \tilde{V}_1 = I.$$

(2)
$$(U_1 U_2)^+ (U_1 U_2) = U_2^+ U_1^+ U_1 U_2 = I.$$
$$(U_1 U_2)(U_1 U_2)^+ = U_1 U_2 U_2^+ U_1^+ = I.$$

The physical significance of the above is that if we perform a rotation from x to x' followed by another rotation from x' to x'', we can accomplish the same resultant rotation by going directly from x to x'' by a different rotation, i.e. if $x' = V_1 x$ and $x'' = V_2 x'$, then $x'' = V_2 V_1 x \Rightarrow x'' = V_3 x$, and $V_3 = V_2 V_1$.

4.3.2 Special Relativity

Special Relativity — a subject originally stimulated by a series of experiments carried out by Michelson and Morley, in which the velocity of light was found to be the same in all reference systems.

Consider the case in Newtonian mechanics where system Σ' moves with velocity υ with respect to system Σ (Fig. 4-6).

Since time is considered absolute, the Newtonian transformation is defined by:

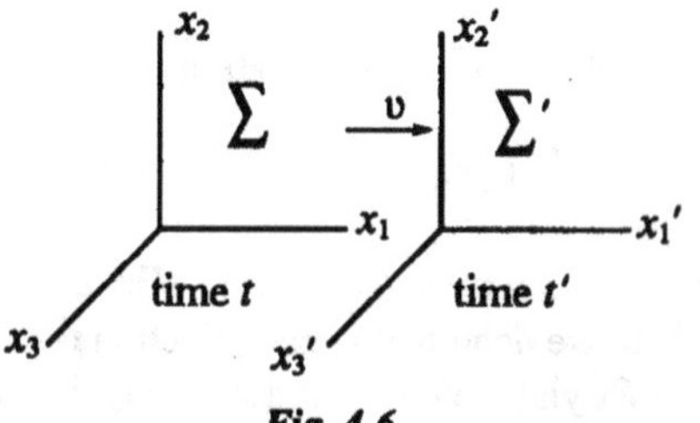

Fig. 4-6

$$t' = t;\ x_2' = x_2;\ x_3' = x_3.$$
$$x_1' = x_1 - \upsilon t.$$

According to Newtonian mechanics, the velocity of a body moving in the Σ' system would appear to be increased by an amount υ when observed in the Σ system. The velocity of light c, however, was observed to have remained the same in all coordinate systems; hence the above transformation is not the proper one. We must inquire into what is meant when we say that c, the velocity of light, remains constant.

Consider the equation of a surface, $x_1^2 + x_2^2 + x_3^2 - c^2t^2 = 0$, which describes wave fronts, or spherical surfaces having radii which increase with time at the rate c. Now if $x_1'^2 + x_2'^2 + x_3'^2 - c^2t'^2 = x_1^2 + x_2^2 + x_3^2 - c^2t^2$, then we are implying that the velocity c is the same in the two systems. Hence, when we say that c is the same in both systems, Σ and Σ', we are demanding that the length of a vector in 4-space remains unchanged, i.e. that space and time obey the following:

$$\sum_{\mu=1}^{4} x_\mu'^2 = \sum_{\mu=1}^{4} x_\mu^2 \quad \rightarrow \quad \tilde{x}_\mu' x_\mu' = \tilde{x}_\mu x_\mu,$$

where $x_4 = ict$, $x_4^2 = -c^2t^2$, and $x_4'^2 = -c^2t'^2$. This implies that the transformation given by $x' = Vx$ should be an orthogonal transformation, where

$$x = \begin{pmatrix} x_1 \\ x_2 \\ x_3 \\ ict \end{pmatrix}; \quad x' = \begin{pmatrix} x_1' \\ x_2' \\ x_3' \\ ict' \end{pmatrix}.$$

Since we would like V to obey $x_2' = x_2$ and $x_3' = x_3$, we proceed as follows:

$$\begin{aligned}
x_2' &= V_{21}x_1 + V_{22}x_2 + V_{23}x_3 + V_{24}x_4 \rightarrow V_{22} = 1,\ V_{21} = 0 = V_{23} = V_{24}. \\
x_3' &= V_{31}x_1 + V_{32}x_2 + V_{33}x_3 + V_{34}x_4 \rightarrow V_{33} = 1,\ V_{31} = 0 = V_{32} = V_{34}. \\
x_1' &= V_{11}x_1 + V_{12}x_2 + V_{13}x_3 + V_{14}x_4 \rightarrow V_{12} = V_{13} = 0, V_{11} \ \&\ V_{14} \neq 0. \\
x_4' &= V_{41}x_1 + V_{42}x_2 + V_{43}x_3 + V_{44}x_4 \rightarrow V_{42} = V_{43} = 0, V_{41} \ \&\ V_{44} \neq 0.
\end{aligned}$$

Finally, then, we have:

$$x_1' = V_{11}x_1 + ictV_{14}. \qquad x_4' = ict' = V_{41}x_1 + ictV_{44}.$$

In the $x_1 - t$ example illustrated in Fig. 4-6, where we do not rotate axes x_2 and x_3, V constitutes a 2×2 matrix:

$$V = \begin{pmatrix} V_{11} & V_{14} \\ V_{41} & V_{44} \end{pmatrix} = \begin{pmatrix} \cos\varphi & \sin\varphi \\ -\sin\varphi & \cos\varphi \end{pmatrix},$$

where $\det|V| = 1$ and where φ is complex; i.e. we have here a rotation in 2-dimensional complex space. To show this, note the following:

x_1' is real while x_4' is pure imaginary, so that from $x_1' = V_{11}x_1 + ictV_{14}$ it follows that V_{11} must be real and V_{14} must be pure imaginary, while from $x_4' = V_{41}x_1 + ictV_{44}$, it follows that V_{41} must be pure imaginary and V_{44} must be real. Hence, $\cos\varphi$ is real and $\sin\varphi$ is pure imaginary. [In general, $\varphi = \alpha + i\theta$, but here, since φ is pure imaginary, $\varphi = i\theta$ (θ real).]

Consider the following trigonometric identities:

$$\cos\varphi = \frac{e^{i\varphi} + e^{-i\varphi}}{2}; \quad \sin\varphi = \frac{e^{i\varphi} - e^{-i\varphi}}{2i}.$$

$$\cosh\theta = \frac{e^{\theta} + e^{-\theta}}{2}; \quad \sinh\theta = \frac{e^{\theta} - e^{-\theta}}{2}. \quad \Rightarrow \quad \sinh^2\theta = \cosh^2\theta - 1.$$

Observe that for $\varphi = i\theta$,

$$\cos\varphi = \frac{e^{-\theta} + e^{\theta}}{2} = \cosh\theta; \quad \sin\varphi = \frac{e^{-\theta} - e^{\theta}}{2i} = i\sinh\theta,$$

so that $\cos\varphi$ is real and $\sin\varphi$ is pure imaginary.

Finally, then,

$$V = \begin{pmatrix} \cosh\theta & 0 & 0 & i\sinh\theta \\ 0 & 1 & 0 & 0 \\ 0 & 0 & 1 & 0 \\ -i\sinh\theta & 0 & 0 & \cosh\theta \end{pmatrix}.$$

Defining $\beta \equiv \tanh\theta$, $\cosh^2\theta = 1 + \sinh^2\theta = 1 + \beta^2\cosh^2\theta$, so that

$$\cosh\theta = \frac{1}{\sqrt{1-\beta^2}}; \qquad \sinh\theta = \frac{\beta}{\sqrt{1-\beta^2}},$$

and substituting into the expressions for x_1' and t',

$$x_1' = \cosh\theta x_1 + ict(i\sinh\theta) = \frac{x_1 - \beta ct}{\sqrt{1-\beta^2}}.$$

$$t' = \frac{1}{ic}[-i\sinh\theta x_1 + ict(\cosh\theta)] = \frac{t - \beta x_1/c}{\sqrt{1-\beta^2}}.$$

The above two equations define the Lorentz Transformation.

[The most general form of the 4-dimensional orthogonal transformation is obtained by a 3-dimensional spatial rotation performed upon V.]

If $x' = Vx$, then $x = \tilde{V}x'$, where

$$\tilde{V} = \begin{pmatrix} \cosh\theta & 0 & 0 & -i\sinh\theta \\ 0 & 1 & 0 & 0 \\ 0 & 0 & 1 & 0 \\ i\sinh\theta & 0 & 0 & \cosh\theta \end{pmatrix}.$$

Hence, the inverse transformation is defined by

$$x_1 = \frac{x_1' + \beta c t'}{\sqrt{1-\beta^2}}. \qquad t = \frac{t' + \beta x_1'/c}{\sqrt{1-\beta^2}}.$$

Consider a point $P(x_1'x_2'x_3't')$ fixed in Σ'. What will be the velocity of point P as observed in Σ?

$$\left(\frac{\partial x_1}{\partial t'}\right)_{x_1'} = \frac{\beta c}{\sqrt{1-\beta^2}}. \qquad \left(\frac{\partial t}{\partial t'}\right)_{x_1'} = \frac{1}{\sqrt{1-\beta^2}}.$$

Hence,

$$\left(\frac{dx_1}{dt}\right)_{\Sigma \text{ for } P} = \frac{\left(\frac{\partial x_1}{\partial t'}\right)_{x_1'}}{\left(\frac{\partial x_1}{\partial t'}\right)_{x_1'}} = \beta c = \upsilon.$$

Thus we see that $\beta c = \upsilon$ is the relative velocity of Σ' with respect to Σ.

We rewrite the Lorentz Transformation and its inverse, substituting for β:

$$x_1' = \frac{x_1 - \upsilon t}{\sqrt{1-\upsilon^2/c^2}}. \qquad t' = \frac{t - \upsilon x_1/c^2}{\sqrt{1-\upsilon^2/c^2}}.$$

$$x_1 = \frac{x_1' + \upsilon t'}{\sqrt{1-\upsilon^2/c^2}}. \qquad t = \frac{t' + \upsilon x_1'/c^2}{\sqrt{1-\upsilon^2/c^2}}.$$

Consequences:

(1) Consider two events occurring in Σ' a time interval $t_B' - t_A'$ apart. What will be the observed time interval in Σ? We must evaluate $t_B - t_A$ where $x_1'(A) = x_1'(B)$; $x_2'(A) = x_2'(B)$; $x_3'(A) = x_3'(B)$. Then,

$$t_B - t_A = \frac{t_B' - t_A'}{\sqrt{1-\beta^2}} \geq t_B' - t_A'.$$

Example: The apparent increased lifetime of a meson in the lab system.

(2) Consider a length $L' = x_1'(A) - x_1'(B)$ in Σ' (Fig. 4-7). What will be the observed length L in the Σ system? $x_A' - x_B' = L'$ in Σ', and is independent of t_B' and t_A' since the ruler is fixed. When we ask, what is L in Σ, we must specify that $t_A = t_B$, otherwise any length is possible when the measurement is made, since the bar is moving with respect to Σ.

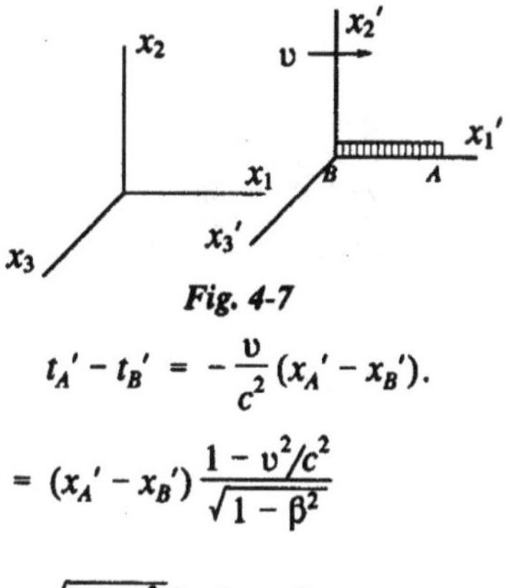

Fig. 4-7

At $t_A = t_B$:

$$t_A' + \frac{\upsilon}{c^2}x_A' = t_B' + \frac{\upsilon}{c^2}x_B' \quad \Rightarrow \quad t_A' - t_B' = -\frac{\upsilon}{c^2}(x_A' - x_B').$$

$$x_A - x_B = \frac{(x_A' - x_B') + \upsilon(t_A' - t_B')}{\sqrt{1-\beta^2}} = (x_A' - x_B')\frac{1 - \upsilon^2/c^2}{\sqrt{1-\beta^2}}$$

$$= \sqrt{1-\beta^2}\,(x_A' - x_B').$$

Thus, $x_A - x_B = \sqrt{1-\beta^2}\,(x_A' - x_B') \leq (x_A' - x_B')$ and therefore represents a contraction.

Example: The electron's electrostatic field for $\upsilon = 0$ yields equipotential surfaces, i.e. spherical surfaces about the electron. When in motion at very high velocities, the equipotential surfaces appear to be ellipsoidal (Fig. 4-8).

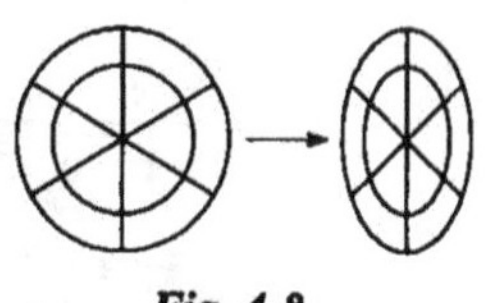
Fig. 4-8

Note that the expression $x_A - x_B = \sqrt{1-\beta^2}\,(x_A' - x_B') \leq (x_A' - x_B')$ could have been obtained from the inverse transformation, i.e. $(x_A' - x_B') = [(x_A - x_B) + \upsilon(t_A - t_B)]/\sqrt{1-\beta^2}$, where $t_A = t_B$.

4.4 Tensor Analysis in N-Dimensional Space

4.4.1 Tensors and Pseudo Tensors

Definition: A $\begin{Bmatrix} \text{tensor} \\ \text{pseudo tensor} \end{Bmatrix}$ of the nth rank, $T_{i_1 i_2 \ldots i_n}$ $(i \to 1,2,3,\ldots,N)$ is one which, under an orthogonal transformation $x' = Vx$, transforms like

$$T'_{i_1 i_2 \ldots i_n} = \begin{cases} \sum_{j_i=1}^{N} V_{i_1 j_1} V_{i_2 j_2} \ldots V_{i_n j_n} T_{j_1 j_2 \ldots j_n}. \\ \det|V| \sum_{j_i=1}^{N} V_{i_1 j_1} V_{i_2 j_2} \ldots V_{i_n j_n} T_{j_1 j_2 \ldots j_n}. \end{cases}$$

Notes:

(1) If $\det|V| = +1$, there is no distinction between psudo tensor and tensor.

(2) The set formed of coordinates $x_{i_1}\, x_{i_2} \ldots x_{i_n}$ is a tensor of the nth rank, since each coordinate transforms like a tensor.

(3) δ_{ij} is an invariant tensor of the 2nd rank, since

$$\sum_{j_1 j_2=1}^{N} V_{i_1 j_1} V_{i_2 j_2} \delta_{j_1 j_2} = \sum_{j_1=1}^{N} V_{i_1 j_1} V_{i_2 j_1} = (V\tilde{V})_{i_1 i_2} = \delta_{i_1 i_2}.$$

(4) $\varepsilon_{i_1 i_2 \ldots i_n}$, which $= \begin{Bmatrix} +1 \\ -1 \end{Bmatrix}$ if $i_1 \ldots i_n$ is an $\begin{matrix}\text{even} \\ \text{odd}\end{matrix}$ permutation of $1,2,\ldots,n$ and $= 0$ otherwise, is an invariant pseudo tensor of the nth rank. [Note that n must $= N$ for the case of ε.]

Proof: Consider

$$\sum_{j_i=1}^{N} V_{i_1 j_1} V_{i_2 j_2} \ldots V_{i_n j_n} \varepsilon_{j_1 j_2 \ldots j_n} = \det \begin{vmatrix} V_{i_1 1} & V_{i_1 2} & V_{i_1 3} & \cdot & V_{i_1 n} \\ V_{i_2 1} & V_{i_2 2} & \cdot & \cdot & V_{i_2 n} \\ \cdot & \cdot & \cdot & \cdot & \cdot \\ \cdot & \cdot & \cdot & \cdots & \cdot \\ \cdot & \cdot & \cdot & \cdot & \cdot \\ V_{i_n 1} & V_{i_n 2} & \cdot & \cdot & V_{i_n n} \end{vmatrix}$$

But

$$\det\begin{vmatrix} V_{i_1 1} & V_{i_1 2} & V_{i_1 3} & & \cdot & V_{i_1 n} \\ V_{i_2 1} & V_{i_2 2} & \cdot & & \cdot & V_{i_2 n} \\ \cdot & \cdot & \cdot & & \cdot & \cdot \\ \cdot & \cdot & \cdot & \cdots\cdots & \cdot & \cdot \\ \cdot & \cdot & \cdot & & \cdot & \cdot \\ V_{i_n 1} & V_{i_n 2} & \cdot & & \cdot & V_{i_n n} \end{vmatrix}$$

$$= \det|V| \times \begin{cases} +1 & \text{if } i_1 \ldots i_n \text{ is an even permutation of } 1 \ldots n\,. \\ -1 & \text{if } i_1 \ldots i_n \text{ is an even permutation of } 1 \ldots n\,. \\ 0 & \text{otherwise}\,. \end{cases}$$

Hence, multiplying through by $\det|V|$, recalling that $(\det|V|)^2 = +1$, we have

$$\det|V| \sum_{j_i=1}^{N} V_{i_1 j_1} V_{i_2 j_2} \ldots V_{i_n j_n} \varepsilon_{j_1 j_2 \ldots j_n}\, \varepsilon_{j_1 j_2 \ldots j_n} = \varepsilon_{i_1 i_2 \ldots i_n}\,.$$

Examples in 3-dimensional space

(a) If **A** and **B** are vectors (tensors of the first rank), then **A**×**B** is a pseudo vector, or a pseudo tensor of the first rank.

Proof: By definition,

$$\begin{aligned} (\mathbf{A}\times\mathbf{B})_i' &= \sum_{jk} \varepsilon_{ijk}' A_j' B_k' \\ &= \sum_{jk} \det|V| \sum_{lmn} V_{il} V_{jm} V_{kn} \varepsilon_{lmn} \sum_{o} V_{jo} A_o \sum_{p} V_{kp} B_p \\ &= \det|V| \sum_{lmnop} \left[\sum_{j} V_{jm} V_{jo} \sum_{k} V_{kn} V_{kp} \right] V_{il} \varepsilon_{lmn} A_o B_p \\ &= \det|V| \sum_{lmnop} \delta_{mo} \delta_{np} V_{il} \varepsilon_{lmn} A_o B_p = \det|V| \sum_{lmn} V_{il} \varepsilon_{lmn} A_m B_n \\ &= \det|V| \sum_{l} V_{il} \left[\sum_{mn} \varepsilon_{lmn} A_m B_n \right] = \det|V| \sum_{l} V_{il} (\mathbf{A}\times\mathbf{B})_l\,. \end{aligned}$$

Hence (**A**×**B**) is a pseudo vector by the law of transformation.

(b) If **A** is a pseudo vector and **B** is a good vector, then (**A**×**B**) is a good vector.

Proof:

$$(\mathbf{A}\times\mathbf{B})_i' = \sum_{jk} \varepsilon_{ijk}' A_j' B_k'$$

$$= \sum_{jk} \det|V| \sum_{lmn} V_{il} V_{jm} V_{kn} \varepsilon_{lmn} \sum_{o} V_{jo} A_o \sum_{p} V_{kp} B_p$$

$$= (\det|V|)^2 \sum_{lmnop} \left[\sum_{jk} V_{jm} V_{jo} V_{kn} V_{kp}\right] V_{il} \varepsilon_{lmn} A_o B_p ,$$

which, as above, reduces to $+1 \sum_l V_{il} \left[\sum_{mn} \varepsilon_{lmn} A_m B_n\right]$. Thus,

$$(\mathbf{A}\times\mathbf{B})_i' = \sum_l V_{il} (\mathbf{A}\times\mathbf{B})_l ,$$

satisfying the law of transformation of a good vector.

Consider the inversion transformation in 3 dimensions expressed by $x_1' = -x_1;\ x_2' = -x_2;\ x_3' = -x_3$. The inversion transformation matrix V is given by:

$$V = \begin{pmatrix} -1 & 0 & 0 \\ 0 & -1 & 0 \\ 0 & 0 & -1 \end{pmatrix};\ \det|V| = -1.$$

[Note that for the case of 3 dimensions, an inversion is different from a rotation for which $\det|V|_{\text{rot}} = +1$, while for the case of 4 dimensions, an inversion is the same as a rotation, i.e. for odd dimensionality, an inversion is different from a rotation. A reflection, which changes the sign of one coordinate only, is <u>always</u> different from a rotation, regardless of the dimensionality.]

Under the 3-dimensional inversion above, $\mathbf{r} \to -\mathbf{r}$ and $\mathbf{v} \to -\mathbf{v}$ (derivatives of coordinates must also change sign). Hence **r** and **v**, which change sign, behave like good vectors. Therefore, $(\mathbf{r}\times\mathbf{v})_i = \sum_{jk} \varepsilon_{ijk} r_j \upsilon_k \ \to\ (\mathbf{r}\times\mathbf{v})_i'$, which doesn't change sign because it is the product of two sign-changing components — hence, a pseudo vector.

Consider the graphical interpretation of an inversion (Fig. 4-9), where we have constructed the vector cross product (r×v) formed of two vectors **r** and **v**. We see that the angular momentum vector (r×v) is a pseudo vector, since (r×v)′ = (r×v).

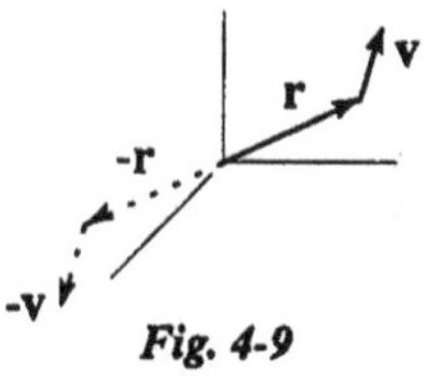

Fig. 4-9

Vectors and pseudo vectors in electromagnetic theory

In electromagnetic theory (restricted to 3 dimensions), the following Maxwell equations apply:

$$\nabla\cdot\mathbf{E} = 4\pi\rho.$$

$$\nabla\cdot\mathbf{H} = 0.$$

$$\nabla\times\mathbf{E} = -\frac{1}{c}\dot{\mathbf{H}}.$$

$$\nabla\times\mathbf{H} = \frac{1}{c}\dot{\mathbf{E}} + \frac{4\pi\mathbf{j}}{c}.$$

We perform an inversion and investigate the behavior of **E** and **H**, i.e. see if they are vectors or pseudo vectors. Before we can do this, we must know what happens to **j** and ρ under an inversion.

We adopt the usual convention that charge remains unchanged under an inversion (non-conjugation of charge), i.e. if positive, it remains positive. We also require that Maxwell's equations be invariant to all transformations, as is necessary by physical law.

We have that $\mathbf{j} = \rho\mathbf{v}$, and since **v** changes sign, **j** must change sign; hence, **j** is a good vector. Furthermore, since $\partial/\partial x' \to -\partial/\partial x$, the gradient operator, $\nabla \equiv \partial/\partial x$, changes sign and is also a good vector. Thus, under an inversion, $\nabla\cdot \to -\nabla\cdot$ and $\nabla\times \to -\nabla\times$.

Since $\nabla\cdot\mathbf{E} = 4\pi\rho$ and ρ doesn't change sign but $\nabla\cdot$ does, **E** must change sign; hence **E** is a good vector. Since $\nabla\times\mathbf{E}$ remains unchanged, $\dot{\mathbf{H}}$ and therefore **H** remain unchanged; hence **H** is a pseudo vector. Thus, if system Σ undergoes an inversion to system Σ', we have the following results:

$$\left\{\begin{matrix}\rho \to \rho \\ \mathbf{H} \to \mathbf{H}\text{ pseudo vector}\end{matrix}\right\} \qquad \left\{\begin{matrix}\mathbf{j} \to -\mathbf{j} \\ \mathbf{E} \to -\mathbf{E}\end{matrix}\right\}\text{ vectors.}$$

According to Einstein, all physical laws must be invariant to 4-dimensional orthogonal transformations (i.e. Lorentz transformations). Hence, we must be able to express physical laws as good tensor relations.

In 4 dimensions we have $x_1,\ x_2,\ x_3,\ x_4 = ict$, where the x_μ are good vector components, as are the dx_μ, commonly called the 4-vector.

According to Newton, $\mathbf{F} = \dot{\mathbf{p}}$, which is a good vector relationship in 3 dimensions. However, it doesn't hold for 4 dimensions, i.e. the velocity vector is no longer a good vector in 4 dimensions. Which vector should we replace it with?

Define

$$ds^2 \equiv -\sum_{\mu=1}^{4} dx_\mu^2 = -(dx_1^2 + dx_2^2 + dx_3^2) + c^2dt^2.$$

Under a Lorentz orthogonal transformation V, ds is a scalar and remains invariant; hence $\frac{dx_\mu}{ds} : \left(\frac{dx_1}{ds}, \frac{dx_2}{ds}, \frac{dx_3}{ds}, \frac{dx_4}{ds}\right)$, behaves as a good vector, and transforms as $\left(\frac{dx_\mu}{ds}\right)' = \sum_{\nu=1}^{4} V_{\mu\nu} \frac{dx_\nu}{ds}$.

Note that $ds/dt = \sqrt{-(dx_1^2 + dx_2^2 + dx_3^2) + c^2dt^2}/dt$, so that defining the 3-dimensional velocity υ_i as

$$\upsilon_1 \equiv dx_1/dt\,; \quad \upsilon_2 \equiv dx_2/dt\,; \quad \upsilon_3 \equiv dx_3/dt\,,$$

we have $\upsilon \equiv \sqrt{\upsilon_1^2 + \upsilon_2^2 + \upsilon_3^2}$ and $ds/dt = \sqrt{c^2 - \upsilon^2}$. Then

$$\frac{dx_1}{ds} = \frac{dx_1/dt}{ds/dt} = \frac{\upsilon_1}{\sqrt{c^2 - \upsilon^2}} = \frac{\upsilon_1}{c}\frac{1}{\sqrt{1 - \beta^2}}. \quad (\beta = \upsilon/c)$$

$$\frac{dx_4}{ds} = ic\frac{dt}{ds} = \frac{ic}{\sqrt{c^2 - \upsilon^2}} = \frac{i}{\sqrt{1 - \beta^2}}.$$

Hence,

$$\frac{dx_\nu}{ds} : \left(\frac{\upsilon_1}{c\sqrt{1 - \beta^2}}, \frac{\upsilon_2}{c\sqrt{1 - \beta^2}}, \frac{\upsilon_3}{c\sqrt{1 - \beta^2}}, \frac{i}{\sqrt{1 - \beta^2}}\right).$$

We now introduce the 4-momentum

$$p_\nu = m_o c \frac{dx_\nu}{ds},$$

which we see represents good vectors. We have

$$p_i = \frac{m_o \upsilon_i}{\sqrt{1-\beta^2}}\ (i = 1,2,3);\quad p_4 = \frac{i m_o c}{\sqrt{1-\beta^2}} = \frac{iE}{c},\ E = \frac{m_o c^2}{\sqrt{1-\beta^2}},$$

where the p_i are the generalized momenta of the particle and p_4 is the generalized energy of the particle.

For small velocities ($\beta << 1$), the momenta behave like 3-dimensional momenta, since

$$p_1 = m_o \upsilon_1 + \ldots,\ p_2 = m_o \upsilon_2 + \ldots,\ p_3 = m_o \upsilon_3,$$

while

$$p_4 \cong i\left(m_o c + m_o \upsilon^2/2c\right) = iE/c \ \rightarrow\ E = m_o c^2 + m_o \upsilon^2/2 + \ldots$$

For a system at rest, $\upsilon = 0$, $p_1 = p_2 = p_3 = 0$, $E_o = m_o c^2$ (rest energy), while for a system moving with velocity $\upsilon > 0$, $p > m_o \upsilon$, $E > E_o$. Let us define

$$m = \frac{m_o}{\sqrt{1-\beta^2}},$$

where m_o is the rest mass and m the moving mass. Then,

$$p_1 = m\upsilon_1,\ p_2 = m\upsilon_2,\ p_3 = m\upsilon_3,\ p_4 = imc,\ E = mc^2.$$

4.4.2 Similarity Transformations

Consider a linear transformation in Σ which transforms a point $P(x)$ into another point $Q(y)$ by the transformation A, i.e. $y = Ax$.

Now consider a transformation S which transforms Σ into Σ', such that $y' = Sy$ and $x' = Sx$.

We ask! What transformation A' in Σ' will accomplish the same transformation in Σ' as A did in Σ, i.e. we seek an A' such that $P(x') \rightarrow Q(y')$, where $y' = A'x'$.

Theorem: $A' = SAS^{-1}$ defines a similarity transformation.

Proof: If $y' = A'x'$,

$$\begin{Bmatrix} y' = Sy \\ x' = Sx \end{Bmatrix} \Rightarrow Sy = A'Sx \Rightarrow S^{-1}Sy = S^{-1}A'Sx.$$

Hence, $y = S^{-1}A'Sx$. But $y = Ax \Rightarrow Ax = S^{-1}A'Sx \Rightarrow A = S^{-1}A'S$. Therefore, multiplying on the left by S and on the right S^{-1}, we have:

$$SAS^{-1} = SS^{-1}A'SS^{-1} = A'.$$

If S is a unitary transformation, it means that we have used a unitary matrix S ($S^+ = S^{-1}$). If we use an orthogonal matrix S ($\tilde{S} = S^{-1}$), then the similarity transformation is called an orthogonal transformation.

Example in 2-dimensional real space (Fig. 4-10)

Let A be a rotation in Σ of 90° and let S be a reflection from Σ to Σ'. Find A'.

Under A: $x \to y$. Under S: $x \to x'$ and $y \to y'$. Under A': $x' \to y'$

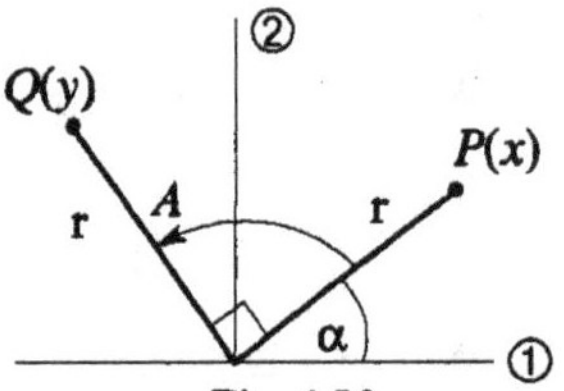

Fig. 4-10

Under A:

$$\begin{Bmatrix} y_1 = -r\sin\alpha = -x_2. \\ y_2 = r\cos\alpha = x_1. \end{Bmatrix} \Rightarrow A = \begin{pmatrix} 0 & -1 \\ 1 & 0 \end{pmatrix}.$$

Reflecting about the ② axis:

$$\begin{Bmatrix} x_1' = -x_1. \\ x_2' = x_2. \end{Bmatrix} \Rightarrow S = \begin{pmatrix} -1 & 0 \\ 0 & 1 \end{pmatrix} = S^{-1}.$$

$$S: \begin{Bmatrix} y_1' = -y_1 \\ y_2' = y_2 \end{Bmatrix} \quad A: \begin{Bmatrix} y_1 = -x_2 \\ y_2 = x_1 \end{Bmatrix} \quad S: \begin{Bmatrix} x_2 = x_2' \\ x_1 = -x_1' \end{Bmatrix}$$

$$\begin{Bmatrix} y_1' = -y_1 = x_2 = x_2' \\ y_1' = -y_1 = x_2 = x_2' \end{Bmatrix} \Rightarrow \begin{Bmatrix} y_1' = x_2' \\ y_2' = -x_1' \end{Bmatrix} \Rightarrow A' = \begin{pmatrix} 0 & 1 \\ -1 & 0 \end{pmatrix}.$$

$$SAS^{-1} = \begin{pmatrix} -1 & 0 \\ 0 & 1 \end{pmatrix}\begin{pmatrix} 0 & -1 \\ 1 & 0 \end{pmatrix}\begin{pmatrix} -1 & 0 \\ 0 & 1 \end{pmatrix} = \begin{pmatrix} 0 & 1 \\ 1 & 0 \end{pmatrix}\begin{pmatrix} -1 & 0 \\ 0 & 1 \end{pmatrix} = \begin{pmatrix} 0 & 1 \\ -1 & 0 \end{pmatrix} = A'.$$

Definition: If $H = H^+$ (i.e. $H_{ij} = H_{ji}^*$), then H is a Hermitian matrix.

[Note: Every real symmetric matrix is Hermitian].

Theorem: If U is a unitary matrix and H is Hermitian, then UHU^{-1} is also Hermitian.

Proof:

$$(UHU^{-1})^+ = (UHU^+)^+ = (U^+)^+H^+U^+ = UHU^+ = UHU^{-1}.$$

We have thereby shown that the Hermitian property is retained for unitary similarity transformations.

[Note: since we have shown that the product of unitary matrices yields another unitary matrix, then the unitary property is retained under a unitary similarity transformation.]

4.5 Matrices in N-dimensional Space

4.5.1 Diagonalization of a Matrix

Theorem: If A is either a Hermitian matrix or a unitary matrix, then there exists a unitary matrix U, such that

$$UAU^+ = UAU^{-1} = \begin{pmatrix} \lambda_1 & 0 & 0 & \cdot & \cdot & \cdot & \cdot & \cdot & 0 \\ 0 & \lambda_2 & 0 & \cdot & \cdot & \cdot & \cdot & \cdot & \cdot \\ 0 & 0 & \lambda_3 & \cdot & \cdot & \cdot & \cdot & \cdot & \cdot \\ \cdot & 0 & \cdot & \cdot & \cdot & \cdot & \cdot & \cdot & \cdot \\ \cdot & \cdot & \cdot & \cdot & \cdot & \cdot & \cdot & \lambda_{n-1} & \cdot \\ 0 & 0 & 0 & \cdot & \cdot & \cdot & \cdot & \cdot & \lambda_n \end{pmatrix}.$$

[We have already dealt with diagonalizations for real symmetric tensors T_{ij} in 3 dimensions, which are merely special cases of a Hermitian matrix.]

Proof (i): For A Hermitian, i.e. $A = A^+$.

Consider the characteristic equation $A\psi = \lambda\psi$, where ψ is an $n\times 1$ matrix, an eigenvector of A, and where λ is a number, the eigenvalue of A (at this stage not necessarily real).

If the above is true, then we have that $(A - \lambda I)\psi = 0$, the only solution of which, for $\psi \neq 0$, is given by

$$\det|A - \lambda I| = \begin{vmatrix} A_{11}-\lambda & A_{12} & A_{13} & \cdots & A_{1n} \\ A_{21} & A_{22}-\lambda & A_{23} & \cdots & A_{2n} \\ \cdot & \cdot & A_{33}-\lambda & \cdots & \cdot \\ \cdot & \cdot & \cdot & \cdots & \cdot \\ \cdot & \cdot & \cdot & \cdots & \cdot \\ A_{n1} & A_{n2} & \cdot & \cdots & A_{nn}-\lambda \end{vmatrix} = 0,$$

an nth order equation in λ with solutions $\lambda_1\lambda_2\ldots\lambda_n$, the n eigenvalues of A.

Corresponding to each eigenvalue λ_i there corresponds an eigenvector ψ^i. Hence our characteristic equation can now be written as the set of n^2 equations

(1) $\quad A\psi^i = \lambda_i\psi^i.$

For each λ_i we have a set of n equations. Choose, for example, λ_7 with its corresponding ψ^7. We then have

$$A_{11}\psi_1^7 + A_{12}\psi_2^7 + \ldots + A_{1n}\psi_n^7 = \lambda_7\psi_1^7.$$
$$A_{21}\psi_1^7 + A_{22}\psi_2^7 + \ldots + A_{2n}\psi_n^7 = \lambda_7\psi_1^7.$$
$$\vdots$$
$$A_{n1}\psi_1^7 + A_{n2}\psi_2^7 + \ldots + A_{nn}\psi_n^7 = \lambda_7\psi_1^7.$$

From these equations we can solve for ψ^7. Repeating the above procedure for each λ_i, we build up the $n\times n$ matrix

$$\psi = \begin{pmatrix} \psi_1^1 & \psi_1^2 & & \psi_1^n \\ \psi_2^1 & \psi_2^2 & & \psi_2^n \\ \cdot & \cdot & \cdots & \cdot \\ \cdot & \cdot & & \cdot \\ \cdot & \cdot & & \psi_n^n \end{pmatrix}.$$

We now wish to normalize ψ^i, i.e. choose ψ^i such that $(\psi^i)^+\psi^i = 1$. To proceed, we will have need of the following Lemmas:

Lemma (1): All eigenvalues λ_i are real.

We multiply the characteristic equation (1) on the left by ψ^{i+} to get $\psi^{i+}A\psi^i = \lambda_i\psi^{i+}\psi^i = \lambda_i$. Since λ_i is just a number, it follows that $\lambda_i^+ = \lambda_i^*$. Therefore,

$$\left[\psi^{i+}A\psi^i\right]^* = \left[\psi^{i+}A\psi^i\right]^+ = \psi^{i+}A^+\psi^i = \psi^{i+}A\psi^i = \lambda_i.$$

Lemma (2): Two eigenvectors belonging to two different eigenvalues must be orthogonal.

Consider:

(2) $\quad A\psi^i = \lambda_i\psi^i.$

(3) $\quad A\psi^j = \lambda_j\psi^j.$

Taking the adjoint of (2), $\psi^{i+}A^+ = \lambda_i\psi^{i+}$. But since $A = A^+$, we have

(4) $\quad \psi^{i+}A = \lambda_i\psi^{i+}.$

Multiplying (3) by ψ^{i+} on the left and (4) by ψ^j on the right:

(5) $\quad \psi^{i+}A\psi^j = \lambda_j\psi^{i+}\psi^j.$

(6) $\quad \psi^{i+}A\psi^j = \lambda_i\psi^{i+}\psi^j.$

Subtracting:

(7) $\quad (\lambda_i - \lambda_j)(\psi^{i+}\psi^j) = 0.$

If $\lambda_i \neq \lambda_j$ then $\psi^{i+}\psi^j = 0$, which implies that the two particular eigenvectors are orthogonal.

Note that every $\psi^{i'} = e^{i\theta}\psi^i$ obeys the choice $\psi^{i+}\psi^i = 1$. This means that the phase of the eigenvector is variable. [This comes under consideration when dealing with gauge transformations in Quantum Mechanics.]

Lemma (3): If $\lambda_i = \lambda_j$, the Schmidt orthogonalization method can be used to find a set of orthogonal vectors from the set of unit vectors φ^i.

Let $\lambda_1 = \lambda_2 = \lambda$ and let the corresponding eigenvectors be φ^1 and φ^2.

Then $A\varphi^1 = \lambda_1\varphi^1 = \lambda\varphi^1$ and $A\varphi^2 = \lambda_2\varphi^2 = \lambda\varphi^2$, where $\varphi^2 \neq c\varphi^1$, otherwise φ^2 and φ^1 would satisfy the same equation and essentially be the same eigenvector.

Define $\psi^1 \equiv \varphi^1$ and choose $\varphi^{1+}\varphi^1 = 1$.

Define $\psi^2 \equiv \left[\varphi^2 - \left(\psi^{1+}\varphi^2\right)\varphi^1\right]N_2$, where N_2 is chosen to normalize ψ^2.

Note then that

$$\psi^{1+}\psi^1 = \varphi^{1+}\varphi^1 = 1.$$

$$\begin{aligned}\psi^{2+}\psi^2 &= \left[\varphi^2 - \left(\psi^{1+}\varphi^2\right)\varphi^1\right]^+\left[\varphi^2 - \left(\psi^{1+}\varphi^2\right)\varphi^1\right]N_2^2 \\ &= \left[\varphi^{2+} - \varphi^{1+}\varphi^{2+}\psi^1\right]\left[\varphi^2 - \psi^{1+}\varphi^2\varphi^1\right]N_2^2 \\ &= \left[\varphi^{2+}\varphi^2 + \varphi^{1+}\varphi^{2+}\varphi^1\varphi^{1+}\varphi^2\varphi^1 - \varphi^{2+}\varphi^{1+}\varphi^2\psi^1\right]N_2^2 \\ &\quad -\left[\varphi^{1+}\varphi^{2+}\varphi^1\varphi^2\right]N_2^2 = 1,\end{aligned}$$

since $\psi^{2+}\psi^2 \neq 0$ and N_2 is chosen to normalize ψ^2.

For any set of $\lambda_i = \lambda$, say $\lambda_1 = \lambda_2 = \lambda_3 = \lambda$, we have a set of corresponding eigenvectors $\varphi_1\ \varphi_2\ \varphi_3$. First define $\psi^1 \equiv \varphi_1$ and choose $\varphi_1{}^+\varphi_1 = 1$.

Then define

$$\psi^2 \equiv \left[\varphi_2 - \left(\psi^{1+}\varphi_2\right)\psi^1\right]N_2,$$

where N_2 is chosen to normalize ψ^2. Then, as shown above, $\psi^{2+}\psi^2 = 1$, while $\psi^{1+}\psi^2 \equiv \left[\overset{+}{\varphi_1}\varphi_2 - \left(\overset{+}{\varphi_1}\varphi_2\right)\overset{+}{\varphi_1}\varphi_1\right]N_2 \equiv 0.$

Then define

$$\psi^3 \equiv \left[\varphi_3 - \left(\psi^{2+}\varphi_3\right)\psi^2 - \left(\psi^{1+}\varphi_3\right)\varphi^1\right]N_3,$$

where N_3 is chosen to normalize ψ^3. Then, $\psi^{3+}\psi^3 = 1$, while $\psi^{1+}\psi^3 \equiv$ $\left[\overset{+}{\varphi_1}\varphi_3 - \left(\psi^{2+}\varphi_3\right)\psi^{1+}\psi^2 - \left(\psi^{1+}\varphi_3\right)\psi^{1+}\psi^1\right]N_3 \equiv 0$, and $\psi^{2+}\psi^3 \equiv$ $\left[\psi^{2+}\varphi_3 - \left(\psi^{2+}\varphi_3\right)\psi^{2+}\psi^2 - \left(\psi^{1+}\varphi_3\right)\psi^{2+}\psi^1\right]N_3 \equiv 0.$

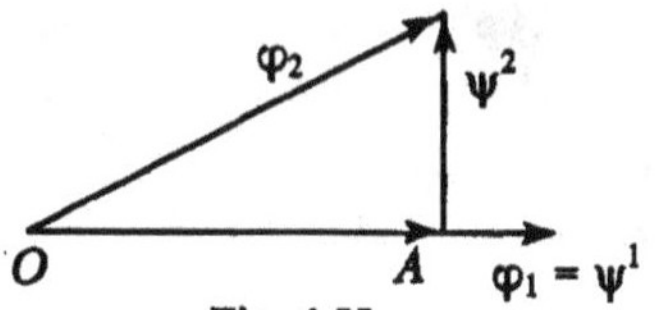

Fig. 4-11

To show how the vectors ψ^2 and ψ^3 were created geometrically:

We first construct ψ^2 to be perpendicular to φ_1 (or ψ^1) (Fig. 4-11). Then since $\psi^2 = \varphi_2 - \vec{OA}$ and since $\vec{OA}$ = projection of φ_2 on φ_1, $\vec{OA} = \varphi_2^+\varphi_1 = \varphi_1^+\varphi_2$, we have

$$\psi^2 = \left[\varphi_2 - \left(\varphi_1^+\varphi_2\right)\varphi_1\right]N.$$

Returning to our theorem, we have shown that, for any set of eigenvalues, the n eigenvectors can be so chosen that $\psi^{i+}\psi^j = \delta_{ij}$.

Let us choose U such that

$$U^+ = \begin{pmatrix} \psi_1^1 & \psi_1^2 & & \psi_1^n \\ \psi_2^1 & \psi_2^2 & & \psi_2^n \\ \cdot & \cdot & \cdots & \cdot \\ \cdot & \cdot & & \cdot \\ \psi_n^1 & \psi_n^2 & & \psi_n^n \end{pmatrix}.$$

It then follows that

$$(U^+)^+ = U = \tilde{U}^* = \begin{pmatrix} \psi_1^{1*} & \psi_2^{1*} & & \psi_n^{1*} \\ \psi_1^{2*} & \psi_2^{2*} & & \psi_n^{2*} \\ \cdot & \cdot & \cdots & \cdot \\ \cdot & \cdot & & \cdot \\ \psi_1^{n*} & \psi_2^{n*} & & \psi_n^{n*} \end{pmatrix} = \begin{pmatrix} \psi_1^1 & \psi_2^1 & & \psi_n^1 \\ \psi_1^2 & \psi_2^2 & & \psi_n^2 \\ \cdot & \cdot & \cdots & \cdot \\ \cdot & \cdot & & \cdot \\ \psi_1^n & \psi_2^n & & \psi_n^n \end{pmatrix}.$$

[Note then that

$$UU^+ = \begin{pmatrix} \psi^{1+}\psi^1 & \psi^{1+}\psi^2 & & \psi^{1+}\psi^n \\ \psi^{2+}\psi^1 & \psi^{2+}\psi^2 & & \psi^{2+}\psi^n \\ \cdot & \cdot & \cdots & \cdot \\ \cdot & \cdot & & \cdot \\ \psi^{n+}\psi^1 & \psi^{n+}\psi^2 & & \psi_n^+\psi^n \end{pmatrix} = \begin{pmatrix} 1 & 0 & & 0 \\ 0 & 1 & & 0 \\ 0 & 0 & & 0 \\ \cdot & \cdot & \cdots & \cdot \\ \cdot & \cdot & & \cdot \\ \cdot & \cdot & & 1 \end{pmatrix},$$

illustrating that U is a unitary matrix.]

To show that $UAU^+ = \lambda_i \delta_{ij}$:

$$AU^+ = \begin{pmatrix} A\psi^1 & A\psi^2 & & A\psi^n \\ \lambda_1\psi_1^1 & \lambda_2\psi_1^2 & & \lambda_n\psi_1^n \\ \lambda_1\psi_2^1 & \lambda_2\psi_2^2 & \cdots & \lambda_n\psi_2^n \\ \cdot & \cdot & & \cdot \\ \cdot & \cdot & & \cdot \\ \lambda_1\psi_n^1 & \lambda_2\psi_n^2 & & \lambda_n\psi_n^n \end{pmatrix}, \text{ and}$$

$$UAU^+ = \begin{pmatrix} \lambda_1\psi^{1+}\psi^1 & \lambda_2\psi^{i+}\psi^2 & & \lambda_n\psi^{i+}\psi^n \\ \lambda_1 & 0 & & 0 \\ 0 & \lambda_2 & & 0 \\ 0 & 0 & \cdots & 0 \\ \cdot & \cdot & & \cdot \\ \cdot & \cdot & & \cdot \\ \cdot & \cdot & & \cdot \\ \cdot & \cdot & & \lambda_n \end{pmatrix}.$$

We have already shown that if A is Hermitian and U is unitary, then UAU^+ is still Hermitian, thus implying that the eigenvalues λ_i are real. For A real and Hermitian, since the λ_i are real, the ψ^i can be anything. We will always choose the ψ^i to be real.

Proof (ii): For A unitary, i.e. $AA^+ = 1$, $A^+ = A^{-1}$.

Consider the same characteristic equations as in Proof (i):

(8) $\quad A\psi^i = \lambda_i\psi^i.$

(9) $\quad A\psi^j = \lambda_j\psi^j.$

Although we can choose the ψ^i such that $\psi^{i+}\psi^i = 1$, the problem is to show that for $i \neq j$, $\psi^{i+}\psi^j = \delta_{ij}$. This can be done as follows:

Take the adjoint of (8) and apply it to (8) on the left: $\psi^{i+}A^+A\psi^i = \psi^{i+}\psi^i\lambda_i^*\lambda_i$. Thus, $\lambda_i^*\lambda_i = 1 \Rightarrow |\lambda_i|^2 = 1 \Rightarrow \lambda_i = e^{i\theta}$, θ real. Hence the eigenvalues of a unitary matrix are complex, with magnitude equal to unity.

(10) $A\psi^i = e^{i\theta}\psi^i.$

(11) $\psi^{j+}A^+ = e^{-i\varphi}\psi^{j+}.$

Applying (11) to the left side of (10):

(12) $\psi^{j+}A^+A\psi^i = e^{i(\theta-\varphi)}\psi^{j+}\psi^i. \quad \rightarrow \quad \psi^{j+}\psi^i\left(e^{i(\theta-\varphi)} - 1\right) = 0.$

But for $i \neq j$, $\theta \neq \varphi$, $\lambda_i \neq \lambda_j$ so that $\psi^{j+}\psi^i = 0$. The procedure follows now exactly as for Proof (i).

For the case that A is unitary and real, since $A\psi^i = \lambda_i\psi^i$:

$$\begin{cases} \text{If } \theta = 0 \ (\lambda \text{ real}),\ \psi^i \text{ may be complex, but can also be real.} \\ \text{If } \theta \neq 0 \ (\lambda \text{ complex}),\ \psi^i \text{ must be complex.} \end{cases}$$

4.5.2 Applications of Matrix Diagonalization

We have shown that for every matrix A (whether unitary or Hermitian) we can find a unitary matrix U such that

$$UAU^+ = \Lambda = \begin{pmatrix} \lambda_1 & 0 & & 0 \\ 0 & \lambda_2 & & 0 \\ 0 & 0 & & 0 \\ \cdot & \cdot & \cdots & \cdot \\ \cdot & \cdot & & \cdot \\ \cdot & \cdot & & \lambda_n \end{pmatrix},$$

where $UU^+ = U^+U = 1$, i.e. $U^+ = U^{-1}$.

Example 1: Diagonalization of a real $2nd$ rank symmetric tensor T_{ij}. [An example in real 3-dimensional space would be the moment of inertia tensor.]

$$T = \begin{pmatrix} T_{11} & T_{12} & \cdots & T_{1n} \\ T_{21} & T_{22} & \cdots & T_{2n} \\ T_{13} & T_{23} & \cdots & T_{3n} \\ \cdot & \cdot & \cdots & \cdot \\ \cdot & \cdot & \cdots & \cdot \\ T_{1n} & T_{2n} & \cdots & T_{nn} \end{pmatrix}.$$

Since $T_{ij} = T^*_{ji}$, $T = T^+$ and T is Hermitian. For the characteristic equation, $T\psi^i = \lambda_i\psi^i$, all the λ are real; hence, we choose ψ to be real. We then have

$$U^+ = \begin{pmatrix} \psi_1^1 & \psi_1^2 & & \psi_1^n \\ \psi_2^1 & \psi_2^2 & & \psi_2^n \\ \cdot & \cdot & \cdots & \cdot \\ \cdot & \cdot & & \cdot \\ \psi_n^1 & \psi_n^2 & & \psi_n^n \end{pmatrix} \text{ is real } \rightarrow U = (U^+)^+ = (\tilde{U}^+) \rightarrow U^+ = \tilde{U},$$

so that

$$UTU^+ = \begin{pmatrix} \psi_1^1 & \psi_2^1 & & \psi_n^1 \\ \psi_1^2 & \psi_2^2 & & \psi_n^2 \\ \cdot & \cdot & \cdots & \cdot \\ \cdot & \cdot & & \cdot \\ \psi_1^n & \psi_2^n & & \psi_n^n \end{pmatrix} \begin{pmatrix} \lambda_1\psi_1^1 & \lambda_2\psi_1^2 & & \lambda_n\psi_1^n \\ \lambda_1\psi_2^1 & \cdot & & \cdot \\ \cdot & \cdot & \cdots & \cdot \\ \cdot & \cdot & & \cdot \\ \lambda_1\psi_n^1 & \lambda_2\psi_n^2 & & \lambda_n\psi_n^n \end{pmatrix} = \begin{pmatrix} \lambda_1 & 0 & & 0 \\ 0 & \lambda_2 & & 0 \\ \cdot & \cdot & \cdots & \cdot \\ \cdot & \cdot & & \cdot \\ 0 & 0 & & \lambda_n \end{pmatrix},$$

where $U\tilde{U} = \tilde{U}U = 1$ and U is real.

Hence we have shown that by performing a real rotation (U) in real space, any 2*nd* rank symmetric tensor can be diagonalized.

Example 2: Small vibrations and normal modes in Mechanics

Consider a system of n particles acting through each other by means of a potential in quadratic form.

If $q_1\, q_2 \ldots q_n$ are the generalized coordinates of a system of particles and $V = \sum_{ij} \upsilon_{ij} q_i q_j$ is the potential, then the equation of motion is given by

$$m_i \ddot{q}_i = -\sum_j \upsilon_{ij} q_j \quad (i = 1,2,\ldots,n),$$

a system of n partial differential equations. m_i is a scalar (for a complex system, m would correspond to some inertial property). In cartesian coordinates, m = mass.

The problem of solving these n differential equations is alleviated by noting that we can form a matrix equation from the above if we define $x_i \equiv \sqrt{m_i}\, q_i$.

Then $\sqrt{m_i}\sqrt{m_i}\,\ddot{q}_i = -\sum_j \upsilon_{ij} q_j \sqrt{m_j}/\sqrt{m_j} \quad \rightarrow \quad \ddot{x}_i = -\sum_j \dfrac{\upsilon_{ij} x_j}{\sqrt{m_j m_i}}.$

The matrix equation becomes $\ddot{x} = -Tx$, where

$$\ddot{x} = \begin{pmatrix} \ddot{x}_1 \\ \ddot{x}_2 \\ \cdot \\ \cdot \\ \ddot{x}_n \end{pmatrix} \quad \text{and} \quad T_{ij} = \frac{\upsilon_{ij}}{\sqrt{m_i m_j}}.$$

Note first that T_{ij} is real. Since V is real, let us choose υ_{ij} to be symmetric. Actually, υ_{ij} is not uniquely defined, since if one considers $V = x_1 x_2$, we could have $\upsilon_{12} = 1$ and $\upsilon_{21} = 0$, or $\upsilon_{12} = 1/2$ and $\upsilon_{21} = 1/2$. We choose this latter symmetric case. Thus, T_{ij} is real and symmetric ($T^*_{ij} = T_{ji} = T^*_{ji}$), so that T is Hermitian. Thus, there exists a unitary matrix U such that $UTU^+ = \Lambda$.

Let $y = Ux$, where

$$y = \begin{pmatrix} y_1 \\ y_2 \\ \cdot \\ \cdot \\ y_n \end{pmatrix}.$$

Then, $\ddot{x} = -Tx \rightarrow U\ddot{x} = -UT(U^+U)x \rightarrow \ddot{y} = -\Lambda y \rightarrow \ddot{y}_i = -\lambda_i y_i$ for every i.

[Note that since υ_{ij} doesn't depend on time, and m_i doesn't depend on time, T_{ij} is independent of time and U is independent of time, so that $\dot{U} = 0$, which is what we have implied above.]

The differential equations are now much simpler, and their solutions, representing oscillating functions, are the normal modes of vibration

$$y_i = A_i e^{+i\sqrt{\lambda_i}t} + B_i e^{-i\sqrt{\lambda_i}t},$$

where the A_i and B_i depend on the initial conditions of position and velocity.

Using the inverse transformation,

$$x = U^{-1}y = U^+y \rightarrow x \text{ in terms of } t,$$

and

$$q_i = \frac{x_i}{\sqrt{m_i}} \rightarrow q_i \text{ in terms of } t.$$

Example 3: Rotation in 2-dimensional real space.

The rotation matrix is given by

$$A = \begin{pmatrix} \cos\theta & -\sin\theta \\ \sin\theta & \cos\theta \end{pmatrix},$$

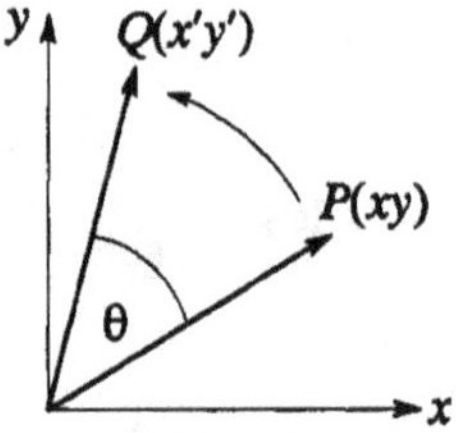

Fig. 4-12

which can be seen by realizing that we can achieve the above rotation by a rotation of axis through an angle $-\theta$ (Fig. 4-12). Then since the 2-dimensional real space rotation matrix is given by

$$A' = \begin{pmatrix} \cos\varphi & \sin\varphi \\ -\sin\varphi & \cos\varphi \end{pmatrix},$$

we merely set $\varphi = -\theta$ to get A above. A is clearly an orthogonal matrix, thus a real unitary matrix ($U^+ = \tilde{U} = U^{-1}$).

$x' = Ax$, where x' and x refer to coordinates in the primed and unprimed systems, respectively. We seek the eigenvalues and eigenvectors of A. We let $z = x + iy$ and investigate the following ($x' = x\cos\theta - y\sin\theta$ and $y' = x\sin\theta + y\cos\theta$):

$$z' = (x' + iy') = e^{i\theta}(x + iy) = e^{i\theta}z.$$

$$z'^* = (x' - iy') = e^{-i\theta}(x - iy) = e^{-i\theta}z^*.$$

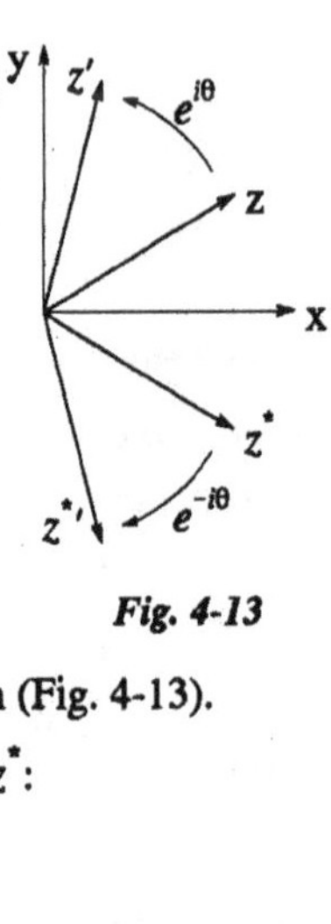

Fig. 4-13

In complex notation $z' = e^{i\theta}z$ represents a rotation of vector z, while $z'^* = e^{-i\theta}z^*$ represents a rotation of the complex conjugate of this vector in the opposite direction (Fig. 4-13).

We construct the following two eigenvectors from z and z^*:

$$\varphi_1 = \begin{pmatrix} \frac{1}{\sqrt{2}} \\ \frac{i}{\sqrt{2}} \end{pmatrix}. \quad \varphi_2 = \begin{pmatrix} \frac{1}{\sqrt{2}} \\ -\frac{i}{\sqrt{2}} \end{pmatrix}.$$

Note that

$$A\varphi_1 = \begin{pmatrix} \frac{\cos\theta - i\sin\theta}{\sqrt{2}} \\ \frac{\sin\theta + i\cos\theta}{\sqrt{2}} \end{pmatrix} = \begin{pmatrix} \frac{e^{-i\theta}}{\sqrt{2}} \\ \frac{ie^{-i\theta}}{\sqrt{2}} \end{pmatrix} = e^{-i\theta}\begin{pmatrix} \frac{1}{\sqrt{2}} \\ \frac{i}{\sqrt{2}} \end{pmatrix} \quad \Rightarrow \quad A\varphi_1 = e^{-i\theta}\varphi_1.$$

Similarly,

$$A\varphi_2 = \begin{pmatrix} \dfrac{\cos\theta + i\sin\theta}{\sqrt{2}} \\ \dfrac{\sin\theta - i\cos\theta}{\sqrt{2}} \end{pmatrix} = \begin{pmatrix} \dfrac{e^{i\theta}}{\sqrt{2}} \\ \dfrac{-ie^{i\theta}}{\sqrt{2}} \end{pmatrix} = e^{i\theta}\begin{pmatrix} \dfrac{1}{\sqrt{2}} \\ \dfrac{-i}{\sqrt{2}} \end{pmatrix} \Rightarrow A\varphi_2 = e^{i\theta}\varphi_2 .$$

We see that φ_1 and φ_2 do indeed constitute eigenvectors. [Note also that $\varphi_1^* = \varphi_2$.] We now construct the matrix U^+ from the eigenvectors φ_1 and φ_2.

$$U^+ = \begin{pmatrix} \dfrac{1}{\sqrt{2}} & \dfrac{1}{\sqrt{2}} \\ \dfrac{i}{\sqrt{2}} & \dfrac{-i}{\sqrt{2}} \end{pmatrix} \Rightarrow U = \begin{pmatrix} \dfrac{1}{\sqrt{2}} & \dfrac{-i}{\sqrt{2}} \\ \dfrac{1}{\sqrt{2}} & \dfrac{i}{\sqrt{2}} \end{pmatrix} \Rightarrow UU^+ = 1.$$

Then,

$$AU^+ = \begin{pmatrix} \dfrac{e^{-i\theta}}{\sqrt{2}} & \dfrac{e^{i\theta}}{\sqrt{2}} \\ \dfrac{ie^{-i\theta}}{\sqrt{2}} & \dfrac{-ie^{i\theta}}{\sqrt{2}} \end{pmatrix} \Rightarrow UAU^+ = \begin{pmatrix} e^{-i\theta} & 0 \\ 0 & e^{i\theta} \end{pmatrix},$$

and the eigenvalues are seen to be $e^{-i\theta}$ and $e^{i\theta}$.

Converse: any matrix B, with eigenvalues (diagonal elements) $e^{-i\theta}$ and $e^{i\theta}$ can be considered to be equivalent to a rotation in 2-dimensional real space.

Proof: If $B = \begin{pmatrix} e^{-i\theta} & 0 \\ 0 & e^{i\theta} \end{pmatrix}$, then we can find an A (and a U) such that

$$B = UAU^+ = \begin{pmatrix} e^{-i\theta} & 0 \\ 0 & e^{i\theta} \end{pmatrix} \Rightarrow A = \begin{pmatrix} \cos\theta & -\sin\theta \\ \sin\theta & \cos\theta \end{pmatrix}.$$

Theorem: Any finite 3-dimensional rotation can be achieved by a single 2-dimensional rotation, if we can find the proper axis.

Proof: Let the rotation in 3-dimensional space be specified by a 3×3 real, unitary matrix A. Then we can find a matrix U such that

$$UAU^+ = \Lambda = \begin{pmatrix} \lambda_1 & 0 & 0 \\ 0 & \lambda_2 & 0 \\ 0 & 0 & \lambda_3 \end{pmatrix}, \text{ where } \det|UAU^+| = \det|A| = \lambda_1\lambda_2\lambda_3 = +1.$$

Since A is unitary, its eigenvalues are complex, such that $|\lambda_i^2| = 1$. A possible choice is $\lambda_i = e^{i\theta_i}$ or $e^{-i\theta_i}$.

If we let $\lambda_1 = e^{i\theta}$ ($\theta \neq 0$), then $A\psi^1 = e^{i\theta}\psi^1$, where ψ^1 is an eigenvector corrsponding to the eigenvalue $\lambda_1 = e^{i\theta}$. It then follows that $A\psi^{1*} = e^{-i\theta}\psi^{1*}$, from which we see that $e^{-i\theta}$ is also an eigenvalue of A.

Hence, $\lambda_2 = e^{-i\theta}$, $\lambda_3 = 1$ and $A\psi^3 = 1\cdot\psi^3$, so that

$$UAU^{+} = \begin{pmatrix} \lambda_1 & 0 & 0 \\ 0 & \lambda_2 & 0 \\ 0 & 0 & \lambda_3 \end{pmatrix} = \begin{pmatrix} e^{i\theta} & 0 & 0 \\ 0 & e^{-i\theta} & 0 \\ 0 & 0 & 1 \end{pmatrix} = \begin{pmatrix} \begin{pmatrix} e^{i\theta} & 0 \\ 0 & e^{-i\theta} \end{pmatrix} & 0 \\ 0 & 1 \end{pmatrix}.$$

Since A is real, we choose ψ^3 to be real, and let it be the z-axis, i.e.,

$$\psi^3 = \begin{pmatrix} 0 \\ 0 \\ 1 \end{pmatrix}.$$

But we have previously shown that any matrix with elements $e^{i\theta}$ and $e^{-i\theta}$ is equivalent to a 2-dimensional rotation in real space, so that

$$UAU^{+} = \begin{pmatrix} \cos\theta & -\sin\theta & 0 \\ \sin\theta & \cos\theta & 0 \\ 0 & 0 & 1 \end{pmatrix},$$

where θ refers to the particular axis about which we are performing the 2-dimensional rotation.

Any 4-dimensional rotation in real space can be written as

$$\begin{pmatrix} e^{i\theta} & 0 & 0 & 0 \\ 0 & e^{-i\theta} & 0 & 0 \\ 0 & 0 & e^{i\theta} & 0 \\ 0 & 0 & 0 & e^{-i\theta} \end{pmatrix},$$

which is equivalent to two 2-dimensional rotations.

The eigenvalues operate in pairs, since if $A\psi^i = e^{i\theta}\psi^i$, where ψ^i is the eigenvector for $e^{i\theta}$, then $A\psi^{i*} = e^{-i\theta}\psi^{i*}$ and $e^{-i\theta}$ is the other eigenvalue.

For 5 dimensions, the rotation matrix becomes

$$\begin{pmatrix} 1 & 0 & 0 & 0 & 0 \\ 0 & e^{i\theta} & 0 & 0 & 0 \\ 0 & 0 & e^{-i\theta} & 0 & 0 \\ 0 & 0 & 0 & e^{i\theta} & 0 \\ 0 & 0 & 0 & 0 & e^{-i\theta} \end{pmatrix}.$$

Any rigid body can move from position A to position B by a translation plus a 3-dimensional rotation. We have just seen that we can accomplish this by moving point P to point P' by a translation, then accomplish the 3-dimensional rotation by choosing a proper axis through point P' and performing the equivalent 2-dimensional rotation.

4.6 Linear Independence and Completeness

4.6.1 Orthonormality and Linear Independence

Definition: Let $x^1x^2x^3 \cdots x^m$ represent m $n\times 1$ matrices, where m can be less than n, the dimensionality of space.

$$\text{If } \sum_{i=1}^{m} a_i x^i = 0 \begin{Bmatrix} \text{only if all } a_i = 0 \\ \text{and not all } a_i = 0 \end{Bmatrix}, \text{then } x^1x^2 \cdots x^m \text{ are defined be}$$

$$\text{linearly} \begin{Bmatrix} \text{independent} \\ \text{dependent} \end{Bmatrix}.$$

Theorem: If H is Hermitian and $n\times n$ $(H = H^{+})$, and its eigenvectors $\psi^1\psi^2 \cdots \psi^n$ obey $\psi^{i+}\psi^j = \delta_{ij}$, then $\sum_{i=1}^{n} a_i\psi^i = 0$ will be true only when all $a_i = 0$, i.e. these eigenvectors are necessarily independent.

Proof: Multiplying $\sum_{i=1}^{n} a_i\psi^i = 0$ on the left by ψ^{j+}, we have

$$\psi^{j+}\sum_{i=1}^{n} a_i\psi^i = \sum_{i=1}^{n} \delta_{ij}a_i = a_j = 0.$$

Since the above holds for any j, we have proved the theorem, namely that orthonormality implies linear independence.

Note that linear independence does not necessarily imply orthonormality. We can, however, construct an orthonormal set from a set of m linearly independent vectors using the Schmidt orthogonalization method, as follows:

Let $\varphi_1\ \varphi_2 \ldots \varphi_m$ be a linearly independent set. We then set $\varphi_1^+\varphi_1 = 1$ (i.e. we normalize φ_1), and let $\psi_1 = \varphi_1$. We now set $\psi_2 = (\varphi_2 - \psi_1^+\varphi_2\psi_1)N_2$, where N_2 is chosen to normalize ψ_2, i.e. $\psi_2^+\psi_2 = 1$. Observe that

$$\psi_1^+\psi_2 = (\varphi_1^+\varphi_2 - \psi_1^+\varphi_2\psi_1^+\psi_1)N_2 = 0.$$

We now set $\psi_3 = (\varphi_3 - \psi_2^+\varphi_3\psi_2 - \psi_1^+\varphi_3\psi_1)N_3$, where N_3 is chosen to normalize ψ_3. Observe that

$$\psi_1^+\psi_3 = (\psi_1^+\varphi_3 - \psi_2^+\varphi_3\psi_1^+\psi_2 - \psi_1^+\varphi_3\psi_1^+\psi_1)N_3 = 0.$$

$$\psi_2^+\psi_3 = (\psi_2^+\varphi_3 - \psi_2^+\varphi_3\psi_2^+\psi_2 - \psi_1^+\varphi_3\psi_2^+\psi_1)N_3 = 0.$$

4.6.2 Completeness

Definition: The set of vectors $\psi^1\psi^2 \cdots \psi^n$ forms a complete set if *every* ψ can be expressed as

$$\psi = \sum_{i=1}^{n} a_i\psi^i,$$

where a_i is a number.

Theorem: Completeness implies linear independency.

Proof: Define a matrix B and a vector a as follows:

$$B = \begin{pmatrix} \psi_1^1 & \cdots & \psi_1^n \\ \cdot & & \\ \cdot & & \\ \cdot & & \\ \psi_n^1 & \cdots & \psi_n^n \end{pmatrix}; \qquad a = \begin{pmatrix} a_1 \\ a_2 \\ \cdot \\ \cdot \\ \cdot \\ a_n \end{pmatrix}.$$

Then $\psi = \sum_{i=1}^{n} a_i\psi^i = a_1\psi^1 + a_2\psi^2 + \cdots + a_n\psi^n$, and

$$\left.\begin{matrix} \psi_1 = a_1\psi_1^1 + a_2\psi_1^2 + \cdots + a_n\psi_1^n. \\ \cdot \\ \cdot \\ \cdot \\ \psi_n = a_1\psi_n^1 + a_2\psi_n^2 + \cdots + a_n\psi_n^n. \end{matrix}\right\} \Rightarrow \psi = B \cdot a.$$

If ψ is given, in order for the a_i to exist, $\det|B| \neq 0$.

Assume the set ψ^i $(i \to n)$ is a linear dependent set. Then, $\sum_{i=1}^{n} b_i\psi^i = 0$, where not all $b_i = 0$.

If we let

$$b = \begin{pmatrix} b_1 \\ \cdot \\ \cdot \\ b_n \end{pmatrix},$$

then $\sum_{i=1}^{n} b_i\psi^i = b_1\psi^1 + b_2\psi^2 \cdots + b_n\psi^n = 0$, or the set of equations,

$$\left.\begin{matrix} b_1\psi_1^1 + b_2\psi_1^2 + \cdots + b_n\psi_1^n. = 0 \\ b_1\psi_2^1 + b_2\psi_2^2 + \cdots + b_n\psi_2^n. = 0 \\ \cdot \\ \cdot \\ \cdot \\ b_n\psi_n^1 + b_n\psi_n^2 + \cdots + b_n\psi_n^n. = 0 \end{matrix}\right\} \Rightarrow B \cdot b = 0.$$

But if not all $b_i = 0$, then $\det|B| = 0$, which is contrary to the completeness hypothesis. Hence, all $b_i = 0$, and the set $\psi^1\psi^2 \cdots \psi^n$ constitutes a linear independent set. Furthermore, the a_i are uniquely determined by $a_i = \psi^{i+}\psi$.

Converse: Any set of n linearly independent vectors (n = dimensionality of space) also form a complete set.

Summary:

Orthonormality (in n dimensions) $\supset$ linear independence $\supset$ completeness.

Completeness $\supset$ linear independence (in n dimensions) $\supset$ orthonormality (which can be constructed by the method of 4.6.1)

Chapter 5

Hilbert Space

5.1 Definitions

Hilbert space is the space of functions defined on the interval $a \le x \le b$. The algebra of Hilbert space is "almost" identical with that of N-dimensional space, but is concerned with functions. We will limit ourselves to functions which are continuous and single-valued.

Normalization of a function $f_1(x)$: In Hilbert space, if $\int_a^b f_1^* f_1 dx = 1$, then $f_1(x)$ is normalized, where the range of x is given by $a \le x \le b$, where a and b may or may not include ∞.

In N-dimensional space, $\psi^+\psi = 1$ defines normalization. The similarity between ψ and f is seen as follows: choose the interval between a and b to be divided (Fig. 5-1) into N sections. Choose some point, x_i, within each division. Then,

$$\int_a^b f_1^* f_1 dx = \lim_{N\to\infty} \sum_{i=1}^{N} f_1^*(x_i) f_1(x_i).$$

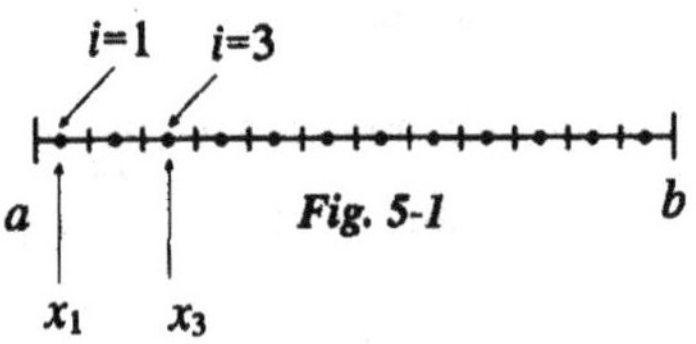

Fig. 5-1

Regarding $f_1(x_i)$ as a vector,

$$f_1(x_i) = \begin{pmatrix} f_1(x_1) \\ f_1(x_2) \\ \cdot \\ \cdot \\ \cdot \\ f_1(x_n) \end{pmatrix}$$

is an $N \times 1$ matrix, or a vector in N-dimensional space.

The concept of a function as a vector in Hilbert space is used in defining the definite integral

$$\int_a^b f(x)g(x)dx = \lim_{N\to\infty}\sum_{i=1}^{N} f(x_i)g(x_i).$$

From this definition of the integral, the transition from N-dimensional space to Hilbert space is made by $\sum_1^N \to \int_a^b dx$. Thus, for normalization,

$$\psi^+\psi = 1 \to \int_a^b f^* f dx = 1.$$

Orthogonality: In Hilbert space, if $\int_a^b f_1^* f_2 dx = 0$, then f_1 and f_2 are orthogonal. In N-dimensional space, $\psi^+\varphi = 0$ defines orthogonalization, so that $\psi^+\varphi = 0 \to \int_a^b f^* g dx = 0$.

Orthonormality of a set:

$$\psi^1 \cdots \psi^n \ni \psi^{i+}\psi^j = \delta_{ij} \quad \to \quad f_1 \cdots f_n \ni \int_a^b f_i^* f_j dx = \delta_{ij}.$$

Linear dependence in Hilbert space: If $\sum_{\nu=1}^{n} a_\nu f_\nu = 0$ and not all $a_\nu = 0$, for all x, then the set of f_ν are linearly dependent.

Theorem: Orthonormality implies linear independence.

Proof: The orthonormality condition is given by $\int_a^b f_i^* f_j dx = \delta_{ij}$.
If $\sum_{\nu=1}^{n} a_\nu f_\nu = 0$ and we multiply through by f_ρ^*, we obtain

$$f_\rho^*\left[a_1 f_1 + a_2 f_2 + \cdots + a_\nu f_\nu + \cdots + a_n f_n\right] = 0.$$

Integrating,

$$\int_a^b f_\rho^* \left(\sum_{\nu=1}^{n} a_\nu f_\nu\right) d\nu = a_\rho = 0.$$

Since this holds for all ρ, all coefficients $a_\nu = 0$.

Completeness in Hilbert space: $f_1 f_2 \cdots f_N$ form a complete set for the interval $a \le x \le b$ if and only if any single-valued function $f(x)$ can be written as

$$f(x) = \lim_{N\to\infty} \sum_{\nu=1}^{N} a_\nu f_\nu(x).$$

It is clear that, given an $\varepsilon > 0$, there exists an $N(x)$ such that

$$(1) \qquad \left| f(x) - \sum_{\nu=1}^{N(x)} a_\nu f_\nu(x) \right| < \varepsilon.$$

If there exists a maximum finite N for all x in $a \le x \le b$, the series is said to converge uniformly. In general, our approximating series will not satisfy this condition, but will satisfy the weaker condition, that of convergence in the mean:

$$(2) \qquad \lim_{N\to\infty} \int_a^b \left| f(x) - \sum_{\nu=1}^{N(x)} a_\nu f_\nu(x) \right|^2 dx = 0.$$

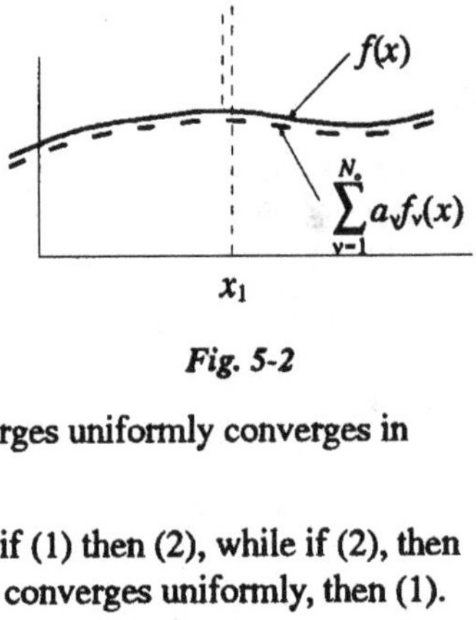

Fig. 5-2

The difference between the two modes of convergence may be seen graphically (Fig. 5-2). If the width of the pip $\to 0$ as $N \to \infty$, we have mean convergence, for the value is unchanged by changing the value of the integrand at isolated points; but we do not have uniform convergence since $N(x_1) = \infty$. Note that any series which converges uniformly converges in the mean.

Condition (2) is less restrictive than (1), i.e., if (1) then (2), while if (2), then not necessarily (1), but if the limit as $N \to \infty$ converges uniformly, then (1).

Completeness does not necessarily imply orthonormality, but an orthonormal set $g_1 g_2 \ldots g_N$ can be constructed from a set of complete functions by the trivial extension of the Schmidt method. If $f_1 f_2 \ldots f_N$ be the complete set, we set $g_1 = f_1$,

$$g_2 = \left\{ f_2 - \left[\int_a^b g_1^* f_2 dx \right] \right\} N_2 \cdots \cdot g_n = \left\{ f_n - \sum_{m=1}^{n-1} \left[\int_a^b g_1^* f_2 dx \right] \right\} N_n \cdots.$$

5.2 Weierstrass' Theorem

The series $1\ x\ x^2 x^3 \ldots$ form a complete (though not orthonormal) set for the definite interval $a \le x \le b$ (Fig. 5-3) (∞ not included).

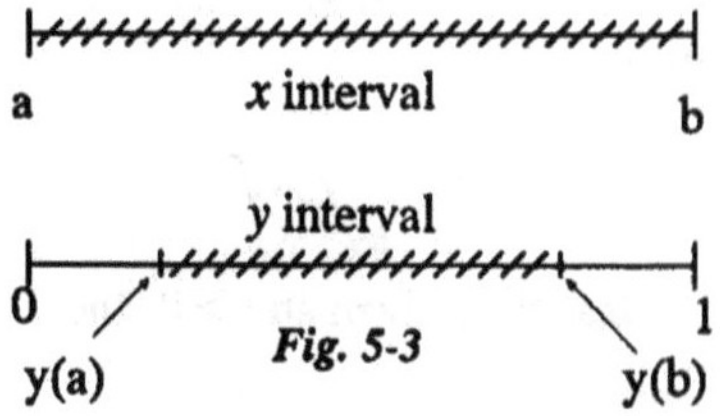

Fig. 5-3

Let $y = (x - a + \varepsilon)/A$, where $A > b - a$ and $\varepsilon > 0$ (Fig. 5-3).

Due to the linear relationship between x and y, if the set $1\ y\ y^2 y^3 \ldots$ forms a complete set, then the set $1\ x\ x^2 x^3 \ldots$ is also a complete set. Hence, without loss of generality, we will show the set of x's to be complete within the interval $0 < a \le x \le b < 1$.

Proof: Define a function $D_n(x - x_0) \equiv N_n\left[1 - (x - x_0)^2\right]^n$, a polynomial in x of the power $2n$, where $0 \le x \le 1$ and $0 < a \le x_0 \le b < 1$.

This function becomes very sharply peaked as $n \to \infty$. N_n is merely a normalizing factor (Fig. 5-4).

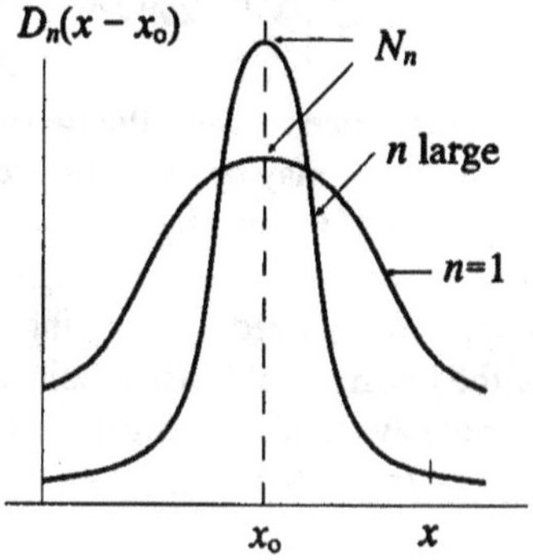

Fig. 5-4

In order to normalize the curve so as to resemble the Dirac δ function, we first set

$$|x - x_0| \ge \delta \begin{pmatrix} \ne 0 \\ < 1 \end{pmatrix},$$

so that, for $\delta \begin{pmatrix} \ne 0 \\ < 1 \end{pmatrix}$,

$$\left[1 - (x - x_0)^2\right]^n \le \left[1 - \delta^2\right]^n \underset{n \to \infty}{\to} 0.$$

Normalizing $D_n(x - x_0)$,

$$\int_0^1 \left[1 - (x - x_0)^2\right]^n dx = \left\{ \int_0^{x_0 - \delta} + \int_{x_0 - \delta}^{x_0 + \delta} + \int_{x_0 + \delta}^{1} \right\} \left[1 - (x - x_0)^2\right]^n dx$$

$$= \int_{x_0 - \delta}^{x_0 + \delta} \left[1 - (x - x_0)^2\right]^n dx + O\left[1 - \delta)^2\right]^n,$$

where O denotes "to the order of".

Letting $y = x - x_0$, so that the limits become $\pm\delta$,

$$\int_0^1 \left[1 - (x - x_0)^2\right]^n dx = \int_{-\delta}^{+\delta} (1 - y^2)^n dy + O(1 - \delta^2)^n.$$

Consider the integral from -1 to $+1$ as broken up into portions -1 to $-\delta$, $-\delta$ to $+\delta$, $+\delta$ to $+1$. If we do this for our peaked curve, we get

$$\int_{-\delta}^{+\delta} (1 - y^2)^n dy + O(1 - \delta^2)^n = \int_{-1}^{+1} (1 - y^2)^n dy + O(1 - \delta^2)^n,$$

since the regions -1 to $-\delta$ and $+\delta$ to $+1$ do not include $x = x_0$ (i.e. $y = 1$). [Note also that the above expression doesn't depend upon x.]

Hence we set $N_n^{-1} = \int_{-1}^{+1} (1 - y^2)^n dy$, so that

$$D_n(x - x_0) = \frac{\left[1 - (x - x_0)^2\right]^n}{\int_{-1}^{+1} (1 - y^2)^n dy} \quad \Rightarrow \quad \int_0^1 D_n(x - x_0) dx = 1,$$

i.e. $D_n(x - x_0)$ is normalized to unity.

Consider the denominator term $I(n) \equiv \int_{-1}^{+1} (1 - y^2)^n dy$:

$$\int_{-1}^{+1} (1 - y^2)^n dy = y(1 - y^2)^n \Big|_{-1}^{+1} - \int_{-1}^{+1} (-2ny^2)(1 - y^2)^{n-1} dy$$

$$= -2n \int_{-1}^{+1} (1 - y^2 - 1)(1 - y^2)^{n-1} dy = -2n[I(n) - I(n-1)].$$

Hence,

$$I(n) = \frac{2n}{2n+1} I(n-1).$$

$$I(0) = \int_{-1}^{+1} dy = 2 \quad \Rightarrow \quad I(n) = \frac{2n}{2n+1} \cdot \frac{2(n-1)}{2(n-1)+1} \cdots 2$$

$$= \frac{2^n n!}{(2n+1)(2n-1)\cdots} = \frac{(2^n n!)^2}{(2n+1)!}$$

Employing Stirling's formula for large n (to be proven in Math Physics II), $\ln n! \approx n\ln n - n$, we have:

$$\begin{aligned}\ln I(n) &\approx 2n\ln 2 + 2n(\ln n - 1) - (2n+1)[\ln(2n+1) - 1] \\ &= 2n\ln 2n - 2n - (2n+1)[\ln(2n+1) - 1] \\ &\approx \ln(2n)! - \ln(2n+1)! = \ln\frac{(2n)!}{(2n+1)!} = \ln\frac{1}{2n+1}.\end{aligned}$$

Thus, $I(n) \approx \dfrac{1}{(2n+1)}$.

We have that

$$D_n(x \neq x_o) \approx n\left[1 - (x - x_o)^2\right]^n \rightarrow 0,$$

since if we define K as $0 < 1 - (x - x_o)^2 \equiv K < 1$, then $\ln K < 0$. In addition,

$$n\left[1 - (x - x_o)^2\right]^n = n\exp(n\ln K) = \frac{n}{\exp(n|\ln K|)}.$$

$$\lim_{n\to\infty}\frac{n}{\exp(n|\ln K|)} = \lim_{n\to\infty}\frac{1}{|\ln K|\exp(n|\ln K|} \propto \frac{1}{e^\infty} = 0.$$

(l'Hospital's rule was used for evaluating the above expression.)

We also had that $\int_0^1 D_n(x - x_o)dx \rightarrow 1$. Hence,

$$D_n(x - x_o) \underset{n\to\infty}{\rightarrow} \delta(x - x_o).$$

Defining a function $F_n(x_o) \equiv \int_0^1 D_n(x - x_o)f(x)dx$, where x_o is a variable and $f(x)$ is continuous, then $F_n(x_o)$ is a polynomial in x_o of the power $2n$, and $\lim_{n\to\infty} F_n(x_o) = \int_0^1 \delta(x - x_o)f(x)dx = f(x_o)$. The requirement for completeness is that $f(x) = \lim_{n\to\infty}\sum_{\nu=1}^{n} a_\nu f_\nu(x)$, so that if we set $F_n(x_o) = \sum_{\nu=1}^{n} a_\nu f_\nu(x_o)$, then $f(x_o) = \lim_{n\to\infty} F_n(x_o)$. But this is just what we have shown above, where $F_n(x_o)$ is a polynomial in x_o of power $2n$.

Corollary: The set of functions $1\ x\ x^2 \ldots x^n \ldots y\ y^2\ y^3 \ldots$ forms a complete set, i.e. for any function $f(x_o\, y_o)$,

$$f(x_o y_o) = \lim_{n\to\infty} F_n(x_o y_o),$$

where $F_n(x_o y_o)$ is a polynomial in $x_o y_o$.

Proof: Define

$$F_n(x_o y_o) \equiv \int_0^1\!\!\int_0^1 D_n(x - x_o) D_n(y - y_o) f(xy)\, dx\, dy,$$

so that

$$\lim_{n\to\infty} F_n(x_o y_o) = \int_0^1\!\!\int_0^1 \delta(x - x_o)\delta(y - y_o) f(xy)\, dx\, dy = f(x_o y_o),$$

where $F_n(x_o y_o)$ is a polynomial in x_o and y_o of power $2n$.

5.3 Examples of Complete Orthonormal Sets

5.3.1 Fourier Series

Consider some function continuous in the region shown in Fig. 5-5. We already know that any function $f(x\,y)$ can be represented in a power series in x and y, as follows:

$$f(xy) = \sum_{nm=1}^{\infty} A_{nm}\, x^n y^m.$$

Consider that particular $f(xy) = \rho F(\theta)$. Since $x = \rho\cos\theta$ and $y = \rho\sin\theta$, we have

$$\rho F(\theta) = \sum_{nm=1}^{\infty} A_{nm} \rho^{n+m} \cos^n\theta \sin^m\theta.$$

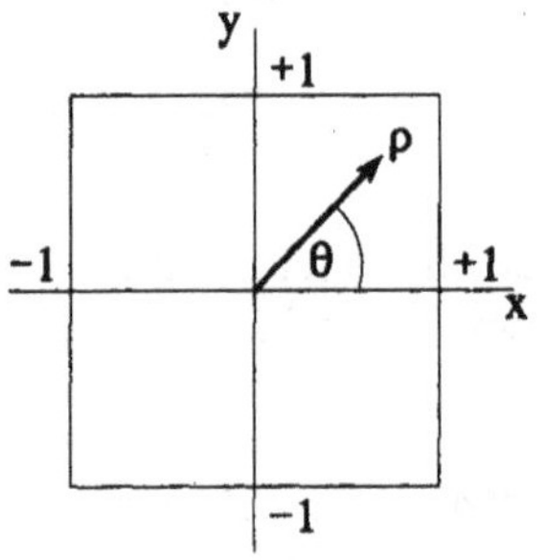

Fig. 5-5

Since the above is true in the squared region (Fig. 5-5), it will be true on a circle of radius $\rho = 1$. Thus, for $\rho = 1$,

$$F(\theta) = \sum_{nm=1}^{\infty} A_{nm} \cos^n\theta \sin^m\theta.$$

We now make a change of form, in order to satisfy the orthogonality conditions.

From trigonometry we recall that

$$\sin r\theta = \frac{e^{ir\theta} - e^{-ir\theta}}{2i}; \qquad \cos r\theta = \frac{e^{ir\theta} + e^{-ir\theta}}{2}.$$

Expanding $\sin^n\theta = \left((e^{i\theta} - e^{-i\theta})/2i\right)^n$ and $\cos^n\theta = \left((e^{i\theta} + e^{-i\theta})/2\right)^n$, and grouping terms in $e^{ir\theta} \pm e^{-ir\theta}$, we have

$$\begin{Bmatrix} \cos^n\theta \\ \sin^n\theta \end{Bmatrix} = A\begin{Bmatrix} \cos\theta \\ \sin\theta \end{Bmatrix} + B\begin{Bmatrix} \cos 2\theta \\ \sin 2\theta \end{Bmatrix} + \cdots + R\begin{Bmatrix} \cos n\theta \\ \sin n\theta \end{Bmatrix}.$$

Hence, by a rearrangement of terms, $F(\theta)$ = series in $\cos n\theta$, $\sin n\theta$. Thus, any continuous function $F(\theta)$ can be represented by a series in $\cos n\theta$ and $\sin n\theta$, as follows:

$$F(\theta) = \frac{a_0}{2} + \sum_{n=1}^{\infty} a_n \cos n\theta + \sum_{n=1}^{\infty} b_n \sin n\theta,$$

and holds for the circular variable, i.e. $F(\theta + 2\pi) = F(\theta)$. But any function of x that satisfies $f(x + 2\pi) = f(x)$ and is continuous, can be represented by

$$f(x) = \frac{a_0}{2} + \sum_{n=1}^{\infty} a_n \cos nx + \sum_{n=1}^{\infty} b_n \sin nx,$$

and it holds for the linear variable x for functions whose periodicity is equivalent to that of the circular variable θ.

Orthogonality relations

$$\int_{-\pi}^{+\pi} \cos nx \cos mx\, dx = \int_{-\pi}^{+\pi} \frac{1}{2}\left[\cos(n+m)x + \cos(n-m)x\right] dx = \pi\delta_{nm}.$$

$$\int_{-\pi}^{+\pi} \sin nx \sin mx\, dx = \int_{-\pi}^{+\pi} \frac{1}{2}\left[\cos(n-m)x - \cos(n+m)x\right] dx = \pi\delta_{nm}.$$

$$\int_{-\pi}^{+\pi} \sin nx \cos mx\, dx = \int_{-\pi}^{+\pi} \frac{1}{2}\left[\sin(n+m)x + \sin(n-m)x\right] dx = 0\begin{Bmatrix} n \neq m \\ n = m \end{Bmatrix}.$$

Hence,

$$\int_{-\pi}^{+\pi} f(x)dx = \int_{-\pi}^{+\pi} \frac{a_o}{2}\,dx = \frac{a_o}{2}2\pi = \pi a_o \;\rightarrow\; a_o = \frac{1}{\pi}\int_{-\pi}^{+\pi} f(x)dx,$$

and since the functions are orthogonal,

$$\int_{-\pi}^{+\pi} f(x)\cos nx\,dx = \int_{-\pi}^{+\pi}\sum_{\nu=1}^{\infty} a_\nu \cos\nu x \cos nx\,dx = \pi a_n.$$

$$\int_{-\pi}^{+\pi} f(x)\sin nx\,dx = \int_{-\pi}^{+\pi}\sum_{\nu=1}^{\infty} b_\nu \sin\nu x \sin nx\,dx = \pi b_n.$$

Hence,

$$a_n = \int_{-\pi}^{+\pi} f(x)\cos nx\,dx. \qquad b_n = \int_{-\pi}^{+\pi} f(x)\sin nx\,dx.$$

Thus, knowing $f(x)$ we can find the Fourier series that represents it.

If we now wish to extend the interval, we make the following scale change: Set $y = \pi x/L$, so that when $-\pi \le y \le \pi$, $-L \le x \le L$. It then follows that

$$f(y) = \frac{a_o}{2} + \sum_{n=1}^{\infty} a_n \cos ny + \sum_{n=1}^{\infty} b_n \sin ny.$$

$$f(x) = \frac{a_o}{2} + \sum_{n=1}^{\infty} a_n \cos\frac{n\pi x}{L} + \sum_{n=1}^{\infty} b_n \sin\frac{n\pi x}{L}. \quad (-L \le x \le L)$$

If $f(x)$ is not periodic, then the expression is only good within the given interval. If $f(x)$ is periodic, then the expression is good everywhere.

Solving for the coefficients a_n and b_n, we have:

$$\int_{-L}^{+L} f(x)\cos\frac{n\pi x}{L}dx = \int_{-L}^{+L}\sum_{m=1}^{\infty} a_m \cos\frac{m\pi x}{L}\cos\frac{n\pi x}{L}dx = a_n L.$$

$$\int_{-L}^{+L} f(x)\sin\frac{n\pi x}{L}dx = \int_{-L}^{+L}\sum_{m=1}^{\infty} b_m \sin\frac{m\pi x}{L}\sin\frac{n\pi x}{L}dx = b_n L.$$

Thus,

$$a_n = \frac{1}{L}\int_{-L}^{+L} f(x)\cos\frac{n\pi x}{L}dx. \qquad b_n = \frac{1}{L}\int_{-L}^{+L} f(x)\sin\frac{n\pi x}{L}dx.$$

The requirement that $f(x)$ be continuous, i.e. that for any $\varepsilon > 0$, there exists an η_o such that for all $\eta > \eta_o$, $|f(x) - f(x-\eta)| < \varepsilon$, presupposes that $f(x)$ be bounded, since the above continuity statement says that $f(x)$ cannot be infinite and therefore must be bounded.

5.3.2 Generalization of a Hermitian Operator in Hilbert Space

Consider the complete and orthonormal set of functions $\cos n\pi x/L$ and $\sin n\pi x/L$. Can we relate these to eigenfunctions of a "functional" Hermitian operator? We recall, by analogy, that the eigenvectors of a Hermitian operator in vector space were orthonormal.

Definition: If

$$\int_a^b f_1^* \mathrm{O} f_2 dx = \left[\int_a^b f_2^* \mathrm{O} f_1 dx\right]^*$$

for any f_1 and f_2 in Hilbert space, then O is a Hermitian operator in Hilbert space. O may contain integrals, derivatives, functions, etc.

In n-dimensional space, if H is Hermitian (i.e. $H = H^+ \rightarrow H_{ij} = H_{ji}^*$), then

$$(\psi_1^+ H\psi_2) = \sum_{ij} (\psi_1^*)_i H_{ij} (\psi_2)_j,$$

where ψ_1 and ψ_2 are arbitrary vectors in n-dimensional space.

$$(\psi_2^+ H\psi_1) = \sum_{ij} (\psi_2^*)_i H_{ij} (\psi_1)_j = \sum_{ij} (\psi_2^*)_j H_{ji} (\psi_1)_i,$$

where we have interchanged dummy indices i and j. Thus,

$$(\psi_2^+ H\psi_1)^* = \sum_{ij} (\psi_2)_j H_{ji}^* (\psi_1^*)_i = \sum_{ij} (\psi_2)_j H_{ij} (\psi_1^*)_i = (\psi_1^+ H\psi_2).$$

Hence, $\psi_1^+ H\psi_2 = (\psi_2^+ H\psi_1)^*$ is the expression which becomes generalized into that of a Hermitian operator in Hilbert space.

To show that if $\psi_1^+ H\psi_2 = (\psi_2^+ H\psi_1)^*$ for all ψ_i, then H is Hermitian, we choose the ψ_i such that ψ_2 has only its 4th element (=1) and such that ψ_1 has only its 3rd element (=1).

Thus,

$$\psi_2 = \begin{pmatrix} 0 \\ 0 \\ 0 \\ 1 \\ \cdot \\ \cdot \\ \cdot \end{pmatrix}, \quad \psi_1 = \begin{pmatrix} 0 \\ 0 \\ 1 \\ 0 \\ \cdot \\ \cdot \\ \cdot \end{pmatrix}.$$

Then we would have

$$\psi_1^{\dagger} H \psi_2 = (0010\,..\,0) \begin{pmatrix} & & \\ & H_{ij} & \\ & & \end{pmatrix} \begin{pmatrix} 0 \\ 0 \\ 0 \\ 1 \\ \cdot \\ \cdot \\ \cdot \end{pmatrix} = (0010\,..\,0) \begin{pmatrix} H_{14} \\ H_{24} \\ H_{34} \\ \cdot \\ \cdot \\ \cdot \\ H_{n4} \end{pmatrix} = H_{34}.$$

Similarly,

$$(\psi_2^{\dagger} H \psi_1)^* = (0001\,..\,0) \begin{pmatrix} & & \\ & H_{ij}^* & \\ & & \end{pmatrix} \begin{pmatrix} 0 \\ 0 \\ 1 \\ 0 \\ \cdot \\ \cdot \\ \cdot \end{pmatrix} = (0001\,..\,0) \begin{pmatrix} H_{13}^* \\ H_{23}^* \\ H_{33}^* \\ \cdot \\ \cdot \\ \cdot \\ H_{n3}^* \end{pmatrix} = H_{43}^*.$$

Thus, by hypothesis, $H_{34} = H_{43}^* \Rightarrow H = H^{\dagger}$.

Let us now consider some familiar operators and check whether they are good Hermitian functional operators.

(1) $\mathrm{O} = \dfrac{d}{dx}$:

$$\int_a^b f_1^* \frac{df_2}{dx}\, dx = \left[f_1^* f_2 \right]_a^b - \int_a^b f_2 \frac{df_1^*}{dx}\, dx = -\left[\int_a^b f_2^* \frac{df_1}{dx}\, dx \right]^*,$$

where f_1 and f_2 are chosen to vanish at the limits. Thus, due to the presence of the minus sign, O is not Hermitian.

(2) $\mathrm{O} = \frac{1}{i}\frac{d}{dx}$:

$$\frac{i}{i}\int_a^b f_1^* \frac{df_2}{dx}\,dx = -\left[\frac{1}{i}\int_a^b f_2 \frac{df_1^*}{dx}\,dx\right] = \left[\frac{1}{i}\int_a^b f_2^* \frac{df_1}{dx}\,dx\right]^*,$$

where again we have chosen f_1 and f_2 to vanish at the boundary. Thus, we see that due to the presence of the factor i, O is Hermitian. [In quantum mechanics $\frac{\hbar}{i}\frac{\partial}{\partial x}$ is associated with the momentum of a particle.]

(3) Any real function of x, $\mathrm{O}_r(x)$, is a good Hermitian operator, since

$$\int_a^b f_1^* \mathrm{O}_r(x) f_2 dx = \left[\int_a^b f_2^* \mathrm{O}_r(x) f_1 dx\right]^*.$$

We now proceed to show that the complete and orthonormal set comprising $\cos\frac{n\pi x}{l}$ and $\sin\frac{n\pi x}{l}$ is connected with a specific Hermitian operator.

Let Hilbert space be formed only of continuous periodic functions of period $2L$, i.e. $f(x + 2L) = f(x)$, and let us study the algebra of such a space.

Consider the operator $\mathrm{O} = \frac{d^2}{dx^2}$:

$$\int_{-L}^{+L} f_1^* \frac{d^2}{dx^2} f_2 dx = \left[f_1^* \frac{df_2}{dx}\right]_{-L}^{+L} - \int_{-L}^{+L} \frac{df_1^*}{dx}\frac{df_2}{dx}\,dx$$

$$= \left[f_1^* \frac{df_2}{dx}\right]_{-L}^{+L} - \left[\frac{df_1^*}{dx} f_2\right]_{-L}^{+L} + \int_{-L}^{+L} f_2 \frac{d^2}{dx^2} f_1^* dx = 0.$$

Since the functions are periodic the first two expressions on the right vanish. Hence,

$$\int_{-L}^{+L} f_1^* \frac{d^2}{dx^2} f_2 dx = \left[\int_{-L}^{+L} f_2^* \frac{d^2}{dx^2} f_1 dx\right]^*$$

and $\mathrm{O} = \frac{d^2}{dx^2}$ is a good functional Hermitian operator.

To evaluate the eigenfunctions of $\frac{d^2}{dx^2}$, we set up the "eigenvector" equation, $Of^i = \lambda_i f^i$, which in this case becomes $\frac{d^2f}{dx^2} = -k^2 f$, where $\lambda = -k^2$. Solutions are easily seen to be $f = \begin{Bmatrix} \cos kx \\ \sin kx \end{Bmatrix}$. Since $f(x+2L) = f(x)$, we must have $\sin kx = \sin k(x+2L)$, so that $2kL = 2\pi n$, or $k = n\pi/L$, $n = 0, 1 \ldots$ Note that k is real, and therefore k^2 and $-k^2$ are real. Hence, the eigenvalues of $\frac{d^2}{dx^2}$ are real and negative. We have thus shown that the eigenfunctions of the Hermitian operator $\frac{d^2}{dx^2}$ form a complete and orthonormal set.

5.3.3 Fourier Integral Theorem; Fourier Transforms

Since

$$\cos\frac{n\pi x}{L} = \frac{e^{in\pi x/L} + e^{-in\pi x/L}}{2} \quad \text{and} \quad \sin\frac{n\pi x}{L} = \frac{e^{in\pi x/L} - e^{-in\pi x/L}}{2i},$$

any function that is expressible in a series of sines and cosines can be expressed as a linear combination of $e^{in\pi x/L}$ and $e^{-in\pi x/L}$. Thus,

$$f(x) = \sum_{n=-\infty}^{+\infty} A_n e^{in\pi x/L} \cdot \frac{1}{2L}.$$

For $n \neq n_o$ and for periodic functions of period $2L$,

$$\frac{1}{2L}\int_{-L}^{+L} e^{i(n-n_o)\pi x/L}dx = \frac{1}{2L}\left[\frac{e^{i(n-n_o)\pi x/L}}{i(n-n_o)\pi/L}\right]_{-L}^{+L}$$

$$= \frac{1}{2i(n-n_o)\pi L}\left[e^{i(n-n_o)\pi} - e^{-i(n-n_o)\pi}\right] = 0,$$

while for $n = n_o$, $\frac{1}{2L}\int_{-L}^{+L} dx = 1$. Hence,

$$\int_{-L}^{+L} f(x)e^{-in\pi x/L}dx = \int_{-L}^{+L}\sum_{m=-\infty}^{+\infty} A_m \frac{1}{2L} e^{i(m-n)\pi x/L}dx = A_n.$$

Let us introduce $k_n \equiv n\pi/L$, so that as Δn varies in steps of 1, Δk_n varies in steps of π/L. Then,

$$f(x) = \sum_{k_n=-\infty}^{k_n=+\infty} A(k_n)e^{ik_n x} \cdot \left(\frac{\Delta k_n}{2\pi}\right),$$

where

$$A(k_n) = \int_{-L}^{+L} f(x)e^{-ik_n x}dx.$$

As $L \to \infty$, Δk_n gets smaller and smaller, so that k_n practically maps out a continuum. We would then have

$$A(k) = \int_{-L}^{+L} f(x)e^{-ikx}dx \quad \text{and} \quad f(x) = \frac{1}{2\pi}\int_{-\infty}^{+\infty} A(k)e^{ikx}dk.$$

If we define $g(k) \equiv \frac{1}{\sqrt{2\pi}}A(k)$, we then have the Fourier Integral Theorem:

$$f(x) = \frac{1}{\sqrt{2\pi}}\int_{-\infty}^{+\infty} g(k)e^{ikx}dk.$$

$$g(k) = \frac{1}{\sqrt{2\pi}}\int_{-\infty}^{+\infty} f(x)e^{-ikx}dx.$$

$g(k)$ is called the Fourier transform of $f(x)$, while $f(x)$ is called the Fourier transform of $g(k)$. In the above treatment, x and k were dimensionless variables running from $-\infty$ to $+\infty$.

We want the expression for the Fourier Integral Theorem in 3-dimensional space.

Consider a function of 2 variables, $f(xy)$, where y is regarded as a parameter. Applying the Fourier Integral Theorem to $f(xy)$ as a function of x, we get:

(1) $$f(xy) = \frac{1}{\sqrt{2\pi}}\int_{-\infty}^{+\infty} g(yk_1)e^{ik_1 x}dk_1.$$

(2) $$g(yk_1) = \frac{1}{\sqrt{2\pi}}\int_{-\infty}^{+\infty} f(xy)e^{-ik_1 x}dx.$$

Regarding k_1 as a parameter in $g(yk_1)$:

$$(5) \qquad g(yk_1) = \frac{1}{\sqrt{2\pi}}\int_{-\infty}^{+\infty} h(k_1k_2)e^{ik_2y}dk_2.$$

$$(6) \qquad h(k_1k_2) = \frac{1}{\sqrt{2\pi}}\int_{-\infty}^{+\infty} g(yk_1)e^{-ik_2y}dy.$$

Substituting (3) into (1) and (2) into (4):

$$(3) \qquad f(xy) = \frac{1}{2\pi}\iint_{-\infty}^{+\infty} h(k_1k_2)e^{i(k_1x + k_2y)}dk_1dk_2.$$

$$(4) \qquad h(k_1k_2) = \frac{1}{(2\pi)^{3/2}}\iint_{-\infty}^{+\infty} f(xy)e^{-i(k_1x + k_2y)}dy.$$

By a similar extension to 3 dimensions (3 variables), the Fourier Integral Theorem becomes:

$$f(xyz) = \frac{1}{(2\pi)^{3/2}}\int_{-\infty}^{+\infty} g(k_1k_2k_3)e^{i(\mathbf{k}\cdot\mathbf{r})}dk_1dk_2dk_3.$$

$$g(k_1k_2k_3) = \frac{1}{(2\pi)^{3/2}}\int_{-\infty}^{+\infty} f(xyz)e^{-i(\mathbf{k}\cdot\mathbf{r})}dxdydz.$$

Examples of use of Fourier transforms

[1] *Dirac δ-function*

In 1 dimension,

$$\delta(x - x_o) = \frac{1}{\sqrt{2\pi}}\int_{-\infty}^{+\infty} g(k)e^{ikx}dk \rightarrow g(k) = \frac{1}{\sqrt{2\pi}}\int_{-\infty}^{+\infty} \delta(x - x_o)e^{-ikx}dx$$

$$= \frac{e^{-ikx_o}}{\sqrt{2\pi}}.$$

Thus, $\delta(x - x_o) = \frac{1}{2\pi}\int_{-\infty}^{+\infty} g(k)e^{ik(x - x_o)}dk.$

In 3 dimensions, $\delta^3(\mathbf{r} - \mathbf{r}_o) = \frac{1}{(2\pi)^3}\iiint_{-\infty}^{+\infty} e^{i\mathbf{k}\cdot(\mathbf{r} - \mathbf{r}_o)}d^3\mathbf{k}$, $(d^3\mathbf{k} \equiv dk_1dk_2dk_3)$.

[2] *Wave equation:* $\Delta\varphi - \frac{1}{c^2}\frac{\partial^2\varphi}{\partial t^2} = 0$, where $\varphi = \varphi(\mathbf{r}t)$.

Regarding t as a parameter, $\varphi(\mathbf{r}t) = \frac{1}{(2\pi)^{3/2}}\int\limits_{\substack{\text{all } k\\ \text{space}}} g(\mathbf{k}t)e^{i(\mathbf{k}\cdot\mathbf{r})}d^3\mathbf{k}$, so that

$$\Delta\varphi(\mathbf{r}t) = \frac{1}{(2\pi)^{3/2}}\int\limits_{\substack{\text{all } k\\ \text{space}}} g(\mathbf{k}t)\Delta e^{i(\mathbf{k}\cdot\mathbf{r})}d^3\mathbf{k},$$

where we note that Δ operates on $f(\mathbf{r})$, while $g(\mathbf{k}t)$ is independent of $\mathbf{r}$. Thus, $\frac{\partial}{\partial x}e^{i\mathbf{k}\cdot\mathbf{r}} = e^{i\mathbf{k}\cdot\mathbf{r}}\cdot ik_x \Rightarrow \frac{\partial^2}{\partial x^2}e^{i\mathbf{k}\cdot\mathbf{r}} = (-k_x^2)e^{i\mathbf{k}\cdot\mathbf{r}}$, so that $\Delta e^{i\mathbf{k}\cdot\mathbf{r}} = -k^2 e^{i\mathbf{k}\cdot\mathbf{r}}$, where $k^2 = |\mathbf{k}|^2 = -(k_x^2 + k_y^2 + k_z^2)$. Hence,

$$\Delta\varphi(\mathbf{r}t) = \frac{1}{(2\pi)^{3/2}}\int\limits_{\substack{\text{all } k\\ \text{space}}} g(\mathbf{k}t)e^{i(\mathbf{k}\cdot\mathbf{r})}(-k^2)d^3\mathbf{k}.$$

$$\ddot{\varphi}(\mathbf{r}t) = \frac{1}{(2\pi)^{3/2}}\int\limits_{\substack{\text{all } k\\ \text{space}}} \ddot{g}(\mathbf{k}t)e^{i(\mathbf{k}\cdot\mathbf{r})}d^3\mathbf{k}.$$

Substituting into the wave equation,

$$-\frac{1}{(2\pi)^{3/2}}\iiint\limits_{k} e^{i(\mathbf{k}\cdot\mathbf{r})}\left(k^2 g(\mathbf{k}t) + \frac{1}{c^2}\ddot{g}(\mathbf{k}t)\right)d^3\mathbf{k} = 0.$$

Setting $d^3\mathbf{r} = dxdydz$, we perform the following operation:

$$\iiint\limits_{-\infty}^{+\infty} e^{-i(\mathbf{k}_o\cdot\mathbf{r})}d^3\mathbf{r}\iiint\limits_{k} e^{i(\mathbf{k}\cdot\mathbf{r})}\left(k^2 g(\mathbf{k}t) + \frac{1}{c^2}\ddot{g}(\mathbf{k}t)\right)d^3\mathbf{k} = 0,$$

where $\mathbf{k}_o$ is any fixed vector in k space. We previously had that $\delta^3(\mathbf{k}-\mathbf{k}_o) = \frac{1}{(2\pi)^3}\iiint\limits_{-\infty}^{+\infty} e^{i\mathbf{r}\cdot(\mathbf{k}-\mathbf{k}_o)}d^3\mathbf{r}$, so that our expression above reduces to $\iiint\limits_{k}\delta^3(\mathbf{k}-\mathbf{k}_o)\left(k^2 g(\mathbf{k}t) + \frac{1}{c^2}\ddot{g}(\mathbf{k}t)\right)d^3\mathbf{k} = k_o^2 g(\mathbf{k}_o t) + \frac{1}{c^2}\ddot{g}(\mathbf{k}_o t) = 0$. But since $\mathbf{k}_o$ was an arbitrary vector, we have that $\ddot{g}(\mathbf{k}t) = -k^2c^2 g(\mathbf{k}t) = -\omega^2 g(\mathbf{k}t)$, where $\omega = kc$.

Solutions of the above equation are easily seen to be $g(\mathbf{k}t) = A_{\mathbf{k}}e^{i\omega t} + B_{\mathbf{k}}e^{-i\omega t}$, so that

$$\varphi(\mathbf{r}t) = \frac{1}{(2\pi)^{3/2}}\iiint_{-\infty}^{+\infty}\left[A_{\mathbf{k}}e^{i(\mathbf{k}\cdot\mathbf{r}+\omega t)} + B_{\mathbf{k}}e^{i(\mathbf{k}\cdot\mathbf{r}-\omega t)}\right]d^3\mathbf{k},$$

which we recognize as an expression for plane waves. It represents the most general solution of the wave equation, where $A_{\mathbf{k}}$ and $B_{\mathbf{k}}$ are any two functions of **k**.

5.3.4 Piecewise Continuous Functions

Definition: A function $f(x)$ is piecewise continuous if it is discontinuous only at a finite number of points, and if the amount of discontinuity is finite.

Theorem: A piecewise continuous function can be represented by a Fourier series which converges to the function at all points where the function is continuous and which, at a point of discontinuity, x_0, converges to the value $1/2\left[f(x_0^+) + f(x_0^-)\right]$.

Proof: Let a function $f(x)$ be piecewise continuous in a manner as shown in Fig. 5-6. At every point of discontinuity we can draw a curve connecting the points of discontinuity (Fig. 5-7).

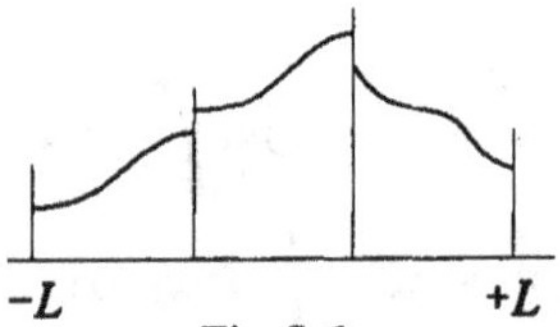

Fig. 5-6

Since a Fourier series can be used for piecewise continuous functions, except at points of discontinuity, we let

$$S_n(x) = \frac{a_0}{2} + \sum_{m=1}^{n} a_m \cos\frac{m\pi x}{L} + b_m \sin\frac{m\pi x}{L}$$

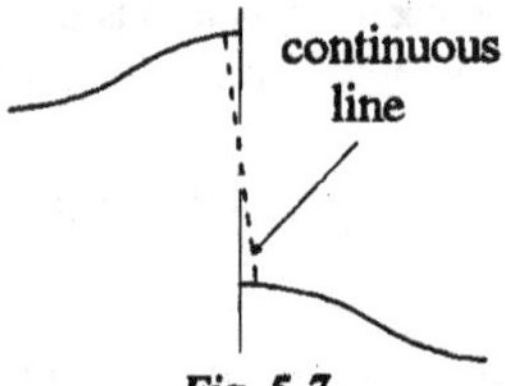

Fig. 5-7

represent a continuous function $S_n(x)$, which we make approach $f(x)$ as close as we want over the continuous line above. Thus,

$$S_n(x) \underset{n\to\infty}{\to} f(x)$$

except at the discontinuous point, while at the point of discontinuity,

$$S_n(x) \underset{n\to\infty}{\to} 1/2\left[f(x^+) + f(x^-)\right].$$

Observe that

$$\left\{\begin{aligned} a_n &= \frac{1}{L}\int_{-L}^{+L} f(x)\cos\frac{n\pi x}{L}\,dx \\ b_n &= \frac{1}{L}\int_{-L}^{+L} f(x)\sin\frac{n\pi x}{L}\,dx \end{aligned}\right\}$$

are well defined since we can consider the integrals as divided into separate regions of integration.

Define a function $D(x) = D(x+2L)$ to be an odd function (Fig. 5-8) $[D(-x) = -D(x)]$, where

$$D(x) = \frac{1}{2L}(L-x) \text{ for } 0 < x \le L.$$

$$D(x) = -\frac{1}{2L}(L-|x|) \text{ for } -L \le x < 0.$$

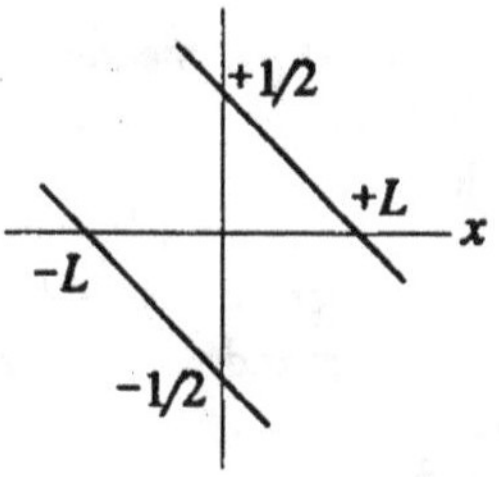

Fig. 5-8

Discontinuities exist at $x = 0, \pm 2L, \pm 4L$ etc. We have then that $D(0^+) = 1/2 = -D(0^-)$. Note that

$$a_n = \frac{1}{L}\int_{-L}^{+L} D(x)\cos\frac{n\pi x}{L}\,dx = 0$$

[the cosine function is even while $D(x)$ is odd, so that the product is odd and the integral vanishes], while

$$b_n = \frac{1}{L}\int_{-L}^{+L} D(x)\sin\frac{n\pi x}{L}\,dx \neq 0.$$

Let us call that $S_n(x)$ which represents $D(x)$ above, $S_n^D(x)$, i.e.

$$S_n^D(x) = \frac{a_0}{2} + \sum_{m=1}^{n} a_m\cos\frac{m\pi x}{L} + \sum_{m=1}^{n} b_m\sin\frac{m\pi x}{L},$$

where

$$a_n = \frac{1}{L}\int_{-L}^{+L} D(x)\cos\frac{n\pi x}{L}\,dx = 0. \qquad a_0 = \frac{1}{L}\int_{-L}^{+L} D(x)dx = 0.$$

$$b_n = \frac{1}{L}\int_{-L}^{+L} D(x)\sin\frac{n\pi x}{L}\,dx \neq 0.$$

Thus, $S_n^D(0) = 0$ for any n. Hence, $\lim_{n\to\infty} S_n^D(0) = 0 = \frac{1}{2}\left[D(0^+) + D(0^+)\right]$, and holds for any odd function with a finite jump at $x = 0$.

To show this to be true for any function, odd or even, consider the function $f(x)$ shown in Fig. 5-9.

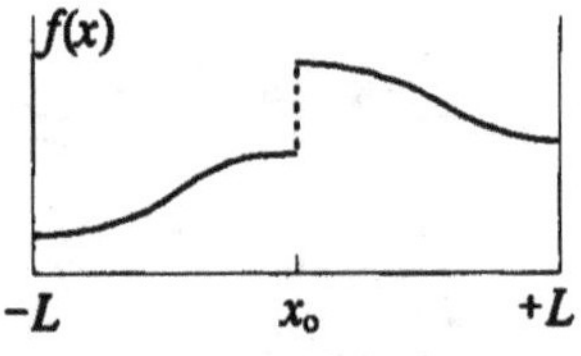

Fig. 5-9

Define

$$F(x) \equiv f(x) - \left[f(x_o^+) - f(x_o^-)\right]D(x - x_o),$$

where $D(x - x_o)$ is the same D function defined above.

$f(x)$ is obviously discontinuous at $x = x_o$. However, what we want to show is that $F(x)$ <u>is</u> continuous at $x = x_o$. Hence, compare $F(x_o^+)$ and $F(x_o^-)$.

$$F(x_o^+) = f(x_o^+) - \left[f(x_o^+) - f(x_o^-)\right]D(x_o^+ - x_o).$$

But, since $D(x_o^+ - x_o) = D(0^+) = 1/2$,

$$F(x_o^+) = f(x_o^+) - (1/2)\left[f(x_o^+) - f(x_o^-)\right] = (1/2)\left[f(x_o^+) + f(x_o^-)\right].$$

$$F(x_o^-) = f(x_o^-) - \left[f(x_o^+) - f(x_o^-)\right]D(x_o^- - x_o).$$

But, since $D(x_o^- - x_o) = D(0^+) = -1/2$,

$$F(x_o^-) = f(x_o^-) + (1/2)\left[f(x_o^+) - f(x_o^-)\right] = (1/2)\left[f(x_o^+) + f(x_o^-)\right] = F(x_o^+).$$

Thus, $F(x_o^+) = F(x_o^-)$ and $F(x_o)$ is continuous at $x = x_o$.

We now let $\Sigma_n(x)$, $S_n(x)$ and $S_n^D(x)$ represent the Fourier series of $F(x)$, $f(x)$ and $D(x - x_o)$, respectively. Then, from the definition of $F(x)$, it follows that

$$\frac{1}{L}\int_{-L}^{+L} F(x)\cos\frac{n\pi x}{L}\,dx = \frac{1}{L}\int_{-L}^{+L} f(x)\cos\frac{n\pi x}{L}\,dx$$
$$- [[\ \ ]]\,\frac{1}{L}\int_{-L}^{+L} D(x - x_o)\cos\frac{n\pi x}{L}\,dx.$$

$$\frac{1}{L}\int_{-L}^{+L} F(x)\sin\frac{n\pi x}{L}\,dx = \frac{1}{L}\int_{-L}^{+L} f(x)\sin\frac{n\pi x}{L}\,dx$$
$$- [[\ \ ]]\,\frac{1}{L}\int_{-L}^{+L} D(x - x_o)\sin\frac{n\pi x}{L}\,dx.$$

Hence, $a_{Fm} = a_{fm} - [[\ ]]a_{Dm}$ and $b_{Fm} = b_{fm} - [[\ ]]b_{Dm}$, where $[[\ ]] = \left[f(x_o^+) - f(x_o^-)\right]$. It then follows that

$$\sum_{m=1}^{n} a_{Fm} \cos\frac{m\pi x}{L} = \sum_{m=1}^{n} a_{fm} \cos\frac{m\pi x}{L} - [[\ ]]\sum_{m=1}^{n} a_{Dm} \cos\frac{m\pi x}{L}.$$

$$\sum_{m=1}^{n} b_{Fm} \sin\frac{m\pi x}{L} = \sum_{m=1}^{n} b_{fm} \sin\frac{m\pi x}{L} - [[\ ]]\sum_{m=1}^{n} b_{Dm} \sin\frac{m\pi x}{L}.$$

Thus,

$$\begin{aligned}
&\frac{a_o}{2} + \sum_{m=1}^{n} a_{Fm} \cos\frac{m\pi x}{L} + \sum_{m=1}^{n} b_{Fm} \sin\frac{m\pi x}{L} \\
&\quad = \frac{a_o}{2} + \sum_{m=1}^{n} a_{fm} \cos\frac{m\pi x}{L} + \sum_{m=1}^{n} b_{fm} \sin\frac{m\pi x}{L} \\
&\quad\quad - [[\ ]]\left\{\frac{a_o}{2} + \sum_{m=1}^{n} a_{Dm} \cos\frac{m\pi x}{L} + \sum_{m=1}^{n} b_{Dm} \sin\frac{m\pi x}{L}\right\}.
\end{aligned}$$

Collecting,

$$\Sigma_n(x) = S_n(x) - \left[f(x_o^+) - f(x_o^-)\right]S_n^D(x - x_o)$$

But for any value of n, at $x = x_o$, we have shown that $S_n^D(0) = 0$. Thus,

$$\lim_{n\to\infty} S_n(x_o) = \lim_{n\to\infty} \Sigma_n(x_o) = F(x_o) = (1/2)\left[f(x_o^+) + f(x_o^-)\right],$$

and the theorem is proved.

If there is more than 1 point of discontinuity present, we define an $F(x)$:

$$F(x) = f(x) - \sum_i \left[f(x_i^+) - f(x_i^-)\right]D(x_i - x).$$

We have thus shown that any piecewise continuous function in the interval $-L \le x \le +L$ can be expressed as $S_n(x) = \frac{a_o}{2} + \sum_{m=1}^{n} a_m \cos\frac{m\pi x}{L} + b_m \sin\frac{m\pi x}{L}$, with $\lim_{n\to\infty} S_n(x) = (1/2)\left[f(x_o^+) + f(x_o^-)\right]$, where the point of discontinuity is $x_o = x$.

Theorem: For certain $f(x)$ in the range $0 \le x \le L$ we can have

$$(1)\ f(x) = \frac{A_0}{2} + \sum_{n=1}^{\infty} A_n \cos\frac{n\pi x}{L}$$

or

$$(2)\ f(x) = \sum_{n=1}^{\infty} B_n \sin\frac{n\pi x}{L}$$

Proof: Define a function $F(x)$ to be an even function (Fig. 5-10), i.e. define

$F(x) = f(x)$ for $x > 0$.
$F(x) = f(|x|)$ for $x < 0$.

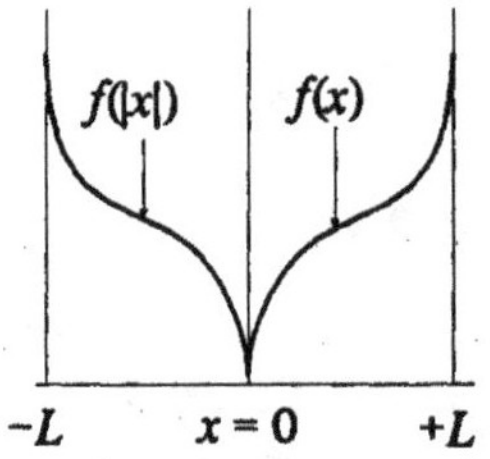

Fig. 5-10

Expanding $F(x)$ in the interval from $-L$ to $+L$,

$$F(x) = \frac{c_0}{2} + \sum_{n=1}^{\infty} c_n \cos\frac{n\pi x}{L} + \sum_{n=1}^{\infty} d_n \sin\frac{n\pi x}{L},$$

where

$$c_n = \frac{1}{L}\int_{-L}^{0} F(x)\cos\frac{n\pi x}{L}\,dx + \frac{1}{L}\int_{0}^{L} F(x)\cos\frac{n\pi x}{L}\,dx$$

and

$$d_n = \frac{1}{L}\int_{-L}^{0} F(x)\sin\frac{n\pi x}{L}\,dx + \frac{1}{L}\int_{0}^{L} F(x)\sin\frac{n\pi x}{L}\,dx.$$

Replacing x by $-x$ in the first integral of d_n, we get

$$\int_{-L}^{0} F(x)\sin\frac{n\pi x}{L}\,dx \Rightarrow \int_{-x=+L}^{-x=0} F(-x)\sin\frac{-n\pi x}{L}\,d(-x)$$

$$= \int_{L}^{0} F(x)\left(-\sin\frac{n\pi x}{L}\right)(-dx) = \int_{L}^{0} F(x)\left(\sin\frac{n\pi x}{L}\right)dx$$

$$= -\int_{0}^{L} F(x)\sin\frac{n\pi x}{L}\,dx,$$

which exactly cancels the second term of d_n. Hence, when $F(x)$ is an even function, there remains only a cosine series.

But between 0 and L, $F(x) = f(x)$, so that in the interval $0 \le x \le L$,

$$f(x) = \frac{A_o}{2} + \sum_{n=1}^{\infty} A_n \cos\frac{n\pi x}{L},$$

where

$$A_n = \frac{2}{L}\int_0^L f(x)\cos\frac{n\pi x}{L}\,dx.$$

If we now define $F(x)$ to be an odd function (Fig. 5-11), such that $F(x) = f(x)$ for $x > 0$, $F(x) = -f(|x|)$ for $x < 0$, then $F(x)$ is expressed only in terms of a sine series (similar argument to that above), and since $F(x) = f(x)$ in the interval $0 \le x \le L$, then

$$f(x) = \sum_{n=1}^{\infty} B_n \sin\frac{n\pi x}{L},$$

where

$$B_n = \frac{2}{L}\int_0^L f(x)\sin\frac{n\pi x}{L}\,dx.$$

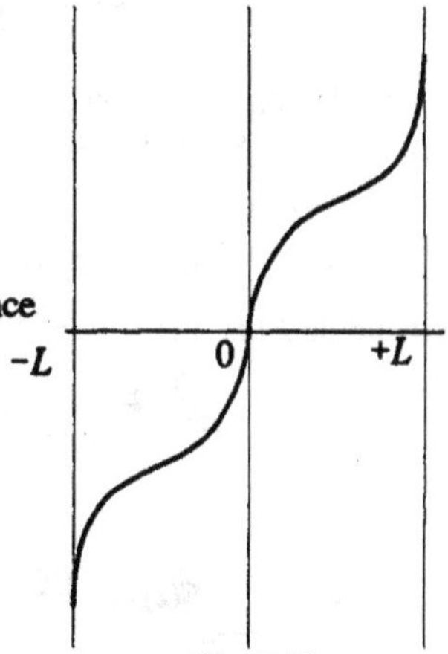

Fig. 5-11

5.3.5 Legendre Polynomials

We have already seen (Section 5.2) that the set $1, x, x^2 \ldots$ forms a complete but not necessarily orthonormal set. Consider the interval $-1 \le x \le +1$ and let us use the Schmidt orthogonalization method to construct an orthonormal set from the above complete one. To proceed, we define

$$\psi^o = \frac{1}{\sqrt{2}}P_o = \frac{1}{\sqrt{2}}, \text{i.e. } P_o = 1. \rightarrow \int_{-1}^{+1}(\psi^o)^2 dx = \frac{1}{2}\int_{-1}^{+1} dx = \left[\frac{x}{2}\right]_{-1}^{+1} = 1.$$

Noting that $\int_{-1}^{+1} x^2 dx = \left[\frac{x^3}{3}\right]_{-1}^{+1} = \frac{2}{3}$, we set

$$\psi^1 = \sqrt{3/2}\,P_1 = \sqrt{3/2}\,x, \text{ i.e. } P_1 = x. \rightarrow \int_{-1}^{+1}\sqrt{3/2}\,x\cdot\sqrt{1/2}\,dx = 0,$$

and P_o and P_1 are orthogonal within the given range.

Using the Schmidt method,

$$\psi^2 = \left[\varphi^2 - \left\{\int_{-1}^{+1}\psi^1\varphi^2\right\}\psi^1 - \left\{\int_{-1}^{+1}\psi^o\varphi^2\right\}\psi^o\right]N^{(2)}.$$

Since $\psi^1 = \sqrt{(3/2)}x$, $\psi^o = \sqrt{(1/2)}$ and $\int_{-1}^{+1}\sqrt{(3/2)}x \cdot x^2 dx = 0$, we take $\varphi^2 = x^2$, so that

$$\psi^2 = \left[x^2 - \left\{\int_{-1}^{+1}\sqrt{1/2}\cdot x^2 dx\right\}\sqrt{1/2}\right]N^{(2)} = \left[x^2 - \frac{x^3}{3}\right]N^{(2)} = \frac{N^{(2)}}{3}(3x^2 - 1).$$

Since $\int_{-1}^{+1}(3x^2 - 1)^2 dx = \int_{-1}^{+1}(9x^4 - 6x^2 + 1)dx = \frac{8}{5}, \frac{N^{(2)}}{3} = \sqrt{5/8} \rightarrow N^{(2)} = \frac{3\sqrt{5/2}}{2}$.
Hence, we take

$$\psi^2 = \sqrt{5/2}P_2 = \sqrt{5/2}\cdot(1/2)\cdot(3x^2 - 1) \rightarrow P_2 = (1/2)\cdot(3x^2 - 1).$$

Theorem: The general member of a series of Legendre polynomials is given by $[(2n + 1)/2]^{1/2}P_n(x)$ $n = 0, 1, 2, \ldots \infty$, where

$$P_n(x) = \frac{1}{2^n n!}\frac{d^n}{dx^n}(x^2 - 1)^n; \quad \int_{-1}^{+1}P_n(x)P_m(x)dx = \frac{2}{2n + 1}\delta_{nm}.$$

[Note that $P_o = 1$, $P_1 = x$, $P_2 = (3x^2 - 1)/2$, etc.]

If the above is true, then $[(2n + 1)/2]^{1/2}P_n(x)$ represents a complete (a linear combination of the x's) and orthonormal set.

Proof: Consider the following integral:

$$\int_{-1}^{+1}\left\{\frac{d^n}{dx^n}(x^2 - 1)^n\right\}\left\{\frac{d^m}{dx^m}(x^2 - 1)^m\right\}dx$$

$$= \left[\frac{d^{n-1}}{dx^{n-1}}(x^2 - 1)^n \frac{d^m}{dx^m}(x^2 - 1)^m\right]_{-1}^{+1}$$

$$- \int_{-1}^{+1}\left\{\frac{d^{n-1}}{dx^{n-1}}(x^2 - 1)^n\right\}\left\{\frac{d^{m+1}}{dx^{m+1}}(x^2 - 1)^m\right\}dx.$$

For $s < n$,

$$\frac{d^s}{dx^s}(x^2-1)^n = \frac{d^{s-1}}{dx^{s-1}}n(x^2-1)^{n-1}(2x)$$

$$= \frac{d^{s-2}}{dx^{s-2}}\left[n(n-1)(x^2-1)^{n-2}(2x)^2 + 2n(x^2-1)^{n-1}\right].$$

Hence $\frac{d^s}{dx^s}(x^2-1)^n$ goes as $(x^2-1)^{n-s}$, which = 0 at $x = \pm 1$. Therefore,

$\left[\frac{d^{n-1}}{dx^{n-1}}(x^2-1)^m\right]_{-1}^{+1} = 0$, and the first term in the integral vanishes. Thus,

$$\int_{-1}^{+1}\left\{\frac{d^n}{dx^n}(x^2-1)^n\right\}\left\{\frac{d^m}{dx^m}(x^2-1)^m\right\}dx$$

$$= -\int_{-1}^{+1}\left\{\frac{d^{n-1}}{dx^{n-1}}(x^2-1)^n\right\}\left\{\frac{d^{m+1}}{dx^{m+1}}(x^2-1)^m\right\}dx$$

$$= \left[\frac{d^{n-2}}{dx^{n-2}}(x^2-1)^n\frac{d^{m+1}}{dx^{m+1}}(x^2-1)^m\right]_{-1}^{+1}$$

$$+\int_{-1}^{+1}\left\{\frac{d^{n-2}}{dx^{n-2}}(x^2-1)^n\frac{d^{m+2}}{dx^{m+2}}(x^2-1)^m\right\}dx.$$

As above and for the same reasons, the first term in the integral vanishes. If we continue the process, we finally get:

$$\int_{-1}^{+1}\left\{\frac{d^n}{dx^n}(x^2-1)^n\frac{d^m}{dx^m}(x^2-1)^m\right\}dx = (-1)^n\int_{-1}^{+1}(x^2-1)^n\frac{d^{m+n}}{dx^{m+n}}(x^2-1)^m dx.$$

Consider the following two cases:

(a) For $n \neq m$ (say $n > m$), then $n + m > 2m$ and $\frac{d^{n+m}}{dx^{n+m}}(x^2-1)^m = 0$.

(b) For $n = m$, $\frac{d^{2n}}{dx^{2n}}(x^2-1)^n = (2n)!$

Hence,

$$(2^{2n})(n!)^2\int_{-1}^{+1}P_n(x)P_m(x)dx = \delta_{nm}(-1)^n(2n)!\int_{-1}^{+1}(1-x^2)^n dx.$$

To evaluate $\int_{-1}^{+1}(1-x^2)^n dx$, first set $x = \cos\theta$, obtaining:

$$\int_{-1}^{+1}(1-x^2)^n dx = \int_{-1}^{+1}\sin^{2n}\theta(d\cos\theta) = \Big[\cos\theta\sin^{2n}\theta\Big]_\pi^0 - \int_\pi^0 \cos\theta\, d(\sin^{2n}\theta)$$

$$= -\int_\pi^0 2n\cos^2\theta\sin^{2n-1}\theta\, d\theta = -\int_\pi^0 2n\sin^{2n-1}\theta\, d\theta + \int_\pi^0 2n\sin^{2n+1}\theta\, d\theta$$

$$= 2n\int_0^\pi \sin^{2n-1}\theta\, d\theta - \int_0^\pi 2n\sin^{2n+1}\theta\, d\theta.$$

We also have that

$$\int_{-1}^{+1}(1-x^2)^n dx = \int_{-1}^{+1}\sin^{2n}\theta(d\cos\theta) = -\int_\pi^0 \sin^{2n+1}\theta\, d\theta = \int_0^\pi \sin^{2n+1}\theta\, d\theta.$$

Hence, $\int_0^\pi \sin^{2n+1}\theta\, d\theta = 2n\int_0^\pi \sin^{2n-1}\theta\, d\theta - \int_0^\pi 2n\sin^{2n+1}\theta\, d\theta$, or

$$\int_0^\pi \sin^{2n+1}\theta\, d\theta = \frac{2n}{2n+1}\int_0^\pi \sin^{2n-1}\theta\, d\theta = \left(\frac{2n}{2n+1}\right)\left(\frac{2n-2}{2n-1}\right)\int_0^\pi \sin^{2n-3}\theta\, d\theta,$$

i.e.,

$$\int_0^\pi \sin^{2n+1}\theta\, d\theta = \frac{2n(2n-2)\ldots 2}{(2n+1)(2n-1)\ldots 3}\int_0^\pi \sin\theta d\theta$$

$$= \frac{2\cdot 2^n\cdot n!}{(2n+1)(2n-1)\ldots 3}.$$

Hence,

$$\int_{-1}^{+1}P_n(x)P_m(x)dx = \delta_{nm}\frac{(2n)!2^{-2n}}{(n!)^2}\frac{2\cdot 2^n\cdot n!}{(2n+1)(2n-1)\ldots 3}$$

$$= \delta_{nm}\frac{2}{2n+1}.$$

Consider the following polynomial:

$$P_r'(x) = \frac{(2r)!}{2^r(r!)^2}\left\{x^r - \frac{r(r-1)}{2(2r-1)}x^{r-2} + \frac{r(r-1)(r-2)(r-3)}{2\cdot4\cdot(2r-1)(2r-3)}x^{r-4} - \dots\right\}.$$

We would like to show that $P_r'(x)$ is identical to $P_r(x) = \frac{1}{2^2 r!}\frac{d^r}{dx^r}(x^2-1)^r$.

Evaluating:

$$\frac{d^r}{dx^r}(x^2-1)^r = \frac{d^r}{dx^r}\left(x^{2r} - rx^{2r-2} + \frac{r(r-1)x^{2r-4}}{2!} - \frac{r(r-1)(r-2)x^{2r-6}}{3!} + \dots\right)$$

$$= (2r)(2r-1)(2r-2)\dots(r+1)x^r - r(2r-2)(2r-3)\dots(r-1)x^{r-2}$$

$$+ \frac{r(r-1)(2r-4)(2r-5)\dots(r-3)x^{r-4}}{2!}$$

$$- \frac{r(r-1)(r-2)(2r-6)(2r-7)\dots(r-5)x^{r-6}}{3!} + \dots$$

$$= \frac{(2r)!x^r}{r!} - \frac{r(2r-2)!x^{r-2}}{(r-2)!} + \frac{r(r-1)(2r-4)!x^{r-4}}{2!(r-4)!}$$

$$- \frac{r(r-1)(r-2)(2r-6)!x^{r-6}}{3!(r-6)!} + \dots$$

$$= \frac{(2r)!}{r!}\left\{x^r - \frac{r^2(r-1)x^{r-2}}{2r(2r-1)} + \frac{r^2(r-1)^2(r-2)(r-3)x^{r-4}}{2(2r)(2r-1)(2r-2)(2r-3)} + \dots\right\}$$

$$= \frac{(2r)!}{r!}\left\{x^r - \frac{r(r-1)x^{r-2}}{2(2r-1)} + \frac{r(r-1)(r-2)(r-3)x^{r-4}}{2\cdot4(2r-1)(2r-3)} + \dots\right\}$$

Hence,

$$P_r(x) = \frac{(2r)!}{2^r(r!)^2}\left\{x^r - \frac{r(r-1)x^{r-2}}{2(2r-1)} + \frac{r(r-1)(r-2)(r-3)x^{r-4}}{2\cdot4(2r-1)(2r-3)} + \dots\right\}$$

$$= P_r'(x).$$

Properties of Legendre polynomials

1] Since the set of Legendre polynomials was defined in the interval $-1 \le x \le +1$, and since $\cos\theta$ is also defined in this interval, we can express any function of $\cos\theta$ in a series of Legendre polynomials, i.e.:

$$f(\cos\theta) = \sum_{n=0}^{\infty} f_n P_n(\cos\theta), \text{ where}$$

$$\int_{-1}^{+1} f(\cos\theta)P_m(\cos\theta)d(\cos\theta) = \sum_{n=0}^{\infty}\int_{-1}^{+1} f_n P_n(\cos\theta)P_m(\cos\theta)d(\cos\theta)$$

$$= f_m \frac{2}{2m+1)}.$$

Hence,

$$f_n = \frac{2n+1)}{2}\int_{-1}^{+1} f(\cos\theta)P_n(\cos\theta)d(\cos\theta).$$

2] $P_n(1) = 1$ for any n.

Let $x = 1 + \varepsilon$, where $\varepsilon \to 0$. Then $x^2 - 1 = (x+1)(x-1) = \varepsilon\cdot(2+\varepsilon)$.

$$P_n(x) = \frac{1}{2^n n!}\frac{d^n}{dx^n}(x^2-1)^n = \frac{1}{2^n n!}\frac{d^n}{dx^n}\varepsilon^n(2+\varepsilon)^n.$$

$$P_n(1) = \frac{1}{2^n n!}\frac{d^n}{dx^n}\left[\varepsilon^n(2+\varepsilon)^n\right]_{\varepsilon\to 0}.$$

Taking derivatives of the bracketed expression on the right will always involve derivatives of ε^n multiplied by some factor and derivatives of $(2+\varepsilon)^n$ multiplied by ε raised to some power. This last set vanishes as $\varepsilon \to 0$.

Hence, since $\frac{d^n}{dx^n}\left[\varepsilon^n(2+\varepsilon)^n\right]_{\varepsilon\to 0} \Rightarrow n!2^n$, we have

$$P_n(1) = \frac{1}{2^n n!}n!2^n = 1.$$

We may have occasion to observe a complete and orthonormal set and may want to know whether this set is a set of Legendre polynomials (e.g. the two sets differ only in their normalization factors). We investigate $P_n(1)$.

$$\left\{\begin{matrix}\text{If } P_n(1) = 1 \text{ the set is a set of Legendre polynomials.} \\ \text{If } P_n(1) \neq 1 \text{ the set is not a set of Legendre polynomials.}\end{matrix}\right\}$$

5.3.6 Multipole Expansion — (example involving the use of Legendre polynomials in electrostatics)

Consider a charge distribution $\rho(\mathbf{r})$ in the region shown (Fig. 5-12). It is desired to find the potential at a point, P, far from and outside this charge distribution. The potential is given by

$$V_R = \int \frac{\rho(\mathbf{r})}{|\mathbf{R}-\mathbf{r}|}\, d^3\mathbf{r}$$

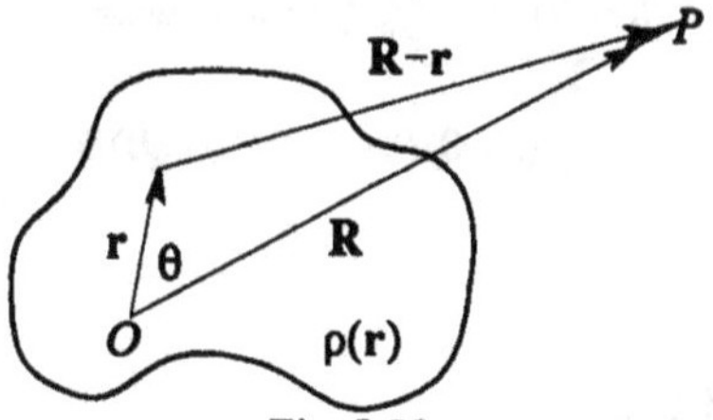

Fig. 5-12

for all $R > r$ in the distribution.

It is desired to have an expression of the form $V_R = \sum_{n=0}^{\infty} \frac{\sigma_n}{R^{n+1}}$. Therefore, consider

$$|\mathbf{R}-\mathbf{r}| = \left(R^2 + r^2 - 2rR\cos\theta\right)^{1/2} = R\left(1 + \frac{r^2}{R^2} - \frac{2r}{R}\cos\theta\right)^{1/2}.$$

If we set $\xi = r/R$ and $x = \cos\theta$, then $|\mathbf{R}-\mathbf{r}| = R\left(1 + \xi^2 - 2\xi x\right)^{1/2}$. Using Taylor's theorem, we expand in powers of ξ about point $\xi = 0$.

$$\left(1 + \xi^2 - 2\xi x\right)^{-1/2} = \sum_{n=0}^{\infty} \frac{\xi^n}{n!}\left\{\frac{\partial^n}{\partial\xi^n}\left(1 + \xi^2 - 2\xi x\right)^{-1/2}\right\}_{\xi=0}.$$

n	$\frac{1}{n!}\left\{\frac{\partial^n}{\partial\xi^n}\left(1+\xi^2-2\xi x\right)^{-1/2}\right\}_{\xi=0}$
0	1
1	$-\frac{1}{2}\left(1+\xi^2-2\xi x\right)^{-3/2}(2\xi-2x)\Big]_{\xi=0} = x$
2	$\frac{1}{2!}\cdot\frac{3}{2}\cdot\frac{1}{2}\left(1+\xi^2-2\xi x\right)^{-5/2}(2\xi-2x)^2 - \frac{1}{2}\left(1+\xi^2-2\xi x\right)^{-3/2} 2\Big]_{\xi=0} = \frac{(3x^2-1)}{2}$

Observe that

$$\left(1+\xi^2-2\xi x\right)^{-1/2} = \left(P_o(x)+P_1(x)+P_2(x)+\ldots\right)\frac{r^n}{R^n}.$$

We strongly believe that $\dfrac{1}{|\mathbf{R}-\mathbf{r}|} = \sum_{n=o}^{\infty}\dfrac{P_n(\cos\theta)r^n}{R^{n+1}}$ for all n, where the $P_n(\cos\theta)$ are the Legendre polynomials. [Thus far, we have only shown this to be true for $n = 0, 1, 2$.] If the above is true, then

$$V_R = \int\rho(\mathbf{r})\sum_{n=o}^{\infty}\frac{P_n(\cos\theta)r^n}{R^{n+1}}\,d^3\mathbf{r}$$

$$= \int\frac{\rho(\mathbf{r})}{R}\,d^3\mathbf{r} + \int\frac{r\cos\theta}{R^2}\,\rho(\mathbf{r})\,d^3\mathbf{r} + \int\frac{r^2(3\cos^2\theta-1)}{2R^3}\,\rho(\mathbf{r})\,d^3\mathbf{r} + \ldots$$

$$= \frac{\sigma_o}{R}+\frac{\sigma_1}{R^2}+\frac{\sigma_2}{R^3}+\ldots = \sum_{n=o}^{\infty}\frac{\sigma_n}{R^{n+1}},$$

where

$\sigma_o = \int\rho(\mathbf{r})d^3\mathbf{r}$, the total charge,

$\sigma_1 = \int r\cos\theta\rho(\mathbf{r})d^3\mathbf{r}$, the dipole moment, and

$\sigma_2 = \int\frac{r^2}{2}(3\cos^2\theta-1)\rho(\mathbf{r})d^3\mathbf{r}$, the quadrupole moment, etc.

Theorem: $\dfrac{1}{|\mathbf{R}-\mathbf{r}|} = \sum_{n=o}^{\infty}\dfrac{P_n(x)r^n}{R^{n+1}}$, where the $P_n(x)$ are identically the Legendre polynomials.

Proof: $\dfrac{1}{|\mathbf{R}-\mathbf{r}|} = \dfrac{1}{\left\{(X-x)^2+(Y-y)^2+(Z-z)^2\right\}^{1/2}} = \dfrac{1}{\{\ \}^{1/2}}.$

$$\frac{\partial^2}{\partial x^2}\frac{1}{|\mathbf{R}-\mathbf{r}|} = \frac{\partial}{\partial x}\left[-\frac{\{\ \}^{-3/2}}{2}2(X-x)(-1)\right] = \frac{\partial}{\partial x}\frac{X-x}{\{\ \}^{3/2}}$$

$$= -\frac{3}{2}\{\ \}^{-5/2}(-2)(X-x)^2 - \frac{1}{\{\ \}^{3/2}} = \frac{3(X-x)^2-\{\ \}}{\{\ \}^{5/2}}.$$

Hence,

$$\Delta_r \frac{1}{|\mathbf{R}-\mathbf{r}|} = \left\{\frac{\partial^2}{\partial x^2} + \frac{\partial^2}{\partial y^2} + \frac{\partial^2}{\partial z^2}\right\} \frac{1}{|\mathbf{R}-\mathbf{r}|}$$

$$= \frac{3(X-x)^2 + 3(Y-y)^2 + 3(Z-z)^2 - 3\{\ \}}{\{\ \}^{5/2}} = 0.$$

Taking the Laplacian of $\frac{1}{|\mathbf{R}-\mathbf{r}|} = \sum_{n=0}^{\infty} \frac{P_n(x) r^n}{R^{n+1}}$ and setting $x = \cos\theta$, we have

$$\sum_{n=0}^{\infty} \frac{1}{R^{n+1}} \Delta_r \left[P_n(\cos\theta) r^n\right] = 0,$$

true for any value of R. We now multiply through by R and then let $R \to \infty$ (so that only the $n = 0$ term remains). We then get $\Delta_r \left[P_0(\cos\theta) r^0\right] = 0$. Multiplying through by R^2 and then letting $R \to \infty$ (so that only the $n = 1$ term remains) yields $\Delta_r \left[P_1(\cos\theta) r^1\right] = 0$. Repeating the process we ultimately obtain $\Delta_r \left[P_n(\cos\theta) r^n\right] = 0$.

Consider the Laplacian operator in spherical coordinates:

$$\Delta = \frac{1}{r^2}\frac{\partial}{\partial r}\left(r^2 \frac{\partial}{\partial r}\right) + \frac{1}{r^2 \sin\theta}\frac{\partial}{\partial \theta}\left(\sin\theta \frac{\partial}{\partial \theta}\right) + \frac{1}{r^2 \sin^2\theta}\frac{\partial^2}{\partial \varphi^2},$$

of which only the first two terms operate on $P_n(\cos\theta) r^n$.

Observe that $\Delta_r \frac{1}{|\mathbf{R}-\mathbf{r}|}$ only referred to Cartesian coordinate Laplacian operator Δ, i.e. with respect to Cartesian coordinates of r. However, we are now going to refer $\Delta_r \left[P_n(\cos\theta) r^n\right]$ to spherical coordinates. We are allowed to do this since $\Delta\varphi$ is a scaler, hence the same in all coordinate systems.

Operating on r^n:

$$\frac{1}{r^2}\frac{\partial}{\partial r}\left(r^2 \frac{\partial r^n}{\partial r}\right) = \frac{1}{r^2}\frac{\partial}{\partial r}\left(n r^{n+1}\right) = \frac{n}{r^2}(n+1) r^n = n(n+1) r^{n-2}.$$

Hence, operating on $P_n(\cos\theta)r^n$:

$$\Delta_r\left[P_n(\cos\theta)r^n\right] = n(n+1)r^{n-2}P_n(\cos\theta)$$

$$+\frac{r^{n-2}}{\sin\theta}\frac{\partial}{\partial\theta}\left[\sin\theta\frac{\partial P_n(\cos\theta)}{\partial\theta}\right] = 0.$$

$$\Downarrow$$

[1] $$r^{n-2}\left\{n(n+1)P_n(\cos\theta)+\frac{1}{\sin\theta}\frac{\partial}{\partial\theta}\left[\sin\theta\frac{\partial P_n(\cos\theta)}{\partial\theta}\right]\right\} = 0.$$

If $x = \cos\theta$, $\sin^2\theta = 1 - \cos^2\theta = 1 - x^2$, $dx = -\sin\theta d\theta$, and $\dfrac{d}{\sin\theta d\theta} = -\dfrac{d}{dx}$. Substituting into [1], we have:

$$n(n+1)P_n(x)+\frac{d}{dx}\left[(1-x^2)\frac{dP_n(x)}{dx}\right] = 0.$$

Rewriting,

[2] $$\frac{d}{dx}\left[(1-x^2)\frac{dP_n(x)}{dx}\right] = -n(n+1)P_n(x).$$

Let us define a new operator $H \equiv \dfrac{d}{dx}\left[(1-x^2)\dfrac{d}{dx}\right]$, so that equation [2] becomes

[3] $$HP_n(x) = -n(n+1)P_n(x),$$

and the $P_n(x)$ are seen to be the eigensolutions of the operator H. We now show that H is intimately connected with the Laplacian operator. In fact,

$$H = \frac{1}{\sin\theta}\frac{\partial}{\partial\theta}\left(\sin\theta\frac{\partial}{\partial\theta}\right) = \frac{d}{dx}\left[(1-x^2)\frac{d}{dx}\right].$$

Theorem:

$$\int_{-1}^{+1}\varphi^* H\psi dx = \left\{\int_{-1}^{+1}\psi^* H\varphi dx\right\}^*,$$

where $H = \dfrac{d}{dx}\left[(1-x^2)\dfrac{d}{dx}\right]$.

Proof:

$$\int_{-1}^{+1}\varphi^*\frac{d}{dx}\left[(1-x^2)\frac{d\psi}{dx}\right]dx = \left[\varphi^*(1-x^2)\frac{d\psi}{dx}\right]_{-1}^{+1} - \int_{-1}^{+1}(1-x^2)\frac{d\psi}{dx}\frac{d\varphi^*}{dx}dx$$

$$= 0 - \int_{-1}^{+1}(1-x^2)\frac{d\psi}{dx}\frac{d\varphi^*}{dx}dx.$$

$$\left\{\int_{-1}^{+1}\psi^* H\varphi dx\right\}^* = -\left\{\int_{-1}^{+1}(1-x^2)\frac{d\psi}{dx}\frac{d\varphi^*}{dx}dx\right\}^* = -\int_{-1}^{+1}(1-x^2)\frac{d\psi}{dx}\frac{d\varphi^*}{dx}dx.$$

Thus, H is Hermitian.

Theorem: If H is Hermitian for the interval $a \leq x \leq b$ and if $H\psi = \lambda\psi$ and $H\varphi = \lambda\varphi$, then if $\lambda \neq \lambda'$,

$$\int_a^b \psi^*\varphi dx = 0,$$

i.e. the eigensolutions are orthogonal.

Proof:

(a) $$\int_a^b \varphi^* H\psi dx = \lambda\int_a^b \varphi^*\psi dx.$$

(b) $$\left\{\int_a^b \psi^* H\varphi dx\right\}^* = \lambda'\left\{\int_a^b \psi^*\varphi dx\right\}^* = \lambda'\int_a^b \varphi^*\psi dx.$$

It follows from (a) and (b) that $(\lambda - \lambda'^*)\int_a^b \varphi^*\psi dx = 0$. If H is real, then it follows from $\int_a^b \psi^* H\psi dx = \lambda\int_a^b \psi^*\psi dx$ that since the left side is real and the integral on the right is real, λ must be real. Thus, since λ' is real, $\lambda' = \lambda'^*$. Hence, from the above, $(\lambda - \lambda'^*)\int_a^b \varphi^*\psi dx = 0$. But by hypothesis, $\lambda \neq \lambda'$, so that $\int_a^b \varphi^*\psi dx = 0$ and the eigensolutions of a Hermitian operator for $\lambda \neq \lambda'$ are orthogonal.

[Note the similarity to the properties of a Hermitian matrix in n-dimensional space.]

As an example of another Hermitian operator, let us define the following:

$$(-L^2) \equiv \frac{1}{\sin\theta}\frac{\partial}{\partial\theta}\left(\sin\theta\frac{\partial}{\partial\theta}\right) + \frac{1}{\sin^2\theta}\frac{\partial^2}{\partial\varphi^2}.$$

Theorem:

$$\int\Psi^*(-L^2)\Phi d\Omega = \int\Psi^*(-L^2)\Phi\sin\theta d\theta d\varphi = \left\{\int\Phi^*(-L^2)\Psi\sin\theta d\theta d\varphi\right\}^*.$$

Proof: We have already shown (Sect. 5.3.2) that

$$\int_0^{2\pi}\Psi^*\frac{\partial^2}{\partial\varphi^2}\Phi d\varphi = \int_0^{2\pi}\Phi\frac{\partial^2}{\partial\varphi^2}\Psi^* d\varphi.$$

Hence, $(-L^2)$ is the sum of two Hermitian operators and is therefore itself a Hermitian operator.

Summary:

(a) $$\frac{1}{(1+\xi^2-2\xi x)^{1/2}} = \sum_{n=0}^{\infty}P_n(x)\xi^n,$$
where the $P_n(x)$ are real polynomials in x.

(b) The $P_n(x)$ are the eigensolutions of the Hermitian operator H, i.e.
$HP_n(x) = -n(n+1)P_n(x)$, such that $\int_{-1}^{+1}P_n(x)P_m(x)dx = 0$ for $n \neq m$.

(c) For $x = 1$, $\frac{1}{(1-\xi)} = \sum_{n=0}^{\infty}P_n(1)\xi^n$. But $\frac{1}{(1-\xi)} = \sum_{n=0}^{\infty}\xi^n \Rightarrow P_n(1) = 1$.

(d) The $P_n(x)$ are polynomials in x of degree n.

(e) The $P_n(x)$ are orthogonal sets of polynomials, i.e. $\int_{-1}^{+1}P_n(x)P_m(x)dx = 0$;
their normalizing property is sustained by showing that $P_n(1) = 1$.

5.3.7 Recursion Formulae for Legendre Polynomials

Consider

$$\frac{1}{(1+\xi^2-2\xi x)^{1/2}} = \sum_{n=0}^{\infty} P_n(x)\xi^n,$$

where the $P_n(x)$ are the Legendre polynomials.

Taking ln of both sides,

$$-\frac{1}{2}\ln(1+\xi^2-2\xi x) = \ln\left(\sum_{n=0}^{\infty} P_n(x)\xi^n\right).$$

We now perform two differentiations.

Differentiating with respect to ξ:

$$\frac{\partial}{\partial\xi}\left(-(1/2)\ln(1+\xi^2-2\xi x)\right) = \frac{-\xi+x}{(1+\xi^2-2\xi x)} = \frac{\sum_{n=0}^{\infty} P_n(x)n\xi^{n-1}}{\sum_{n=0}^{\infty} P_n(x)\xi^n}.$$

Equating coefficients of ξ^n,

$$\sum_{n=0}^{\infty}\xi^n(xP_n - P_{n-1}) = \sum_{n=0}^{\infty}\xi^n\left[(n+1)P_{n+1} + (n-1)P_{n-1} - 2xnP_n\right],$$

and, since this must be true for any $\xi = \frac{r}{R} < 1$,

$$(n+1)P_{n+1} + (n-1)P_{n-1} = 2xnP_n + xP_n - P_{n-1}.$$

$$\Downarrow$$

(I) $\quad (n+1)P_{n+1} + nP_{n-1} = x(2n+1)P_n.$

Differentiating with respect to x:

$$\frac{\partial}{\partial x}\left(-(1/2)\ln(1+\xi^2-2\xi x)\right) = \frac{\xi}{(1+\xi^2-2\xi x)} = \frac{\sum_{n=0}^{\infty} P_n'(x)n\xi^n}{\sum_{n=0}^{\infty} P_n(x)\xi^n}.$$

We now equate coefficients of ξ^{n+1}.

$$\sum_{n=0}^{\infty}\xi^{n+1}P_n = \sum_{n=0}^{\infty}\xi^{n+1}(P'_{n+1} + P'_{n-1} - 2xP'_n).$$

Since this must be true for any $\xi = \frac{r}{R} < 1$,

(II) $\quad P'_{n+1} + P'_{n-1} - 2xP'_n - P_n = 0.$

All other recursion formulae found in the literature are obtained by linear combinations of appropriate derivatives of the above two formulae, I and II.

5.3.8 Spherical Harmonics (another example of a complete and orthonormal set)

We have already shown in Section 5.3.3 that on the interval $(-\pi, \pi)$ any function of φ can be expressed as

$$f(\varphi) = \sum_{m=-\infty}^{\infty} A_m e^{\pm im\varphi},$$

where the functions $e^{\pm im\varphi}$ are related to the eigenfunctions of the Hermitian operator $\partial^2/\partial\varphi^2$.

We shall see that when we consider functions, $F(\theta\varphi)$, defined on a spherical surface, we will get a series of spherical harmonics which are eigenfunctions of the Hermitian operator $(-L^2)$, where

$$-L^2 \equiv \frac{1}{\sin\theta}\frac{\partial}{\partial\theta}\left(\sin\theta\frac{\partial}{\partial\theta}\right) + \frac{1}{\sin^2\theta}\frac{\partial^2}{\partial\varphi^2}.$$

Note the following:

<u>In 1 dim:</u> $1, x, x^2 \ldots$ form a complete set in the interval $a \le x \le b$.

<u>In 2 dim:</u> $x^t y^u$ form a complete set, where t and u are positive integers and

where $a \le \begin{pmatrix} x \\ y \end{pmatrix} \le b$.

<u>In 3 dim:</u> $x^t y^u z^v$ (where t, u and v are positive integers) form a complete set

for $a \le \begin{pmatrix} x \\ y \\ z \end{pmatrix} \le b$.

From the transformation to spherical coordinates, given by

$$z = r\cos\theta, \quad x + iy = r\sin\theta\, e^{+i\varphi}, \quad x - iy = r\sin\theta\, e^{-i\varphi},$$

it follows that

$$(x+iy)^t(x-iy)^u z^v = (x+iy)^{t-u}(x+iy)^u(x-iy)^u z^v$$

$$= (x+iy)^{t-u}(x^2+y^2)^u z^v = r^{t-u}\sin^{t-u}\theta\, e^{i(t-u)\varphi} r^{2u}\sin^{2u}\theta\, z^v$$

$$= r^{t+u+v}\sin^{t+u}\theta\cos^v\theta\, e^{i(t-u)\varphi}.$$

Since $x^t y^u z^v$ form a complete set, $(x+iy)^t(x-iy)^u z^v$ will also form complete set, since it is a linear combination of xy and t. Hence,

$$\sum_{tuv} A_{tuv}\, r^{t+u+v}\sin^{t+u}\theta\cos^v\theta\, e^{i(t-u)\varphi}$$

form a complete set in 3 dimensions, and any function can be represented by such a series.

For the particular class of functions, $F(\theta\varphi)$, which lie on the surface of a sphere for which $r = 1$,

$$F(\theta\varphi) = \sum_{tuv} A_{tuv}\cos^v\theta\sin^{t+u}\theta\, e^{i(t-u)\varphi}.$$

We define an $m \equiv |t-u| \quad \Rightarrow \quad t + u = m + \begin{Bmatrix} 2u \text{ if } t > u \\ 2t \text{ if } t < u \end{Bmatrix}$, so that

$$\sin^{t+u}\theta = \sin^m\theta \cdot \begin{Bmatrix} \sin^{2u}\theta \\ \sin^{2t}\theta \end{Bmatrix}.$$

Note that

$$\sin^{2u}\theta = \sin^2\theta\cdot\sin^2\theta\cdot\sin^2\theta\cdot\cdots\cdot\sin^2\theta \quad (u \text{ terms})$$

$$= (1-\cos^2\theta)(1-\cos^2\theta)\cdots(1-\cos^2\theta) \quad (u \text{ factors}),$$

i.e. $\begin{Bmatrix} \sin^{2u}\theta \\ \sin^{2t}\theta \end{Bmatrix}$ are polynomials in $\cos^2\theta$. Hence,

$$F(\theta\varphi) = \sum_{nm} A_{nm}\cos^n\theta\sin^m\theta\, e^{im\varphi}.$$

Setting $n + m = l$, we have for $l = 0, 1, 2$:

l	m	n	No. of States	Function (Unnormalized)
0	0	0	1	1
1	0	1	1	$\cos\theta$
	1	0	2	$\sin\theta e^{+i\theta}$, $\sin\theta e^{-i\theta}$
2	0	2	1	$\cos^2\theta - 1/3$ (see below)
	1	1	2	$\sin\theta\cos\theta e^{+i\theta}$, $\sin\theta\cos\theta e^{-i\theta}$
	2	0	2	$\sin^2\theta e^{+2i\theta}$, $\sin^2\theta e^{-2i\theta}$

To investigate for orthonormality over the surface of the sphere, we note first that $\cos\theta$ ($l = 1$) is orthogonal to 1 ($l = 0$) on the surface of the sphere, since

$$\int_0^{2\pi}\int_0^{\pi}\cos\theta\cdot 1\cdot\sin\theta d\theta d\varphi = 2\pi\int_0^{\pi}\sin\theta\cdot\cos\theta\cdot d\theta = \pi\int_0^{\pi}\sin 2\theta d\theta$$

$$= -\left[\frac{\pi}{2}\cos 2\theta\right]_0^{\pi} = 0.$$

We also note that $\sin\theta e^{\pm i\varphi}$ ($l = 1$) are orthogonal to 1 ($l = 0$) by virtue of the $e^{i\varphi}$, since

$$\int_0^{\pi}\sin^2\theta d\theta\int_0^{2\pi}e^{\pm i\varphi}d\varphi = \int_0^{\pi}\sin^2\theta d\theta\left[(\mp i)e^{i\varphi}\right]_0^{2\pi} = 0.$$

Similarly, $\sin\theta e^{\pm i\varphi}$ ($l = 1$) are orthogonal to $\cos\theta$ ($l = 1$). In fact, any exponential term will be orthogonal to any other function not having the same exponential power.

Note that $\cos^2\theta$ ($l = 2$) is orthogonal to $\cos\theta$ ($l = 1$), since

$$\int_0^{\pi}\cos^3\theta\sin\theta d\theta = \left[-\frac{\cos^4\theta}{4}\right]_0^{\pi} = 0.$$

However, when we investigate the orthogonality poroperty for $\cos^2\theta$ ($l = 2$) and 1 ($l = 0$), we get

$$\int_0^{\pi}\cos^2\theta\sin\theta d\theta = \left[-\frac{\cos^3\theta}{3}\right]_0^{\pi} = -\frac{1}{3}(-1-1) = \frac{2}{3}.$$

Furthermore,

$$\int_0^\pi 1\cdot\sin\theta d\theta = \Big[-\cos\theta\Big]_0^\pi = -(-2) = +2.$$

Hence, to get orthogonality, we take the term in $\cos^2\theta$ to be $\left(\cos^2\theta - 1/3\right)$. We now investigate $\sin\theta\cos\theta e^{\pm i\varphi}$ and $\sin\theta e^{\pm i\varphi}$.

$$\int_0^{2\pi}\int_0^\pi \sin^3\theta\cos\theta \begin{Bmatrix} e^{\pm 2i\varphi} \\ e^0 \end{Bmatrix} d\theta d\varphi = \begin{cases} \displaystyle\int_0^{2\pi}\int_0^\pi \sin^3\theta\cos\theta e^{\pm 2i\varphi} d\theta d\varphi \\ \displaystyle\int_0^{2\pi}\int_0^\pi \sin^3\theta\cos\theta d\theta d\varphi \end{cases}$$

$$= \begin{cases} \displaystyle\int_0^{2\pi} e^{\pm 2i\varphi} d\varphi \left[\frac{\sin^4\theta}{4}\right]_0^\pi = 0. \\ \displaystyle 2\pi\int_0^\pi \sin^3\theta\cos\theta d\theta = 2\pi\left[\frac{\sin^4\theta}{4}\right]_0^\pi = 0. \end{cases}$$

Hence we see that our functions represent a complete and orthonormal set. For each l there are $2l + 1$ states, i.e. for $l = 0$, we have 1 state, for $l = 1$, we have 3 states, for $l = 2$, we have 5 states, etc.

The spherical harmonics are given by

$$Y_{l,m_l}(\theta\varphi) = \left[\frac{(2l+1)(l-m)!}{4\pi(l+m)!}\right]^{1/2} P_l^m(\cos\theta)e^{im\varphi},$$

where

$$P_l^m(x) = (1-x^2)^{m/2}\frac{d^m}{dx^m}P_l(x)$$

are the Associated Legendre polynomials and $P_l(x)$ are the Legendre polynomials. $m_l = \pm m = -l, -l+1, \ldots +l$, i.e. $2l+1$ values of m_l for every l value, and $|m_l| = m$.

Note that if $x = \cos\theta$,

$$P_l^m(\cos\theta) = \sin^m\theta\frac{d^m}{dx^m}P_l(\cos\theta) \rightarrow \sin^m\theta\frac{d^m}{d(\cos\theta)^m}P_l(\cos\theta).$$

Since $\frac{d^m}{dx^m} P_l(\cos\theta)$ yields polynomials in $(\cos\theta)$ of degree $l - m = n$, $P_l^m(\cos\theta)e^{im_l\varphi} = \sin^m\theta \frac{d^m}{d(\cos\theta)^m} P_l(\cos\theta)e^{\pm im\varphi}$ yields polynomials of the form

$\cos^n\theta \cdot \sin^m\theta \cdot e^{\pm im\varphi}$, and these are seen to be the (required spherical harmonics, or) complete orthogonal set discussed. Hence, the $Y_{l,m_l}(\theta\varphi)$ as given above are a complete set of polynomials in 3 dimensions. We have yet to show they are normalized.

Theorem:

$$\int_0^{2\pi} d\varphi \int_0^{\pi} \sin\theta d\theta \left[Y^*_{l,m_l}(\theta\varphi) Y_{l',m'_{l'}}(\theta\varphi) \right] = \delta_{ll'}\delta_{m_l m'_{l'}}.$$

Note that for $m_l \neq m'_{l'}$,

$$\int_0^{\pi} \sin\theta d\theta\, F(lml'm') \int_0^{\pi} P_{l'}^{m'}(\cos\theta) P_l^m(\cos\theta) e^{i(m'_{l'}-m_l)\varphi} d\varphi$$

$$= \int_0^{\pi} G(\theta) \sin\theta d\theta\, F(lml'm') \int_0^{2\pi} e^{i(m'_{l'}-m_l)\varphi} d\varphi = 0;$$

hence the theorem need only be shown true for $m_l = m'_{l'}$, i.e. we have to show that

$$-2\pi \int_0^{\pi} \frac{d(\cos\theta)}{4\pi} \left[\frac{(2l+1)(l-m)!(2l'+1)(l'-m)!}{(l+m)!(l'+m)!} \right]^{1/2} P_l^m(\cos\theta) P_{l'}^m(\cos\theta) = \delta_{ll'}$$

$$\Downarrow$$

$$\int_{-1}^{+1} P_l^m(x) P_{l'}^m(x) = \frac{2}{2l+1} \frac{(l+m)!}{(l-m)!} \delta_{ll'}.$$

Proof: The proof will be rather lengthy and will involve several other proofs along the way.

Let us define $y_l^m \equiv \frac{d^m}{dx^m} P_l(x)$, noting that $y_l^0 = P_l(x)$ and $\frac{d}{dx} y_l^m = y_l^{m+1}$.

Consider the equation:

$$(1) \qquad (1 - x^2) \frac{d^2}{dx^2} y_l^m - 2x(m + 1) \frac{d}{dx} y_l^m = -(l - m)(l + m + 1) y_l^m.$$

Equation (1) will be verified by induction, i.e. we will show to be true for $m = 0$ [(a) below], and then show that if true for m, it is true for $m + 1$ [(b) below].

(a) For $m = 0$, $y_l^m = P_l(x)$, so that

$$(1 - x^2)\frac{d^2}{dx^2}P_l(x) - 2x\frac{d}{dx}P_l(x) = -l(l+1)P_l(x).$$

This is seen to be merely

$$\frac{d}{dx}\left[(1 - x^2)\frac{d}{dx}P_l(x)\right] = -l(l+1)P_l(x),$$

which is the differential equation satisfied by the Legendre polynomials (Sect. 5.3.6).

(b) Assume equation (1) is true for m:

Differentiating (1), using $y_l^{m+1} = \frac{d}{dx}y_l^m$, we have

$$(1 - x^2)\frac{d^2}{dx^2}y_l^{m+1} - 2x\frac{d^2}{dx^2}y_l^m - 2x(m+1)\frac{d^2}{dx^2}y_l^m - 2(m+1)\frac{d}{dx}y_l^m$$
$$= -(l-m)(l+m+1)y_l^{m+1}$$
$$\Downarrow$$
$$(1 - x^2)\frac{d^2}{dx^2}y_l^{m+1} - 2x(m+2)\frac{d^2}{dx^2}y_l^m - 2(m+1)\,y_l^{m+1}$$
$$= -(l-m)(l+m+1)y_l^{m+1}$$
$$\Downarrow$$
$$(1 - x^2)\frac{d^2}{dx^2}y_l^{m+1} - 2x(m+2)\frac{d}{dx}y_l^{m+1} - 2(m+1)\,y_l^{m+1}$$
$$= -(l-m)(l+m+1)y_l^{m+1}$$

Observe that

$$\begin{aligned}(l-m-1)(l+m+2) &= [(l-m)-1][(l+m+1)+1] \\ &= (l-m)(l+m+1) + (l-m) - (l+m+1) - 1 \\ &= (l-m)(l+m+1) - 2m - 2.\end{aligned}$$

Hence our equation becomes

$$(1-x^2)\frac{d^2}{dx^2}y_l^{m+1} - 2x(m+2)\frac{d}{dx}y_l^{m+1} = -(l-m-1)(l+m+2)y_l^{m+1},$$

which is merely equation (1) with m replaced by $m + 1$; hence (1) is verified.

Now consider the equation

$$(2) \qquad \frac{d}{dx}\left[(1-x^2)\frac{d}{dx}\right]P_l^m(x) + \left[l(l+1) - \frac{m^2}{(1-x^2)}\right]P_l^m(x) = 0.$$

Note that for $m = 0$, $P_l^0(x) = P_l(x)$ and equation (2) becomes

$$\frac{d}{dx}\left[(1-x^2)\frac{d}{dx}\right]P_l(x) + l(l+1)P_l(x) = 0,$$

which is again the differential equation satisfied by the Legendre polynomials; hence, equation (2) is true for $m = 0$.

By direct differentiation of equation (2),

$$\frac{dP_l^m(x)}{dx} = \frac{d}{dx}\left[(1-x^2)^{m/2}y_l^m\right] = -mx(1-x^2)^{m/2-1}y_l^m + (1-x^2)^{m/2}y_l^{m+1}.$$

$$(1-x^2)\frac{dP_l^m(x)}{dx} = -mx(1-x^2)^{m/2}y_l^m + (1-x^2)^{m/2+1}\frac{dy_l^m}{dx}.$$

$$\frac{d}{dx}\left\{(1-x^2)\frac{dP_l^m(x)}{dx}\right\} = -m(1-x^2)^{m/2}y_l^m + m^2x^2(1-x^2)^{m/2-1}y_l^m$$

$$-mx(1-x^2)^{m/2}\frac{dy_l^m}{dx} - (m+2)x(1-x^2)^{m/2}\frac{dy_l^m}{dx} + (1-x^2)^{m/2+1}\frac{d^2y_l^m}{dx}.$$

Collecting terms,

$$\frac{d}{dx}\left\{(1-x^2)\frac{dP_l^m(x)}{dx}\right\} = (1-x^2)^{m/2}\left[(1-x^2)\frac{d^2y_l^m}{dx^2} - 2x(m+1)\frac{dy_l^m}{dx}\right]$$

$$-(1-x^2)^{m/2}\left[my_l^m - \frac{m^2x^2}{(1-x^2)}y_l^m\right].$$

But, from equation (1), since

$$(1-x^2)\frac{d^2y_l^m}{dx^2} - 2x(m+1)\frac{dy_l^m}{dx} = -(l-m)(l+m+1)y_l^m,$$

We have that

$$\frac{d}{dx}\left\{(1-x^2)\frac{dP_l^m(x)}{dx}\right\} = -(1-x^2)^{m/2}(l-m)(l+m+1)y_l^m$$

$$-(1-x^2)^{m/2}\left[m-\frac{m^2x^2}{(1-x^2)}\right]y_l^m.$$

Noting that

$$\begin{aligned}-(l-m)(l+m+1) &= -l(l+1) - ml + m(l+1) + m^2\\ &= -l(l+1) + m + m^2,\end{aligned}$$

$$\frac{d}{dx}\left\{(1-x^2)\frac{dP_l^m(x)}{dx}\right\} = \left[-l(l+1)+\frac{m^2}{(1-x^2)}\right](1-x^2)^{m/2}y_l^m,$$

thus verifying equation (2).

Rewriting equation (2) using $x = \cos\theta$ and $dx = -\sin\theta d\theta$, we have

$$\frac{1}{\sin\theta}\frac{\partial}{\partial\theta}\left[\sin\theta\frac{\partial}{\partial\theta}\right]P_l^m(\cos\theta) + \left[l(l+1)-\frac{m^2}{\sin^2\theta}\right]P_l^m(\cos\theta) = 0,$$

which, since $Y_{lm_l}(\theta\varphi) = f(lm)P_l^m(\cos\theta)e^{im\varphi}$, is the same as

$$\frac{1}{\sin\theta}\frac{\partial}{\partial\theta}\left[\sin\theta\frac{\partial}{\partial\theta}\right]Y_{lm_l}(\theta\varphi) = -\left[l(l+1)-\frac{m^2}{\sin^2\theta}\right]Y_{lm_l}(\theta\varphi).$$

Note also that $\frac{\partial^2}{\partial\varphi^2}Y_{lm_l}(\theta\varphi) = -m_l^2Y_{lm_l}(\theta\varphi) = -m^2Y_{lm_l}(\theta\varphi)$. Hence our differential equation for Y_{lm_l} becomes:

$$\frac{1}{\sin\theta}\frac{\partial}{\partial\theta}\left[\sin\theta\frac{\partial Y_{lm_l}}{\partial\theta}\right] + \frac{1}{\sin^2\theta}\frac{\partial^2 Y_{lm_l}}{\partial\varphi^2} = -L^2Y_{lm_l} = -l(l+1)Y_{lm_l}.$$

We have already seen (Sect. 5.3.6) that $-L^2$ is a good Hermitian operator. Hence, $Y_{lm_l}(\theta\varphi)$ are the eigensolutions of Hermitian operator $-L^2$ with eigenvalues $-l(l+1)$.

Properties of the eigensolutions:

1. The eigenvalues are functions of l only. Hence, for each l, there are $2l+1$ degenerate states ($2l+1$ values of m_l), each state with eigenvalue $-l(l+1)$.
2. For $l \neq l'$, i.e. for two different eigenvalues, the eigenfunctions are orthogonal. [This general property of Hermitian operators was discussed in Sect. 5.3.6.] Hence, for $l \neq l'$,

$$\int_0^{2\pi} d\varphi \int_0^{\pi} \sin\theta d\theta \left[Y^*_{lm_l}(\theta\varphi) Y_{l'm_l}(\theta\varphi)\right] = 0.$$

We have already shown the functions to be orthogonal for $m_l \neq m_l'$ (p. 143). Now we have shown it for $l \neq l'$. It only remains to show that

$$\int_{-1}^{+1} \left[P_l^m(x)\right]^2 dx = \frac{2}{(2l+1)} \frac{(l+m)!}{(l-m)!},$$

i.e. that the $P_l^m(x)$ are a normalized set.

Theorem: If the set ψ_s are the eigensolutions of a Hermitian operator H, and if the set of ψ_s constitute a complete set (they are already necessarily orthogonal), then these ψ_s are the only eigensolutions of H (i.e. they are unique); furthermore, there exist only the particular eigenvalues associated with $H\psi_s = \lambda_s\psi_s$.

Proof: $H\psi_s = \lambda_s\psi_s$ and we have $\Psi = \sum_s a_s\psi_s$ by completeness property. Then if $H\Psi = \lambda'\Psi$ so that λ' constitutes a diferent eigenvalue with a different eigensolution Ψ, we have

$$\left\{\begin{aligned} H\Psi &= \sum_s a_s H\psi_s = \sum_s a_s\lambda_s\psi_s \\ H\Psi &= \lambda'\Psi = \sum_s \lambda' a_s\psi_s \end{aligned}\right\} \quad \rightarrow \quad \sum_s (\lambda_s - \lambda')a_s\psi_s = 0.$$

But if the ψ_s are complete, $a_s \neq 0$, so that $\lambda_s = \lambda'$, the eigenvalues are distinct and therefore the ψ_s are distinct.

Reverting to our problem, to show that the set $Y_{lm_l}(\theta\varphi)$ are a normalized set, we have to prove the following theorem:

Theorem:

$$\int_{-1}^{+1}\left[P_l^m(x)\right]^2 dx = \frac{2(l+m)!}{(2l+1)(l-m)!},$$

where

$$P_l^m(x) = (1-x^2)^{m/2}\frac{d^m P_l(x)}{dx^m} = (1-x^2)^{m/2}y_l^m = (1-x^2)^{m/2}\frac{dy_l^{m-1}}{dx},$$

so that

$$\left[P_l^m(x)\right]^2 = (1-x^2)^m\left\{\frac{dy_l^{m-1}}{dx}\right\}^2.$$

Proof:

$$\int_{-1}^{+1}\left[P_l^m(x)\right]^2 dx = \int_{-1}^{+1}(1-x^2)^m\left\{\frac{dy_l^{m-1}}{dx}\right\}^2 dx.$$

By partial integration,

$$= \left[(1-x^2)^m\frac{dy_l^{m-1}}{dx}y_l^{m-1}\right]_{-1}^{+1} - \int_{-1}^{+1}y_l^{m-1}(-2mx)(1-x^2)^{m-1}\frac{dy_l^{m-1}}{dx}dx$$

$$-\int_{-1}^{+1}y_l^{m-1}(1-x^2)^m\frac{d^2y_l^{m-1}}{dx^2}dx$$

$$= -\int_{-1}^{+1}y_l^{m-1}(1-x^2)^{m-1}\left[(-2mx)\frac{dy_l^{m-1}}{dx} + (1-x^2)^m\frac{d^2y_l^{m-1}}{dx^2}\right]dx.$$

Setting $m = m-1$ in equation (1) yields

$$(1-x^2)\frac{d^2y_l^{m-1}}{dx^2} - 2xm\frac{dy_l^{m-1}}{dx} = -(l+m)(l-m+1)Y_l^{m-1}.$$

Substituting this result into the above equation:

$$\int_{-1}^{+1}\left[P_l^m(x)\right]^2 dx = \int_{-1}^{+1} y_l^{m-1}(1-x^2)^{m-1}\left[(l+m)(l-m+1)Y_l^{m-1}\right]dx$$

$$= (l+m)(l-m+1)\int_{-1}^{+1}\left(y_l^{m-1}\right)^2(1-x^2)^{m-1}dx$$

$$= (l+m)(l-m+1)\int_{-1}^{+1}\left(P_l^{m-1}(x)\right)^2 dx$$

$$= (l+m)(l-m+1)\cdot(l+m-1)(l-m+2)\int_{-1}^{+1}\left(P_l^{m-2}(x)\right)^2 dx\,.$$

Hence,

$$\int_{-1}^{+1}\left[P_l^m(x)\right]^2 dx = (l+m)(l+m-1)\cdots(l+1)(l-m+1)(l-m+2)\cdots$$

$$\cdots l\int_{-1}^{+1}\left(P_l^0(x)\right)^2 dx$$

$$= (l+m)(l+m-1)\cdots(l+1)l(l-m+1)(l-m+2)\cdots$$

$$\cdots(l-1)\int_{-1}^{+1}\left(P_l^0(x)\right)^2 dx\,.$$

Substituting

$$\int_{-1}^{+1}\left(P_l^0(x)\right)^2 dx = \frac{2}{2l+1}\,,\ \text{and}$$

$$\frac{(l+m)!}{(l-m)!} = \frac{(l+m)(l+m-1)\cdots l(l-1)\cdots(l-m+1)(l-m)!}{(l-m)!}\,,$$

we have the result:

$$\int_{-1}^{+1}\left(P_l^m(x)\right)^2 dx = \frac{2}{2l+1}\frac{(l+m)!}{(l-m)!}\,.$$

In summary, we have shown the following:

(1) $Y_{lm_l}(\theta\varphi)$ are a complete set of functions on the surface of a sphere.

(2) $Y_{lm_l}(\theta\varphi)$ form an orthogonal set for $l \neq l'$, $m_l \neq m'_{l'}$.

(3) $Y_{lm_l}(\theta\varphi)$ form a normalized set.

Hence, the $Y_{lm_l}(\theta\varphi)$ form a complete and orthonormal set, and any function $F(\theta\varphi)$ defined within the given interval can be expressed as a sum of spherical harmonics, i.e.

$$F(\theta\varphi) = \sum_{l=0}^{\infty}\sum_{m_l=-l}^{+l} A_{lm_l} Y_{lm_l}(\theta\varphi),$$

where

$$A_{lm_l} = \int_0^{2\pi} d\varphi \int_0^{\pi} Y^*_{lm_l}(\theta\varphi) F(\theta\varphi) \sin\theta d\theta .$$

This result will prove useful in the study of electromagnetic, electron and proton waves. The procedure is to analyze the wave into spherical harmonics. Those waves for which $l = 0$ are called s waves, those for which $l = 1$ are called p waves, those for which $l = 2$ are called d waves, etc. A_{lm_l} represents the amplitude for these particular waves.

Since the $Y_{lm_l}(\theta\varphi)$ are complete and are eigenfunctions of a Hermitian operator (the spherical part of ∇^2), the $Y_{lm_l}(\theta\varphi)$ are unique, i.e. they are the <u>only</u> functions satisfying $H\psi_s = \lambda_s\psi_s$.

Summary of Polynomials and Related Properties

Legendre Polynomials: $P_l(x) = \dfrac{1}{2^l l!}\dfrac{d^l}{dx^l}(x^2-1)^l.$

The $P_l(x)$ are the eigensolutions of

$$\frac{d}{dx}\left[(1-x^2)\frac{dP_l(x)}{dx}\right] = -l(l+1)P_l(x).$$

$$\frac{1}{\sin\theta}\frac{\partial}{\partial\theta}\left[\sin\theta\frac{\partial P_l(\cos\theta)}{\partial\theta}\right] = -l(l+1)P_l(\cos\theta).$$

Associated Legendre Polynomials: $P_l^m(x) = (1-x^2)^{m/2}\dfrac{d^m}{dx^m}P_l^m(x).$

The $P_l^m(x)$ are the eigensolutions of

$$\frac{d}{dx}\left[(1-x^2)\frac{dP_l^m(x)}{dx}\right] + \left[l(l+1) - \frac{m^2}{(1-x^2)}\right]P_l^m(x) = 0.$$

$$\frac{1}{\sin\theta}\frac{\partial}{\partial\theta}\left[\sin\theta\frac{\partial P_l^m(\cos\theta)}{\partial\theta}\right] + \left[l(l+1) - \frac{m^2}{\sin^2\theta}\right]P_l^m(\cos\theta) = 0.$$

Spherical Harmonics: $\mathrm{Y}_{lm}(\theta\varphi) = \left[\dfrac{(2l+1)(l-m)!}{4\pi(l+m)!}\right]^{1/2} P_l^m(\cos\theta)e^{im\varphi}.$

The $\mathrm{Y}_{lm}(\theta\varphi)$ are the eigensolutions of

$$\frac{d}{dx}\left[(1-x^2)\frac{d\mathrm{Y}(\theta\varphi)}{dx}\right] + \left[l(l+1) - \frac{m^2}{(1-x^2)}\right]\mathrm{Y}(\theta\varphi) = 0.$$

$$\frac{1}{\sin\theta}\frac{\partial}{\partial\theta}\left[\sin\theta\frac{\partial \mathrm{Y}(\theta\varphi)}{\partial\theta}\right] + \frac{1}{\sin^2\theta}\frac{\partial^2\mathrm{Y}(\theta\varphi)}{\partial\varphi^2} = -l(l+1)\mathrm{Y}(\theta\varphi).$$

Appendix A

Problems and Solutions

Problem Set A

I. Using tensor notation, prove the following:

(i) $\mathbf{A}\times(\mathbf{B}\times\mathbf{C}) = \mathbf{B}(\mathbf{A}\cdot\mathbf{C}) - \mathbf{C}(\mathbf{A}\cdot\mathbf{B})$.

$$\{\mathbf{A}\times(\mathbf{B}\times\mathbf{C})\}_i = \sum_{jk}\varepsilon_{ijk}(\mathbf{B}\times\mathbf{C})_k A_j = \sum_{jklm}\varepsilon_{ijk}A_j\varepsilon_{klm}B_lC_m$$
$$= \sum_{jklm}\varepsilon_{ijk}\varepsilon_{lmk}A_jB_lC_m \overset{*}{=} \sum_{jlm}(\delta_{il}\delta_{jm} - \delta_{im}\delta_{jl})A_jB_lC_m$$
$$= \sum_j B_iA_jC_j - \sum_l A_lB_lC_i = B_i\sum_j A_jC_j - C_i\sum_l A_lB_l.$$

Hence, $\mathbf{A}\times(\mathbf{B}\times\mathbf{C}) = \mathbf{B}(\mathbf{A}\cdot\mathbf{C}) - \mathbf{C}(\mathbf{A}\cdot\mathbf{B})$.

(ii) $(\mathbf{A}\times\mathbf{B})\cdot(\mathbf{C}\times\mathbf{D}) = (\mathbf{A}\cdot\mathbf{C})(\mathbf{B}\cdot\mathbf{D}) - (\mathbf{A}\cdot\mathbf{D})(\mathbf{B}\cdot\mathbf{C})$.

$$(\mathbf{A}\times\mathbf{B})\cdot(\mathbf{C}\times\mathbf{D}) = \sum_i(\mathbf{A}\times\mathbf{B})_i(\mathbf{C}\times\mathbf{D})_i = \sum_{ijklm}\varepsilon_{ijk}A_jB_k\varepsilon_{ilm}C_lD_m$$
$$= \sum_{jklm}(\delta_{jl}\delta_{km} - \delta_{jm}\delta_{kl})A_jB_kC_lD_m$$
$$= \sum_{jk}(A_jB_kC_jD_k - A_jB_kC_kD_j)$$
$$= \sum_{jk}(A_jC_j)(B_kD_k) - (A_jD_j)(B_kC_k)$$

Hence, $(\mathbf{A}\times\mathbf{B})\cdot(\mathbf{C}\times\mathbf{D}) = (\mathbf{A}\cdot\mathbf{C})(\mathbf{B}\cdot\mathbf{D}) - (\mathbf{A}\cdot\mathbf{D})(\mathbf{B}\cdot\mathbf{C})$.

*This step is proven in Problem Set B.

(iii) $\quad (\mathbf{A}\times\mathbf{B})\times(\mathbf{C}\times\mathbf{D}) = \mathbf{C}[(\mathbf{A}\times\mathbf{B})\cdot\mathbf{D}] - \mathbf{D}[(\mathbf{A}\times\mathbf{B})\cdot\mathbf{C}]$.

$$
\begin{aligned}
[(\mathbf{A}\times\mathbf{B})\times(\mathbf{C}\times\mathbf{D})]_i &= \sum_{jk}\varepsilon_{ijk}\sum_{lm}\varepsilon_{jlm}A_lB_m\sum_{no}\varepsilon_{kno}C_nD_o \\
&= \sum_{jkl}\sum_{mno}\varepsilon_{ijk}\varepsilon_{jlm}\varepsilon_{kno}A_lB_mC_nD_o \\
&= \sum_{jlmno}(\delta_{in}\delta_{jo} - \delta_{io}\delta_{jn})\varepsilon_{jlm}A_lB_mC_nD_o \\
&= \sum_{jlm}\varepsilon_{jlm}\left[A_lB_mC_iD_j - A_lB_mC_jD_i\right] \\
&= C_i\,[\mathbf{D}\cdot(\mathbf{A}\times\mathbf{B})] - D_i\,[\mathbf{C}\cdot(\mathbf{A}\times\mathbf{B})].
\end{aligned}
$$

Hence, $(\mathbf{A}\times\mathbf{B})\times(\mathbf{C}\times\mathbf{D}) = \mathbf{C}[(\mathbf{A}\times\mathbf{B})\cdot\mathbf{D}] - \mathbf{D}[(\mathbf{A}\times\mathbf{B})\cdot\mathbf{C}]$.

II. If $m\ddot{\mathbf{r}} = (e/c)\mathbf{v}\times\mathbf{H}$, show that $\mathrm{v} \equiv |\dot{\mathbf{r}}|$ is constant in time.

Since $\dot{\mathbf{r}} = \mathbf{v} = \upsilon\boldsymbol{t}_o$, where $\boldsymbol{t}_o$ is the unit tangent vector, we have

$$m\ddot{\mathbf{r}} = m(\dot{\upsilon}\boldsymbol{t}_o + \upsilon\dot{\boldsymbol{t}}_o) = (e/c)\upsilon\boldsymbol{t}_o\times\mathbf{H}.$$

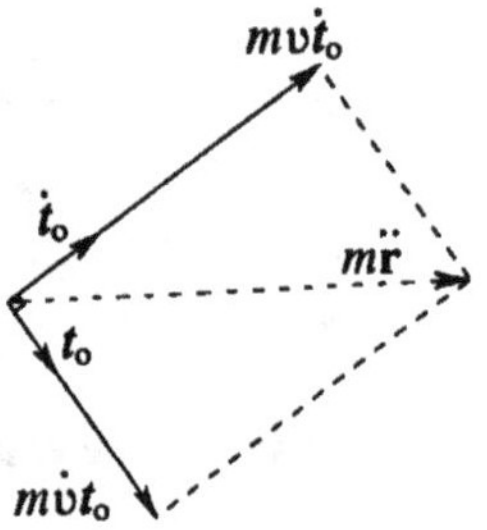

Fig. A-1

Consider $m(\dot{\upsilon}\boldsymbol{t}_o + \upsilon\dot{\boldsymbol{t}}_o)$ and $m(\upsilon\boldsymbol{t}_o\times\mathbf{H})$:

$\boldsymbol{t}_o\times\mathbf{H}$ is a vector perpendicular to $\boldsymbol{t}_o$, $\dot{\boldsymbol{t}}_o$ is a vector perpendicular to $\boldsymbol{t}_o$, and $(\dot{\upsilon}\boldsymbol{t}_o + \upsilon\dot{\boldsymbol{t}}_o)$ determines a plane (Fig. A-1). The only way that $m\ddot{\mathbf{r}} = (e/c)\upsilon\boldsymbol{t}_o\times\mathbf{H}$ can hold and still be perpendicular to $\boldsymbol{t}_o$ is for $\dot{\upsilon} = 0$.

III. If in the above problem $\mathbf{H}$ is constant, so that $\dot{\mathbf{H}} = 0$, show that the motion of a particle can be resolved into that of a uniform velocity plus a constant rotation.

Define a new vector $\mathbf{c} \equiv \mathbf{r}\times\mathbf{H} + \mathbf{d}$, where $\mathbf{d}$ is a constant vector. It then follows that $\dot{\mathbf{c}} = \dot{\mathbf{r}}\times\mathbf{H} + \mathbf{r}\times\dot{\mathbf{H}} + \dot{\mathbf{d}}$. But, since by hypothesis and by definition, $\dot{\mathbf{H}} = 0$ and $\dot{\mathbf{d}} = 0$, $\dot{\mathbf{c}} = \dot{\mathbf{r}}\times\mathbf{H} = \mathbf{v}\times\mathbf{H} = (mc/e)\ddot{\mathbf{r}} \;\rightarrow\; \mathbf{c} = (mc/e)\dot{\mathbf{r}} + \mathbf{E}$, where $\mathbf{E}$ is a constant vector. Hence, $(mc/e)\dot{\mathbf{r}} = \mathbf{r}\times\mathbf{H} + \mathbf{d} - \mathbf{E}$, or

$$\dot{\mathbf{r}} = (e/mc)[\mathbf{r}\times\mathbf{H} + (\mathbf{d} - \mathbf{E})].$$

Setting $\dot{\mathbf{r}}_o = (e/mc)(\mathbf{d} - \mathbf{E})$, we have $\dot{\mathbf{r}} = (e/mc)(\mathbf{r}\times\mathbf{H}) + \dot{\mathbf{r}}_o$. If $\dot{\mathbf{r}}_o$ is parallel to $\mathbf{H}$ and the z-axis is chosen along $\mathbf{H}$, then $\dot{\mathbf{r}}_o = \mathbf{v}_o = (00\upsilon_o)$ and $\mathbf{H} = (00H)$. Thus,

$$\dot{\mathbf{r}} = i_o\frac{eHy}{mc} + j_o\frac{-eHx}{mc} + k_o\upsilon_o.$$

Since $\dot{\mathbf{r}} = i_o\dfrac{dx}{dt} + j_o\dfrac{dy}{dt} + k_o\dfrac{dz}{dt}$ and $\mathbf{r} = i_o x + j_o y + k_o z$, we have

$$\frac{dx}{dt} = \frac{eHy}{mc};\quad \frac{dy}{dt} = -\frac{eHx}{mc}.\quad \rightarrow\quad -\frac{eH}{mc}\frac{dx}{dt} = \frac{d^2y}{dt^2} = -\left(\frac{eH}{mc}\right)^2 y,$$

or,

$$\frac{d^2y}{dt^2} + \left(\frac{eH}{mc}\right)^2 y = 0,$$

with solutions $y = \begin{Bmatrix} A\sin[(eH/mc)t + \varphi_o] \\ A\cos[(eH/mc)t + \varphi_o] \end{Bmatrix}$. Also, $\dfrac{dz}{dt} = \upsilon_o \rightarrow z = \upsilon_o t + z_o$.

We also have that $x = -\dfrac{mc}{eH}\dfrac{dy}{dt} = -\left(\dfrac{mc}{eH}\right)A\left(-\dfrac{eH}{mc}\right)\sin\left(\dfrac{eHt}{mc} + \varphi_o\right)$, resulting in

$$x = A\sin\left(\frac{eHt}{mc} + \varphi_o\right).$$
$$y = A\cos\left(\frac{eHt}{mc} + \varphi_o\right).$$
$$z = \upsilon_o t + z_o.$$

$\mathbf{r}$ then is given by the expression

$$\mathbf{r} = i_o A\sin\left(\frac{eHt}{mc} + \varphi_o\right) + j_o A\cos\left(\frac{eHt}{mc} + \varphi_o\right) + k_o(\upsilon_o t + \varphi_o),$$

which we see is the sum of a constant rotation (first two terms on the right) and a uniform velocity (last term).

IV. The Radius of Curvature R is defined as $R^2 = \left|\dfrac{dt_o}{ds}\right|^{-2}$.
For the problem above, find expressions for R, Torsion $C_2 = \left|\dfrac{db_o}{ds}\right|$ and Total Curvature C, where $C^2 = R^{-2} + C_2^2$.

From $\dot{t}_o = \upsilon \dfrac{dt_o}{ds}$, we have that $\left|\dfrac{dt_o}{ds}\right|^2 = \dfrac{|\dot{t}_o|^2}{\upsilon^2}$. But $t_o = \dfrac{\dot{\mathbf{r}}}{\upsilon} \rightarrow \dot{t}_o = \dfrac{\ddot{\mathbf{r}}}{\upsilon}$ (υ is constant in time), so that $\left|\dfrac{dt_o}{ds}\right|^2 = \dfrac{|\ddot{\mathbf{r}}|^2}{\upsilon^4}$. Hence, $R^2 = \dfrac{\upsilon^4}{|\ddot{\mathbf{r}}|^2}$.

$$\dot{\mathbf{r}} = i_o\left[\frac{eHA}{mc}\cos\left(\frac{eHt}{mc} + \varphi_o\right)\right] + j_o\left[\frac{-eHA}{mc}\sin\left(\frac{eHt}{mc} + \varphi_o\right)\right] + k_o\upsilon_o.$$

$$\ddot{\mathbf{r}} = -\left(\frac{eH}{mc}\right)^2 A\left[i_o \sin\left(\frac{eHt}{mc} + \varphi_o\right) + j_o \cos\left(\frac{eHt}{mc} + \varphi_o\right)\right].$$

$$|\dot{\mathbf{r}}|^2 = \upsilon^2 = A^2\left(\frac{eH}{mc}\right)^2 + \upsilon_o^2.$$

$$|\ddot{\mathbf{r}}|^2 = a^2 = A^2\left(\frac{eH}{mc}\right)^4 = \left(\frac{eH}{mc}\right)^2\left(\upsilon^2 - \upsilon_o^2\right).$$

Thus,

$$R^2 = \left(\frac{mc}{eH}\right)^2 \frac{\upsilon^4}{\left(\upsilon^2 - \upsilon_o^2\right)}.$$

Torsion $C_2 = \left|\dfrac{db_o}{ds}\right|$, where $\dot{b}_o = \upsilon \dfrac{db_o}{ds} \rightarrow \left|\dfrac{db_o}{ds}\right| = \dfrac{|\dot{b}_o|}{\upsilon}$. We also have that $b_o = t_o \times n_o$ and $\dot{b}_o = \dot{t}_o \times n_o + t_o \times \dot{n}_o$. Since $n_o = \dfrac{\dot{t}_o}{|\dot{t}_o|}$, $\dot{n}_o = \dfrac{\ddot{t}_o}{|\dot{t}_o|}$, so that $\dot{b}_o = t_o \times \dot{n}_o$ ($\dot{t}_o \times n_o = 0$).

Setting $\varphi = (eH/mc)t + \varphi_o$, we have

$$t_o = \frac{\dot{\mathbf{r}}}{\upsilon} = \frac{1}{\upsilon}\left[i_o \frac{eH}{mc} A\cos\varphi + j_o \frac{-eH}{mc} A\sin\varphi + k_o\upsilon_o\right].$$

$$\dot{t}_o = n_o|\dot{t}_o| = -\frac{1}{\upsilon}\left[i_o(eH/mc)^2 A\sin\varphi + j_o(eH/mc)^2 A\cos\varphi\right].$$

$$\ddot{t}_o = \dot{n}_o|\dot{t}_o| = \frac{1}{\upsilon}\left[-i_o(eH/mc)^3 A\cos\varphi + j_o(eH/mc)^3 A\sin\varphi\right].$$

$$\dot{b}_o = i_o\left[t_2\dot{n}_3 - t_3\dot{n}_2\right] + j_o\left[t_3\dot{n}_1 - t_1\dot{n}_3\right] + k_o\left[t_1\dot{n}_2 - t_2\dot{n}_1\right],$$

where $\dot{n}_3 = 0$.

$$|\dot{t}_o|\dot{b}_o = i_o\left[-\frac{\upsilon_o}{\upsilon^2}\left(\frac{eH}{mc}\right)^3 A\sin\varphi\right] + j_o\left[-\frac{\upsilon_o}{\upsilon^2}\left(\frac{eH}{mc}\right)^3 A\cos\varphi\right]$$

$$+ k_o\left[\left(\frac{eH}{mc}\right)^4 \frac{A^2}{\upsilon^2}(\cos\varphi\sin\varphi - \cos\varphi\sin\varphi)\right].$$

$$\dot{b}_o = \frac{-\upsilon_o/\upsilon^2}{\frac{A}{\upsilon}\left(\frac{eH}{mc}\right)^3}\left[i_o\left(\frac{eH}{mc}\right)^3 A\sin\varphi + j_o\left(\frac{eH}{mc}\right)^3 A\cos\varphi\right].$$

$$|\dot{b}_o| = \left(\frac{eH}{mc}\right)\frac{\upsilon_o}{\upsilon}; \quad \left|\frac{db_o}{ds}\right| = \frac{|\dot{b}_o|}{\upsilon} = \left(\frac{eH}{mc}\right)\frac{\upsilon_o}{\upsilon^2}.$$

$$C_2^2 = \left|\frac{db_o}{ds}\right|^2 = \left(\frac{eH\upsilon_o}{mc\upsilon^2}\right)^2.$$

Total Curvature

$$C^2 = C_1^2 + C_2^2 = \frac{1}{R^2} + C_2^2 = \left(\frac{eH}{mc}\right)^2\frac{(\upsilon^2-\upsilon_o^2)}{\upsilon^4} + \left(\frac{eH}{mc}\right)^2\frac{\upsilon_o^2}{\upsilon^4}$$

$$= \left(\frac{eH}{mc}\right)^2\frac{1}{\upsilon^4}(\upsilon^2 - \upsilon_o^2 + \upsilon_o^2) = \left(\frac{eH}{mc}\right)^2\frac{1}{\upsilon^2}.$$

$$C^2 = \left(\frac{eH}{mc\upsilon}\right)^2.$$

Note: $C = \left|\frac{dn_o}{ds}\right| = \frac{|\dot{n}_o|}{\upsilon}$. But $|\dot{t}_o||\dot{n}_o| = \frac{1}{\upsilon}\left(\frac{eH}{mc\upsilon}\right)^3 A$.

Hence,

$$|\dot{n}_o| = \frac{A\left(\frac{eH}{mc}\right)^3}{\upsilon\frac{1}{\upsilon}A\left(\frac{eH}{mc}\right)^2} = \frac{eH}{mc} \quad\Rightarrow\quad C = \frac{eH}{mc}.$$

Problem Set B

I. **If λ_1 λ_2 λ_3 are the eigenvalues of a symmetrical tensor of 2nd rank, T_{ij}, prove the following:**

(a) $\sum_i T_{ii} = \sum_i \lambda_i$.

(b) $\sum_{ij} T_{ij}T_{ij} = \sum_i \lambda_i^2$.

(c) $\sum_{ijk} T_{ij}T_{jk}T_{ki} = \sum_i \lambda_i^3$.

(c) $\det|T_{ij}| = \lambda_1\lambda_2\lambda_3$.

(a) If λ_1 λ_2 λ_3 are the eigenvalues of a symmetric 2nd rank tensor, T_{ij}, then in a particular reference frame,

$$T_{ij} \rightarrow T_{ij}' = \begin{pmatrix} \lambda_1 & 0 & 0 \\ 0 & \lambda_2 & 0 \\ 0 & 0 & \lambda_3 \end{pmatrix}.$$

$$T_{ij} = \sum_{kl} u_{ki}u_{lj}T_{kl}',$$

so that

$$T_{ii} = \sum_{ikl} u_{ki}u_{li}T_{kl}' = \sum_{kl} \delta_{kl}T_{kl}' = \sum_k T_{kk}' = \sum_k \lambda_k.$$

Actually, $\sum_i T_{ii}$ is a contraction of a 2nd rank tensor, hence a scalar, and therefore a scalar in all coordinate systems and in particular in the system for which $\sum_k T_{kk} = \sum_k \lambda_k$.

(b) $$\sum_{ij} T_{ij}T_{ij} = \sum_{ij}\sum_{klmn} u_{ki}u_{lj}u_{mi}u_{nj}T_{kl}'T_{mn}' = \sum_{klmn} \delta_{km}\delta_{ln}T_{kl}'T_{mn}'$$

$$= \sum_{klmn} \delta_{km}\delta_{ln}\delta_{kl}\delta_{mn}\lambda_k\lambda_m = \sum_{lmn} \delta_{ml}\delta_{ln}\delta_{mn}\lambda_m^2 = \sum_{mn} \delta_{mn}^2\lambda_m^2 = \sum_m \lambda_m^2.$$

Hence, $\sum_{ij} T_{ij}T_{ij} = \sum_m \lambda_m^2$.

We have here performed a double contraction on a 4th-rank tensor, yielding a scalar, hence a scalar in the primed system, in which $\sum_{ij} T'_{ij} T'_{ij} = \sum_{ij} \lambda_i \delta_{ij} = \sum_i \lambda_i^2$.

(c) Following the procedure above, $\sum_{ijk} T_{ij}T_{jk}T_{ki}$ is essentially a triple contraction of a 6th-rank tensor, yielding a scalar. Hence,

$$\sum_{ijk} T_{ij}T_{jk}T_{ki} = \sum_{ijk} T'_{ij}\,T'_{jk}\,T'_{ki} = \sum_{ijk} \lambda_i\delta_{ij}\lambda_j\delta_{jk}\lambda_k\delta_{ki}$$
$$= \sum_{jk} \lambda_j^2\lambda_k\delta_{jk}\lambda_k\delta_{kj} = \sum_k \lambda_k^3.$$

(d) Consider

$$\sum_{lmn} T_{il}T_{jm}T_{kn}\varepsilon_{lmn} = \begin{vmatrix} T_{i1} & T_{i2} & T_{i3} \\ T_{j1} & T_{j2} & T_{j3} \\ T_{k1} & T_{k2} & T_{k3} \end{vmatrix},$$

and

$$\sum_{ijk}\left[\sum_{lmn} T_{il}T_{jm}T_{kn}\varepsilon_{lmn}\right]\varepsilon_{ijk} = \sum_{ijk} \begin{vmatrix} T_{i1} & T_{i2} & T_{i3} \\ T_{j1} & T_{j2} & T_{j3} \\ T_{k1} & T_{k2} & T_{k3} \end{vmatrix} \varepsilon_{ijk} = 6\det|T_{ij}|.$$

Note that for any interchange of jk, the det changes sign, as does ε_{ijk}; hence the product $|\ |\cdot\varepsilon$ retains its sign and value. Therefore the sum $\Sigma|\ |\cdot\varepsilon$ yields 6 terms, each of which is $\det|T_{ij}|$. Thus, $\det|T_{ij}| = \frac{1}{6}\sum_{ijklmn} T_{il}T_{jm}T_{kn}$, a 0th-rank tensor, i.e. a scalar, in every coordinate system. Hence, by the invariant property, $\det|T_{ij}| = \det|T'_{ij}| = \lambda_1\,\lambda_2\,\lambda_3$.

II. Prove the following:

(a) $$\varepsilon_{ijk}\varepsilon_{lmn} = \delta_{il}\delta_{jm}\delta_{kn} + \delta_{im}\delta_{jn}\delta_{kl} + \delta_{in}\delta_{jl}\delta_{km} - \delta_{im}\delta_{jl}\delta_{kn} - \delta_{il}\delta_{jn}\delta_{km} - \delta_{in}\delta_{jm}\delta_{kl}.$$

(b) $$\sum_k \varepsilon_{ijk}\varepsilon_{lmk} = \delta_{il}\delta_{jm} - \delta_{im}\delta_{jl}.$$

(c) $\sum_{jk} \varepsilon_{ijk}\varepsilon_{ljk} = 2\delta_{il}$.

(d) $\sum_{ijk} \varepsilon_{ijk}\varepsilon_{ijk} = 6$.

(a) Consider

$$\varepsilon_{ijk}\varepsilon_{lmn} = \delta_{il}\delta_{jm}\delta_{kn} + \delta_{im}\delta_{jn}\delta_{kl} + \delta_{in}\delta_{jl}\delta_{km} - \delta_{im}\delta_{jl}\delta_{kn} - \delta_{il}\delta_{jn}\delta_{km} - \delta_{in}\delta_{jm}\delta_{kl}$$

$$= \begin{vmatrix} \delta_{il} & \delta_{im} & \delta_{in} \\ \delta_{jl} & \delta_{jm} & \delta_{jn} \\ \delta_{kl} & \delta_{km} & \delta_{kn} \end{vmatrix}.$$

1. For lmn arranged in the order ijk, the product is + 1 (and both sides = + 1).
2. For lmn arranged in such an order that an even permutation is required to bring it into the order ijk, the product is + 1 (and both sides = + 1).
3. For lmn arranged in such an order that an odd permutation is required to bring it into the order ijk, the product is − 1 (and both sides = − 1).

The 6 possible combinations of $\varepsilon_{ijk}\varepsilon_{lmn}$ corresponding to the possible arrangements of lmn are:

[ijk, ijk]	[ijk, jki]	[ijk, kij]	[ijk, ikj]	[ijk, kji]	[ijk, jik]
↓	↓	↓	↓	↓	↓
+1	+1	+1	−1	−1	−1
$i = l$	$i = n$	$i = m$	$i = l$	$i = n$	$i = m$
$j = m$	$j = l$	$j = n$	$j = n$	$j = m$	$j = l$
$k = n$	$k = m$	$k = l$	$k = m$	$k = l$	$k = n$
$(\delta_{il}\delta_{jm}\delta_{kn})$	$(\delta_{in}\delta_{jl}\delta_{km})$	$(\delta_{im}\delta_{jn}\delta_{kl})$	$-(\delta_{il}\delta_{jn}\delta_{km})$	$-(\delta_{in}\delta_{jm}\delta_{kl})$	$-(\delta_{in}\delta_{jl}\delta_{kn})$

Noting the determinant above, we see that no two values lmn or ijk may be the same. Hence, only one of the 6 possibilities above can exist for given values of ijk lmn. Hence the sum of the 6 terms yields the desired expression.

(b) Consider $\sum_k \varepsilon_{ijk}\varepsilon_{lmk} = \delta_{il}\delta_{jm} - \delta_{im}\delta_{jl}$.

Setting $n = k$ and summing over k in (a) above,

$$\sum_k \varepsilon_{ijk}\varepsilon_{lmk} = \sum_k \delta_{il}\delta_{jm}\delta_{kk} + \sum_k \delta_{im}\delta_{jk}\delta_{kl} + \sum_k \delta_{ik}\delta_{jl}\delta_{km}$$
$$- \sum_k \delta_{im}\delta_{jl}\delta_{kk} - \sum_k \delta_{il}\delta_{jk}\delta_{km} - \sum_k \delta_{ik}\delta_{jm}\delta_{kl}$$
$$= 3\delta_{il}\delta_{jm} + \delta_{im}\delta_{jl} + \delta_{im}\delta_{jl} - 3\delta_{im}\delta_{jl} - \delta_{il}\delta_{jm} - \delta_{il}\delta_{jm}$$
$$= \delta_{il}\delta_{jm} - \delta_{im}\delta_{jl}.$$

Note that for any i and j we have 2 possible terms,

$\varepsilon_{ijk}\varepsilon_{ijk}$	$\varepsilon_{ijk}\varepsilon_{jik}$
⇓	⇓
$+1$	-1
$i = l$	$i = m$
$j = m$	$j = l$
$(\delta_{il}\delta_{jm})$	$-(\delta_{im}\delta_{jl})$

.

(c) Contracting expression (b) above with respect to j, i.e. setting $m = j$ and summing over j, we have:

$$\sum_{jk} \varepsilon_{ijk}\varepsilon_{ljk} = \sum_j \delta_{il}\delta_{jj} - \sum_j \delta_{ij}\delta_{jl} = 3\delta_{il} - \delta_{il} = 2\delta_{il}.$$

Note that for $i \neq l$, both sides vanish, while for $i = l$, we get 2 terms, $(\varepsilon_{ijk})^2 + (\varepsilon_{ikj})^2 = 2$.

(d) Contracting expression (c) above with respect to i, i.e setting $l = i$ and summing over i, we have

$$\sum_{ijk} \varepsilon_{ijk}\varepsilon_{ijk} = 2\sum_i \delta_{ii} = 2\cdot 3 = 6.$$

Note that we have 6 terms to sum over:

$$\sum (\varepsilon_{123})^2 + (\varepsilon_{132})^2 + (\varepsilon_{213})^2 + (\varepsilon_{231})^2 + (\varepsilon_{312})^2 + (\varepsilon_{321})^2 = 6.$$

III. Prove the following:

(a) $\left[\mathbf{A}\times(\mathbf{B}\times\mathbf{C})\right]_i = \sum_{jklm} \varepsilon_{ijk}\varepsilon_{klm} A_j B_l C_m$.

(b) $\mathbf{A}\times(\mathbf{B}\times\mathbf{C}) = \mathbf{B}(\mathbf{A}\cdot\mathbf{C}) - \mathbf{C}(\mathbf{A}\cdot\mathbf{B})$.

(a) $$\left[\mathbf{A}\times(\mathbf{B}\times\mathbf{C})\right]_i = \sum_{jk}\varepsilon_{ijk}A_j(\mathbf{B}\times\mathbf{C})_k = \sum_{jk}\varepsilon_{ijk}A_j\sum_{lm}\varepsilon_{klm}B_lC_m$$
$$= \sum_{jklm}\varepsilon_{ijk}\varepsilon_{klm}A_jB_lC_m\,.$$

(b) From (a) above, $$\left[\mathbf{A}\times(\mathbf{B}\times\mathbf{C})\right]_i = \sum_{jklm}\varepsilon_{ijk}\varepsilon_{klm}A_jB_lC_m$$
$$= \sum_{jklm}\varepsilon_{ijk}\varepsilon_{lmk}A_jB_lC_m\,.$$

But since $\sum_k \varepsilon_{ijk}\varepsilon_{lmk} = \delta_{il}\delta_{jm} - \delta_{im}\delta_{jl}$,

$$\left[\mathbf{A}\times(\mathbf{B}\times\mathbf{C})\right]_i = \sum_{jlm}(\delta_{il}\delta_{jm} - \delta_{im}\delta_{jl})A_jB_lC_m$$
$$= \sum_{lm}(\delta_{il}A_mC_mB_l - \delta_{im}A_lB_lC_m) = \sum_m(A_mC_mB_i - A_mB_mC_i)$$
$$= B_i(\mathbf{A}\cdot\mathbf{C}) - C_i(\mathbf{A}\cdot\mathbf{B})\,.$$

IV.

(a) Find the moment of inertia tensor for a cube of side a and total mass m in any cartesian coordinate system with the center of the cube as its origin. The density of the cube is assumed to be uniform.
[*Hint:* Find I_{ij} for the simplest coordinate system.]

(b) Use the above tensor to find the moment of inertia about any axis which passes through the center of the cube.

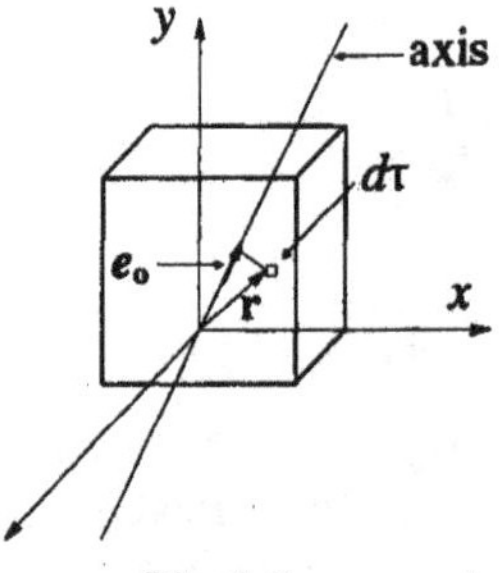

Fig. A-2

(a) From Fig. A-2, $I_{ij} = \int \rho d\tau\,[r^2\delta_{ij} - r_ir_j]$,
where $\rho = m/a^3$ and $r^2 = x^2 + y^2 + z^2$.

$$I_{11} = \frac{m}{a^3}\iiint(y^2 + z^2)dxdydz$$
$$= \frac{m}{a^3}\iint\left[(y^2+z^2)\right]_{-a/2}^{+a/2} xdydz = \frac{m}{a^2}\iint(y^2+z^2)dydz$$
$$= \frac{m}{a^2}\int\left[(y^3 + z^2y)\right]_{-a/2}^{+a/2}dz = \frac{m}{a}\left[\frac{a^2z}{12} + \frac{z^3}{3}\right]_{-a/2}^{+a/2} = \frac{m}{a}\left[\frac{a^3}{12} + \frac{a^3}{12}\right] = \frac{ma^2}{6}.$$

$$I_{12} = -\frac{m}{a^3}\iiint xydxdydz = -\frac{m}{a^3}\iint\left[\frac{x^2y}{2}\right]_{-a/2}^{+a/2}dydz = 0.$$

$$I_{13} = -\frac{m}{a^3}\iiint xzdxdydz = 0.$$

$$I_{12} = I_{21} = I_{23} = I_{32} = I_{13} = I_{31} = 0.$$

By symmetry, $I_{22} = \frac{m}{a^3}\iiint(x^2 + z^2)dxdydz = \frac{ma^2}{6} = I_{33}$, i.e. the system shows no preference for x, y or z axes. Hence, $I_{ij} = \frac{ma^2}{6}\delta_{ij}$.

(b) To find the moment of inertia about an arbitrary axis through the origin, we have $I = \sum_{ij} e_ie_jI_{ij}$, where e_ie_j are the direction cosines of a unit vector e_o lying on the axis. Thus, $I = \sum_{ij} e_ie_j\frac{ma^2}{6}\delta_{ij} = \frac{ma^2}{6}\sum_i e_i^2 = \frac{ma^2}{6}$.

V.

(a) Prove that if A_i is an invariant tensor of the 1st rank, then $A_i = 0$.

(b) Prove that if S_{ij} is an invariant anti-symmetric tensor of the 2nd rank, then $S_{ij} = 0$.

(c) Prove that any invariant symmetrical tensor of the 2nd rank must be of the form $c\delta_{ij}$.

(d) Prove that the tensor $c\delta_{ij}$ is also the only form for any invariant tensor of the 2nd rank.

(a) by geometrical argument

If A_i is a 1st rank invariant tensor, it is a vector whose components do not change under a rotation. The only vector whose components do not change under a rotation is the null vector $A_i = 0$.

(a) by analytic argument

The invariant property is expressed as $A_i = A_i'$, or $A_i' - A_i = 0$.
But since $A_i' = \sum_j u_{ij}A_j$ and $A_i = \sum_j \delta_{ij}A_j$, it follows that $\sum_j (u_{ij} - \delta_{ij})A_j = 0$.

Thus we have a series of 3 homogeneous equations in the 3 unknowns A_j:

$$(u_{11}-1)A_1 + u_{12}A_2 + u_{13}A_3 = 0.$$
$$u_{21}A_1 + (u_{22}-1)A_2 + u_{23}A_3 = 0.$$
$$u_{31}A_1 + u_{32}A_2 + (u_{33}-1)A_3 = 0.$$

For a solution of these homogeneous equations such that $A_j \neq 0$, the $\det|\upsilon_{ij}| = 0$, where $\upsilon_{ij} = (u_{ij} - \delta_{ij})$. But we can always choose the rotation such that $(u_{ij} - \delta_{ij}) \neq 0$. Hence, the only solution of these equations is $A_j = 0$.

(b) <u>by geometrical argument</u>

If S_{ij} is an anti-symmetrical invariant tensor of the 2nd rank, it can be represented by a vector, similarly invariant, where

$$A_k = \sum_{ij} \varepsilon_{kij} S_{ij} \qquad \mathbf{A}:(2S_{23}, 2S_{13}, 2S_{12}).$$

The only invariant vector is the one whose components vanish, as in (a) above. Hence, $S_{ij} = 0$ for $i \neq j$. However, by the anti-symmetry property, $S_{ii} = 0$. Thus, $S_{ij} = 0$ for all i, j.

(b) <u>by analytic argument</u>

By the invariant property, $S_{ij} = S_{ij}' \;\rightarrow\; S_{ij}' - S_{ij} = 0$. But we also have that $S_{ij}' = \sum_{kl} u_{ik}u_{jl}S_{kl}$ and $S_{ij} = \sum_{kl} \delta_{ik}\delta_{jl}S_{kl}$. Thus, $\sum_{kl}(u_{ik}u_{jl} - \delta_{ik}\delta_{jl})S_{kl} = 0$, yielding 9 equations in the 9 unknowns S_{kl}; i and j are free to be separately 1, 2 and 3. For example, if $i = 1$ and $j = 2$, one of the 9 terms would look like the following:

$$u_{11}u_{21}S_{11} + u_{12}u_{21}S_{21} + u_{13}u_{21}S_{31} + (u_{11}u_{22} - 1)S_{12} + u_{12}u_{22}S_{22}$$
$$+ u_{13}u_{22}S_{32} + u_{11}u_{23}S_{13} + u_{12}u_{23}S_{23} + u_{13}u_{23}S_{33} = 0;$$

and we get 8 more terms like this.

The condition that any number n of homogeneous equations in n unknowns yield a non-zero solution for these equations is that the det $|C_{ijkl}| = 0$, where $C_{ijkl} = u_{ik}u_{jl} - \delta_{ik}\delta_{jl}$.

But, since we have performed arbitrary rotations, we can always find a rotation such that $\det|C_{ijkl}| \neq 0$. Therefore, the only solution of the homogeneous equations is for $S_{kl} = 0$.

(c)(d) by geometrical argument

Any invariant symmetrical tensor of the 2nd rank can be represented by a quadratic surface $\sum_{ij} T_{ij} x_i x_j = \pm 1$. The only invariant surface is a sphere, $x^2 + y^2 + z^2 = \text{const.}$, which is represented in ternsor notation by the surface $\sum_{ij} c\delta_{ij} x_i x_j = \pm 1$. Hence, the invariant symmetrical tensor must be of the form $c\delta_{ij}$, where c is a constant.

(c)(d) by analytic argument

We already know that any 2nd rank symmetrical tensor can be diagonalized. Hence, $T_{ij}' = \lambda^i \delta_{ij}$ can be formed by a rotation from T_{ij}''. The invaraent property then states that $T_{ij}'' - T_{ij}' = 0$. We would like to show that $\lambda^i = c = \lambda^j$.

$T_{ij}' = \sum_{kl} u_{ik} u_{jl} T_{kl}'$, and therefore, since $T_{ij}'' = \lambda^i \delta_{ij}$ or $T_{kl}' = \lambda^k \delta_{kl}$, we have

$$\lambda^i \delta_{ij} = \sum_{kl} u_{ik} u_{jl} \lambda^k \delta_{kl}$$

$$\Downarrow$$

$$\sum_l u_{il} u_{jl} \lambda^i = \sum_{kl} u_{ik} u_{jl} \lambda^k \delta_{kl} = \sum_l u_{il} u_{jl} \lambda^l \rightarrow \sum_l u_{il} u_{jl} (\lambda^l - \lambda^i) = 0.$$

As before, since the transformation U is arbitrary, this implies that $\lambda^i = \lambda^l$.

Problem Set C

I. If φ and ψ are both scalar functions, show that

(a) $\nabla(\varphi + \psi) = \nabla\varphi + \nabla\psi$.

(b) $\nabla(\varphi\psi) = \varphi\nabla\psi + \psi\nabla\varphi$.

(a) $[\nabla(\varphi + \psi)]_i = \dfrac{\partial}{\partial x_i}(\varphi + \psi) = \dfrac{\partial\varphi}{\partial x_i} + \dfrac{\partial\psi}{\partial x_i} \;\rightarrow\; \nabla(\varphi + \psi) = \nabla\varphi + \nabla\psi$.

(b) $[\nabla(\varphi\psi)]_i = \dfrac{\partial(\varphi\psi)}{\partial x_i} = \psi\dfrac{\partial\varphi}{\partial x_i} + \varphi\dfrac{\partial\psi}{\partial x_i} \;\rightarrow\; \nabla(\varphi\psi) = \varphi\nabla\psi + \psi\nabla\varphi$.

II. If φ is a scalar function and **A** and **B** are both vector functions, show that

(a) $\nabla\cdot(\mathbf{A} + \mathbf{B}) = \nabla\cdot\mathbf{A} + \nabla\cdot\mathbf{B}$.

(b) $\nabla\times(\mathbf{A} + \mathbf{B}) = \nabla\times\mathbf{A} + \nabla\times\mathbf{B}$.

(c) $\nabla\cdot(\varphi\mathbf{A}) = \mathbf{A}\cdot\nabla\varphi + \varphi\nabla\cdot\mathbf{A}$.

(d) $\nabla\times(\varphi\mathbf{A}) = -\mathbf{A}\times\nabla\varphi + \varphi\nabla\times\mathbf{A}$.

(e) $\nabla(\mathbf{A}\cdot\mathbf{B}) = (\mathbf{A}\cdot\nabla)\mathbf{B} + (\mathbf{B}\cdot\nabla)\mathbf{A} + \mathbf{A}\times(\nabla\times\mathbf{B}) + \mathbf{B}\times(\nabla\times\mathbf{A})$.

(f) $\nabla\cdot(\mathbf{A}\times\mathbf{B}) = \mathbf{B}\cdot(\nabla\times\mathbf{A}) - \mathbf{A}\cdot(\nabla\times\mathbf{B})$.

(g) $\nabla\times(\mathbf{A}\times\mathbf{B}) = \mathbf{A}(\nabla\cdot\mathbf{B}) - \mathbf{B}(\nabla\cdot\mathbf{A}) + (\mathbf{B}\cdot\nabla)\mathbf{A} - (\mathbf{A}\cdot\nabla)\mathbf{B}$.

(a) $$\nabla\cdot(\mathbf{A} + \mathbf{B}) = \sum_i \frac{\partial}{\partial x_i}(\mathbf{A} + \mathbf{B})_i = \sum_i \frac{\partial}{\partial x_i}(A_i + B_i)$$

$$= \sum_i \left(\frac{\partial A_i}{\partial x_i} + \frac{\partial B_i}{\partial x_i}\right) = \nabla\cdot\mathbf{A} + \nabla\cdot\mathbf{B}.$$

(b) $$[\nabla\times(\mathbf{A} + \mathbf{B})]_i = \sum_{jk} \varepsilon_{ijk}\frac{\partial}{\partial x_j}(A_k + B_k) = \sum_{jk} \varepsilon_{ijk}\left(\frac{\partial A_k}{\partial x_j} + \frac{\partial B_k}{\partial x_j}\right)$$

$$= [\nabla\times\mathbf{A}]_i + [\nabla\times\mathbf{B}]_i.$$

(c) $$\nabla\cdot\varphi\mathbf{A} = \sum_i \frac{\partial}{\partial x_i}(\varphi\mathbf{A})_i = \sum_i \left(A_i \frac{\partial\varphi}{\partial x_i} + \varphi\frac{\partial A_i}{\partial x_i}\right) = \mathbf{A}\cdot\nabla\varphi + \varphi\nabla\cdot\mathbf{A}.$$

(d) $$\nabla\times(\varphi\mathbf{A})_i = \sum_{jk}\varepsilon_{ijk}\frac{\partial}{\partial x_j}(\varphi A_k) = \sum_{jk}\varepsilon_{ijk}\left(\varphi\frac{\partial A_k}{\partial x_j} + A_k\frac{\partial\varphi}{\partial x_j}\right)$$

$$= \varphi(\nabla\times\mathbf{A})_i + (\nabla\varphi\times\mathbf{A})_i.$$

(e) $$[\nabla(\mathbf{A}\cdot\mathbf{B})]_i = \frac{\partial}{\partial x_i}\sum_j A_jB_j = \sum_j \frac{\partial A_j}{\partial x_i}B_j + \sum_j \frac{\partial B_j}{\partial x_i}A_j$$

$$= \sum_j\left[B_j\frac{\partial A_j}{\partial x_i} + B_j\frac{\partial A_i}{\partial x_j} - B_j\frac{\partial A_i}{\partial x_j} + A_j\frac{\partial B_j}{\partial x_i} + A_j\frac{\partial B_i}{\partial x_j} - A_j\frac{\partial B_i}{\partial x_j}\right]$$

$$= \sum_j\left[B_j\left(\frac{\partial A_j}{\partial x_i} - \frac{\partial A_i}{\partial x_j}\right) + A_j\left(\frac{\partial B_j}{\partial x_i} - \frac{\partial B_i}{\partial x_j}\right) + B_j\frac{\partial A_i}{\partial x_j} + A_j\frac{\partial B_i}{\partial x_j}\right]$$

$$= \sum_{jpq}\left[(\delta_{pj}\delta_{qi} - \delta_{pi}\delta_{qj})\left(B_j\frac{\partial A_p}{\partial x_q} + A_j\frac{\partial B_p}{\partial x_q}\right)\right] + (\mathbf{B}\cdot\nabla)A_i + (\mathbf{A}\cdot\nabla)B_i$$

$$= \sum_{jpqm}\varepsilon_{pqm}\varepsilon_{jim}\left(B_j\frac{\partial A_p}{\partial x_q} + A_j\frac{\partial B_p}{\partial x_q}\right) + (\mathbf{B}\cdot\nabla)A_i + (\mathbf{A}\cdot\nabla)B_i$$

$$= \sum_{jpqm}\left[\varepsilon_{mpq}\frac{\partial A_p}{\partial x_q}\varepsilon_{mji}B_j + \varepsilon_{mpq}\frac{\partial B_p}{\partial x_q}\varepsilon_{mji}A_j\right] + (\mathbf{B}\cdot\nabla)A_i + (\mathbf{A}\cdot\nabla)B_i$$

$$= \sum_{jm}\varepsilon_{ijm}B_j\sum_{pq}\varepsilon_{mqp}\frac{\partial A_p}{\partial x_q} + \sum_{jm}\varepsilon_{ijm}A_j\sum_{pq}\varepsilon_{mqp}\frac{\partial B_p}{\partial x_q}$$

$$+ (\mathbf{B}\cdot\nabla)A_i + (\mathbf{A}\cdot\nabla)B_i$$

$$= [\mathbf{B}\times(\nabla\times\mathbf{A})]_i + [\mathbf{A}\times(\nabla\times\mathbf{B})]_i + (\mathbf{B}\cdot\nabla)A_i + (\mathbf{A}\cdot\nabla)B_i$$

(f) $$\nabla\cdot(\mathbf{A}\times\mathbf{B}) = \sum_i\frac{\partial}{\partial x_i}(\mathbf{A}\times\mathbf{B})_i = \sum_{ijk}\frac{\partial}{\partial x_i}(\varepsilon_{ijk}A_jB_k) = \sum_{ijk}\varepsilon_{ijk}\frac{\partial}{\partial x_i}(A_jB_k)$$

$$= \sum_{ijk}\varepsilon_{ijk}\frac{\partial A_j}{\partial x_i}B_k + \sum_{ijk}\varepsilon_{ijk}\frac{\partial B_k}{\partial x_i}A_j.$$

Interchanging subscripts,

$$\begin{aligned}\nabla\cdot(\mathbf{A}\times\mathbf{B}) &= \sum_{ijk}\varepsilon_{kij}\frac{\partial A_j}{\partial x_i}B_k - \sum_{ijk}\varepsilon_{jik}\frac{\partial B_k}{\partial x_i}A_j\\ &= \sum_k(\nabla\times\mathbf{A})_k B_k - \sum_j(\nabla\times\mathbf{B})_j A_j = \mathbf{B}\cdot(\nabla\times\mathbf{A}) - \mathbf{A}\cdot(\nabla\times\mathbf{B}).\end{aligned}$$

(g)
$$\begin{aligned}[\nabla\times(\mathbf{A}\times\mathbf{B})]_i &= \sum_{jk}\varepsilon_{ijk}\frac{\partial}{\partial x_j}(\mathbf{A}\times\mathbf{B})_k = \sum_{jklm}\varepsilon_{ijk}\frac{\partial}{\partial x_j}\varepsilon_{klm}A_lB_m\\ &= \sum_{jklm}\varepsilon_{ijk}\varepsilon_{lmk}\frac{\partial}{\partial x_j}A_lB_m = \sum_{jlm}(\delta_{il}\delta_{jm} - \delta_{im}\delta_{jl})\frac{\partial}{\partial x_j}A_lB_m\\ &= \sum_{lm}\left[\delta_{il}\frac{\partial(A_lB_m)}{\partial x_m} - \delta_{im}\frac{\partial(A_lB_m)}{\partial x_l}\right]\\ &= \sum_m\frac{\partial(A_iB_m)}{\partial x_m} - \sum_l\frac{\partial(A_lB_i)}{\partial x_l}\\ &= (\nabla\cdot\mathbf{B})A_i + (\mathbf{B}\cdot\nabla)A_i - (\nabla\cdot\mathbf{A})B_i - (\mathbf{A}\cdot\nabla)B_i\,.\end{aligned}$$

III. If φ is the electrostatic potential and $\mathbf{E}$ the electric field ($\mathbf{E} = -\nabla\varphi$), the electric force $\mathbf{F}$ acting on any volume V bounded by a closed surface S is $\mathbf{F} = \int_V \rho\mathbf{E}d\tau$, where ρ is the electric charge density inside V ($\nabla^2\varphi = -4\pi\rho$). Show that $\mathbf{F}$ can also be written as $F_i = \int_S\sum_j T_{ij}dS_j$, where T_{ij}, the stress tensor, is given by

$$T_{ij} = \frac{1}{4\pi}\left[E_iE_j - \frac{E^2}{2}\delta_{ij}\right].$$

Solution:

$$\begin{aligned}\int_V\rho E_i d\tau &= -\frac{1}{4\pi}\int_V\nabla^2\varphi E_i d\tau = -\frac{1}{4\pi}\int_V\nabla\cdot(\nabla\varphi)E_i d\tau = \frac{1}{4\pi}\int_V(\nabla\cdot\mathbf{E})E_i d\tau\\ &= \frac{1}{4\pi}\int\left(\sum_j\frac{\partial E_j}{\partial x_j}\right)E_i d\tau = \frac{1}{4\pi}\int\left(\sum_j\frac{\partial(E_iE_j)}{\partial x_j} - E_j\frac{\partial E_i}{\partial x_j}\right)d\tau\,.\end{aligned}$$

Consider now $\sum_j E_j \frac{\partial E_i}{\partial x_j} = (\mathbf{E}\cdot\nabla)E_i$:

We have shown previously that

$$\nabla(\mathbf{A}\cdot\mathbf{B}) = (\mathbf{A}\cdot\nabla)\mathbf{B} + (\mathbf{B}\cdot\nabla)\mathbf{A} + \mathbf{A}\times(\nabla\times\mathbf{B}) + \mathbf{B}\times(\nabla\times\mathbf{A}).$$

Setting $\mathbf{B} = \mathbf{A}$, $\nabla\mathbf{A}^2 = 2(\mathbf{A}\cdot\nabla)\mathbf{A} + 2\mathbf{A}\times(\nabla\times\mathbf{A})$, while for $\mathbf{A} = \mathbf{E}$, using $\nabla\times\mathbf{E} = -\nabla\times(\nabla\varphi) = 0$, we have

$$\nabla\mathbf{E}^2 = 2(\mathbf{E}\cdot\nabla)\mathbf{E} \rightarrow (\mathbf{E}\cdot\nabla)E_i = \frac{1}{2}\nabla_i\mathbf{E}^2 = \frac{1}{2}\frac{\partial}{\partial x_i}\sum_j E_j E_j$$
$$= \sum_j \frac{1}{2}\frac{\partial}{\partial x_j}\delta_{ij}E^2$$

Thus, $\sum_j E_j \frac{\partial E_i}{\partial x_j} = \sum_j \frac{1}{2}\frac{\partial}{\partial x_j} E^2\delta_{ij}$, and

$$F_i = \frac{1}{4\pi}\int_V \sum_j \left(\frac{\partial(E_j E_i)}{\partial x_j} - \frac{1}{2}\frac{\partial(E^2\delta_{ij})}{\partial x_j}\right) d\tau$$
$$= \frac{1}{4\pi}\int_V \sum_j \frac{\partial}{\partial x_j}\left[E_j E_i - \frac{1}{2}E^2\delta_{ij}\right) d\tau.$$

Finally, by Gauss' theorem,

$$F_i = \int_V \sum_j \frac{\partial T_{ij}}{\partial x_j} d\tau = \sum_j \int_S T_{ij} dS_j,$$

where $4\pi T_{ij} = E_i E_j - \frac{E^2}{2}\delta_{ij}$.

IV. Verify that at any point, the direction of **E** is along a principal axis of the above stress tensor T_{ij}. Calculate the eigenvalues of this stress tensor and show that it represents a tension $(+E^2/8\pi)$ along the direction of **E** and a pressure $(-E^2/8\pi)$ normal to this direction.

Solution:

Consider $4\pi T_{ij} = E_i E_j - \frac{E^2}{2}\delta_{ij}$ from Problem III above. We can rotate to principal axes, since we know that any symmetrical tensor of the 2nd rank can be so diagonalized.

$$4\pi T_{ij}' = 4\pi \sum_{kl} u_{ik}u_{jl}T_{kl} = \sum_{kl} u_{ik}u_{jl}\left(E_kE_l - \frac{E^2}{2}\delta_{kl}\right)$$
$$= \sum_k u_{ik}E_k \sum_l u_{jl}E_l - \frac{1}{2}\sum_k u_{ik}u_{jk}E^2 ,$$

and

$$4\pi T_{ij}' = E_i'E_j' - \frac{E^2}{2}\delta_{ij},$$

where $E_i'E_j'$ refer to the components of **E** in the new system of coordinates. Since for $i \neq j$, $T_{ij}' = 0$, it follows that $E_i'E_j' = 0$.

For $E_i'E_j' = 0$ the following are possible: $E_1E_2 = 0$; $E_1E_3 = 0$; $E_2E_3 = 0$.

If $E_1 \neq 0$, then $E_2 = E_3 = 0$.

If $E_2 \neq 0$, then $E_1 = E_3 = 0$.

If $E_3 \neq 0$, then $E_1 = E_2 = 0$.

Hence we see that along the principal axes, **E** has its component $|\mathbf{E}|$ along one of the axes, with its other two components equal to 0.

We choose principal axes such that $\mathbf{E}:(E,0,0)$. Then,

$$i = j = 1 \quad 4\pi T_{11}' = 4\pi\lambda_1 = E^2 - E^2/2 = E^2/2 .$$
$$i = j = 2 \quad 4\pi T_{22}' = 4\pi\lambda_2 = 0 - E^2/2 = -E^2/2 .$$
$$i = j = 3 \quad 4\pi T_{33}' = 4\pi\lambda_3 = 0 - E^2/2 = -E^2/2 .$$

Thus,

$$T_{ij}' = \begin{pmatrix} \frac{E^2}{8\pi} & 0 & 0 \\ 0 & -\frac{E^2}{8\pi} & 0 \\ 0 & 0 & -\frac{E^2}{8\pi} \end{pmatrix}$$

We see that $\lambda_1 = E^2/8\pi$ represents a tension along E_1, or along **E**. To show that T_{ij}' represents a compression perpendicular to **E** we have to show that for any direction perpendicular to **E**, λ_2 and λ_3 do not change and are equal to $-E^2/8\pi$.

Let us, therefore, perform a rotation about the **E** axis, and go from the ijk system to the $i'j'k'$ system by a rotation through an angle α.

$T_{ij}'' = \sum_{lm} u_{il}u_{jm}T_{lm}'$, which only has meaning for $l = m$. For $i \neq j$,

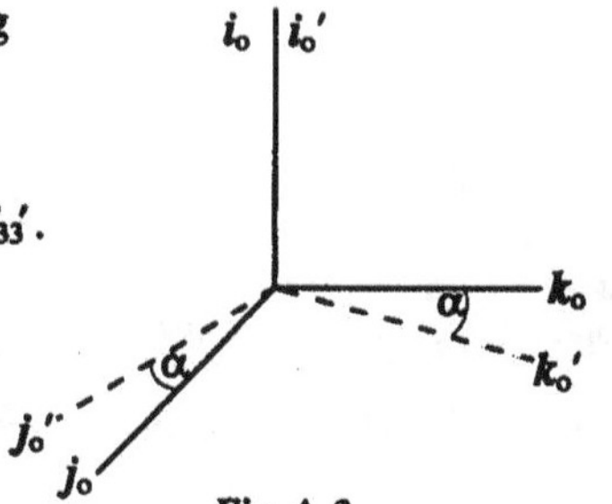

Fig. A-3

$$T_{ij}'' = u_{i1}u_{j1}T_{11}' + u_{i2}u_{j2}T_{22}' + u_{i3}u_{j3}T_{33}'.$$

Considering the rotation shown in Fig. A-3,

$$j_o \cdot j_o' = \cos\alpha = u_{22}.$$
$$j_o \cdot k_o' = \cos(90 - \alpha) = \sin\alpha = u_{32}.$$
$$k_o \cdot k_o' = \cos\alpha = u_{33}.$$
$$j_o' \cdot k_o = \cos(90 + \alpha) = -\sin\alpha = u_{23}.$$
$$i_o \cdot i_o' = u_{11} = 1.$$
$$i_o \cdot j_o = i_o \cdot j_o' = i_o \cdot k_o = i_o \cdot k_o' = 0; \quad u_{21} = u_{31} = 0.$$

By virtue of these relations, for $i \neq j$,

$$T_{ij}'' = u_{i2}u_{j2}T_{22}' + u_{13}u_{j3}T_{33}'.$$
$$T_{12}'' = T_{21}'' = 0; T_{13}'' = T_{31}'' = 0.$$
$$T_{23}'' = T_{32}'' = u_{22}u_{32}T_{22}' + u_{23}u_{33}T_{33}'$$
$$= \sin\alpha\cos\alpha(-E^2/8\pi) - \sin\alpha\cos\alpha(-E^2/8\pi) = 0.$$

For $i = j$,

$$T_{11}'' = u_{11}^2 T_{11}' = \lambda_1.$$
$$T_{22}'' = u_{22}^2 T_{22}' + u_{23}^2 T_{33}' = \lambda_2 = \lambda_2\cos^2\alpha + \lambda_3\sin^2\alpha = \lambda_2(=\lambda_3).$$
$$T_{33}'' = u_{32}^2 T_{22}' + u_{33}^2 T_{33}' = \lambda_3 = \lambda_2\sin^2\alpha + \lambda_3\cos^2\alpha = \lambda_3(=\lambda_2).$$

Hence under a rotation, λ_2 and λ_3 don't change, and represent a compression $(-E^2/8\pi)$ perpendicular to **E**.

V. Plot the equipotential surfaces and the lines of force for the electrostatic field generated by two equal point charges located at the points $x = \pm 1$.

Solution: A line of force is defined by

$$ds = \lambda \mathbf{E} \rightarrow \frac{dx}{E_x} = \frac{dy}{E_y} = \frac{dz}{E_z}.$$

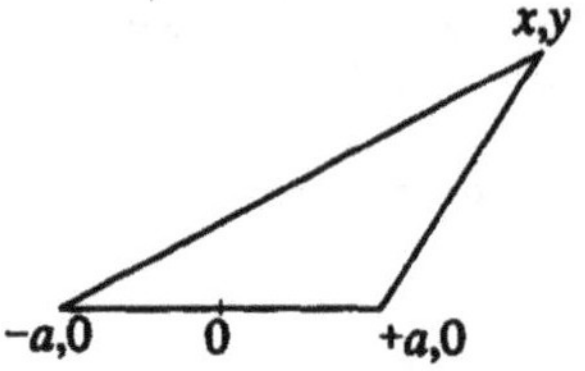

Fig. A-4

By symmetry, any section of the field by a plane including the x-axis will result in the same appearance. Let us therefore take that section made by the xy-plane. We will plot the result for $x > 0$ only. The rest of the picture is then obviously apparent.

From Fig. A-4, we have

$$E_{x_1} = \frac{q}{y^2 + (x+a)^2} \frac{(x+a)}{\left[y^2 + (x+a)^2\right]^{1/2}} = \frac{q(x+a)}{\left[y^2 + (x+a)^2\right]^{3/2}}.$$

$$E_{x_2} = \frac{q(x-a)}{\left[y^2 + (x-a)^2\right]^{3/2}}.$$

Defining $u \equiv (x+a)/y$ and $\upsilon \equiv (x-a)/y$, we have

$$E_x = \frac{q(x+a)}{\left[y^2 + (x+a)^2\right]^{3/2}} + \frac{q(x-a)}{\left[y^2 + (x-a)^2\right]^{3/2}}$$

$$= \frac{qu}{y^2(1+u^2)^{3/2}} + \frac{q\upsilon}{y^2(1+\upsilon^2)^{3/2}}.$$

$$E_y = \frac{qy}{\left[y^2 + (x+a)^2\right]^{3/2}} + \frac{qy}{\left[y^2 + (x-a)^2\right]^{3/2}}$$

$$= \frac{q}{y^2(1+u^2)^{3/2}} + \frac{q}{y^2(1+\upsilon^2)^{3/2}}.$$

$$\frac{dy}{dx} = \frac{E_y}{E_x} = \frac{\dfrac{(1+u^2)^{3/2} + (1+\upsilon^2)^{3/2}}{y^2(1+u^2)^{3/2}(1+\upsilon^2)^{3/2}}}{\dfrac{\upsilon(1+u^2)^{3/2} + u(1+\upsilon^2)^{3/2}}{y^2(1+u^2)^{3/2}(1+\upsilon^2)^{3/2}}} = \frac{(1+u^2)^{3/2} + (1+\upsilon^2)^{3/2}}{\upsilon(1+u^2)^{3/2} + u(1+\upsilon^2)^{3/2}}.$$

From $x + a = uy$ and $x - a = \upsilon y$, we have $y = \dfrac{2a}{u - \upsilon}$ and $x = \dfrac{a(u + \upsilon)}{u - \upsilon}$, so that

$$dy = \frac{-2a}{(u-\upsilon)^2}(du - d\upsilon).$$

$$dx = \frac{a}{(u-\upsilon)^2}\left[-(u+\upsilon)(du - d\upsilon) + (u - \upsilon)(du + d\upsilon)\right]$$

$$= \frac{2a}{(u-\upsilon)^2}(ud\upsilon - \upsilon du).$$

Hence,

$$\frac{dy}{dx} = \frac{d\upsilon - du}{ud\upsilon - \upsilon du}.$$

Combining the two expressions for dy/dx:

$$\frac{d\upsilon - du}{ud\upsilon - \upsilon du} = \frac{1 - du/d\upsilon}{u - \upsilon du/d\upsilon} = \frac{(1+u^2)^{3/2} + (1+\upsilon^2)^{3/2}}{\upsilon(1+u^2)^{3/2} + u(1+\upsilon^2)^{3/2}}.$$

$$\frac{du}{d\upsilon} = \frac{\upsilon(1+u^2)^{3/2} - u(1+u^2)^{3/2}}{u(1+\upsilon^2)^{3/2} - \upsilon(1+\upsilon^2)^{3/2}} = \frac{(\upsilon - u)(1+u^2)^{3/2}}{(u-\upsilon)(1+\upsilon^2)^{3/2}}$$

$$= -\frac{(1+u^2)^{3/2}}{(1+\upsilon^2)^{3/2}}.$$

Separating variables and integrating:

$$\frac{du}{(1+u^2)^{3/2}} = -\frac{d\upsilon}{(1+\upsilon^2)^{3/2}} \quad \Rightarrow \quad u(1+u^2)^{-1/2} + \upsilon(1+\upsilon^2)^{-1/2} = C.$$

In terms of x and y:

$$(x+a)\,[(x+a)^2 + y^2]^{-1/2} + (x-a)\,[(x-a)^2 + y^2]^{-1/2} = C.$$

Setting $a = 1$, we have:

(a) $$\frac{x+1}{\sqrt{(x+1)^2 + y^2}} + \frac{x-1}{\sqrt{(x-1)^2 + y^2}} = C.$$

The equipotential surfaces are defined by $\varphi = C'$:

$$\varphi = \frac{q}{[(x+a)^2+y^2]^{1/2}} + \frac{q}{[(x-a)^2+y^2]^{1/2}} = C',$$

and, since $a = 1$,

$$\text{(b)} \qquad \frac{q}{C'}\left[\frac{1}{[(x+1)^2+y^2]^{1/2}} + \frac{1}{[(x-1)^2+y^2]^{1/2}}\right] = 1.$$

When values of C and C' are substituted into equations (a) and (b) above, we get the following plot of equipotential surfaces and lines of force. In Fig. A-5, solid lines represent lines of force while dotted lines represent equipotential surfaces.

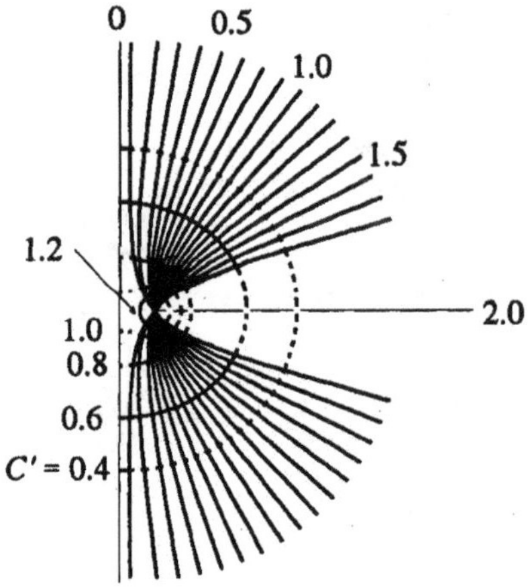

Fig. A-5

VI. If φ is any scalar function and $\mathbf{A}(\mathbf{r})$ any vector function, prove that:

$$(a) \quad \int_V \nabla\varphi d\tau = \int_S \varphi d\mathbf{S}.$$

$$(b) \quad \int_V (\nabla\times\mathbf{A}) d\tau = -\int_S \mathbf{A}\times d\mathbf{S}.$$

(V is any volume bounded by the closed surface S.)

(a) $\int_V (\nabla\varphi)_i d\tau = \int_V \frac{\partial\varphi}{\partial x_i} d\tau.$

Gauss' theorem states that $\sum_i \int_V \frac{\partial}{\partial x_i} T_{j_1j_2\ldots i\ldots j_n} d\tau = \int_S \sum_i T_{j_1j_2\ldots i\ldots j_n} dS_i$. If we want the case that index i not appear in the tensor $T_{j_1j_2\ldots i\ldots j_n}$ we have to find $\int_V \frac{\partial}{\partial x_j} T_{i_1i_2\ldots i_n} d\tau$. We proceed by defining $R_{iji_1i_2\ldots i_n} = \delta_{ij} T_{i_1i_2\ldots i_n}$. Then

$$\sum_i \frac{\partial}{\partial x_i} R_{iji_1i_2\ldots i_n} = \sum_i \frac{\partial}{\partial x_i} \delta_{ij} T_{i_1i_2\ldots i_n},$$

and

$$\int_V \frac{\partial}{\partial x_j} T_{i_1i_2\ldots i_n} d\tau = \sum_i \int_V \frac{\partial}{\partial x_i} \delta_{ij} T_{i_1i_2\ldots i_n} d\tau = \sum_i \int_V \frac{\partial}{\partial x_i} R_{iji_1i_2\ldots i_n} d\tau,$$

which, by Gauss' theorem

$$= \sum_i \int_S R_{iji_1i_2\ldots i_n} dS_i = \sum_i \delta_{ij} \int_S T_{i_1i_2\ldots i_n} dS_i = \int_S T_{i_1i_2\ldots i_n} dS_j.$$

For the current problem, we set $T = \varphi$, yielding

$$\int_V \frac{\partial\varphi}{\partial x_i} d\tau = \int_S \varphi dS_i \quad \rightarrow \quad \int_V \nabla\varphi d\tau = \int_S \varphi d\mathbf{S}.$$

(b) $\int_V (\nabla\times\mathbf{A})_i d\tau = \int_V \sum_{jk} \varepsilon_{ijk} \frac{\partial A_k}{\partial x_j} d\tau = \int_V \sum_{jk} \frac{\partial}{\partial x_j} \varepsilon_{ijk} A_k \, d\tau.$

Defining $R_{ij} = \sum_k \varepsilon_{ijk} A_k$, we have

$$\int_V (\nabla\times\mathbf{A})_i d\tau = \int_V \sum_j \frac{\partial}{\partial x_j} \sum_k \varepsilon_{ijk} A_k \, d\tau = \int_V \sum_j \frac{\partial}{\partial x_j} R_{ij} d\tau,$$

which, by Gauss' theorem $= \int_S R_{ij} dS_j = \int_S \sum_k \varepsilon_{ijk} A_k dS_j$

$$= -\int_S \sum_k \varepsilon_{ikj} A_k dS_j = -\int_S (\mathbf{A}\times d\mathbf{S})_i.$$

$$\Downarrow$$

$$\int_V (\nabla\times\mathbf{A}) d\tau = -\int_S \mathbf{A}\times d\mathbf{S}.$$

Problem Set D

I. Show that if a fluid is incompressible, then $\nabla\cdot\mathbf{v} = 0$, where $\mathbf{v}$ is the velocity function of the fluid.

Solution: By virtue of the conservation of mass (which says nothing about homogeneity or incompressibility of the medium), we have

$$(1) \qquad \frac{\partial \rho}{\partial t} + \nabla\cdot\rho\mathbf{v} = \frac{\partial \rho}{\partial t} + \rho\nabla\cdot\mathbf{v} + \nabla\rho\cdot\mathbf{v} = 0.$$

If, in addition, the fluid is incompressible, then by definition, $\frac{d\rho}{dt} = 0$, so that

$$(2) \qquad \frac{d\rho}{dt} = \frac{\partial \rho}{\partial t} + \frac{\partial \rho}{\partial x}\frac{dx}{dt} + \frac{\partial \rho}{\partial y}\frac{dy}{dt} + \frac{\partial \rho}{\partial z}\frac{dz}{dt} = 0 \rightarrow \frac{\partial \rho}{\partial t} + \nabla\rho\cdot\mathbf{v} = 0.$$

Substituting (2) into (1) we have $\nabla\cdot\mathbf{v} = 0$.

II. The stress tensor for a viscous fluid is known to be of the form

$$T_{ij} = -p\delta_{ij} - \frac{2\mu}{3}(\nabla\cdot\mathbf{v})\delta_{ij} + \mu\left(\frac{\partial v_i}{\partial x_j} + \frac{\partial v_j}{\partial x_i}\right),$$

where p is the pressure and μ the coefficient of viscosity.

Show that the equation of motion can be expressed as

$$\rho\left[\frac{\partial \mathbf{v}}{\partial t} + (\mathbf{v}\cdot\nabla)\mathbf{v}\right] = \mathbf{F} - \nabla p + \frac{\mu}{3}\nabla(\nabla\cdot\mathbf{v}) + \mu\nabla^2\mathbf{v},$$

where ρ is the fluid density and $\mathbf{F}$ the body force per unit volume.

Solution: We saw in Chapter 3 that the equation of motion for deformable bodies is given by

$$\rho a_i = F_i + \sum_j \frac{\partial}{\partial x_j} T_{ij},$$

where T_{ij} is the stress tensor.

$$\mathbf{a} = \frac{d\mathbf{v}}{dt} = \frac{\partial \mathbf{v}}{\partial t} + \sum_i \frac{\partial \mathbf{v}}{\partial x_i}\frac{dx_i}{dt} = \frac{\partial \mathbf{v}}{\partial t} + \sum_i v_i \frac{\partial \mathbf{v}}{\partial x_i} = \frac{\partial \mathbf{v}}{\partial t} + (\mathbf{v}\cdot\nabla)\mathbf{v}.$$

$$\rho a_i = \rho\left[\frac{\partial \upsilon_i}{\partial t} + (\mathbf{v}\cdot\nabla)\upsilon_i\right]$$

$$= F_i + \sum_j \frac{\partial}{\partial x_j}\left[-p\delta_{ij} - \frac{2\mu}{3}(\nabla\cdot\mathbf{v})\delta_{ij} + \mu\left(\frac{\partial \upsilon_i}{\partial x_j} + \frac{\partial \upsilon_j}{\partial x_i}\right)\right].$$

$$= F_i - \frac{\partial p}{\partial x_i} - \frac{2}{3}\left(\mu\frac{\partial}{\partial x_i}(\nabla\cdot\mathbf{v})\right) + \mu\sum_j\left(\frac{\partial^2 \upsilon_i}{\partial x_j^2} + \frac{\partial}{\partial x_i}\frac{\partial \upsilon_j}{\partial x_j}\right)$$

$$= F_i - (\nabla p)_i - (2\mu/3)\nabla_i(\nabla\cdot\mathbf{v}) + \mu\nabla^2\upsilon_i + \mu\nabla_i(\nabla\cdot\mathbf{v})$$

$$= F_i - (\nabla p)_i + (\mu/3)\nabla_i(\nabla\cdot\mathbf{v}) + \mu\nabla^2\upsilon_i.$$

Collecting terms,

$$\rho\left[\frac{\partial \mathbf{v}}{\partial t} + (\mathbf{v}\cdot\nabla)\mathbf{v}\right] = \mathbf{F} - \nabla p + \frac{\mu}{3}\nabla(\nabla\cdot\mathbf{v}) + \mu\nabla^2\mathbf{v}.$$

III. Prove that for any general tensor function of the nth rank:

$$(a)\quad \sum_{ijk}\int_S \varepsilon_{ijk}\frac{\partial}{\partial x_j}T_{ki_1i_2\ldots i_{n-1}}dS_i = \sum_i \oint_s T_{ii_1i_2\ldots i_{n-1}}ds_i,$$

where S is a 2-dimensional surface bounded by the closed curve s. The indices $i_1\ldots i_{n-1}$ may vary arbitrarily from 1 to 3; i.e. prove Stokes' theorem for an arbitrary general tensor of the nth rank.

$$(b)\quad \sum_{ij}\int_S \varepsilon_{ijk}\frac{\partial}{\partial x_j}T_{i_1i_2\ldots i_n}dS_i = \oint_s T_{i_1i_2\ldots i_n}ds_k,$$

where $ki_1\ldots i_n$ can vary arbitrarily from 1 to 3.

$$(c)\quad \int_S \nabla\varphi\times d\mathbf{S} = \oint_s \varphi d\mathbf{s},$$

where φ is a scalar function.

(a) Consider the rectangle of sides δx_3 and $\sqrt{\delta x_1^2 + \delta x_2^2}$, shown in Fig. A-6.

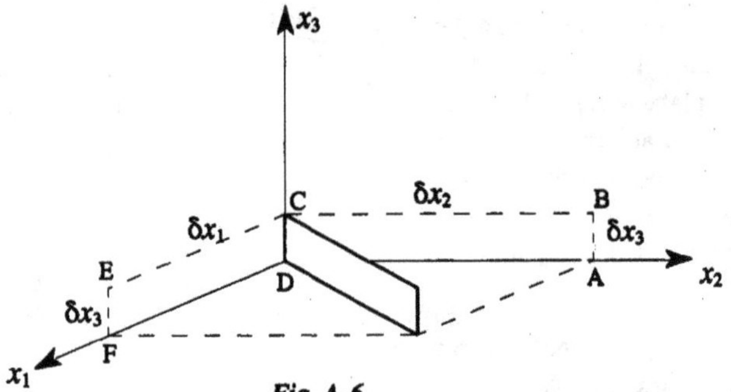

Fig. A-6

$$\sum_{ijk}\int_S \varepsilon_{ijk}\frac{\partial}{\partial x_j}T_{ki_1i_2\dots i_{n-1}}dS_i = \int_S\left(\frac{\partial}{\partial x_2}T_{3i_1i_2\dots i_{n-1}} - \frac{\partial}{\partial x_3}T_{2i_1i_2\dots i_{n-1}}\right)dS_1$$

$$+\int_S\left(\frac{\partial}{\partial x_3}T_{1i_1i_2\dots i_{n-1}} - \frac{\partial}{\partial x_1}T_{3i_1i_2\dots i_{n-1}}\right)dS_2$$

$$+\int_S\left(\frac{\partial}{\partial x_1}T_{2i_1i_2\dots i_{n-1}} - \frac{\partial}{\partial x_2}T_{1i_1i_2\dots i_{n-1}}\right)dS_3 .$$

$$= \int_S\left[\left|T_{3\dots}\right|_{x_2}^{x_2+\delta x_2}dx_3 - \left|T_{2\dots}\right|_{x_3}^{x_3+\delta x_3}dx_2\right] + \int_S\left[\left|T_{1\dots}\right|_{x_3}^{x_3+\delta x_3}dx_1 - \left|T_{3\dots}\right|_{x_1}^{x_1+\delta x_1}dx_3\right]$$

$$= \int_S\left[\left|T_{3\dots}\right|_{x_2+\delta x_2} - \left|T_{3\dots}\right|_{x_2}\right]dx_3 - \int_S\left[\left|T_{2\dots}\right|_{x_3+\delta x_3} - \left|T_{2\dots}\right|_{x_3}\right]dx_2$$

$$+ \int_S\left[\left|T_{1\dots}\right|_{x_3+\delta x_3} - \left|T_{1\dots}\right|_{x_3}\right]dx_1 - \int_S\left[\left|T_{3\dots}\right|_{x_1+\delta x_1} - \left|T_{3\dots}\right|_{x_1}\right]dx_3$$

$$= \int_A^B T_{3\dots}dx_3 + \int_C^D T_{3\dots}dx_3 + \int_B^C T_{2\dots}dx_2 + \int_D^A T_{2\dots}dx_2$$

$$+ \int_C^E T_{1\dots}dx_1 + \int_F^D T_{1\dots}dx_1 + \int_E^F T_{3\dots}dx_3 + \int_D^C T_{3\dots}dx_3$$

$$= \oint T_{3\dots}dx_3 + \oint T_{1\dots}dx_1 + \oint T_{2\dots}dx_2 = \sum_i \oint_s T_{ii_1\dots i_{n-1}}ds_i .$$

The arbitrary, unclosed surface, shown in Fig. A-7, can always be so subdivided to get plane rectangles by passing planes parallel to the x-y plane and parallel to the x-z plane. Each rectangle, then, will be of the type discussed above, i.e. an infinitesimal rectangle of sides δx_3 and $\sqrt{\delta x_1^2 + \delta x_2^2}$.

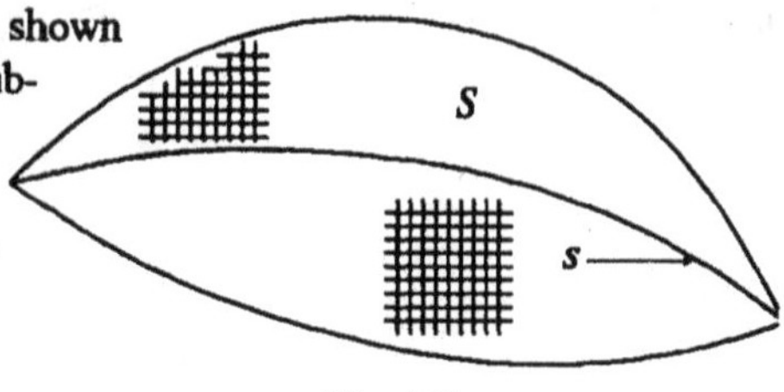

Fig. A-7

For any two adjacent rectangles, the line integral over the common side vanishes (Fig. A-8). Hence, the sum of the line integrals over all rectangles comprising S reduces to the line integral over the small, remaining triangles of the type shown in Fig. A-9.

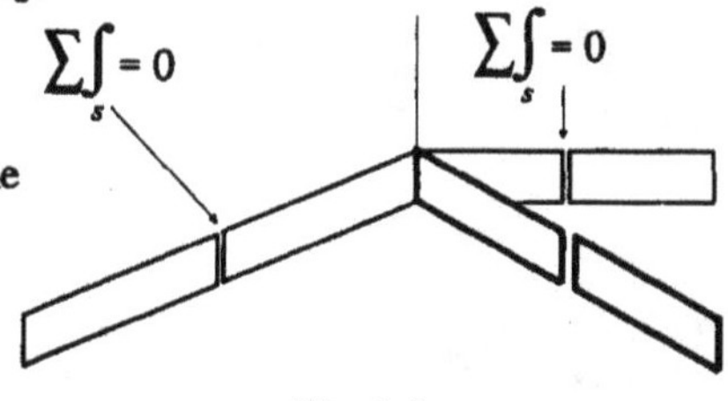

Fig. A-8

It only remains to show that for those triangles whose hypotenuses lie along curve s,

$$\int T_{1i_1\ldots i_{n-1}} dx_1 + \int T_{2i_1\ldots i_{n-1}} dx_2 = \sum_i \int_s T_{ii_1\ldots i_{n-1}} ds_i \, .$$

Fig. A-9

But for infinitesimally small partitions of S, ds:(dx_1, dx_2) (Fig. A-10). Hence,

$$\sum_i T_{ii_1\ldots i_{n-1}} ds_i = T_{1i_1\ldots i_{n-1}} dx_1 + T_{2i_1\ldots i_{n-1}} dx_2 \, .$$

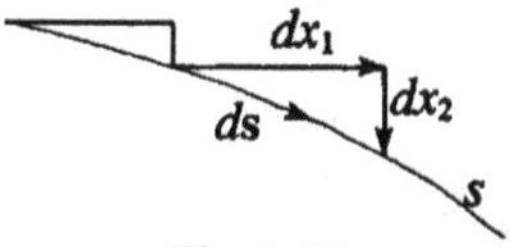

Fig. A-10

(b) Define a tensor $S_{lki_1\ldots i_{n-1}} \equiv \delta_{lk} T_{i_1\ldots i_{n-1}}$, so that

$$\begin{aligned}\int_S \sum_{ij} \varepsilon_{ijk} \frac{\partial}{\partial x_j} T_{i_1 i_2\ldots i_n} dS_i &= \int_S \sum_{ijl} \varepsilon_{ijl} \delta_{lk} \frac{\partial}{\partial x_j} T_{i_1 i_2\ldots i_n} dS_i \\ &= \int_S \sum_{ijl} \varepsilon_{ijl} \frac{\partial}{\partial x_j} S_{lki_1 i_2\ldots i_n} dS_i \rightarrow \sum_i \oint_s S_{iki_1\ldots i_{n-1}} ds_i \\ &= \sum_i \oint_s \delta_{ik} T_{i_1\ldots i_{n-1}} ds_i = \oint_s T_{i_1\ldots i_{n-1}} ds_k \, .\end{aligned}$$

(The last two steps follow from part (a) above.)

(c) $$\int_S [\nabla\varphi \times d\mathbf{S}]_i = \sum_{jk}\int_S \varepsilon_{ijk}\frac{\partial\varphi}{\partial x_j}dS_k$$

$$= -\sum_{jk}\int_S \varepsilon_{kji}\frac{\partial\varphi}{\partial x_j}dS_k \rightarrow -\oint_s \varphi ds_i .$$

(The last step follows from part (b) above.)

IV. If **H** is an arbitrary vector field and $\nabla\cdot\mathbf{H} = 0$ everywhere, prove that

$$\mathbf{H} = \nabla\times\mathbf{A},$$

where

$$A_1(x_1x_2x_3) = \int_0^{x_3} H_2(x_1x_2x_3')dx_3' - \int_0^{x_2} H_3(x_1x_2'0)dx_2'.$$

$$A_2(x_1x_2x_3) = -\int_0^{x_3} H_1(x_1x_2x_3')dx_3'.$$

$$A_3(x_1x_2x_3) = 0.$$

Solution: Let $\int H_1(x_1x_2x_3')dx_3' = G_1(x_1x_2x_3')$, so that $\frac{\partial G_1}{\partial x_3'} = H_1(x_1x_2x_3')$.

Then

$$\int_0^{x_3} H_1(x_1x_2x_3')dx_3' = G_1(x_1x_2x_3) - G_1(x_1x_20),$$

and

$$\frac{\partial}{\partial x_3}\int_0^{x_3} H_1(x_1x_2x_3')dx_3' = H_1(x_1x_2x_3) - 0 = H_1(x_1x_2x_3).$$

It follows then that

$$\frac{\partial A_2}{\partial x_3} = -\frac{\partial}{\partial x_3}\int_0^{x_3} H_1(x_1x_2x_3')dx_3' = -H_1(x_1x_2x_3).$$

$$\frac{\partial A_3}{\partial x_2} = 0.$$

Thus,

$$\frac{\partial A_3}{\partial x_2} - \frac{\partial A_2}{\partial x_3} = H_1(x_1x_2x_3).$$

Similarly:

$$\frac{\partial A_1}{\partial x_3} = \frac{\partial}{\partial x_3}\int_0^{x_3} H_2(x_1x_2x_3')dx_3' - \frac{\partial}{\partial x_3}\int_0^{x_2} H_3(x_1x_2'0)dx_2' = H_2(x_1x_2x_3) - 0.$$

$$\frac{\partial A_3}{\partial x_1} = 0.$$

$$\frac{\partial A_1}{\partial x_3} - \frac{\partial A_3}{\partial x_1} = H_2(x_1x_2x_3).$$

Proceeding similarly once again:

$$\frac{\partial A_2}{\partial x_1} = -\frac{\partial}{\partial x_1}\int_0^{x_3} H_1(x_1x_2x_3')dx_3' = -\int_0^{x_3} \frac{\partial}{\partial x_1} H_1(x_1x_2x_3')dx_3'.$$

$$\frac{\partial A_1}{\partial x_2} = \frac{\partial}{\partial x_2}\int_0^{x_3} H_2(x_1x_2x_3')dx_3' - \frac{\partial}{\partial x_2}\int_0^{x_2} H_3(x_1x_2'0)dx_2'$$

$$= \int_0^{x_3} \frac{\partial}{\partial x_2} H_2(x_1x_2x_3')dx_3' - H_3(x_1x_20).$$

$$\frac{\partial A_2}{\partial x_1} - \frac{\partial A_1}{\partial x_2} = -\int_0^{x_3}\left(\frac{\partial H_1}{\partial x_1} + \frac{\partial H_2}{\partial x_2}\right)dx_3' + H_3(x_1x_20).$$

But $\nabla\cdot\mathbf{H} = 0 \;\rightarrow\; \dfrac{\partial H_1}{\partial x_1} + \dfrac{\partial H_2}{\partial x_2} = -\dfrac{\partial H_3}{\partial x_3}$, so that

$$\frac{\partial A_2}{\partial x_1} - \frac{\partial A_1}{\partial x_2} = \int_0^{x_3} \frac{\partial H_3(x_1x_2x_3')}{\partial x_3} dx_3' + H_3(x_1x_20) = H_3(x_1x_2x_3),$$

and we have established that

$$\left\{\begin{aligned} H_1(x_1x_2x_3) &= \frac{\partial A_3}{\partial x_2} - \frac{\partial A_2}{\partial x_3} \\ H_2(x_1x_2x_3) &= \frac{\partial A_1}{\partial x_3} - \frac{\partial A_3}{\partial x_1} \\ H_3(x_1x_2x_3) &= \frac{\partial A_2}{\partial x_1} - \frac{\partial A_1}{\partial x_2} \end{aligned}\right\} \rightarrow \mathbf{H}(x_1x_2x_3) = \nabla\times\mathbf{A}.$$

Hence, given the fact that $\nabla\cdot\mathbf{H} = 0$ everywhere, one can always find a vector **A** such that $\mathbf{H} = \nabla\times\mathbf{A}$. [This vector **A** is not unique, as will be discussed in Problem VI.]

V. If $\nabla\cdot\mathbf{H} = 0$ everywhere except inside a finite region R and if in this region R, $\int_R \nabla\cdot\mathbf{H}d\tau \neq 0$, then prove that it is impossible to have a vector function **A** such that $\mathbf{H} = \nabla\times\mathbf{A}$ everywhere <u>outside</u> the region R.

Solution: Let region R be bounded by S', within which $\int_R \nabla\cdot\mathbf{H}d\tau \neq 0$. Consider a surface S completely enclosing R, as shown in Fig. A-11. By Gauss' theorem:

$$\int_{\substack{R+R' \\ R'\to\infty}} \nabla\cdot\mathbf{H}d\tau = \int_{S\to} \mathbf{H}\cdot d\mathbf{S}.$$

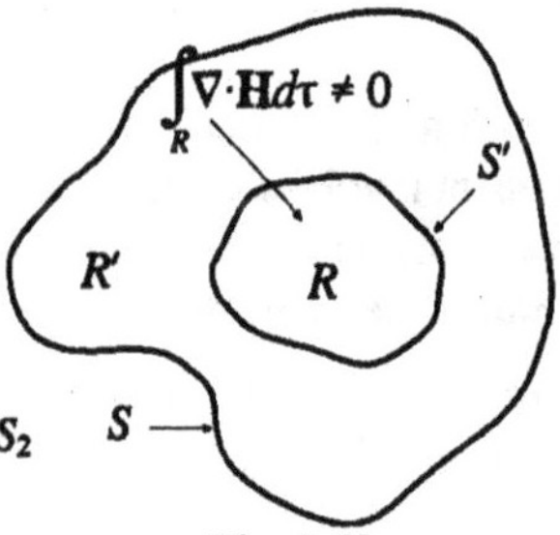

Fig. A-11

For <u>any closed</u> surface S,

$$\oint_S \nabla\times\mathbf{A}\cdot d\mathbf{S} = \int_{S_1} \nabla\times\mathbf{A}\cdot n_o dS_1 + \int_{S_2} \nabla\times\mathbf{A}\cdot n_o dS_2$$

$$= \oint_{\substack{\text{clockwise} \\ s}} \mathbf{A}\cdot d\mathbf{s} + \oint_{\substack{\text{counter clockwise} \\ s}} \mathbf{A}\cdot d\mathbf{s} = 0.$$

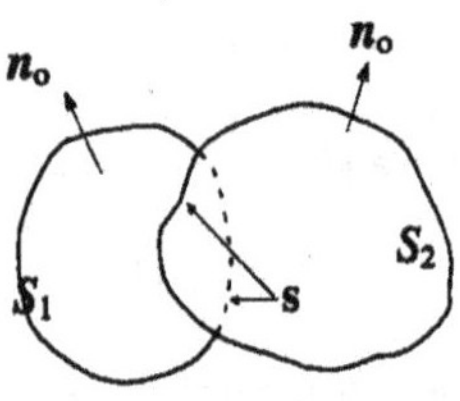

Fig. A-12

This last statement follows since we can always divide S into S_1 and S_2, each of which has the common curve s enclosing it (Fig. A-12).

Hence, for any closed surface S, $\oint_S \nabla\times\mathbf{A}\cdot d\mathbf{S} = 0$.

Program for proof:

1. Assume **H** can be given by $\nabla\times\mathbf{A}$ everywhere in R'.
2. Then show that this leads to a contradiction of the above, i.e. $\oint_S \nabla\times\mathbf{A}\cdot d\mathbf{S} \neq 0$.
3. By this contradiction we know that $\mathbf{H} \neq \nabla\times\mathbf{A}$ in R'.
4. We have demonstrated the proof.

1. $\mathbf{H} = \nabla\times\mathbf{A}$ in R'.

2. $$\int\limits_{\substack{R+R'\\R'\to\infty}} \nabla\cdot\mathbf{H}\,d\tau = \int\limits_{R} \nabla\cdot\mathbf{H}\,d\tau = \int\limits_{S\to\infty} \mathbf{H}\cdot d\mathbf{S} = \oint\limits_{S\to\infty} \nabla\times\mathbf{A}\cdot d\mathbf{S} \neq 0.$$

3. But this is a contradiction, since for any closed surface S, $\oint\limits_{S} \nabla\times\mathbf{A}\cdot d\mathbf{S} = 0.$

4. Hence, $\mathbf{H} \neq \nabla\times\mathbf{A}$ in R'.

VI. Let $\mathbf{H}$ be a constant vector everywhere. Construct a scalar function φ and a vector function $\mathbf{A}$ such that $\mathbf{H} = \nabla\varphi$ and $\mathbf{H} = \nabla\times\mathbf{A}$ with $\nabla\cdot\mathbf{A} = 0$. Are the functions φ and $\mathbf{A}$ unique?

Solution: If $\mathbf{H}$ is to equal $\nabla\varphi$ and be a constant vector, then since the gradient involves the first derivative, φ must be linear in its variables, i.e. representable by $\mathbf{r}$. We can therefore construct from $\mathbf{r}$ and $\mathbf{H}$ the only possible scalar, namely, $\mathbf{r}\cdot\mathbf{H} = \varphi$.

(Note: Since $\nabla\times\mathbf{H} = 0$ everywhere, $\mathbf{H} = \nabla\varphi$, where $d\varphi = \nabla\varphi\cdot d\mathbf{s} = \mathbf{H}\cdot d\mathbf{s}$; hence, $\varphi = \mathbf{H}\cdot\mathbf{r}$.)

It is obvious that φ is unique only to within an additive constant, for if φ_2 be also a solution, $\mathbf{H} = \nabla\varphi_2$ and $\nabla(\varphi_2 - \varphi_1) = 0$, so that $\varphi_2 = \varphi_1 + \text{const.}$

If $\mathbf{H}$ is to equal $\nabla\times\mathbf{A}$ (which again involves first derivatives of $\mathbf{A}$) and still be a constant, $\mathbf{A}$ must be linear in its variables, i.e. in $\mathbf{r}$. We then can form from $\mathbf{r}$ and $\mathbf{H}$ the only possible vector, namely $\mathbf{A} = a(\mathbf{r}\times\mathbf{H})$, so that

$$\mathbf{H} = \nabla\times\mathbf{A} = a\nabla\times(\mathbf{r}\times\mathbf{H}) = a\left[\mathbf{r}(\nabla\cdot\mathbf{H}) - \mathbf{H}(\nabla\cdot\mathbf{r}) + (\mathbf{H}\cdot\nabla)\mathbf{r} - (\mathbf{r}\cdot\nabla)\mathbf{H}\right].$$

Since $\mathbf{H}$ is a constant vector, $\nabla\cdot\mathbf{H} = 0$ and $(\mathbf{r}\cdot\nabla)\mathbf{H} = 0$, reducing the above to

$$\mathbf{H} = a\left[-3\mathbf{H} + \mathbf{H}\right] = -2a\mathbf{H} \Rightarrow a = -1/2.$$

Thus, $\mathbf{A} = -(1/2)\,\mathbf{r}\times\mathbf{H}$. Again, $\mathbf{A}$ is not unique, since we can find another vector $\mathbf{B}$ such that $\mathbf{B} = \mathbf{A} + \nabla\varphi'$. Then, $\mathbf{H} = \nabla\times\mathbf{B} = \nabla\times\mathbf{A}$ and $\nabla\cdot\mathbf{B} = \nabla\cdot\mathbf{A} + \nabla^2\varphi'$. Since $\nabla\cdot\mathbf{A}$ is to be zero, if we choose $\nabla^2\varphi' = 0$, we have completely determined $\mathbf{B}$.

In fact, we can choose $\nabla\varphi' = \nabla\varphi$ (from the first part), i.e. $\mathbf{H} = \nabla\varphi = \nabla\varphi'$. Then, $\mathbf{B} = \mathbf{A} + \mathbf{H}$, $\nabla\times\mathbf{B} = \nabla\times\mathbf{A}$, and $\nabla\cdot\mathbf{B} = \nabla\cdot\mathbf{A}$.

Problem Set E

I. Show that if $\mathbf{F}_1 = \nabla\varphi$ and $\mathbf{F}_2 = \nabla\times\mathbf{A}$, then $\int \mathbf{F}_1\cdot\mathbf{F}_2 d\tau = 0$ provided that as $|\mathbf{r}| \to \infty$, either $|\varphi\mathbf{F}_2|$ or $|\mathbf{A}\times\mathbf{F}_1|$ goes to zero faster than r^{-2}.

Solution:

$$\int \mathbf{F}_1\cdot\mathbf{F}_2 d\tau = \int \nabla\varphi\cdot\nabla\times\mathbf{A}\, d\tau = \int \sum_i \frac{\partial\varphi}{\partial x_i} \sum_{jk} \varepsilon_{ijk} \frac{\partial A_k}{\partial x_j} d\tau$$

$$= \int \sum_i \left[\frac{\partial}{\partial x_i}[\varphi(\nabla\times\mathbf{A})_i] - \varphi\frac{\partial}{\partial x_i}(\nabla\times\mathbf{A})_i\right].$$

But since $\sum_i \frac{\partial}{\partial x_i}(\nabla\times\mathbf{A})_i = \nabla\cdot(\nabla\times\mathbf{A}) = 0$,

$$\int \mathbf{F}_1\cdot\mathbf{F}_2 d\tau = \sum_i \int \frac{\partial}{\partial x_i} \sum_{jk} \varepsilon_{ijk}\, \varphi \frac{\partial A_k}{\partial x_j} d\tau,$$

which, by virtue of Gauss' theorem,

$$= \sum_i \oint_S \varphi(\nabla\times\mathbf{A})_i dS_i = \oint_S \varphi(\nabla\times\mathbf{A})\cdot d\mathbf{S}.$$

Finally, then

$$\int_{\substack{\text{all}\\ \text{space}}} \mathbf{F}_1\cdot\mathbf{F}_2 d\tau = \oint_{S\to\infty} \varphi(\nabla\times\mathbf{A})\cdot d\mathbf{S} = \oint_{S\to\infty} \left[\nabla\times(\varphi\mathbf{A}) + \mathbf{A}\times\nabla\varphi\right]\cdot d\mathbf{S}.$$

We have already shown (Prob. V, Set D) that $\oint_S \nabla\times\mathbf{V}\cdot d\mathbf{S} = 0$, i.e. that the surface integral of the curl of a vector over a closed surface vanishes. Hence,

$$\int_{\substack{\text{all}\\ \text{space}}} \mathbf{F}_1\cdot\mathbf{F}_2 d\tau = \oint_{S\to\infty} (\mathbf{A}\times\nabla\varphi)\cdot d\mathbf{S} = \oint_{S\to\infty} \varphi(\nabla\times\mathbf{A})\cdot d\mathbf{S}.$$

As $|\mathbf{r}| \to \infty$,

$$\left\{\begin{aligned} |\varphi(\nabla\times\mathbf{A})||d\mathbf{S}| &= |\varphi\mathbf{F}_2||d\mathbf{S}| \to \frac{r^2}{r^{2+\varepsilon}} \to 0. \\ |\mathbf{A}\times\nabla\varphi||d\mathbf{S}| &= |\mathbf{A}\times\mathbf{F}_1||d\mathbf{S}| \to \frac{r^2}{r^{2+\varepsilon}} \to 0. \end{aligned}\right\} \Rightarrow \int_{\substack{\text{all}\\ \text{space}}} \mathbf{F}_1\cdot\mathbf{F}_2 d\tau = 0.$$

II. Prove that if $\nabla^2\varphi = 0$ and if as $|\mathbf{r}| \to \infty$, φ goes to zero faster than $r^{-1/2}$, then $\varphi = 0$ everywhere.

Solution: Set $\mathbf{F}_1 = \mathbf{F}_2 = \nabla\varphi$ in Prob. (1) above. Letting $\varphi \to$ const. r^s, where $s < -1/2$, i.e. letting $s = -1/2 - \eta$, where $\eta > 0$,

$$|\varphi\mathbf{F}_2| = |\varphi\nabla\varphi| \underset{|\mathbf{r}|\to\infty}{\to} \frac{1}{r^{\eta+1/2}}\frac{1}{r^{\eta+3/2}} = \frac{1}{r^{2\eta+2}},$$

and since $\eta > 0$, $|\varphi\mathbf{F}_2| \to 0$ faster than r^{-2}. Thus $\int_V \mathbf{F}_1\cdot\mathbf{F}_2 d\tau = \int_V (\nabla\varphi)^2 d\tau$. The quantity $(\nabla\varphi)^2$ is positive definite, so that $\int_V (\nabla\varphi)^2 d\tau$ vanishes only for $\nabla\varphi = 0$, or $\varphi =$ const. But we know that $\varphi \to 0$ as $|\mathbf{r}| \to \infty$; hence, $\varphi = 0$ at infinity. But since $\varphi =$ constant, $\varphi = 0$ everywhere.

III. The wave equation for a meson (satisfied by any particle with mass) is given by

$$\nabla^2\varphi - \frac{1}{c^2}\ddot{\varphi} - \kappa^2\varphi = 0.$$

(*a*) Show that $\varphi = \varphi_o \exp\left[\pm i(\mathbf{k}\cdot\mathbf{r} - \omega t)\right]$ is a solution in which φ_o is an arbitrary constant and $\mathbf{k}$ is related to ω by $c^2(k^2 + \kappa^2) - \omega^2 = 0$.

(*b*) What is the phase velocity of this wave?

Solution:

(a) From $\varphi = \varphi_o \exp\left[\pm i(\mathbf{k}\cdot\mathbf{r} - \omega t)\right]$,

$$\frac{\partial\varphi}{\partial x} = \varphi_o \exp[\ ](\pm i k_x)$$

$$\left.\begin{aligned} \frac{\partial^2\varphi}{\partial x^2} &= \varphi_o \exp[\ ](-k_x^2). \\ \frac{\partial^2\varphi}{\partial y^2} &= \varphi_o \exp[\ ](-k_y^2) \\ \frac{\partial^2\varphi}{\partial z^2} &= \varphi_o \exp[\ ](-k_z^2) \end{aligned}\right\} \to \nabla^2\varphi = -k^2\varphi_o \exp[\ ] = -k^2\varphi.$$

$$\frac{\partial\varphi}{\partial t} = \varphi_o \exp[\;](\mp i\omega); \quad \frac{1}{c^2}\frac{\partial^2\varphi}{\partial t^2} = -\frac{\omega^2}{c^2}\varphi_o \exp[\;] = -\frac{\omega^2\varphi}{c^2}.$$

Hence, $\nabla^2\varphi - \dfrac{1}{c^2}\dfrac{\partial^2\varphi}{\partial t^2} - \kappa^2\varphi = -\left\{k^2 - \dfrac{\omega^2}{c^2} + \kappa^2\right\}\varphi = 0$. Thus,

$\varphi = \varphi_o \exp\left[\pm i(\mathbf{k}\cdot\mathbf{r} - \omega t)\right]$ will be a solution provided $c^2(k^2 + \kappa^2) - \omega^2 = 0$.

(b) To find the phase velocity of the wave, let us construct wave surfaces at a particular instant of time t_o (Fig. A-13). The wave surfaces are given by φ = const., or what amounts to the same thing, $z = \mathbf{k}\cdot\mathbf{r} - \omega t_o$ = const. c'.

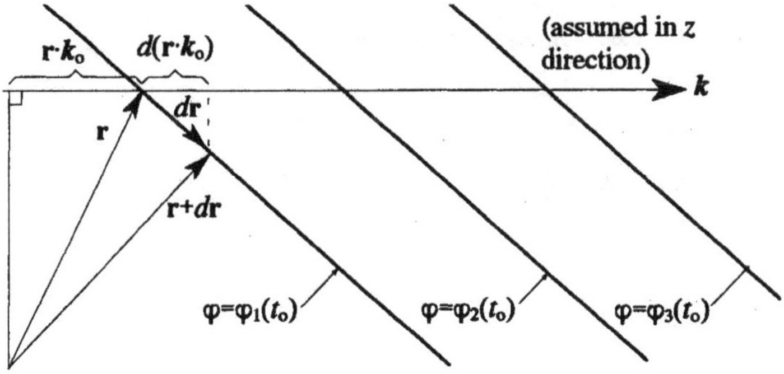

Fig. A-13

These are the equations of planes; the different planes within the family are determined by the parameter c', the constant.

Assume for the moment that $\boldsymbol{k}$ is any vector in space, and let us move a distance $d\mathbf{r}$ along $\varphi = \varphi_1$, where $z = z_1$ (constant). Then, along $z = z_1$, $dz = 0$, i.e. $d(\mathbf{k}\cdot\mathbf{r}) = 0$. We see from Fig. A-13 that the only way that $d(\mathbf{k}\cdot\mathbf{r})$ can equal 0 is for $\mathbf{k}$ to be perpendicular to the wave surfaces. Hence, we correct our diagram to be as shown in Fig. A-14.

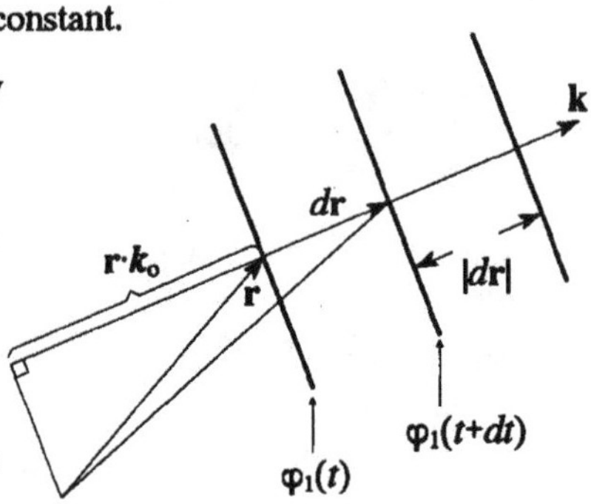

Fig. A-14

The phase velocity, or the velocity with which the wave surface moves, is given by $b = |d\mathbf{r}|/dt$. If now we allow the wave surface to move with time and observe the <u>same wave front</u> (say φ_1) after some time interval dt, then $dz = 0$ again. Thus, $0 = d(\mathbf{k}\cdot\mathbf{r}) - \omega dt$. But, since $d\mathbf{k} = 0$, it follows that $\mathbf{k}\cdot d\mathbf{r} = \omega dt$. From Fig. A-14 we see that $\mathbf{k}\cdot d\mathbf{r} = k|d\mathbf{r}|$. Thus, the phase velocity is given by [using the result from part (a) above]

$$b = \frac{|d\mathbf{r}|}{dt} = \frac{\omega}{k} = c\sqrt{1 + \kappa^2/k^2}.$$

IV. Let $A_x\, A_y\, A_z$ be the components of vector **A** in cartesian coordinates, and $A_\rho\, A_\varphi\, A_z$ the components of **A** in cylindrical coordinates.

(*a*) Express $A_\rho\, A_\varphi\, A_z$ in terms of $A_x\, A_y\, A_z$.

(*b*) Using the above relations together with a direct change of variables in differentiation, show that

$$\frac{\partial A_x}{\partial x} + \frac{\partial A_y}{\partial y} + \frac{\partial A_z}{\partial z} = \frac{1}{\rho}\frac{\partial}{\partial \rho}(\rho A_\rho) + \frac{1}{\rho}\frac{\partial A_\varphi}{\partial \varphi} + \frac{\partial A_z}{\partial z}.$$

(a) From Fig. A-15,

$$\begin{aligned} x &= \rho\cos\varphi. \\ y &= \rho\sin\varphi. \\ z &= z. \end{aligned} \quad \rightarrow \quad \begin{aligned} \rho &= \sqrt{x^2 + y^2}. \\ \varphi &= \tan^{-1} y/x. \\ z &= z. \end{aligned}$$

From the above:

$$\begin{aligned} \partial\rho/\partial x &= x/\rho = \cos\varphi. \\ \partial\rho/\partial y &= y/\rho = \sin\varphi. \\ \partial\rho/\partial z &= 0. \end{aligned}$$

$$\begin{aligned} \partial\varphi/\partial x &= -y/\rho^2 = -\sin\varphi/\rho. \\ \partial\varphi/\partial y &= x/\rho^2 = \cos\varphi/\rho. \\ \partial\varphi/\partial z &= 0. \end{aligned}$$

$$\begin{aligned} \partial z/\partial x &= 0. \\ \partial z/\partial y &= 0. \\ \partial z/\partial z &= 1. \end{aligned}$$

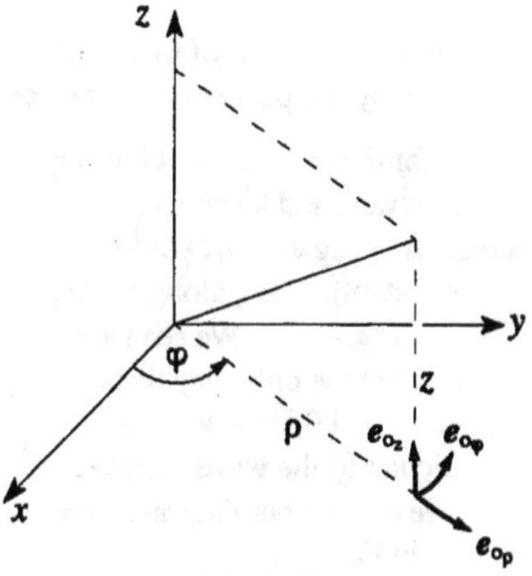

Fig. A-15

In curvilinear coordinates a vector is expressed by $(\mathbf{A})_i = \mathbf{A}\cdot\boldsymbol{e}_{oi}$, where $\boldsymbol{e}_{oi}$ is a unit vector in the ξ_i direction, i.e. $A_\rho = \mathbf{A}\cdot\boldsymbol{e}_{o\rho}$, etc. But, since

$$\boldsymbol{e}_{o\rho} = (\cos\varphi, \sin\varphi, 0), \quad \boldsymbol{e}_{o\varphi} = (-\sin\varphi, \cos\varphi, 0), \quad \boldsymbol{e}_{oz} = (0, 0, 1),$$

$$\begin{aligned} A_\rho &= A_x\cos\varphi + A_y\sin\varphi. \\ A_\varphi &= -A_x\sin\varphi + A_y\cos\varphi. \\ A_z &= A_z. \end{aligned} \quad \rightarrow \quad \begin{aligned} A_\rho &= \frac{xA_x}{\sqrt{x^2+y^2}} + \frac{yA_y}{\sqrt{x^2+y^2}}. \\ A_\varphi &= \frac{-yA_x}{\sqrt{x^2+y^2}} + \frac{xA_y}{\sqrt{x^2+y^2}}. \\ A_z &= A_z. \end{aligned}$$

(b) We also have:

$$\frac{\partial A_x}{\partial x} = \frac{\partial A_x}{\partial \rho}\frac{\partial \rho}{\partial x} + \frac{\partial A_x}{\partial \varphi}\frac{\partial \varphi}{\partial x} + \frac{\partial A_x}{\partial z}\frac{\partial z}{\partial x}.$$

$$\frac{\partial A_y}{\partial y} = \frac{\partial A_y}{\partial \rho}\frac{\partial \rho}{\partial y} + \frac{\partial A_y}{\partial \varphi}\frac{\partial \varphi}{\partial y} + \frac{\partial A_y}{\partial z}\frac{\partial z}{\partial y}.$$

$$\frac{\partial A_z}{\partial z} = \frac{\partial A_z}{\partial \rho}\frac{\partial \rho}{\partial z} + \frac{\partial A_z}{\partial \varphi}\frac{\partial \varphi}{\partial z} + \frac{\partial A_z}{\partial z}\frac{\partial z}{\partial z} = \frac{\partial A_z}{\partial z}.$$

From solution (a):

$$\left\{\begin{aligned} A_\rho\sin\varphi &= A_x\sin\varphi\cos\varphi + A_y\sin^2\varphi. \\ A_\varphi\cos\varphi &= -A_x\sin\varphi\cos\varphi + A_y\cos^2\varphi. \end{aligned}\right\} \rightarrow A_y = A_\rho\sin\varphi + A_\varphi\cos\varphi.$$

$$\left\{\begin{aligned} A_\rho\cos\varphi &= A_x\cos^2\varphi + A_y\sin\varphi\cos\varphi. \\ A_\varphi\cos\varphi &= -A_x\sin^2\varphi + A_y\sin\varphi\cos\varphi. \end{aligned}\right\} \rightarrow A_x = A_\rho\cos\varphi - A_\varphi\sin\varphi.$$

$$\begin{aligned} \frac{\partial A_x}{\partial x} &= \left(\cos\varphi\frac{\partial A_\rho}{\partial \rho} - \sin\varphi\frac{\partial A_\varphi}{\partial \rho}\right)\cos\varphi \\ &\quad + \left(-A_\rho\sin\varphi + \cos\varphi\frac{\partial A_\rho}{\partial \varphi} - A_\varphi\cos\varphi - \sin\varphi\frac{\partial A_\varphi}{\partial \varphi}\right)\left(-\frac{\sin\varphi}{\rho}\right) \\ &= \frac{\partial A_\rho}{\partial \rho}\cos^2\varphi - \frac{\partial A_\varphi}{\partial \rho}\sin\varphi\cos\varphi + \frac{A_\rho\sin^2\varphi}{\rho} - \frac{\sin\varphi\cos\varphi}{\rho}\frac{\partial A_\rho}{\partial \varphi} \\ &\quad + \frac{\sin\varphi\cos\varphi A_\varphi}{\rho} + \frac{\sin^2\varphi}{\rho}\frac{\partial A_\varphi}{\partial \varphi}. \end{aligned}$$

$$\frac{\partial A_y}{\partial y} = \left(\sin\varphi\frac{\partial A_\rho}{\partial\rho} + \cos\varphi\frac{\partial A_\varphi}{\partial\rho}\right)\sin\varphi$$

$$+\left(A_\rho\cos\varphi + \sin\varphi\frac{\partial A_\rho}{\partial\varphi} - A_\varphi\sin\varphi + \cos\varphi\frac{\partial A_\varphi}{\partial\varphi}\right)\left(\frac{\cos\varphi}{\rho}\right)$$

$$= \frac{\partial A_\rho}{\partial\rho}\sin^2\varphi + \frac{\partial A_\varphi}{\partial\rho}\sin\varphi\cos\varphi + \frac{A_\rho\cos^2\varphi}{\rho} + \frac{\sin\varphi\cos\varphi}{\rho}\frac{\partial A_\rho}{\partial\varphi}$$

$$-\frac{\sin\varphi\cos\varphi A_\varphi}{\rho} + \frac{\cos^2\varphi}{\rho}\frac{\partial A_\varphi}{\partial\varphi}.$$

$$\frac{\partial A_z}{\partial z} = \frac{\partial A_z}{\partial z}.$$

$$\frac{\partial A_x}{\partial x} + \frac{\partial A_y}{\partial y} + \frac{\partial A_z}{\partial z} = \frac{\partial A_\rho}{\partial\rho} + \frac{A_\rho}{\rho} + \frac{1}{\rho}\frac{\partial A_\varphi}{\partial\varphi} + \frac{\partial A_z}{\partial z}$$

$$= \frac{1}{\rho}\frac{\partial(\rho A_\rho)}{\partial\rho} + \frac{1}{\rho}\frac{\partial A_\varphi}{\partial\varphi} + \frac{\partial A_z}{\partial z}.$$

V. Express $\nabla^2\Phi$ and $\nabla\times\mathbf{A}$ in cylindrical coordinates.

(a) In cylindrical coordinates,

$$\xi_1 = \rho \quad h_1 = 1$$
$$\xi_2 = \varphi \quad h_2 = \rho$$
$$\xi_3 = z \quad h_3 = 1$$

In general,

$$\nabla\cdot(\nabla\Phi) = \frac{1}{h_1h_2h_3}\left[\frac{\partial(\nabla_1\Phi h_2h_3)}{\partial\xi_1} + \frac{\partial(\nabla_2\Phi h_3h_1)}{\partial\xi_2} + \frac{\partial(\nabla_3\Phi h_1h_2)}{\partial\xi_3}\right].$$

Hence, in cylindrical coordinates,

$$\nabla^2\Phi = \nabla\cdot(\nabla\Phi) = \frac{1}{\rho}\left[\frac{\partial(\rho\nabla_\rho\Phi)}{\partial\rho} + \frac{\partial(\nabla_\varphi\Phi)}{\partial\varphi} + \frac{\partial(\rho\nabla_z\Phi)}{\partial z}\right].$$

But

$$\nabla_\rho\Phi = \frac{\partial\Phi}{\partial\rho},\quad \nabla_\varphi\Phi = \frac{1}{\rho}\frac{\partial\Phi}{\partial\varphi},\quad \nabla_z\Phi = \frac{\partial\Phi}{\partial z}.$$

Hence,

$$\nabla^2\Phi = \frac{1}{\rho}\left[\frac{\partial}{\partial\rho}\left(\rho\frac{\partial\Phi}{\partial\rho}\right) + \frac{\partial}{\partial\varphi}\left(\frac{1}{\rho}\frac{\partial\Phi}{\partial\varphi}\right) + \frac{\partial}{\partial z}\left(\rho\frac{\partial\Phi}{\partial z}\right)\right].$$

$$= \frac{1}{\rho}\left[\frac{\partial}{\partial\rho}\left(\rho\frac{\partial\Phi}{\partial\rho}\right)\right] + \frac{1}{\rho^2}\frac{\partial^2\Phi}{\partial\varphi^2} + \frac{\partial^2\Phi}{\partial z^2}.$$

(b) In general,

$$(\nabla\times\mathbf{A})_i = \frac{1}{h_j h_k}\left[\frac{\partial(A_k h_k)}{\partial\xi_j} - \frac{\partial(A_j h_j)}{\partial\xi_k}\right],$$

so that in cylindrical coordinates:

$$(\nabla\times\mathbf{A})_\rho = \frac{1}{\rho}\left[\frac{\partial A_z}{\partial\varphi} - \frac{\partial(\rho A_\varphi)}{\partial z}\right].$$

$$(\nabla\times\mathbf{A})_\varphi = \left[\frac{\partial A_\rho)}{\partial z} - \frac{\partial(A_z)}{\partial\rho}\right].$$

$$(\nabla\times\mathbf{A})_z = \frac{1}{\rho}\left[\frac{\partial(\rho A_\varphi)}{\partial\rho} - \frac{\partial(A_\rho)}{\partial\varphi}\right].$$

VI. Express $\nabla^2\mathbf{A}$ and $(\mathbf{A}\cdot\nabla)\mathbf{A}$ in spherical coordinates.

(a) $\quad \nabla^2\mathbf{A} = \nabla(\nabla\cdot\mathbf{A}) - \nabla\times(\nabla\times\mathbf{A}).$

For spherical coordinates,

$$\begin{aligned} \xi_1 &= r & h_1 &= 1 \\ \xi_2 &= \theta & h_2 &= r \\ \xi_3 &= \varphi & h_3 &= r\sin\theta \end{aligned}$$

We have shown (p. 60) that

$$\nabla\cdot\mathbf{A} = \frac{1}{r^2}\frac{\partial}{\partial r}(r^2 A_r) + \frac{1}{r\sin\theta}\frac{\partial}{\partial\theta}(\sin\theta A_\theta) + \frac{1}{r\sin\theta}\frac{\partial A_\varphi}{\partial\varphi}.$$

We also know (p. 58) that

$$\left|\nabla(\nabla\cdot\mathbf{A})\right|_r = \frac{\partial}{\partial r}(\nabla\cdot\mathbf{A}).$$

$$\left|\nabla(\nabla\cdot\mathbf{A})\right|_\theta = \frac{1}{r}\frac{\partial}{\partial\theta}(\nabla\cdot\mathbf{A}).$$

$$\left|\nabla(\nabla\cdot\mathbf{A})\right|_\varphi = \frac{1}{r\sin\theta}\frac{\partial}{\partial\varphi}(\nabla\cdot\mathbf{A}).$$

From page 62 we have

$$\left|\nabla\times(\nabla\times\mathbf{A})\right|_r = \frac{1}{r^2\sin\theta}\left\{\frac{\partial}{\partial\theta}r\sin\theta(\nabla\times\mathbf{A})_\varphi - \frac{\partial}{\partial\varphi}r(\nabla\times\mathbf{A})_\theta\right\}$$

$$= \frac{1}{r\sin\theta}\left\{\frac{\partial}{\partial\theta}\sin\theta(\nabla\times\mathbf{A})_\varphi - \frac{\partial}{\partial\varphi}(\nabla\times\mathbf{A})_\theta\right\}.$$

$$\left|\nabla\times(\nabla\times\mathbf{A})\right|_\theta = \frac{1}{r\sin\theta}\left\{\frac{\partial}{\partial\varphi}(\nabla\times\mathbf{A})_r - \frac{\partial}{\partial r}r\sin\theta(\nabla\times\mathbf{A})_\varphi\right\}.$$

$$\left|\nabla\times(\nabla\times\mathbf{A})\right|_\varphi = \frac{1}{r}\left\{\frac{\partial}{\partial r}r(\nabla\times\mathbf{A})_\theta - \frac{\partial}{\partial\theta}(\nabla\times\mathbf{A})_r\right\}.$$

Substituting the appropriate expressions for $(\nabla\times\mathbf{A})_r$, $(\nabla\times\mathbf{A})_\theta$ and $(\nabla\times\mathbf{A})_\varphi$ (page 62) yields

$$\left|\nabla\times(\nabla\times\mathbf{A})\right|_r = \frac{1}{r\sin\theta}\left\{\frac{\partial}{\partial\theta}\left[\frac{\sin\theta}{r}\left(\frac{\partial(rA_\theta)}{\partial r} - \frac{\partial A_r}{\partial\theta}\right)\right]\right\}$$

$$- \frac{1}{r\sin\theta}\left\{\frac{\partial}{\partial\varphi}\left[\frac{1}{r\sin\theta}\left(\frac{\partial A_r}{\partial\varphi} - \frac{\partial(r\sin\theta A_\varphi)}{\partial r}\right)\right]\right\}.$$

$$\left|\nabla\times(\nabla\times\mathbf{A})\right|_\theta = \frac{1}{r\sin\theta}\left\{\frac{\partial}{\partial\varphi}\left[\frac{1}{r\sin\theta}\left(\frac{\partial(\sin\theta A_\varphi)}{\partial\theta} - \frac{\partial A_\theta}{\partial\varphi}\right)\right]\right\}$$

$$- \frac{1}{r\sin\theta}\left\{\frac{\partial}{\partial r}\left[\sin\theta\left(\frac{\partial(rA_\theta)}{\partial r} - \frac{\partial A_r}{\partial\theta}\right)\right]\right\}.$$

$$\left|\nabla\times(\nabla\times\mathbf{A})\right|_\varphi = \frac{1}{r}\left\{\frac{\partial}{\partial r}\left[\frac{1}{\sin\theta}\left(\frac{\partial A_r}{\partial\varphi} - \frac{\partial(r\sin\theta A_\varphi)}{\partial r}\right)\right]\right\}$$

$$-\frac{1}{r}\left\{\frac{\partial}{\partial\theta}\left[\frac{1}{r\sin\theta}\left(\frac{\partial(\sin\theta A_\varphi)}{\partial\theta} - \frac{\partial A_\theta}{\partial\varphi}\right)\right]\right\}.$$

Combining:

$$\left|\nabla^2\mathbf{A}\right|_r = \left|\nabla(\nabla\cdot\mathbf{A})\right|_r - \left|\nabla\times(\nabla\times\mathbf{A})\right|_r .$$

$$\left|\nabla^2\mathbf{A}\right|_\theta = \left|\nabla(\nabla\cdot\mathbf{A})\right|_\theta - \left|\nabla\times(\nabla\times\mathbf{A})\right|_\theta .$$

$$\left|\nabla^2\mathbf{A}\right|_\varphi = \left|\nabla(\nabla\cdot\mathbf{A})\right|_\varphi - \left|\nabla\times(\nabla\times\mathbf{A})\right|_\varphi .$$

The importance of the above result lies in the fact that we cannot compute $\nabla^2\mathbf{A}$ from $\mathbf{A}$ in the case of general orthogonal curvilinear coordinates as we could in cartesian coordinates. For instance, we ask: Is $\nabla^2\mathbf{A}$ given in spherical coordinates by the expressions:

$$\left|\nabla^2\mathbf{A}\right|_r = \nabla^2 A_r\,? \quad \left|\nabla^2\mathbf{A}\right|_\theta = \nabla^2 A_\theta\,? \quad \left|\nabla^2\mathbf{A}\right|_\varphi = \nabla^2 A_\varphi\,?$$

The answer is NO! and we cite a simple example. Consider the vector

$$\mathbf{A} = \boldsymbol{i}_o x + \boldsymbol{j}_o y + \boldsymbol{k}_o z = \mathbf{r}.$$

$$\nabla^2\mathbf{r} = \boldsymbol{i}_o\nabla^2 x + \boldsymbol{j}_o\nabla^2 y + \boldsymbol{k}_o\nabla^2 z = 0,$$

if we use cartesian coordinates and apply $\left|\nabla^2\mathbf{A}\right|_i = \nabla^2 A_i$. On the other hand, if we try the same in spherical coordinates:

$$\left|\nabla^2\mathbf{A}\right|_r = \nabla^2 A_r = \nabla^2 r = \frac{1}{r^2}\left[\frac{\partial}{\partial r}\left(r^2\frac{\partial r}{\partial r}\right)\right] = \frac{1}{r^2}\left[\frac{\partial r^2}{\partial r}\right] = \frac{2r}{r^2} = \frac{2}{r}.$$

Thus, $\nabla^2\mathbf{A} = \dfrac{2\mathbf{r}_o}{r} \neq 0.$

(b) We employ the identity:

$$\nabla(\mathbf{A}\cdot\mathbf{B}) = \mathbf{A}\times(\nabla\times\mathbf{B}) + \mathbf{B}\times(\nabla\times\mathbf{A}) + (\mathbf{A}\cdot\nabla)\mathbf{B} + (\mathbf{B}\cdot\nabla)\mathbf{A}.$$

Setting $\mathbf{A} = \mathbf{B}$, we get $\nabla(A^2) = 2\mathbf{A}\times(\nabla\times\mathbf{A}) + 2(\mathbf{A}\cdot\nabla)\mathbf{A}$, so that

$$(1)\qquad (\mathbf{A}\cdot\nabla)\mathbf{A} = \frac{\nabla(A^2)}{2} - \mathbf{A}\times(\nabla\times\mathbf{A}).$$

$A^2 = \{\text{magnitude of } \mathbf{A}\}^2$, which is invariant to coordinate system transformations. In cartesian coordinates, $A^2 = A_x^2 + A_y^2 + A_z^2$.

One can find $A_r\, A_\theta\, A_\varphi$ in terms of $A_x\, A_y\, A_z$ by using the definition of a vector in the new $r\ \theta\ \varphi$ system, as follows: $(\mathbf{A})_i = (\mathbf{A}\cdot \boldsymbol{e}_{oi})$, where $\boldsymbol{e}_{oi}$ is a unit vector in the ith direction, i.e.

$$\begin{aligned} \boldsymbol{e}_{or} &= (\sin\theta\cos\varphi, \sin\theta\sin\varphi, \cos\theta).\\ \boldsymbol{e}_{o\theta} &= (\cos\theta\cos\varphi, \cos\theta\sin\varphi, -\sin\theta).\\ \boldsymbol{e}_{o\varphi} &= (-\sin\varphi, \cos\varphi, 0).\end{aligned}$$

and

$$\begin{aligned} A_r &= A_x\sin\theta\cos\varphi + A_y\sin\theta\sin\varphi + A_z\cos\theta.\\ A_\theta &= A_x\cos\theta\cos\varphi + A_y\cos\theta\sin\varphi - A_z\sin\theta.\\ A_\varphi &= -A_x\sin\varphi + A_y\cos\varphi.\end{aligned}$$

Solving for $A_x\, A_y\, A_z$:

$$\begin{aligned} A_x &= A_r\sin\theta\cos\varphi + A_\theta\cos\theta\sin\varphi - A_\varphi\sin\varphi.\\ A_y &= A_r\sin\theta\sin\varphi + A_\theta\cos\theta\sin\varphi + A_\varphi\cos\varphi.\\ A_z &= A_r\cos\theta - A_\theta\sin\theta.\end{aligned}$$

Evaluating $A^2 = A_x^2 + A_y^2 + A_z^2$, one finds it equal to $A_r^2 + A_\theta^2 + A_\varphi^2$. This result should not surprise us since, in spherical coordinates,

$$\mathbf{A} = A_r\boldsymbol{e}_{or} + A_\theta\boldsymbol{e}_{o\theta} + A_\varphi\boldsymbol{e}_{o\varphi} \Rightarrow \mathbf{A}\cdot\mathbf{A} = A^2 = A_r^2 + A_\theta^2 + A_\varphi^2.$$

Therefore,

$$\left|\nabla(A^2)\right|_i = \frac{1}{h_i}\frac{\partial}{\partial\xi_i}\left[A_r^2 + A_\theta^2 + A_\varphi^2\right].$$

Regarding the second term in Eq. (1), we can solve for (**A**×**B**) by noting that in spherical coordinates,

$$\mathbf{A} = A_r e_{or} + A_\theta e_{o\theta} + A_\varphi e_{o\varphi}.$$

$$\mathbf{B} = B_r e_{or} + B_\theta e_{o\theta} + B_\varphi e_{o\varphi}.$$

$$\Downarrow$$

$$\mathbf{A}\times\mathbf{B} = (A_\theta B_\varphi - A_\varphi B_\theta)e_{or} + (A_\varphi B_r - A_r B_\varphi)e_{o\theta} + (A_r B_\theta - A_\theta B_r)e_{o\varphi}.$$

Substituting the expressions for the gradient and the curl (pages 58 and 62, respectively) into Eq. (1) above,

$$\left|\mathbf{A}\cdot\nabla)\mathbf{A}\right|_r = \frac{1}{2}\frac{\partial}{\partial r}\left[A_r^2 + A_\theta^2 + A_\varphi^2\right]$$

$$-\left\{\frac{A_\theta}{r}\left(\frac{\partial(rA_\theta)}{\partial r} - \frac{\partial A_r}{\partial\theta}\right) - \frac{A_\varphi}{r\sin\theta}\left(\frac{\partial A_r}{\partial\varphi} - \frac{\partial(r\sin\theta A_\varphi)}{\partial r}\right)\right\}.$$

$$\left|\mathbf{A}\cdot\nabla)\mathbf{A}\right|_\theta = \frac{1}{2r}\frac{\partial}{\partial\theta}\left[A_r^2 + A_\theta^2 + A_\varphi^2\right]$$

$$-\left\{\frac{A_\varphi}{r^2\sin\theta}\left(\frac{\partial(r\sin\theta A_\varphi)}{\partial\theta} - \frac{\partial(rA_\theta)}{\partial\varphi}\right) - \frac{A_r}{r}\left(\frac{\partial(rA_\theta)}{\partial r} - \frac{\partial A_r}{\partial\theta}\right)\right\}.$$

$$\left|\mathbf{A}\cdot\nabla)\mathbf{A}\right|_\varphi = \frac{1}{2r\sin\theta}\frac{\partial}{\partial\varphi}\left[A_r^2 + A_\theta^2 + A_\varphi^2\right]$$

$$-\left\{\frac{A_r}{r\sin\theta}\left(\frac{\partial A_r}{\partial\varphi} - \frac{\partial(r\sin\theta A_\varphi)}{\partial r}\right)\right\}$$

$$+\left\{\frac{A_\theta}{r^2\sin\theta}\left(\frac{\partial(r\sin\theta A_\varphi)}{\partial\theta} - \frac{\partial(rA_\theta)}{\partial\varphi}\right)\right\}.$$

Mid-Term Exam Problem Set

I. Vectors **ABC** are related to three other vectors **abc** through the following relations:

$$A_i = \sum_{j=1}^{3} T_{ij}a_j; \quad B_i = \sum_{j=1}^{3} T_{ij}b_j; \quad C_i = \sum_{j=1}^{3} T_{ij}c_j,$$

where T_{ij} is a symmetrical tensor of the 2nd rank with eigenvalues $\lambda_1\lambda_2\lambda_3$.

Prove that

$$\mathbf{A}\cdot(\mathbf{B}\times\mathbf{C}) = \lambda_1\lambda_2\lambda_3[\mathbf{a}\cdot(\mathbf{b}\times\mathbf{c})].$$

Solution: Note that $\mathbf{A}\cdot(\mathbf{B}\times\mathbf{C})$ is a scalar and hence invariant to transformations. Let us rotate to that particular system in which $T_{ij} = \delta_{ij}\lambda_i$, i.e. the principal axes.

$$\mathbf{A}\cdot(\mathbf{B}\times\mathbf{C}) \rightarrow \mathbf{A}'\cdot(\mathbf{B}'\times\mathbf{C}') = \sum_i A_i' \sum_{jk} \varepsilon_{ijk}B_j'C_k'$$

$$= \sum_{ijk} \varepsilon_{ijk} \sum_l T_{il}'a_l' \sum_m T_{jm}'b_m' \sum_n T_{kn}'c_n'$$

$$= \sum_{ijk} \varepsilon_{ijk} \sum_l \lambda_i\delta_{il}a_l' \sum_m \lambda_j\delta_{jm}b_m' \sum_n \lambda_k\delta_{kn}c_n'$$

$$= \sum_{ijk} \varepsilon_{ijk}\lambda_i a_i'\lambda_j b_j'\lambda_k c_k' = \sum_{ijk} \lambda_i\lambda_j\lambda_k\varepsilon_{ijk}a_i'b_j'c_k'$$

$$= \sum_i \lambda_i a_i' \sum_{jk} \varepsilon_{ijk}\lambda_j b_j\lambda_k' c_k' = \lambda_1\lambda_2\lambda_3\, \mathbf{a}'\cdot(\mathbf{b}'\times\mathbf{c}')$$

$$= \lambda_1\lambda_2\lambda_3\, \mathbf{a}\cdot(\mathbf{b}\times\mathbf{c}).$$

The last statement follows since $\mathbf{a}\cdot(\mathbf{b}\times\mathbf{c})$ is a scalar and thus invariant to the transformation.

II. If e_o is a constant unit vector, prove that $V = \left|\iint_S (e_o\cdot\mathbf{r})d\mathbf{S}\right|$, where V is the volume of a region bounded by the closed surface S.

Solution: Since e_o is a constant unit vector, $V = \left|\int_V e_o d\tau\right|$. But we note that

$$\nabla(e_o \cdot \mathbf{r}) = (e_o \cdot \nabla)\mathbf{r} + (\mathbf{r} \cdot \nabla)e_o + e_o \times (\nabla \times \mathbf{r}) + \mathbf{r} \times (\nabla \times e_o).$$

Noting that e_o is a constant unit vector, $(\mathbf{r} \cdot \nabla)e_o = (\nabla \times \mathbf{r}) = (\nabla \times e_o) = 0$, so that $\nabla(e_o \cdot \mathbf{r}) = (e_o \cdot \nabla)\mathbf{r} = e_o$. Hence, from Gauss' theorem,

$$V = \left|\int_V e_o d\tau\right| = \left|\int_V \nabla(e_o \cdot \mathbf{r}) d\tau\right| = \left|\int_S (e_o \cdot \mathbf{r}) d\mathbf{S}\right|.$$

III. Let S be the area of a surface which lies completely on a plane and let t_o be a constant unit vector lying on the same plane. Prove that

$$S = \left|\oint_s (t_o \cdot \mathbf{r}) ds\right|,$$

where S is bounded by the closed curve s.

Solution: Since $t_o \times d\mathbf{S} = |t_o||d\mathbf{S}| \sin(t_o, d\mathbf{S}) = dS$, and since $\nabla(t_o \cdot \mathbf{r}) = t_o$ from Prob. II above, we have

$$S = \left|\int_S t_o \times d\mathbf{S})\right| = \left|\int_S \nabla(t_o \cdot \mathbf{r}) \times d\mathbf{S})\right| = \left|\int_S \sum_{jk} \varepsilon_{ijk} \frac{\partial}{\partial x_j} (t_o \cdot \mathbf{r}) dS_k\right|$$

$$= \left|\oint_s (t_o \cdot \mathbf{r}) ds_i\right|,$$

by Stokes' theorem.

Problem Set G

I. Pauli spin matrices are defined as:

$$\sigma_1 = \begin{pmatrix} 0 & 1 \\ 1 & 0 \end{pmatrix} \quad \sigma_2 = \begin{pmatrix} 0 & -i \\ i & 0 \end{pmatrix} \quad \sigma_3 = \begin{pmatrix} 1 & 0 \\ 0 & -1 \end{pmatrix}$$

Prove:

$$(a) \quad \sigma_i^{+} = \sigma_i \quad (i = 1, 2, 3)$$

$$(b) \quad \sigma_i\sigma_j = \delta_{ij} + i\sum_k \varepsilon_{ijk}\sigma_k \quad (ij \text{ varies from 1 to 3})$$

$$(c) \quad \sigma_i\sigma_j + \sigma_j\sigma_i = 2\delta_{ij}$$

Solutions:

(a)

$$\sigma_1^{+} = \begin{pmatrix} 0 & 1 \\ 1 & 0 \end{pmatrix} = \sigma_1 .$$

$$\sigma_2^{+} = \begin{pmatrix} 0 & +i^* \\ -i^* & 0 \end{pmatrix} = \begin{pmatrix} 0 & -i \\ i & 0 \end{pmatrix} = \sigma_2 .$$

$$\sigma_3^{+} = \begin{pmatrix} 1 & 0 \\ 0 & -1 \end{pmatrix} = \sigma_3 .$$

(b) We first note that $\sigma_i\sigma_j = \delta_{ij} + i\sum_k \varepsilon_{ijk}\sigma_k$ implies that

$$\sigma_1\sigma_2 = i\sigma_3 . \qquad \sigma_2\sigma_1 = -i\sigma_3 .$$

$$\sigma_2\sigma_3 = i\sigma_1 . \qquad \sigma_3\sigma_2 = -i\sigma_1 .$$

$$\sigma_3\sigma_1 = i\sigma_2 . \qquad \sigma_1\sigma_3 = -i\sigma_2 .$$

$$\sigma_1^2 = 1.$$

$$\sigma_2^2 = 1.$$

$$\sigma_3^2 = 1.$$

This, then, is what we have to show.

$$\sigma_1\sigma_2 = \begin{pmatrix} 0 & 1 \\ 1 & 0 \end{pmatrix}\begin{pmatrix} 0 & -i \\ i & 0 \end{pmatrix} = \begin{pmatrix} i & 0 \\ 0 & -i \end{pmatrix} = i\begin{pmatrix} 1 & 0 \\ 0 & -1 \end{pmatrix} = i\sigma_3.$$

$$\sigma_2\sigma_3 = \begin{pmatrix} 0 & -i \\ i & 0 \end{pmatrix}\begin{pmatrix} 1 & 0 \\ 0 & -1 \end{pmatrix} = \begin{pmatrix} 0 & i \\ i & 0 \end{pmatrix} = i\begin{pmatrix} 0 & 1 \\ 1 & 0 \end{pmatrix} = i\sigma_1.$$

$$\sigma_3\sigma_1 = \begin{pmatrix} 1 & 0 \\ 0 & -1 \end{pmatrix}\begin{pmatrix} 0 & 1 \\ 1 & 0 \end{pmatrix} = \begin{pmatrix} 0 & 1 \\ -1 & 0 \end{pmatrix} = i\begin{pmatrix} 0 & -i \\ i & 0 \end{pmatrix} = i\sigma_2.$$

$$\sigma_2\sigma_1 = \begin{pmatrix} 0 & -i \\ i & 0 \end{pmatrix}\begin{pmatrix} 0 & 1 \\ 1 & 0 \end{pmatrix} = \begin{pmatrix} -i & 0 \\ 0 & i \end{pmatrix} = -i\begin{pmatrix} 1 & 0 \\ 0 & -1 \end{pmatrix} = -i\sigma_3.$$

$$\sigma_3\sigma_2 = \begin{pmatrix} 1 & 0 \\ 0 & -1 \end{pmatrix}\begin{pmatrix} 0 & -i \\ i & 0 \end{pmatrix} = \begin{pmatrix} 0 & -i \\ -i & 0 \end{pmatrix} = -i\begin{pmatrix} 0 & 1 \\ 1 & 0 \end{pmatrix} = -i\sigma_1.$$

$$\sigma_1\sigma_3 = \begin{pmatrix} 0 & 1 \\ 1 & 0 \end{pmatrix}\begin{pmatrix} 1 & 0 \\ 0 & -1 \end{pmatrix} = \begin{pmatrix} 0 & -1 \\ 1 & 0 \end{pmatrix} = -i\begin{pmatrix} 0 & -i \\ i & 0 \end{pmatrix} = -i\sigma_2.$$

$$\sigma_1^2 = \begin{pmatrix} 0 & 1 \\ 1 & 0 \end{pmatrix}\begin{pmatrix} 0 & 1 \\ 1 & 0 \end{pmatrix} = \begin{pmatrix} 1 & 0 \\ 0 & 1 \end{pmatrix} = I.$$

$$\sigma_2^2 = \begin{pmatrix} 0 & -i \\ i & 0 \end{pmatrix}\begin{pmatrix} 0 & -i \\ i & 0 \end{pmatrix} = \begin{pmatrix} 1 & 0 \\ 0 & 1 \end{pmatrix} = I.$$

$$\sigma_3^2 = \begin{pmatrix} 1 & 0 \\ 0 & -1 \end{pmatrix}\begin{pmatrix} 1 & 0 \\ 0 & -1 \end{pmatrix} = \begin{pmatrix} 1 & 0 \\ 0 & 1 \end{pmatrix} = I.$$

Thus, for $i = j$,

$$\sigma_i\sigma_j = \delta_{ij},$$

while for $i \neq j$,

$$\sigma_i\sigma_j = i\sigma_k \begin{cases} +1 \\ -1 \end{cases} \text{for } ijk \text{ an } \begin{cases} \text{even} \\ \text{odd} \end{cases} \text{permutation of } 1, 2, 3.$$

(c) Since for $i \neq j$, $\sigma_i\sigma_j = i\sigma_k$ and $\sigma_j\sigma_i = -i\sigma_k$, we have $\sigma_i\sigma_j + \sigma_j\sigma_i = 0$, while for $i = j$, $\sigma_i\sigma_j = I$, so that $\sigma_i\sigma_j + \sigma_j\sigma_i = 2I$. Therefore the expression $\sigma_i\sigma_j + \sigma_j\sigma_i = 2\delta_{ij}$ is satisfied.

II. Hamilton's generalization of pure imaginaries $\underline{ijk}$ may be defined as

$$\underline{i} = -i\sigma_1. \quad \underline{j} = -i\sigma_2. \quad \underline{k} = -i\sigma_3.$$

Show that $\underline{ijk}$ obey the following rules of multiplication:

(a) $\underline{i}^2 = \underline{j}^2 = \underline{k}^2 = -1.$

(b) $\underline{ij} = \underline{-ji} = \underline{k}. \quad \underline{jk} = \underline{-kj} = \underline{i}. \quad \underline{ki} = \underline{-ik} = \underline{j}.$

Solution:

(a)
$$\underline{i}^2 = (-i\sigma_1)(-i\sigma_1) = -\sigma_1^2 = -I.$$
$$\underline{j}^2 = (-i\sigma_2)(-i\sigma_2) = -\sigma_2^2 = -I.$$
$$\underline{k}^2 = (-i\sigma_3)(-i\sigma_3) = -\sigma_3^2 = -I.$$

Thus we see that Hamilton's imaginaries $\underline{ijk}$ obey the same rules as our "ordinary" imaginaries, i.e. $i^2 = -1$, where by -1 we mean $\begin{pmatrix} -1 & 0 \\ 0 & -1 \end{pmatrix}$.

(b)
$$\underline{ij} = (-i\sigma_1)(-i\sigma_2) = -\sigma_1\sigma_2 = -i\sigma_3 = \underline{k}.$$
$$\underline{ji} = (-i\sigma_2)(-i\sigma_1) = -\sigma_2\sigma_1 = +i\sigma_3 = -\underline{k}.$$
$$\underline{jk} = (-i\sigma_2)(-i\sigma_3) = -\sigma_2\sigma_3 = -i\sigma_1 = \underline{i}.$$
$$\underline{kj} = (-i\sigma_3)(-i\sigma_2) = -\sigma_3\sigma_2 = +i\sigma_1 = -\underline{i}.$$
$$\underline{ki} = (-i\sigma_3)(-i\sigma_1) = -\sigma_3\sigma_1 = -i\sigma_2 = \underline{j}.$$
$$\underline{ik} = (-i\sigma_1)(-i\sigma_3) = -\sigma_1\sigma_3 = +i\sigma_2 = -\underline{j}.$$

III. A quaternion Q is defined to be $Q = t + \underline{i}x + \underline{j}y + \underline{k}z$, where $t\ x\ y\ z$ are real numbers.

(Notice that all real numbers are quaternions.)

Prove that:

(*a*) If Q_1 and Q_2 are quaternions, then $Q_1 + Q_2$ and $Q_1 \cdot Q_2$ are quaternions.

(*b*) Q^+Q is a real number, where Q^+ is called the "complex conjugate" of Q and is given by $Q^+ = t - \underline{i}x - \underline{j}y - \underline{k}z$ while $Q = t + \underline{i}x + \underline{j}y + \underline{k}z$.
(Note that Q^+ is also a quaternion.

(*c*) For every quaternion Q there exists another quaternion Q^{-1} such that $QQ^{-1} = Q^{-1}Q = I$.

Solution:

(a) We have to show that $Q_1 + Q_2$ and $Q_1 \cdot Q_2$ can be expressed in the form $t + \underline{i}x + \underline{j}y + \underline{k}z$.

If $Q_1 = t_1 + \underline{i}x_1 + \underline{j}y_1 + \underline{k}z_1$ and $Q_2 = t_2 + \underline{i}x_2 + \underline{j}y_2 + \underline{k}z_2$, then

$$Q_1 + Q_2 = (t_1 + t_2) + (\underline{i}x_1 + \underline{i}x_2) + (\underline{j}y_1 + \underline{j}y_2) + (\underline{k}z_1 + \underline{k}z_2).$$

But since $(\underline{i}x_1 + \underline{i}x_2)$ is simply the addition of two matrices:

$$(\underline{i}x_1 + \underline{i}x_2) = \begin{pmatrix} 0 & -ix_1 \\ -ix_1 & 0 \end{pmatrix} + \begin{pmatrix} 0 & -ix_2 \\ -ix_2 & 0 \end{pmatrix} = \begin{pmatrix} 0 & -i(x_1 + x_2) \\ -i(x_1 + x_2) & 0 \end{pmatrix}$$

$$= -i\sigma_1(x_1 + x_2) = +\underline{i}(x_1 + x_2).$$

Hence, $Q_1 + Q_2 = (t_1 + t_2) + \underline{i}(x_1 + x_2) + \underline{j}(y_1 + y_2) + \underline{k}(z_1 + z_2)$, which is clearly seen to be a quaternion.

$$Q_1 \cdot Q_2 = (t_1 + \underline{i}x_1 + \underline{j}y_1 + \underline{k}z_1) \cdot (t_2 + \underline{i}x_2 + \underline{j}y_2 + \underline{k}z_2)$$

$$= T + \underline{i}X + \underline{j}Y + \underline{k}Z,$$

where

$$T = t_1t_2 - x_1x_2 - y_1y_2 - z_1z_2 \quad \rightarrow \quad T \text{ is real.}$$

$$X = x_1t_2 + x_2t_1 + y_1z_2 - y_2z_1 \quad \rightarrow \quad X \text{ is real.}$$

$$Y = y_1t_2 + y_2t_1 + z_1x_2 - z_2x_1 \quad \rightarrow \quad Y \text{ is real.}$$

$$Z = z_2t_1 + z_1t_2 + x_1y_2 - x_2y_1 \quad \rightarrow \quad Z \text{ is real.}$$

Hence, $Q_1 \cdot Q_2$ is a quaternion.

(b) If $Q^+ = t - \underline{i}x - \underline{j}y - \underline{k}z$, then

$$Q^+ \cdot Q = (t - \underline{i}x - \underline{j}y - \underline{k}z)(t + \underline{i}x + \underline{j}y + \underline{k}z)$$
$$= t^2(\text{real number}) + (x^2 + y^2 + z^2)(\text{real number}).$$
$$= t^2 + x^2 + y^2 + z^2 \text{ (real number)}.$$

(c) We have seen that $Q^+ \cdot Q$ yields a real number. Let us so choose the quaternion Q' to resemble Q^+ yet yield the real number $Q'Q = 1$. In this way we will have formed Q^{-1}.

Letting $Q' = \dfrac{t - ix - jy - kz}{D}$, where $D = t^2 + x^2 + y^2 + z^2$, we have

$$Q'Q = \frac{(t - ix - jy - kz)(t + ix + jy + kz)}{D} = \frac{t^2 + x^2 + y^2 + z^2}{D} = I.$$

$$Q' = Q^{-1} = \frac{t - ix - jy - kz}{t^2 + x^2 + y^2 + z^2}.$$

Observe also that

$$QQ' = \frac{(t + ix + jy + kz)(t - ix - jy - kz)}{D} = I.$$

Thus, $QQ' = Q'Q = I$, and $Q^{-1} = \dfrac{Q^+}{Q^+Q}$.

IV. Let systems Σ and Σ' both be inertial systems, and let system Σ move with a uniform velocity υ_o with respect to system Σ' along the x_1 direction (Fig. A-16). If a particle moves with a constant velocity υ along the same direction, as observed in system Σ, show that the same particle, if observed in the Σ system, would move with a velocity given by

$$\upsilon' = \frac{\upsilon + \upsilon_o}{1 + \dfrac{\upsilon\upsilon_o}{c^2}},$$

where c is the velocity of light.

Solution: The problem is to find υ', the velocity of point P, in system Σ' (Fig. A-16). The Lorentz transformation between systems Σ and Σ' is given by

$$x' = \frac{x + \upsilon_0 t}{\sqrt{1-\beta^2}}, \quad t' = \frac{t + (\upsilon_0 x/c^2)}{\sqrt{1-\beta^2}},$$

where $\beta = \upsilon_0/c$.

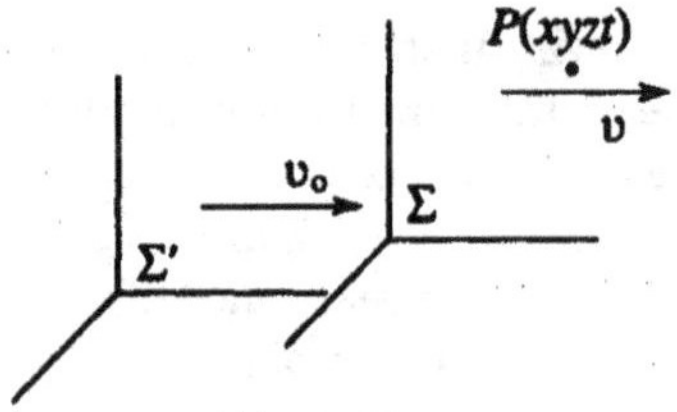

Fig. A-16

In system Σ, $x = x_0 + \upsilon t$, and substituting:

$$x' = \frac{x_0 + (\upsilon + \upsilon_0)t}{\sqrt{1-\beta^2}}, \quad t' = \frac{t + \upsilon_0(x_0 + \upsilon t)/c^2}{\sqrt{1-\beta^2}}.$$

$$\left(\frac{\partial x'}{\partial t}\right)_{\upsilon x_0} = \frac{\upsilon + \upsilon_0}{\sqrt{1-\beta^2}}, \quad \left(\frac{\partial t'}{\partial t}\right)_{\upsilon x_0} = \frac{1 + \upsilon\upsilon_0/c^2}{\sqrt{1-\beta^2}}.$$

Therefore,

$$\upsilon' = \frac{dx'}{dt'} = \frac{\left(\frac{\partial x'}{\partial t}\right)_{\upsilon x_0}}{\left(\frac{\partial t'}{\partial t}\right)_{\upsilon x_0}} = \frac{\upsilon + \upsilon_0}{1 + \upsilon\upsilon_0/c^2}.$$

Note that by combining 2 velocities, each of which is less than c, there results a velocity less than c, for if $\upsilon = c - k$ and $\upsilon_0 = c - \omega$, where both k and ω are greater than 0 but less than c,

$$\upsilon' = \frac{(c - k + c - \omega)c}{c + \frac{c^2 + \omega k + c(-k-\omega)}{c}} = \frac{c(2c - k - \omega)}{(2c - k - \omega) + \frac{\omega k}{c}}$$

$$= \frac{c}{1 + \frac{\omega k}{c(2c - k - \omega)}} < c.$$

Note also that if we combine the velocity of light c with any other velocity less than c, we still get a velocity $\upsilon' = c$:

$$\upsilon' = \frac{c + \upsilon_0}{1 + c\upsilon_0/c^2} = \frac{c + \upsilon_0}{(c + \upsilon_0)/c} = c.$$

Following is an illustration in the use of the matrix method to solve the problem of compounding two velocities, and which therefore represents another method of solving the given problem.

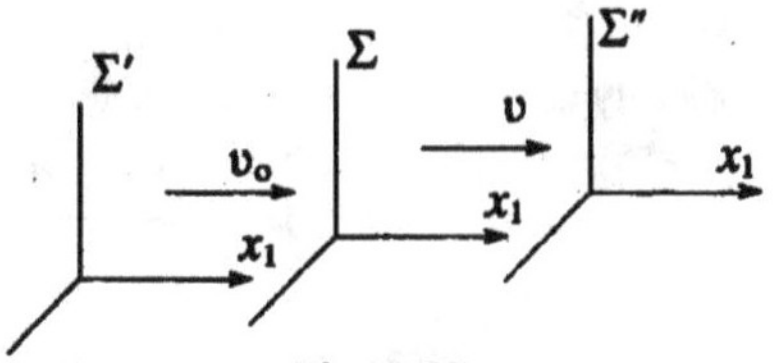

Fig. A-17

Referring to Fig. A-17, $x = V_1 x'$, while $x'' = V_2 x$, so that $x'' = V_2 V_1 x'$, where V_2 and V_1 are the following Lorentz transformations:

$$V_1 = \begin{pmatrix} \frac{1}{\sqrt{1-\beta_1^2}} & 0 & 0 & \frac{i\beta_1}{\sqrt{1-\beta_1^2}} \\ 0 & 1 & 0 & 0 \\ 0 & 0 & 1 & 0 \\ \frac{-i\beta_1}{\sqrt{1-\beta_1^2}} & 0 & 0 & \frac{1}{\sqrt{1-\beta_1^2}} \end{pmatrix}, \quad V_2 = \begin{pmatrix} \frac{1}{\sqrt{1-\beta_2^2}} & 0 & 0 & \frac{i\beta_2}{\sqrt{1-\beta_2^2}} \\ 0 & 1 & 0 & 0 \\ 0 & 0 & 1 & 0 \\ \frac{-i\beta_2}{\sqrt{1-\beta_2^2}} & 0 & 0 & \frac{1}{\sqrt{1-\beta_2^2}} \end{pmatrix}.$$

$\beta_1 = \upsilon_o/c$, $\beta_2 = \upsilon/c$, and the resulting transformation is given by $W = V_2 V_1$, where

$$W_{11} = \frac{1}{\sqrt{(1-\beta_2^2)(1-\beta_1^2)}} + \frac{\beta_1\beta_2}{\sqrt{(1-\beta_2^2)(1-\beta_1^2)}}$$

$$= \frac{1}{\left(\frac{1-\beta_1^2-\beta_2^2+\beta_1^2\beta_2^2}{(1+\beta_1^2\beta_2^2)^2}\right)^{1/2}} = \frac{1}{\sqrt{1-\beta_3^2}}.$$

$$\beta_3 = \frac{\upsilon_3}{c} = \frac{\beta_1+\beta_2}{1+\beta_1\beta_2}.$$

Thus,

$$\upsilon_3 = \frac{\upsilon_o+\upsilon}{1+\upsilon_o\upsilon/c^2}$$

is the velocity of the moving system Σ" as observed in system Σ'.

V. If under a Lorentz transformation one regards the electromagnetic field **E** and **H** as forming an antisymmetric tensor of 2nd rank $F_{\mu\upsilon}$ while the electric current **J** and charge density ρ form a 4-vector j_μ, then show that Maxwell's equations can be written as

$$(a)\quad \sum_{\upsilon=1}^{4} \frac{\partial}{\partial x_\upsilon} F_{\mu\upsilon} = \frac{4\pi j_\mu}{c}. \quad (\mu = 1\ldots4)$$

$$(b)\quad \sum_{\upsilon\kappa\rho} \varepsilon_{\mu\upsilon\kappa\rho} \frac{\partial}{\partial x_\upsilon} F_{\upsilon\rho} = 0,$$

where

$$F_{\mu\upsilon} = -F_{\upsilon\mu},$$
$$F_{12} = H_3, \quad F_{23} = H_1, \quad F_{31} = H_2,$$
$$F_{14} = -iE_1, \quad F_{24} = -iE_2, \quad F_{34} = -iE_3,$$

and where $j_\mu = (\mathbf{J}, i\rho c)$.

Solution:

$$F_{\mu\upsilon} = \begin{pmatrix} 0 & H_3 & -H_2 & -iE_1 \\ -H_3 & 0 & H_1 & -iE_2 \\ H_2 & -H_1 & 0 & -iE_3 \\ iE_1 & iE_2 & iE_3 & 0 \end{pmatrix} \qquad j_\mu = \begin{pmatrix} J_1 \\ J_2 \\ J_3 \\ i\rho c \end{pmatrix} \qquad x_\mu = \begin{pmatrix} x \\ y \\ z \\ ict \end{pmatrix}$$

(a) Consider $\sum_{i=1}^{4} \frac{\partial}{\partial x_i} F_{ji} = \frac{4\pi j_i}{c}$:

For $j = 1$:

$$\frac{\partial F_{12}}{\partial x_2} + \frac{\partial F_{13}}{\partial x_3} + \frac{\partial F_{14}}{\partial x_4} = \frac{4\pi j_1}{c}.$$

$$\left(\frac{\partial H_3}{\partial x_2} - \frac{\partial H_2}{\partial x_3}\right) - \frac{1}{c}\frac{\partial E_1}{\partial t} = \frac{4\pi J_1}{c}. \quad \Rightarrow \quad (\nabla\times\mathbf{H})_1 = \frac{4\pi \mathbf{J}_1}{c} + \frac{1}{c}\frac{\partial \mathbf{E}_1}{\partial t}.$$

For $j = 2$:

$$\frac{\partial F_{21}}{\partial x_1} + \frac{\partial F_{23}}{\partial x_3} + \frac{\partial F_{24}}{\partial x_4} = \frac{4\pi j_2}{c}.$$

$$\left(\frac{\partial H_1}{\partial x_3} - \frac{\partial H_3}{\partial x_1}\right) - \frac{1}{c}\frac{\partial E_2}{\partial t} = \frac{4\pi J_2}{c}. \quad \Rightarrow \quad (\nabla\times\mathbf{H})_2 = \frac{4\pi \mathbf{J}_2}{c} + \frac{1}{c}\frac{\partial \mathbf{E}_2}{\partial t}.$$

For $j = 3$:

$$\frac{\partial F_{31}}{\partial x_1} + \frac{\partial F_{32}}{\partial x_2} + \frac{\partial F_{34}}{\partial x_4} = \frac{4\pi j_3}{c}.$$

$$\left(\frac{\partial H_2}{\partial x_1} - \frac{\partial H_1}{\partial x_2}\right) - \frac{1}{c}\frac{\partial E_3}{\partial t} = \frac{4\pi J_3}{c}. \quad \Rightarrow \quad (\nabla\times\mathbf{H})_3 = \frac{4\pi \mathbf{J}_3}{c} + \frac{1}{c}\frac{\partial \mathbf{E}_3}{\partial t}.$$

For $j = 4$:

$$\frac{\partial F_{41}}{\partial x_1} + \frac{\partial F_{42}}{\partial x_2} + \frac{\partial F_{43}}{\partial x_3} = \frac{4\pi j_4}{c}.$$

$$i\left(\frac{\partial E_1}{\partial x_1} + \frac{\partial E_2}{\partial x_2} + \frac{\partial E_3}{\partial x_3}\right) = ic\frac{4\pi\rho}{c}. \quad \Rightarrow \quad (\nabla\cdot\mathbf{E}) = 4\pi\rho.$$

Thus, for $j = 1, 2, 3$: $(\nabla\times\mathbf{H}) = \dfrac{4\pi\mathbf{J}}{c} + \dfrac{1}{c}\dfrac{\partial\mathbf{E}}{\partial t}$, while for $j = 4$: $(\nabla\cdot\mathbf{E}) = 4\pi\rho$, and we have arrived at 2 of Maxwell's equations.

(b) Consider $\sum_{jkl} \varepsilon_{ijkl}\dfrac{\partial F_{kl}}{\partial x_j} = 0$.

For $i = 1$:

$$\frac{\partial F_{34}}{\partial x_2} - \frac{\partial F_{43}}{\partial x_2} + \frac{\partial F_{42}}{\partial x_3} - \frac{\partial F_{24}}{\partial x_3} + \frac{\partial F_{23}}{\partial x_4} - \frac{\partial F_{32}}{\partial x_4} = 0.$$

$$\Downarrow$$

$$-2i\left(\frac{\partial E_3}{\partial x_2} - \frac{\partial E_2}{\partial x_3}\right) + \frac{2}{ic}\frac{\partial H_1}{\partial t} = 0. \quad \Rightarrow \quad (\nabla\times\mathbf{E})_1 = -\frac{1}{c}\frac{\partial H_1}{\partial t}.$$

For $i = 2$:

$$\frac{\partial F_{41}}{\partial x_3} - \frac{\partial F_{14}}{\partial x_3} + \frac{\partial F_{34}}{\partial x_1} - \frac{\partial F_{43}}{\partial x_1} + \frac{\partial F_{13}}{\partial x_4} - \frac{\partial F_{31}}{\partial x_4} = 0.$$

$$\Downarrow$$

$$-2i\left(\frac{\partial E_3}{\partial x_1} - \frac{\partial E_1}{\partial x_3}\right) + \frac{2i}{c}\frac{\partial H_2}{\partial t} = 0. \quad \Rightarrow \quad (\nabla\times\mathbf{E})_2 = -\frac{1}{c}\frac{\partial H_2}{\partial t}.$$

For $i = 3$:

$$\frac{\partial F_{12}}{\partial x_4} - \frac{\partial F_{21}}{\partial x_4} + \frac{\partial F_{41}}{\partial x_2} - \frac{\partial F_{14}}{\partial x_2} + \frac{\partial F_{24}}{\partial x_1} - \frac{\partial F_{42}}{\partial x_1} = 0.$$

$$\Downarrow$$

$$-2i\left(\frac{\partial E_2}{\partial x_1} - \frac{\partial E_1}{\partial x_2}\right) - \frac{2i}{c}\frac{\partial H_3}{\partial t} = 0. \quad \rightarrow \quad (\nabla\times\mathbf{E})_3 = -\frac{1}{c}\frac{\partial H_3}{\partial t}.$$

For $i = 4$:

$$\frac{\partial F_{23}}{\partial x_1} - \frac{\partial F_{32}}{\partial x_1} + \frac{\partial F_{12}}{\partial x_3} - \frac{\partial F_{21}}{\partial x_3} + \frac{\partial F_{31}}{\partial x_2} - \frac{\partial F_{13}}{\partial x_2} = 0.$$

$$\Downarrow$$

$$2\left(\frac{\partial H_1}{\partial x_1} + \frac{\partial H_2}{\partial x_2} + \frac{\partial H_3}{\partial x_3}\right) = 0. \quad \rightarrow \quad (\nabla\cdot\mathbf{H}) = 0.$$

Thus, for $i = 1, 2, 3$: $\nabla\times\mathbf{E} = -\frac{1}{c}\dot{\mathbf{H}}$, while for $i = 4$: $\nabla\cdot\mathbf{H} = 0$, and we have obtained the 2 remaining Maxwell's equations.

Let us attempt to construct the anti-symmetric tensor $F_{\mu\upsilon}$ and the 4-vector j_μ from Maxwell's equations.

In free space, Maxwell's equations are given by:

(i) $\nabla\cdot\mathbf{E} = 4\pi\rho.$

(ii) $\nabla\cdot\mathbf{H} = 0.$

(iii) $\nabla\times\mathbf{E} = -\frac{1}{c}\dot{\mathbf{H}}.$

(iv) $\nabla\times\mathbf{H} = \frac{4\pi\mathbf{J}}{c} + \frac{1}{c}\dot{\mathbf{E}}.$

Since $\nabla\cdot\mathbf{H} = 0$, we know that $\mathbf{H}$ can be expressed as the curl of a vector, i.e. $\mathbf{H} = \nabla\times\mathbf{A}$. Let us also specify that $\nabla\cdot\mathbf{A} = -\frac{1}{c}\frac{\partial\varphi}{\partial t}$. Substituting $\mathbf{H} = \nabla\times\mathbf{A}$ into (iii):

$$\nabla\times\mathbf{E} = -\frac{1}{c}\frac{\partial(\nabla\times\mathbf{A})}{\partial t} = -\frac{1}{c}\nabla\times\dot{\mathbf{A}} \quad \rightarrow \quad \nabla\times\left(\mathbf{E} + \frac{\dot{\mathbf{A}}}{c}\right) = 0.$$

But if the curl of a vector vanishes, we can write

$$\mathbf{E} + \frac{\dot{\mathbf{A}}}{c} = -\nabla\varphi, \quad \text{with} \quad \nabla\cdot\mathbf{A} = -\frac{1}{c}\frac{\partial\varphi}{\partial t}.$$

Consider the 4-dimensional gradient operator (which we know to be a good 4-vector), having components:

$$\square = \left(\frac{\partial}{\partial x}, \frac{\partial}{\partial y}, \frac{\partial}{\partial z}, \frac{\partial}{ic\,\partial t}\right) \rightarrow \frac{\partial}{\partial x_\mu}.$$

To test for vector validity, we dot the expression into $\square$ and observe whether we get a Lorentz invarient as a result. We proceed to do this for j_μ.

$$\square\cdot j_\mu = \frac{\partial J_1}{\partial x} + \frac{\partial J_2}{\partial y} + \frac{\partial J_3}{\partial z} + \frac{\partial\rho}{\partial t} = \nabla\cdot\mathbf{J} + \frac{\partial\rho}{\partial t} = 0,$$

the last by virtue of the equation of continuity. Since the quantity 0 is Lorentz invarient, j_μ is a good 4-vector. Let us now perform the same test for the "4-vector" $V_\mu = (\mathbf{A}, i\varphi)$.

$$\square\cdot V_\mu = \frac{\partial A_x}{\partial x} + \frac{\partial A_y}{\partial y} + \frac{\partial A_z}{\partial z} + \frac{1}{c}\frac{\partial\varphi}{\partial t} = \nabla\cdot\mathbf{A} + \frac{1}{c}\frac{\partial\varphi}{\partial t} = 0,$$

the last by virtue of our choice of $\nabla\cdot\mathbf{A}$. Once again we see that V_μ is a good 4-vector. From $\mathbf{H}_i = (\nabla\times\mathbf{A})_i$ and $\mathbf{E}_i = -\frac{1}{c}\frac{\partial A_i}{\partial t} - \frac{\partial\varphi}{\partial x_i}$:

$$H_x = \frac{\partial A_z}{\partial y} - \frac{\partial A_y}{\partial z} = \frac{\partial V_3}{\partial x_2} - \frac{\partial V_2}{\partial x_3}.$$

$$H_y = \frac{\partial A_x}{\partial z} - \frac{\partial A_z}{\partial x} = \frac{\partial V_1}{\partial x_3} - \frac{\partial V_3}{\partial x_3}.$$

$$H_z = \frac{\partial A_y}{\partial x} - \frac{\partial A_x}{\partial y} = \frac{\partial V_2}{\partial x_1} - \frac{\partial V_1}{\partial x_2}.$$

$$E_x = -\frac{\partial\varphi}{\partial x} - \frac{1}{c}\frac{\partial A_x}{\partial t} = i\left[\frac{\partial V_4}{\partial x_1} - \frac{\partial V_1}{\partial x_4}\right].$$

$$E_y/i = \frac{\partial V_4}{\partial x_2} - \frac{\partial V_2}{\partial x_4}.$$

$$E_z/i = \frac{\partial V_4}{\partial x_3} - \frac{\partial V_3}{\partial x_4}.$$

Note that the terms on the right are components of an anti-symmetric tensor $F_{\mu\upsilon}$, where $F_{\mu\upsilon} = \frac{\partial V_\upsilon}{\partial x_\mu} - \frac{\partial V_\mu}{\partial x_\upsilon}$, so that

$$F_{11} = F_{22} = F_{33} = F_{44} = 0.$$

$$F_{12} = H_z. \quad F_{21} = -H_z.$$

$$F_{23} = H_x. \quad F_{32} = -H_x.$$

$$F_{13} = -H_y. \quad F_{31} = H_y.$$

$$F_{14} = -iE_x. \quad F_{41} = iE_x.$$

$$F_{24} = -iE_y. \quad F_{42} = iE_y.$$

$$F_{34} = -iE_z. \quad F_{43} = iE_z.$$

Hence,

$$F_{\mu\upsilon} = \begin{pmatrix} 0 & H_z & -H_y & -iE_x \\ -H_z & 0 & H_x & -iE_y \\ H_y & -H_x & 0 & -iE_z \\ iE_x & iE_y & iE_z & 0 \end{pmatrix}.$$

Problem Set H

I. Find the eigenvectors and eigenvalues of the Pauli spin matrices σ_1, σ_2 and σ_3.

Solution:

$$\sigma_1 = \begin{pmatrix} 0 & 1 \\ 1 & 0 \end{pmatrix}. \qquad \sigma_2 = \begin{pmatrix} 0 & -i \\ i & 0 \end{pmatrix}. \qquad \sigma_3 = \begin{pmatrix} 1 & 0 \\ 0 & -1 \end{pmatrix}.$$

$$\det \begin{vmatrix} -\lambda & 1 \\ 1 & -\lambda \end{vmatrix} = 0. \qquad \det \begin{vmatrix} -\lambda & -i \\ i & -\lambda \end{vmatrix} = 0. \qquad \det \begin{vmatrix} 1-\lambda & 0 \\ 0 & -1-\lambda \end{vmatrix} = 0.$$

$$\lambda^2 - 1 = 0 \qquad \lambda^2 - 1 = 0 \qquad 1 - \lambda^2 = 0$$

$$\lambda_1 = +1,\ \lambda_2 = -1 \qquad \lambda_1 = +1,\ \lambda_2 = -1 \qquad \lambda_1 = +1,\ \lambda_2 = -1$$

Hence, for all three spin matrices eigenvalues are +1 and −1. We then have: $\sigma_i \Psi^j = \lambda_j \Psi^j$, where $i = 1, 2, 3$ and $j = 1, 2$, and where

$$\Psi^j = \begin{pmatrix} \Psi^{j1} \\ \Psi^{j2} \end{pmatrix}.$$

$$\sigma_1 \Psi^i = \lambda_i \Psi^i \qquad \sigma_2 \Psi^i = \lambda_i \Psi^i \qquad \sigma_3 \Psi^i = \lambda_i \Psi^i$$

$$\begin{pmatrix} 0 & 1 \\ 1 & 0 \end{pmatrix}\begin{pmatrix} \Psi_1^i \\ \Psi_2^i \end{pmatrix} = \lambda_i \begin{pmatrix} \Psi_1^i \\ \Psi_2^i \end{pmatrix} = \pm 1 \begin{pmatrix} \Psi_1^i \\ \Psi_2^i \end{pmatrix}.$$

$$\begin{pmatrix} 0 & -i \\ i & 0 \end{pmatrix}\begin{pmatrix} \Psi_1^i \\ \Psi_2^i \end{pmatrix} = \pm 1 \begin{pmatrix} \Psi_1^i \\ \Psi_2^i \end{pmatrix}.$$

$$\begin{pmatrix} 1 & 0 \\ 0 & -1 \end{pmatrix}\begin{pmatrix} \Psi_1^i \\ \Psi_2^i \end{pmatrix} = \pm 1 \begin{pmatrix} \Psi_1^i \\ \Psi_2^i \end{pmatrix}.$$

$$\left\{ \begin{matrix} \Psi_2^1 = \Psi_1^1 \\ \Psi_1^1 = \Psi_2^1 \end{matrix} \right\} \frac{\Psi_1^1}{\Psi_2^1} = 1.$$

$$\left\{ \begin{matrix} -i\,\Psi_2^1 = \Psi_1^1 \\ i\,\Psi_1^1 = \Psi_2^1 \end{matrix} \right\} \frac{\Psi_1^1}{\Psi_2^1} = -i.$$

$$\left\{ \begin{matrix} \Psi_1^1 = \Psi_1^1 \\ -\Psi_2^1 = \Psi_2^1 \end{matrix} \right\} \Psi_1^1 = 1,\ \Psi_2^1 = 0.$$

$$\begin{Bmatrix} \Psi_2'^2 = -\Psi_1'^2 \\ \Psi_1'^2 = -\Psi_2'^2 \end{Bmatrix} \frac{\Psi_1'^2}{\Psi_2'^2} = -1.$$

$$\begin{Bmatrix} -i\,\Psi_2'^2 = -\Psi_1'^2 \\ i\,\Psi_1'^2 = -\Psi_2'^2 \end{Bmatrix} \frac{\Psi_1'^2}{\Psi_2'^2} = +i..$$

$$\begin{Bmatrix} \Psi_1'^2 = -\Psi_1'^2 \\ -\Psi_2'^2 = -\Psi_2'^2 \end{Bmatrix} \Psi_1'^2 = 0,\ \Psi_2'^2 = 1.$$

Hence,

$$\Psi'^1 = \begin{pmatrix} \frac{1}{\sqrt{2}} \\ \frac{1}{\sqrt{2}} \end{pmatrix}. \quad \Psi'^2 = \begin{pmatrix} \frac{1}{\sqrt{2}} \\ \frac{-1}{\sqrt{2}} \end{pmatrix}.$$

$$\Psi'^1 = \begin{pmatrix} \frac{i}{\sqrt{2}} \\ \frac{-1}{\sqrt{2}} \end{pmatrix}. \quad \Psi'^2 = \begin{pmatrix} \frac{i}{\sqrt{2}} \\ \frac{1}{\sqrt{2}} \end{pmatrix}.$$

$$\Psi'^1 = \begin{pmatrix} 1 \\ 0 \end{pmatrix}. \quad \Psi'^2 = \begin{pmatrix} 0 \\ 1 \end{pmatrix}.$$

II. The interaction of an electron with a magnetic field **H** is described by a hermitian matrix E,

$$E = \frac{e\hbar}{2mc}\bar{\sigma}\cdot\mathbf{H} = \frac{e\hbar}{2mc}\sum_{i=1}^{3}\sigma_i H_i,$$

where e, m, c, $\hbar$ are, respectively, the charge and mass of the electron, the velocity of light, and (Planck's constant)/2π.

(*a*) Show that σ_i transforms like a vector, i.e. that $\sum_i \sigma_i H_i$ depends only on $|\mathbf{H}|$.

(*b*) Find the eigenvalues of E for arbitrarily oriented magnetic field **H**.

Solutions:

(a) If we can show that $\sum_i \sigma_i H_i$ behaves like a scalar, then we will have proved that σ_i behaves like a vector.

$$\sum_i \sigma_i H_i = H_1\begin{pmatrix}0 & 1\\ 1 & 0\end{pmatrix} + H_2\begin{pmatrix}0 & -i\\ i & 0\end{pmatrix} + H_3\begin{pmatrix}1 & 0\\ 0 & -1\end{pmatrix}$$

$$= \begin{pmatrix} H_3 & H_1 - iH_2 \\ H_1 + iH_2 & -H_3 \end{pmatrix}.$$

Since this is a hermitian matrix, there exists a unitary matrix, U, such that $\sum_i \sigma_i H_i = \Lambda = \lambda_i \delta_{ij}$. Then,

$$\det\begin{vmatrix} H_3 - \lambda & H_1 - iH_2 \\ H_1 + iH_2 & -H_3 - \lambda \end{vmatrix} = 0.$$

$$\Downarrow$$

$$\lambda^2 = H_3^2 + H_1^2 + H_2^2 = H^2 \rightarrow \lambda_1 = +|\mathbf{H}|,\ \lambda_2 = -|\mathbf{H}|.$$

Thus, under the unitary transformation U, $\sum_i \sigma_i H_i \rightarrow \begin{pmatrix}1 & 0\\ 0 & -1\end{pmatrix}|\mathbf{H}|$, and $\sum_i \sigma_i H_i$ depends only on $|\mathbf{H}|$, i.e. σ_i behaves like a vector upon transformation, so that $\overline{\sigma}\cdot\mathbf{H} = \sum_i \sigma_i H_i$.

(b) From (a) above:

$$E = \frac{e\hbar}{2mc}\left[H_1\begin{pmatrix}0 & 1\\ 1 & 0\end{pmatrix} + H_2\begin{pmatrix}0 & -i\\ i & 0\end{pmatrix} + H_3\begin{pmatrix}1 & 0\\ 0 & -1\end{pmatrix}\right]$$

$$= \begin{pmatrix} \dfrac{e\hbar}{2mc}H_3 & \dfrac{e\hbar}{2mc}(H_1 - iH_2) \\[2ex] \dfrac{e\hbar}{2mc}(H_1 + iH_2) & -\dfrac{e\hbar}{2mc}H_3 \end{pmatrix}.$$

det $|E - \lambda\delta_{ij}| = 0$ yields:

$$\lambda^2 - \left(\frac{e\hbar}{2mc}\right)^2 H_3^2 - \left(\frac{e\hbar}{2mc}\right)^2 (H_1^2 + H_2^2) = 0 \quad \rightarrow \quad \begin{array}{l} \lambda_1 = +\dfrac{e\hbar}{2mc}H \\ \lambda_2 = -\dfrac{e\hbar}{2mc}H \end{array},$$

where $H = \sqrt{H_1^2 + H_2^2 + H_3^2}$.

III. Let A and B be either unitary or hermitian matrices. Prove that the necessary and sufficient condition for $AB = BA$ is that there exists a unitary matrix U such that UAU^+ and UBU^+ are both diagonal matrices.

Solution:

Sufficiency Since A and B are both either unitary or hermitian, there exists a unitary matrix U which simultaneously reduces A and B to their respective diagonal forms A' and B', where $A' = UAU^+$ and $B' = UBU^+$.

Since two diagonal matrices commute with each other, $A'B' = B'A'$, or

$$UAU^+UBU^+ = UBU^+UAU^+ \rightarrow UABU^+ = UBAU^+ \rightarrow AB = BA.$$

Necessity: Since $AB = BA$ and both A and B are unitary or hermitian, the product AB remains unitary or hermitian; hence, it is possible to find a unitary matrix U such that

(1) $\quad UABU^+ = \Lambda_1$

is a diagonal matrix. Multiplying both sides of (1) on the left by UA^+U^+, we have $UA^+U^+UABU^+ = UA^+U^+\Lambda_1$, or

(2) $\quad UBU^+ = UA^+U^+\Lambda_1$.

Multiplying equation (1) on the right by UA^+U^+, we have $UABU^+UA^+U^+ = \Lambda_1 UA^+U^+ \rightarrow UBAU^+UA^+U^+ = \Lambda_1 UA^+U^+$ (since $AB = BA$), i.e.

(3) $\quad UBU^+ = \Lambda_1 UA^+U^+$.

Equations (2) and (3) show that UA^+U^+ and Λ_1 commute with each other. But, since Λ_1 is a diagonal matrix, UA^+U^+ is necessarily diagonal (see next page). Thus, $UAU^+ = (UA^+U^+)^+$ is also a diagonal matrix.

If we repeat the same procedure, this time multiplying through by UB^+U^+, we get the result that UB^+U^+ is necessarily diagonal, and that UBU^+ is a diagonal matrix as well.

We need now only prove that if 2 matrices commute with each other and one is a diagonal matrix, then the other is necessarily diagonal.

Set $D_{ij} = \lambda_i \delta_{ij}$ (i.e. D a diagonal matrix), and let $DA = AD$. Then we have:

$$(DA)_{ij} = (AD)_{ij} \quad \rightarrow \quad \sum_p D_{ip}A_{pj} = \sum_p A_{ip}D_{pj}.$$

$$\Downarrow$$

$$\sum_p \lambda_i \delta_{ip} A_{pj} = \sum_p A_{ip}\lambda_p \delta_{pj} \quad \rightarrow \quad \lambda_i A_{ij} = A_{ij}\lambda_j = A_{ij}\delta_{ij}\lambda_i.$$

$$\Downarrow$$

$$A_{ij} = A_{ij}\delta_{ij} \quad \rightarrow \quad A \text{ is diagonal.}$$

IV. Prove that the result of any finite rotation in a real 4-dimensional space can be achieved by two 2-dimensional rotations. Furthermore, show that these two 2-dimensional rotations always commute with each other.

Solution: Let A be the matrix in question. Since A is real and orthogonal, it is also a unitary matrix, and so $AA^+ = 1$, i.e. $A^+ = A^{-1}$. One can therefore find a unitary transformation U such that $UAU^+ = \Lambda$ (diagonal matrix). For proper rotations only, $\det|A| = +1$, so that

$$\det|A| = \det|UAU^+| = \det|\Lambda| = \lambda_1\lambda_2\lambda_3\lambda_4 = +1.$$

The eigenvalues of a unitary matrix are complex, and satisfy the expression $|\lambda_i|^2 = \lambda_i\lambda_i^* = 1$. Setting $\lambda_1 = e^{i\theta}$ (θ real):

$$A\psi^1 = e^{i\theta}\psi^1 \quad \Rightarrow \quad A^*\psi^{1*} = e^{-i\theta}\psi^{1*}.$$

Since A is a real, unitary matrix ($A = A^*$),

$$A\psi^1 = e^{i\theta}\psi^1 \quad \rightarrow \quad A\psi^{1*} = e^{-i\theta}\psi^{1*},$$

and $\lambda_2 = e^{-i\theta}$ is also an eigenvalue.

Setting $\lambda_3 = e^{i\varphi}$ (φ real):

$$A\psi^3 = e^{i\varphi}\psi^3 \rightarrow A^*\psi^{3*} = e^{-i\varphi}\psi^{3*}.$$

Since A is a real, unitary matrix ($A = A^*$),

$$A\psi^3 = e^{i\varphi}\psi^3 \rightarrow A\psi^{3*} = e^{-i\varphi}\psi^{3*},$$

and $\lambda_4 = e^{-i\varphi}$ is also an eigenvalue. Therefore,

$$\Lambda = \begin{pmatrix} e^{i\theta} & 0 & 0 & 0 \\ 0 & e^{-i\theta} & 0 & 0 \\ 0 & 0 & e^{i\varphi} & 0 \\ 0 & 0 & 0 & e^{-i\varphi} \end{pmatrix} = \begin{pmatrix} e^{i\theta} & 0 & 0 & 0 \\ 0 & e^{-i\theta} & 0 & 0 \\ 0 & 0 & 1 & 0 \\ 0 & 0 & 0 & 1 \end{pmatrix} \begin{pmatrix} 1 & 0 & 0 & 0 \\ 0 & 1 & 0 & 0 \\ 0 & 0 & e^{i\varphi} & 0 \\ 0 & 0 & 0 & e^{-i\varphi} \end{pmatrix}.$$

We have already proven that any matrix with diagonal elements $e^{i\theta}$, $e^{-i\theta}$ is equivalent to a 2-dimensional rotation. Hence we see that the 4-dimensional rotation in real space, A, is equivalent to two 2-dimensional rotations.

To show that these rotations commute with each other, we need only note that they are diagonal matrices, and then show that all diagonal matrices commute with each other.

Proof: If A and D are diagonal matrices, $A_{ij} = A_i\delta_{ij}$ and $D_{ij} = D_i\delta_{ij}$. Then,

$$(DA)_{ij} = \sum_p D_i\delta_{ip}A_p\delta_{pj} = D_iA_j\delta_{ij} = D_iA_i\delta_{ij}.$$

$$(AD)_{ij} = \sum_p A_i\delta_{ip}D_p\delta_{pj} = A_iD_j\delta_{ij} = D_iA_i\delta_{ij}.$$

V. An infinitesimal rotation is one that is generated by a unitary (or orthogonal) matrix U, where $U - I = \varepsilon$ is an infinitesimal quantity.

Show that if one neglects second or higher order quantities in ε, then

(*a*) $\varepsilon_{ij} = -\varepsilon_{ji}^*$ ($\varepsilon_{ij} = -\varepsilon_{ji}$ for U orthogonal)

(*b*) All infinitesimal rotations commute with each other.

An infinitesimal rotation is a unitary (or orthogonal) transformation such that the components of a vector are "almost" the same in both sets of axes, i.e. the change is infinitesimal.

Thus, for the 3-dimensional position vector $\mathbf{r}(x_1 x_2 x_3)$:

$$x_1' = x_1 + \varepsilon_{11}x_1 + \varepsilon_{12}x_2 + \varepsilon_{13}x_3 \;\rightarrow\; x_i' = x_i + \sum_{j=1}^{3}\varepsilon_{ij}x_j$$

$$\Downarrow$$

$$x_i' = \sum_{j=1}^{3}(\delta_{ij} + \varepsilon_{ij})x_j = \sum_{j=1}^{3}U_{ij}x_j,$$

where $U_{ij} = \delta_{ij} + \varepsilon_{ij} \;\rightarrow\; U = I + \varepsilon$ defines the infinitesimal rotation.

Solution:

(a)

For U unitary

$U^{-1} = U^{+}$

For U orthogonal

$U^{-1} = \tilde{U}$

[Note that $UU^{-1} = (I+\varepsilon)(I-\varepsilon) = I + \varepsilon I - \varepsilon I - \varepsilon^2 = I$, if we neglect terms of order ε^2.]

For U unitary:

$U = I + \varepsilon$

$$\begin{Bmatrix} U^{-1} = I - \varepsilon \\ U^{+} = I + \varepsilon^{+} \end{Bmatrix} \rightarrow \begin{Bmatrix} \varepsilon = -\varepsilon^{+} \\ \varepsilon_{ij} = -\varepsilon_{ji}^{*} \end{Bmatrix}$$

For U orthogonal:

$U = I + \varepsilon$

$$\begin{Bmatrix} U^{-1} = I - \varepsilon \\ \tilde{U} = I + \tilde{\varepsilon} \end{Bmatrix} \rightarrow \begin{Bmatrix} \varepsilon = -\tilde{\varepsilon} \\ \varepsilon_{ij} = -\varepsilon_{ji} \end{Bmatrix}$$

(b) If we neglect terms of order ε^2:

$$(1+\varepsilon_1)(1+\varepsilon_2) = 1 + \varepsilon_1 + \varepsilon_2 + \varepsilon_1\varepsilon_2 \approx 1 + \varepsilon_1 + \varepsilon_2 .$$

$$(1+\varepsilon_2)(1+\varepsilon_1) = 1 + \varepsilon_2 + \varepsilon_1 + \varepsilon_2\varepsilon_1 \approx 1 + \varepsilon_2 + \varepsilon_1 .$$

Since matrices commute with regard to addition, it follows that infinitesimal matrices commute. This implies that, contrary to finite rotations which do not necessarily commute, infinitesimal rotations do commute. In other words, since a rotation is specified by a direction (axis of rotation) and a magnitude (angle turned through) and since it obeys the commutative property in the addition of rotations (i.e. $U_1U_2 = U_2U_1$), then we can represent infinitesimal rotations by vectors.

VI. Let H be a hermitian matrix and let e^{iH} be another matrix, defined as

$$e^{iH} = \sum_{n=0}^{\infty} \frac{(iH)^n}{n!}.$$

Prove that e^{iH} is a unitary matrix.

Solution:

$$e^{iH} = 1 + \frac{iH}{1!} - \frac{H^2}{2!} - \frac{iH^3}{3!} + \frac{H^4}{4!} + \frac{iH^5}{5!} - \frac{H^6}{6!} + \ldots + \frac{iH^n}{n!} + \ldots$$

$$\left(e^{iH}\right)^+ = 1 - \frac{iH^+}{1!} - \frac{(H^+)^2}{2!} + \frac{i(H^+)^3}{3!} + \frac{(H^+)^4}{4!} - \frac{i(H^+)^5}{5!} + \ldots + \frac{-i(H^+)^n}{n!} + \ldots$$

Setting $H = H^+$:

$$\left(e^{iH}\right)^+ = 1 - \frac{iH}{1!} - \frac{H^2}{2!} + \frac{iH^3}{3!} + \frac{H^4}{4!} - \frac{iH^5}{5!} + \ldots + \frac{(-iH)^n}{n!} + \ldots,$$

so that

$$\left(e^{iH}\right)\left(e^{iH}\right)^+ = 1$$

$$+ H(i - i) + H^2\left(-\frac{1}{2} - \frac{1}{2} + 1\right) + H^3\left(\frac{i}{3!} - \frac{i}{3!} + \frac{i}{2} - \frac{i}{2}\right)$$

$$+ H^4\left(\frac{1}{4!} + \frac{1}{4!} - \frac{1}{3!} - \frac{1}{3!} + \frac{1}{2}\cdot\frac{1}{2}\right) \rightarrow H^4\left(\frac{2}{4!} - \frac{8}{4!} + \frac{6}{4!}\right)$$

$$+ H^5\left(\frac{i}{5!} - \frac{i}{5!} + \frac{i}{4!} - \frac{i}{4!} + \frac{i}{2\cdot 3!} - \frac{i}{2\cdot 3!}\right)$$

$$+ H^6\left(-\frac{2}{6!} + \frac{2}{5!} - \frac{2}{2\cdot 4!} + \frac{1}{3!}\cdot\frac{1}{3!}\right) \rightarrow H^6\left(-\frac{2}{6!} + \frac{12}{6!} - \frac{30}{6!} + \frac{20}{6!}\right) +$$

$$\vdots$$

$$+ H^n\left(\frac{(i)^n}{n!} + \frac{(-i)^n}{n!} + \frac{i(-i)^{n-1}}{(n-1)!} + \frac{(-i)(i)^{n-1}}{(n-1)!} + \ldots\right)$$

For n odd, $(-i)^n = -(i^n)$ and the last term sum vanishes. For n even, we get a series of terms, the sum of which vanishes. The result, therefore, is $[\exp(iH)][\exp(iH)]^+ = 1$, which is the condition for a unitary matrix.

Problem Set I

I. An $n \times n$ cyclic matrix C is defined as:

$$C = \begin{pmatrix} C_1 & C_2 & \cdot & \cdot & \cdot & \cdot & \cdot & C_{n-1} & C_n \\ C_n & C_1 & \cdot & \cdot & \cdot & \cdot & \cdot & C_{n-2} & C_{n-1} \\ C_{n-1} & C_n & \cdot & \cdot & \cdot & \cdot & \cdot & C_{n-3} & C_{n-2} \\ \cdot & \cdot & \cdot & \cdot & \cdot & \cdot & \cdot & \cdot & \cdot \\ \cdot & \cdot & \cdot & \cdot & \cdot & \cdot & \cdot & \cdot & \cdot \\ C_2 & C_3 & \cdot & \cdot & \cdot & \cdot & \cdot & C_n & C_1 \end{pmatrix}.$$

Prove that the eigenequation $C\psi^s = \lambda_s\psi^s$ has n solutions, with

$$\lambda_s = \sum_{m=1}^{n} C_m \exp\,[i(m-1)2s\pi/n]\,.$$

and

$$\psi^s = \begin{pmatrix} 1 \\ \exp\,(i2s\pi/n)\cdot 1 \\ \exp\,(i2s\pi/n)\cdot 2 \\ \cdot \\ \cdot \\ \cdot \\ \exp\,(i2s\pi/n)\cdot[n-1] \end{pmatrix}.$$

Solution: Consider the first of the n equations formed from $C\psi^s = \lambda_s\psi^s$, namely:

$$(1) \qquad C_1 + C_2 \exp\frac{i2\pi s(1)}{n} + C_3 \exp\frac{i2\pi s(2)}{n} + C_4 \exp\frac{i2\pi s(3)}{n} + \ldots + C_{n-1} \exp\frac{i2\pi s(n-2)}{n} + C_n \exp\frac{i2\pi s(n-1)}{n}$$

$$= \sum_{m=1}^{n} C_m \exp\frac{i2\pi s(m-1)}{n} = \lambda_s\psi_1^s,$$

i.e. $C_{1i}\psi_i^s = \lambda_s\psi_1^s \;\rightarrow\;$ true for the first equation.

Multiply (1) on the right by exp $(2i\pi s/n)\cdot 1$ to get

$$\sum_{m=1}^{n} C_m \exp\frac{i(m-1)2s\pi}{n} \exp\frac{i2s\pi}{n},$$

a series in C_m multiplied by some exponential. The coefficients C_{m-1} now have the same exponential powers as the previous set of C_m. What about terms C_1 and C_n?

$$\text{For } C_1 \quad \rightarrow \quad C_1 \exp\frac{i2\pi s}{n}.$$

$$\text{For } C_n \quad \rightarrow \quad C_n \exp\frac{i2\pi s(n)}{n} = C_n \exp(i2\pi s) = C_n,$$

yielding

$$\text{(2)} \qquad C_n + C_1 \exp\frac{i2\pi s(1)}{n} + C_2 \exp\frac{i2\pi s(2)}{n} + C_3 \exp\frac{i2\pi s(3)}{n} + \ldots + C_{n-2} \exp\frac{i2\pi s(n-2)}{n} + C_{n-1} \exp\frac{i2\pi s(n-1)}{n}$$

$$= \sum_{m=1}^{n} C_m \exp\frac{i2\pi s(m-1)}{n} \exp\frac{i2\pi s}{n} = \lambda_s \psi_2^s,$$

i.e. $C_{2i}\psi_i^s = \lambda_s \psi_2^s$ $\rightarrow$ true for the second equation. In fact, every operation of $\exp\frac{i2\pi s(p)}{n}$ upon $\lambda_s = \sum_{m=1}^{n} C_m \exp[i(m-1)2s\pi/n]$ serves to shift the coefficients C_m so that $m \rightarrow m-p$. Setting $p = n-1$ yields an equation of the form:

$$C_2 + C_3 \exp\frac{i2\pi s(1)}{n} + C_4 \exp\frac{i2\pi s(2)}{n} + C_5 \exp\frac{i2\pi s(3)}{n} + \ldots + C_n \exp\frac{i2\pi s(n-2)}{n} + C_1 \exp\frac{i2\pi s(n-1)}{n}$$

$$= \sum_{m=1}^{n} C_m \exp\frac{i2\pi s(m-1)}{n} \exp\frac{i2\pi s(n-1)}{n} = \lambda_s \psi_n^s,$$

which we recognize as the nth equation of $C\psi^s = \lambda_s \psi^s$, i.e. $C_{ni}\psi_i^s = \lambda_s \psi_n^s$.

We have therefore shown that $\lambda_s = \sum_{m=1}^{n} C_m \exp\,[i(m-1)2s\pi/n]$ and

$$\psi^s = \begin{pmatrix} 1 \\ \exp\,(i2s\pi/n)\cdot 1 \\ \exp\,(i2s\pi/n)\cdot 2 \\ \cdot \\ \cdot \\ \cdot \\ \exp\,(i2s\pi/n)\cdot[n-1] \end{pmatrix} \text{ satisfy the equation } C\psi^s = \lambda_s\psi^s.$$

II. Consider a one-dimensional, ring-shaped crystal having n lattice points. Let x_i $(i = 1\ldots n)$ be the distance of the ith lattice point from its equilibrium position. The kinetic energy T and potential energy V of the entire crystal are given by

$$T = \frac{m}{2}\sum_{i=1}^{n}\dot{x}_i^2\,, \qquad V = \frac{k}{2}\sum_{i=1}^{n-1}(x_{i+1} - x_i)^2 + \frac{k}{2}(x_n - x_1)^2\,.$$

(*a*) Find the normal modes of this system.

(*b*) Express each x_i as an explicit function of time.

Solution:

(a) $$V = \frac{k}{2}\sum_{i=1}^{n-1}\left[x_{i+1}^2 - 2x_i x_{i+1} + x_i^2\right] + \frac{k}{2}\left[x_n^2 - 2x_1 x_n + x_1^2\right] = \sum_{ij} \upsilon_{ij} x_i x_j\,,$$

where the matrix υ_{ij} is given by:

$$\upsilon = k\begin{pmatrix} 1 & -1/2 & 0 & 0 & \cdot\ \cdot & \cdot & -1/2 \\ -1/2 & 1 & -1/2 & 0 & \cdot\ \cdot & \cdot & 0 \\ 0 & -1/2 & 1 & -1/2 & \cdot\ \cdot & \cdot & 0 \\ 0 & 0 & -1/2 & 1 & \cdot\ \cdot & \cdot & \cdot \\ \cdot & \cdot & \cdot & \cdot & \cdot\ \cdot & \cdot & \cdot \\ \cdot & \cdot & \cdot & \cdot & \cdot\ \cdot & \cdot & \cdot \\ \cdot & \cdot & \cdot & \cdot & \cdot\ \cdot & \cdot & \cdot \\ -1/2 & 0 & 0 & \cdot & \cdot\ \cdot & -1/2 & 1 \end{pmatrix}.$$

Note that υ is a cyclic matrix and, since it is also real and symmetrical, it is hermitian.

The equations of motion of the system are

$$m\ddot{x}_i = -\frac{\partial V}{\partial x_i} = -\sum_{j=1}^{n} \upsilon_{ij} x_j \quad (i = 1, 2 \ldots n).$$

In matrix form, $\ddot{x} = -Tx$, where $T_{ij} = \dfrac{\upsilon_{ij}}{m}$.

Since T is hermitian, one can perform a real rotation in real space to effect its diagonalization, i.e. one can find a U such that $UT\tilde{U} = \Lambda$.

If we define $q = Ux \rightarrow \ddot{q} = U\ddot{x}$ (U independent of time), then

$$U\ddot{x} = -UTx = -UT(\tilde{U}U)x = -(UT\tilde{U})q = -\Lambda q \rightarrow \ddot{q} = -\Lambda q.$$

Thus, for every s, $\ddot{q}_s = -\lambda_s q_s$, where λ_s are the eigenvalues of matrix T, i.e. of matrix υ/m.

In problem I above, the eigenvalues of the cyclic matrix C, λ_{sC}, were found to be

$$\lambda_{sC} = \sum_{m=1}^{n} C_m \exp\left[i(m-1)2\pi s/n\right].$$

Since the eigenvalues of the hermitian matrix T, λ_{sT}, are real, we take the real parts of λ_{sC} to get:

$$\lambda_{sT} = \frac{k}{m}\sum_{m=1}^{n} C_m \cos 2\pi s(m-1)/n,$$

where $C_1 = 1$, $C_2 = -1/2$, $C_n = -1/2$, and all other $C_i = 0$.

Thus,

$$\lambda_{sT} = \frac{k}{m}\left[1 - \frac{1}{2}\cos\frac{2\pi s}{n} - \frac{1}{2}\cos\frac{2\pi s(n-1)}{n}\right] = \frac{k}{m}\left[1 - \cos\frac{2\pi s}{n}\right].$$

Solutions of $\ddot{q}_s = -\lambda_s q_s$ are therefore given by $q_s = A_s e^{i\omega_s t} + B_s e^{-i\omega_s t}$,

where $\omega_s^2 = k/m\,(1 - \cos 2\pi s/n)$. The normal modes of vibration, then, are:

$$q_s = A_s e^{i\sqrt{k/m\,(1-\cos 2\pi s/n)}\,t} + B_s e^{-i\sqrt{k/m\,(1-\cos 2\pi s/n)}\,t}.$$

(b) The inverse transformation is obtained as follows:

$$q = Ux \quad \rightarrow \quad x = \tilde{U}q \quad \rightarrow \quad x_p = \sum_{s=1}^{n} \tilde{U}_{ps} q_s ,$$

where

$$\tilde{U} = \begin{array}{c} \begin{matrix} s=1 & s=2 & \cdots & & \end{matrix} \\ \begin{pmatrix} 1 & 1 & \cdots & \cdot & 1 \\ \cos 2\pi/n & \cos 4\pi/n & \cdots & \cdot & 1 \\ \cos 4\pi/n & \cos 8\pi/n & \cdots & \cdot & 1 \\ \cdot & \cdot & \cdots & \cos 2\pi s(p-1)/n & 1 \\ \cdot & \cdot & \cdots & \cdot & 1 \\ \cdot & \cdot & \cdots & \cdot & 1 \\ \cos 4\pi/n & \cos 8\pi/n & \cdots & \cdot & 1 \\ \cos 2\pi/n & \cos 4\pi/n & \cdots & \cdot & 1 \end{pmatrix} \end{array},$$

and

$$x_p = \sum_{s=1}^{n} \tilde{U}_{ps} q_s = \sum_{s=1}^{n} q_s \cos 2\pi(p-1)s/n .$$

Hence,

$$x_p = \sum_{s=1}^{n} \cos 2\pi s(p-1)/n \left[A_s e^{i\sqrt{k/m}\,(1-\cos 2\pi s/n)t} + B_s e^{-i\sqrt{k/m}\,(1-\cos 2\pi s/n)t} \right].$$

III. Show that for $0 \le x \le \pi$:

$$(a) \quad \sin x = \frac{2}{\pi} - \frac{4}{\pi}\left(\cos\frac{2x}{1\cdot 3} + \cos\frac{4x}{3\cdot 5} + \cos\frac{6x}{5\cdot 7} + \cdots \right).$$

(b) What function does the series represent in the range $-\pi \le x \le 0$?

Solution:

(a) Define the function $F(x)$ to be an even function, so that $F(-x) = +F(x)$ and

$$\left\{ \begin{array}{ll} F(x) = f(x) & 0 \le x \le \pi \\ F(x) = f(|x|) & -\pi \le x \le 0 \end{array} \right\} \quad \text{where } f(x) = \sin x \text{ in this case.}$$

Expanding, $F(x) = \frac{A_o}{2} + \sum_{n=1}^{\infty} A_n \cos nx + \sum_{n=1}^{\infty} B_n \sin nx,$

where

$$A_n = \frac{1}{\pi}\int_{-\pi}^{+\pi} F(x)\cos nx\,dx. \quad A_o = \frac{1}{\pi}\int_{-\pi}^{+\pi} F(x)\,dx. \quad B_n = \frac{1}{\pi}\int_{-\pi}^{+\pi} F(x)\sin nx\,dx.$$

We can write A_n as

$$A_n = \frac{1}{\pi}\left[\int_{-\pi}^{0} F(x)\cos nx\,dx + \int_{0}^{\pi} F(x)\cos nx\,dx\right].$$

Replacing x by $-x$ in the first integral yields

$$\int_{-\pi}^{0} F(-x)\cos n(-x)\,d(-x) = -\int_{\pi}^{0} f(x)\cos nx\,dx = \int_{0}^{\pi} f(x)\cos nx\,dx.$$

Hence,

$$A_n = \frac{2}{\pi}\int_{0}^{\pi} f(x)\cos nx\,dx. \quad A_o = \frac{2}{\pi}\int_{0}^{\pi} f(x)\,dx. \quad B_n = 0.$$

(The last term follows since $F(x)$ is even while $\sin nx$ is odd.)

Evaluating:

$$A_o = \frac{2}{\pi}\int_{0}^{\pi} \sin x\,dx = \left[-\frac{2\cos x}{\pi}\right]_0^{\pi} = \frac{2}{\pi}(1 - \cos\pi) = \frac{4}{\pi}.$$

$$A_n = \frac{2}{\pi}\int_{0}^{\pi} \sin x \cos nx\,dx = \frac{2}{\pi}\int_{0}^{\pi} \frac{1}{2}\left[\sin(n+1)x - \sin(n-1)x\right]$$

$$= \frac{1}{\pi}\left[\frac{-\cos(n+1)x}{n+1} + \frac{\cos(n-1)x}{n-1}\right]_0^{\pi}$$

$$= \frac{1}{\pi}\left[\frac{1-\cos(n+1)\pi}{n+1} - \frac{1-\cos(n-1)\pi}{n-1}\right].$$

For $n = 1, 3, 5, \ldots$ (i.e. odd), A_n vanishes identically, since $1 - \cos(n\pm1)\pi = 0$.

For $n = 2, 4, 6, \ldots$ (i.e. even),

$$A_n \Rightarrow \frac{2}{\pi}\left[\frac{1}{n+1} - \frac{1}{n-1}\right] = \frac{-4}{n^2-1} = \frac{-4}{(n+1)(n-1)}.$$

Hence,

$$\sum_{n=1}^{\infty} A_n \cos nx = A_2 \cos 2x + A_4 \cos 4x + \ldots$$

$$= \frac{-4}{\pi}\left[\frac{\cos 2x}{1\cdot 3} + \frac{\cos 4x}{3\cdot 5} + \frac{\cos 6x}{5\cdot 7} + \ldots\right].$$

Finally then,

$$f(x) = \sin x = \frac{2}{\pi} - \frac{4}{\pi}\left[\frac{\cos 2x}{1\cdot 3} + \frac{\cos 4x}{3\cdot 5} + \frac{\cos 6x}{5\cdot 7} + \ldots\right].$$

(b) In the range $-\pi \le x \le 0$ this series represents $\sin|x|$ or $-\sin x$, since this is the series for an arbitrarily defined even function $F(x) = F(-x)$.

IV. Find the Fourier transform of the Yukawa function

$$Y(r) = \frac{\exp(-\mu r)}{r},$$

where r is the magnitude of a 3-dimensional radial vector and μ is a constant.

Solution:

$$g(\mathbf{k}) = \frac{1}{(2\pi)^{3/2}}\int_{-\infty}^{+\infty} Y(r) e^{-i\mathbf{k}\cdot\mathbf{r}} d^3\mathbf{r} = \frac{1}{(2\pi)^{3/2}}\int_{-\infty}^{+\infty}\frac{1}{r} e^{-i\mathbf{k}\cdot\mathbf{r} - \mu r} dx dy dz.$$

Since $e^{-i\mathbf{k}\cdot\mathbf{r} - \mu r}/r$ is spherically symmetrical, we choose our axes such that $\mathbf{k}$ lies along the z-axis, so that

$$g(\mathbf{k}) = \frac{1}{(2\pi)^{3/2}}\int_{\varphi=0}^{2\pi}\int_{\theta=0}^{\pi}\int_{r=0}^{\infty}\frac{e^{-\mu r}}{r} e^{-ikr\cos\theta} r^2 \sin\theta \, d\theta \, d\varphi \, dr$$

$$= \frac{1}{(2\pi)^{1/2}}\int_{\theta=0}^{\pi}\int_{r=0}^{\infty}\frac{e^{-\mu r}}{r} e^{-ikr\cos\theta} r^2 \frac{ikr\sin\theta}{ikr} dr \, d\theta$$

$$= \frac{1}{ik(2\pi)^{1/2}}\int_{r=0}^{\infty}\left| e^{-\mu r} e^{-ikr\cos\theta}\right|_0^{\pi} dr = \frac{1}{ik(2\pi)^{1/2}}\int_{r=0}^{\infty}\left[e^{(-\mu+ik)r} - e^{(-\mu-ik)r}\right] dr$$

Continuing,

$$g(\mathbf{k}) = \frac{1}{ik\sqrt{2\pi}}\left[\frac{e^{(-\mu+ik)r}}{ik-\mu} + \frac{e^{(-\mu-ik)r}}{ik+\mu}\right]_0^\infty = \frac{1}{ik\sqrt{2\pi}}\left[-\frac{1}{ik-\mu} - \frac{1}{ik+\mu}\right]$$

$$= \frac{1}{ik\sqrt{2\pi}}\left[-\frac{2ik}{-k^2-\mu^2}\right] = \frac{(2/\pi)^{1/2}}{\mu^2+k^2}.$$

V. Show that the most general solution of the Klein-Gordon equation,

$$(1) \qquad \Delta\varphi - \frac{1}{c^2}\ddot{\varphi} - \mu^2\varphi = 0,$$

can be written as

$$\varphi(\mathbf{r}t) = \int\left(f(\mathbf{k})\exp\left[i(\mathbf{k}\cdot\mathbf{r} - \omega t)\right] + d(\mathbf{k})\exp\left[i(\mathbf{k}\cdot\mathbf{r} + \omega t)\right]\right)dk_1dk_2dk_3,$$

where f and d are two arbitrary functions of $\mathbf{k}$, and $\omega = \sqrt{k^2+\mu^2}$.

Solution: Regarding t as parameter, $\varphi(\mathbf{r}t) = \frac{1}{(2\pi)^{3/2}}\int_{-\infty}^{+\infty} g(\mathbf{k}t)e^{i\mathbf{k}\cdot\mathbf{r}}d^3\mathbf{k}$.

$$\Delta\varphi(\mathbf{r}t) = \frac{1}{(2\pi)^{3/2}}\int_{-\infty}^{+\infty} g(\mathbf{k}t)\Delta e^{i\mathbf{k}\cdot\mathbf{r}}d^3\mathbf{k} = \frac{1}{(2\pi)^{3/2}}\int_{-\infty}^{+\infty} g(\mathbf{k}t)(-k^2)e^{i\mathbf{k}\cdot\mathbf{r}}d^3\mathbf{k}.$$

Substituting into equation (1):

$$\int_{-\infty}^{+\infty}\left[-k^2g(\mathbf{k}t)e^{i\mathbf{k}\cdot\mathbf{r}} - \mu^2g(\mathbf{k}t)e^{i\mathbf{k}\cdot\mathbf{r}} - \frac{1}{c^2}\ddot{g}(\mathbf{k}t)e^{i\mathbf{k}\cdot\mathbf{r}}\right]d^3\mathbf{k} = 0.$$

Let $\mathbf{k}_o$ be an arbitrary vector in $\mathbf{k}$ space and perform the following operation:

$$\int_{\substack{\text{all}\\\text{space}}} e^{-i\mathbf{k}_o\cdot\mathbf{r}}d^3\mathbf{r}\int_{\substack{\text{all k}\\\text{space}}}^{+\infty}\left[-(k^2+\mu^2)g(\mathbf{k}t) - \frac{1}{c^2}\ddot{g}(\mathbf{k}t)\right]e^{i\mathbf{k}\cdot\mathbf{r}}d^3\mathbf{k} = 0.$$

$$\Downarrow$$

$$\int_{\substack{\text{all k}\\\text{space}}}^{+\infty}\left[-(k^2+\mu^2)g(\mathbf{k}t) - \frac{1}{c^2}\ddot{g}(\mathbf{k}t)\right]d^3\mathbf{k}\int_{\substack{\text{all}\\\text{space}}} e^{-i\mathbf{r}\cdot(\mathbf{k}-\mathbf{k}_o)}d^3\mathbf{r} = 0.$$

Substituting, $\int_{-\infty}^{+\infty} e^{-i\mathbf{r}\cdot(\mathbf{k}-\mathbf{k}_o)} d^3\mathbf{r} = (2\pi)^3\, \delta^3(\mathbf{k}-\mathbf{k}_o)$, we have:

$$\int_{-\infty}^{+\infty} \left[-(k^2+\mu^2) g(\mathbf{k}t) - \frac{1}{c^2}\ddot{g}(\mathbf{k}t) \right] \delta^3(\mathbf{k}-\mathbf{k}_o) d^3\mathbf{k} = 0.$$

$$\Downarrow$$

$$-(k_o^2+\mu^2) g(\mathbf{k}_o t) - \frac{1}{c^2}\ddot{g}(\mathbf{k}_o t) = 0 \rightarrow \ddot{g}(\mathbf{k}t) = -c^2(k^2+\mu^2) g(\mathbf{k}t).$$

(The last statement follows since $\mathbf{k}_o$ is arbitrary.)

Solutions are therefore given by:

$$g(\mathbf{k}t) = A_{\mathbf{k}} e^{i\omega t} + B_{\mathbf{k}} e^{-i\omega t},$$

where $\omega = c\sqrt{k^2+\mu^2}$ and $A_{\mathbf{k}}$ and $B_{\mathbf{k}}$ are functions of $\mathbf{k}$ and ω.

Substituting back into the original expression for $\varphi(\mathbf{r}t)$:

$$\varphi(\mathbf{r}t) = \frac{1}{(2\pi)^{3/2}} \int_{-\infty}^{+\infty} \left[A_{\mathbf{k}} e^{i(\mathbf{k}\cdot\mathbf{r}+\omega t)} + B_{\mathbf{k}} e^{i(\mathbf{k}\cdot\mathbf{r}-\omega t)} \right] d^3\mathbf{k}.$$

Absorbing the π factor within $A_{\mathbf{k}}$ and $B_{\mathbf{k}}$, we have:

$$\varphi(\mathbf{r}t) = \int_{-\infty}^{+\infty} \left[f_{\mathbf{k}} e^{i(\mathbf{k}\cdot\mathbf{r}+\omega t)} + d_{\mathbf{k}} e^{i(\mathbf{k}\cdot\mathbf{r}-\omega t)} \right] d^3\mathbf{k}.$$

Final Exam Problem Set

I. The electric and magnetic fields are related through Maxwell's equations:

$$\nabla\times\mathbf{H} = \frac{4\pi\mathbf{j}}{c} + \frac{\dot{\mathbf{E}}}{c}. \qquad \nabla\times\mathbf{E} = -\frac{\dot{\mathbf{H}}}{c}.$$

In the above, $\mathbf{j}$ is the electric current density and c the velocity of light.

Prove the following:

$$-\frac{\partial}{\partial t}\int_V \frac{(E^2+H^2)}{8\pi}\,d\tau = \int_V \mathbf{E}\cdot\mathbf{j}\,d\tau + \frac{c}{4\pi}\int_S (\mathbf{E}\times\mathbf{H})\cdot d\mathbf{S},$$

where V is the volume enclosed by the surface S.

Solution: Note the following:

$$\frac{\partial}{\partial t}\left[E^2+H^2\right] = \frac{\partial}{\partial t}[\mathbf{E}\cdot\mathbf{E}+\mathbf{H}\cdot\mathbf{H}] = 2\,[\mathbf{E}\cdot\dot{\mathbf{E}}+\mathbf{H}\cdot\dot{\mathbf{H}}].$$

$$\nabla\cdot[\mathbf{E}\times\mathbf{H}] = \sum_i \frac{\partial}{\partial x_i}[\mathbf{E}\times\mathbf{H}]_i = \sum_{ijk}\frac{\partial}{\partial x_i}\varepsilon_{ijk}E_jH_k$$

$$= \sum_{ijk}\varepsilon_{ijk}\left[E_j\frac{\partial H_k}{\partial x_i} + H_k\frac{\partial E_j}{\partial x_i}\right]$$

$$= \sum_{ijk}\varepsilon_{kij}\frac{\partial E_j}{\partial x_i}H_k - \sum_{ijk}\varepsilon_{jik}\frac{\partial H_k}{\partial x_i}E_j = (\nabla\times\mathbf{E})\cdot\mathbf{H} - (\nabla\times\mathbf{H})\cdot\mathbf{E}.$$

Thus,

(a) $$-\frac{\partial}{\partial t}\int_V \frac{1}{8\pi}\left[E^2+H^2\right]d\tau = -\frac{1}{4\pi}\int_V[\mathbf{E}\cdot\dot{\mathbf{E}}+\mathbf{H}\cdot\dot{\mathbf{H}}]\,d\tau$$

(b) $$\frac{c}{4\pi}\int_S[\mathbf{E}\times\mathbf{H}]\cdot d\mathbf{S} = \frac{c}{4\pi}\int_V\nabla\cdot[\mathbf{E}\times\mathbf{H}]\,d\tau = \frac{c}{4\pi}\int_V\left(\nabla\times\mathbf{E}\cdot\mathbf{H} - \nabla\times\mathbf{H}\cdot\mathbf{E}\right)d\tau$$

$$= -\frac{1}{4\pi}\int_V[\dot{\mathbf{H}}\cdot\mathbf{H}+\dot{\mathbf{E}}\cdot\mathbf{E}]\,d\tau - \int_V\mathbf{j}\cdot\mathbf{E}\,d\tau.$$

Hence,

$$-\frac{\partial}{\partial t}\int_V\frac{1}{8\pi}\left[E^2+H^2\right]d\tau = \int_V\mathbf{j}\cdot\mathbf{E}\,d\tau + \frac{c}{4\pi}\int_S[\mathbf{E}\times\mathbf{H}]\cdot d\mathbf{S}.$$

II. Let H and H' be any two hermitian matrices of equal rank.

(a) Which of the following matrices are hermitian?

1. HH'
2. $HH' + H'H$
3. $i(HH' - H'H)$
4. H^{-1}
5. e^H
6. $H^\dagger H' H$
7. $e^{H+H'}$
8. $e^H \cdot e^{H'}$

(b) What is the necessary and sufficient condition on H and H' for all the above matrices to be hermitian?

Solution:

(a)

1. $(HH')^\dagger = H'^\dagger H^\dagger = H'H \neq HH'$, hence not hermitian.

2. $(HH' + H'H)^\dagger = H'^\dagger H^\dagger + H^\dagger H'^\dagger = H'H + HH'$, hence hermitian.

3. $i(HH' + H'H)^\dagger = -i(H'^\dagger H^\dagger - H^\dagger H'^\dagger) = -i(H'H - HH')$

   $= i(HH' + H'H)^\dagger$, hence hermitian.

4. $HH^{-1} = 1 \rightarrow (HH^{-1})^\dagger = H^{-1\dagger}H^\dagger = H^{-1\dagger}H = 1.$

   $H^{-1}H = 1 \rightarrow H^{-1\dagger}H = H^{-1}H \rightarrow (H^{-1\dagger} - H^{-1})H = 0.$

   But if H^{-1} exists, $\det H \neq 0$ and $H^{-1\dagger} = H^{-1}$, hence hermitian.

5. $$e^H = 1 + H + \frac{H^2}{2} + \frac{H^3}{3!} + \ldots$$

   $$\left[e^H\right]^\dagger = 1 + H^\dagger + \frac{(H^\dagger)^2}{2} + \frac{(H^\dagger)^3}{3!} + \ldots = 1 + H + \frac{H^2}{2} + \frac{H^3}{3!} + \ldots$$

   $= e^H$, hence hermitian.

6. $(H^+H'H)^+ = H^+H'^+H = H^+H'H$, hence hermitian.

7. Define $H'' \equiv H + H'$ (H'' hermitian).

Then $\left[e^{H''}\right]^+ = e^{H''}$ from 5; hence hermitian.

8. $e^H \cdot e^{H'} \equiv AA'$, where A and A' are hermitian.

But from 1, AA' is not hermitian, so that $e^H \cdot e^{H'}$ is not hermitian.

(b) The necessary and sufficient condition that HH' and $e^H \cdot e^{H'}$ be hermitian is that H and H' commute with each other.

Sufficiency (commutation): $HH' = H'H$.

1. $(HH')^+ = H'^+H^+ = H'H = HH'$.

2. $\left[e^H \cdot e^{H'}\right]^+ = \left[e^{H'}\right]^+\left[e^H\right]^+ = e^{H'} \cdot e^H = e^H \cdot e^{H'}$.

The last follows since each of the two terms is a hermitian matrix. Thus, they commute.

Necessity (hermiticity): $HH' = (HH')^+$

1. $(HH')^+ = HH' = H'^+H^+ = H'H$. Thus, H and H' commute.

2. $\left[e^H \cdot e^{H'}\right]^+ = \left[e^{H'}\right]^+\left[e^H\right]^+ = e^H \cdot e^{H'}$. Thus, $e^{H'} \cdot e^H = e^H \cdot e^{H'}$,

and the two terms commute.

III. Let H be a hermitian matrix and let U be defined by

$$U = (I - iH)^{-1}(I + iH),$$

where I is the unit matrix.

(*a*) Prove that $UH = HU$.

(*b*) Prove that U is unitary.

(*c*) Let V be a unitary matrix such that $VHV^+ = \Lambda$, where Λ is a diagonal matrix with $\Lambda_{ij} = \lambda_i\delta_{ij}$. Express *explicitly* the matrix elements of U, defined above, in terms of λ_i and V_{ij}.

Solution:

(a) Noting that $U = (I - iH)^{-1}(I + iH)$ is a function of H only, i.e. $U = U(H)$, it can be expanded in a power series in H.

$$U = \frac{I + iH}{I - iH} = I + 2iH - 2H^2 - 2iH^3 + 2H^4 + 2iH^5 \ldots,$$

and any function of H commutes with H. In this case, it is fairly obvious that

$$UH = IH + 2iH^2 - 2H^3 - 2iH^4 + 2H^5 \ldots = HU.$$

Alternative (direct) method:

$$UH = (I - iH)^{-1}(I + iH)H = (I - iH)^{-1}H(I + iH).$$

$$HU = H(I - iH)^{-1}(I + iH).$$

To show that H commutes with $(I - iH)^{-1}$, we assume it to be true and investigate the consequences.

Let $H(I - iH)^{-1} = (I - iH)^{-1}H$ and multiply both sides on the right by $(I - iH)$, yielding $H = (I - iH)^{-1}H(I - iH) = (I - iH)^{-1}(I - iH)H = H$, an identity; hence our assumption holds, and $UH = HU$.

(b) In power series notation,

$$U^+ = I - 2iH^+ - 2H^{+2} + 2iH^{+3} + 2H^{+4} - 2iH^{+5} \ldots,$$

$$= I - 2iH - 2H^2 + 2iH^3 + 2H^4 - 2iH^5 \ldots = \frac{I - iH}{I + iH}.$$

Hence,

$$UU^+ = U^+U = \frac{I - iH}{I + iH}\frac{I + iH}{I - iH} = I.$$

Alternative (direct) method:

$U = (I - iH)^{-1}(I + iH) \rightarrow (I - iH)U = (I + iH)$. Since from (a) above we know that U commutes with H, we have $(I - iH)U = U(I - iH)$. Thus, $U(I - iH) = I + iH$. Taking the adjoint of both sides, $(I - iH)^+U^+ = (I + iH)^+$, or $(I + iH)U^+ = I - iH$.

Multiplying the original expression for U on the left by the latter term:

$$(I + iH)U^{\dagger}U = (I - iH)(I - iH)^{-1}(I + iH) \rightarrow (I + iH)[U^{\dagger}U - I] = 0.$$

Since $\det|I + iH| \neq 0$, $U^{\dagger}U = I$. [If $\det|I + iH| = 0$, $(I + iH)^{-1}$ would not exist!]

(c) $VHV^{\dagger} = \Lambda$, where $\Lambda_{ij} = \lambda_i \delta_{ij}$.

Applying the power series method again and using the fact that $V^{\dagger}V = I$:

$$\begin{aligned} VUV^{\dagger} &= I + 2i(VHU^{\dagger}) - 2(VHV^{\dagger})(VHV^{\dagger}) \\ &\quad - 2i(VHV^{\dagger})(VHV^{\dagger})(VHV^{\dagger}) + \ldots \\ &= I + 2i\Lambda - 2i\Lambda^2 - 2i\Lambda^3 + 2\Lambda^4 + \ldots \\ &= \frac{I + i\Lambda}{I - i\Lambda} = (I - i\Lambda)^{-1}(I + i\Lambda). \end{aligned}$$

Define

$$W \equiv VUV^{\dagger} \rightarrow W = (I - i\Lambda)^{-1}(I + i\Lambda).$$

$$\left[(I - i\Lambda)W\right]_{ij} = (I + i\Lambda)_{ij} \rightarrow \sum_p (\delta_{ip} - i\lambda_i\delta_{ip})W_{pj} = \delta_{ij} + i\delta_{ij}\lambda_i$$

$$\Downarrow$$

$$\sum_p \delta_{ip}(1 - i\lambda_i)W_{pj} = \delta_{ij}(1 + i\lambda_i) \rightarrow (1 - i\lambda_i)W_{pj} = \delta_{ij}(1 + i\lambda_i)$$

$$\Downarrow$$

$$W_{ij} = \delta_{ij}\left(\frac{1 + i\lambda_i}{1 - i\lambda_i}\right).$$

$$W = VUV^{\dagger} \rightarrow U = V^{\dagger}WV \rightarrow U_{ij} = \sum_{\sigma\varepsilon} V_{i\sigma}{}^{+}W_{\sigma\varepsilon}V_{\varepsilon j}.$$

$$U_{ij} = \sum_{\sigma\varepsilon} V^{+}_{i\sigma}\,\delta_{\sigma\varepsilon}\left(\frac{1 + i\lambda_i}{1 - i\lambda_i}\right)V_{\varepsilon j} \rightarrow U_{ij} = \sum_{\varepsilon} V^{+}_{i\varepsilon}V_{\varepsilon j}\left(\frac{1 + i\lambda_{\varepsilon}}{1 - i\lambda_{\varepsilon}}\right).$$

IV. The differential equation for the time variation of temperature $T(\mathbf{r}t)$ due to heat flow in an isotropic medium is given by

$$\frac{\partial T}{\partial t} = \mu\nabla^2 T,$$

where μ is a positive constant.

(*a*) Prove that the general solution of this equation is

$$T(\mathbf{r}t) = \int f(\mathbf{k}) \exp(i\mathbf{k}\cdot\mathbf{r} - \mu k^2 t) d^3\mathbf{k},$$

where $f(\mathbf{k})$ is determined by the initial distribution of temperature at $t = 0$ through the relation

$$f(\mathbf{k}) = \frac{1}{(2\pi)^3}\int T(\mathbf{r}, t{=}0)\, e^{-i\mathbf{k}\cdot\mathbf{r}} d^3\mathbf{r}.$$

(*b*) If at $t = 0$, $T(\mathbf{r}, t{=}0) = \delta^3(\mathbf{r})$, show that at all later time $t > 0$:

$$T(\mathbf{r}t) = \frac{1}{(4\pi\mu t)^{3/2}} \exp\left(-\frac{r^2}{4\mu t}\right).$$

Solution:

(a) Taking Fourier transforms:

$$T(\mathbf{r}t) = \frac{1}{(2\pi)^{3/2}}\int g(\mathbf{k}t)\, e^{i\mathbf{k}\cdot\mathbf{r}} d^3\mathbf{k},$$

where

$$g(\mathbf{k}t) = \frac{1}{(2\pi)^{3/2}}\int T(\mathbf{r}t)\, e^{-i\mathbf{k}\cdot\mathbf{r}} d^3\mathbf{r}.$$

$$\nabla^2 T(\mathbf{r}t) = -\frac{1}{(2\pi)^{3/2}}\int k^2 g(\mathbf{k}t)\, e^{i\mathbf{k}\cdot\mathbf{r}} d^3\mathbf{k}.$$

Substituting into the differential equation, we have:

$$\int_{k\ \text{space}} \left[k^2 g(\mathbf{k}t) + \frac{\dot{g}(\mathbf{k}t)}{\mu}\right] e^{i\mathbf{k}\cdot\mathbf{r}} d^3\mathbf{k} = 0.$$

Choose an arbitrary vector $\mathbf{k}_o$ in $\mathbf{k}$ space and perform the following:

$$\int_{\substack{\text{all}\\\text{space}}} e^{-i\mathbf{k}_o\cdot\mathbf{r}} d^3\mathbf{r} \int_{\substack{k\\\text{space}}} \left[k^2 g(\mathbf{k}t) + \frac{\dot{g}(\mathbf{k}t)}{\mu}\right] e^{i\mathbf{k}\cdot\mathbf{r}} d^3\mathbf{k}$$

$$= \int_{\substack{k\\\text{space}}} \left[k^2 g(\mathbf{k}t) + \frac{\dot{g}(\mathbf{k}t)}{\mu}\right] d^3\mathbf{k} \int_{\substack{\text{all}\\\text{space}}} e^{i(\mathbf{k}-\mathbf{k}_o)\cdot\mathbf{r}} d^3\mathbf{r} = 0.$$

$$\Downarrow$$

$$\int_{\substack{k\\\text{space}}} \left[k^2 g(\mathbf{k}t) + \frac{\dot{g}(\mathbf{k}t)}{\mu}\right] \delta^3(\mathbf{k}-\mathbf{k}_o) d^3\mathbf{k} = 0. \quad \rightarrow \quad k_o^2 g(\mathbf{k}_o t) + \frac{\dot{g}(\mathbf{k}_o t)}{\mu} = 0.$$

Since $\mathbf{k}_o$ is arbitrary, we have $\dot{g}(\mathbf{k}t) = -\mu k^2 g(\mathbf{k}t)$. Solutions are given by:

$$g(\mathbf{k}t) = f_1(\mathbf{k}) e^{-\mu k^2 t},$$

where $f_1(\mathbf{k}) = g(\mathbf{k}, t{=}0)$. Defining $f(\mathbf{k}) = f_1(\mathbf{k})/(2\pi)^{3/2}$, we have:

$$T(\mathbf{r}t) = \int f(\mathbf{k}) \exp(i\mathbf{k}\cdot\mathbf{r} - \mu k^2 t) d^3\mathbf{k}.$$

(b) At $t = 0$, we have from (a) above, that

$$T(\mathbf{r}, t{=}0) = \int f(\mathbf{k})\, e^{i\mathbf{k}\cdot\mathbf{r}} d^3\mathbf{k},$$

and

$$f(\mathbf{k}) = \frac{1}{(2\pi)^3}\int T(\mathbf{r}, t{=}0) e^{-i\mathbf{k}\cdot\mathbf{r}} d^3\mathbf{r} \quad \rightarrow \quad \frac{1}{(2\pi)^3}\int \delta^3(\mathbf{r}) e^{-i\mathbf{k}\cdot\mathbf{r}} d^3\mathbf{r} = \frac{1}{(2\pi)^3}$$

upon setting $T(\mathbf{r}, t{=}0) = \delta^3(\mathbf{r})$. Hence,

$$T(\mathbf{r}t) = \frac{1}{(2\pi)^3}\int \exp(i\mathbf{k}\cdot\mathbf{r} - \mu k^2 t) d^3\mathbf{k}$$

$$= \frac{1}{(2\pi)^3}\int_0^{2\pi}\int_0^{\pi}\int_0^{\infty} e^{ikr\cos\theta} e^{-\mu k^2 t} k^2 \sin\theta\, dk\, d\theta\, d\varphi$$

$$= \frac{1}{(2\pi)^2}\int_0^{\pi}\int_0^{\infty} e^{ikr\cos\theta} e^{-\mu k^2 t} k^2 \sin\theta\, dk\, d\theta$$

$$T(\mathbf{r}t) = -\frac{1}{ir(2\pi)^2}\int_0^\infty \left| e^{ikr\cos\theta} e^{-\mu k^2 t} k\, dk \right|_0^\pi$$

$$= \frac{1}{ir(2\pi)^2}\int_0^\infty e^{-\mu k^2 t} k \left(e^{ikr} - e^{-ikr} \right) dk$$

$$= \frac{2}{r(2\pi)^2}\int_0^\infty e^{-\mu k^2 t} k\, (\sin kr) dk = -\frac{2}{r(2\pi)^2}\int_0^\infty \frac{\sin kr}{2\mu t} d\left(e^{-\mu k^2 t} \right)$$

$$= -\frac{2}{2\mu r(2\pi)^2}\left\{ \left| \frac{\sin kr}{t} e^{-\mu k^2 t} \right|_0^\infty - \int_0^\infty \frac{r \cos kr}{t} e^{-\mu k^2 t} dk \right\}$$

$$= \frac{1}{4\mu\pi^2 t}\int_0^\infty e^{-\mu k^2 t} \cos kr\, dk.$$

From integral tables, $\int_0^\infty e^{-a^2 x^2} \cos bx\, dx = \frac{\sqrt{\pi}}{2a} e^{-b^2/4a^2}$.

Setting $a^2 = \mu t$ and $b = r$, $\int_0^\infty e^{-\mu k^2 t} \cos kr\, dk = \frac{\sqrt{\pi}}{2(\mu t)^{1/2}} e^{-r^2/4\mu t}$. Thus,

$$T(\mathbf{r}t) = \frac{1}{4\mu\pi^2 t}\frac{\sqrt{\pi}}{2(\mu t)^{1/2}} e^{-r^2/4\mu t} = \frac{1}{(4\pi\mu t)^{3/2}} e^{-r^2/4\mu t}.$$

Part II

Chapter 6

Theory of Functions of a Complex Variable

6.1 Theory of Complex Variables

6.1.1 Complex Numbers

Defined: $w = a + ib$; $\upsilon = c + id$.

Equality: $w = \upsilon$ if and only if $a = c$ and $b = d$.

Addition: $w \pm \upsilon = (a + ib) \pm (c \pm id) = (a \pm c) + i(b \pm d)$

Division: $\dfrac{w}{\upsilon} = w\upsilon^{-1} = \dfrac{a + ib}{c + id} = \dfrac{a + ib}{c + id}\dfrac{c - id}{c - id} = \dfrac{ac + bd}{c^2 + d^2} + i\dfrac{bc - ad}{c^2 + d^2}$

6.1.2 Complex Variables

A complex variable $\overline{z} = x + iy$ is defined in a z-plane formed by a real axis and an imaginary axis (Fig. 6-1). From the polar form of $\overline{z}$, i.e. from $\overline{z} = |\overline{z}|e^{i\theta} = \rho e^{i\theta} = \rho\cos\theta + i\rho\sin\theta$, and from $\overline{z} = x + iy$, we have that $x = \rho\cos\theta$ and $y = \rho\sin\theta$ satisfy the conditions for the equality of two complex variables.

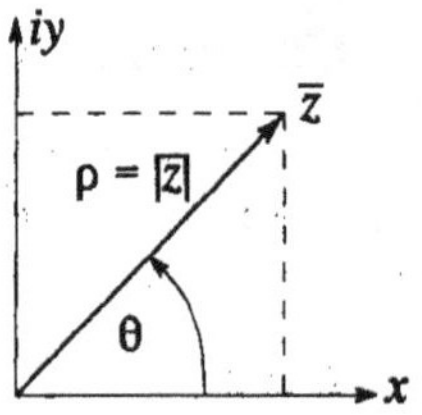

Fig. 6-1

The inverse relationships are given by:

$$\rho = \sqrt{x^2 + y^2}; \quad \theta = \tan^{-1}\frac{y}{x},$$

where ρ = magnitude or absolute value of $\overline{z}$, and θ = modulus of $\overline{z}$.

6.1.3 Addition of Complex Variables

Referring to Fig. 6-2,

$$\bar{R} = \bar{z} + \bar{t} = |\bar{R}|e^{i\theta_1},$$

where $|\bar{R}|^2 = |\bar{z}|^2 + |\bar{t}|^2 - 2|\bar{z}||\bar{t}|\cos\alpha$ and where $\theta_1 = \theta + \gamma = \theta + (180 - \varphi - \alpha)$.

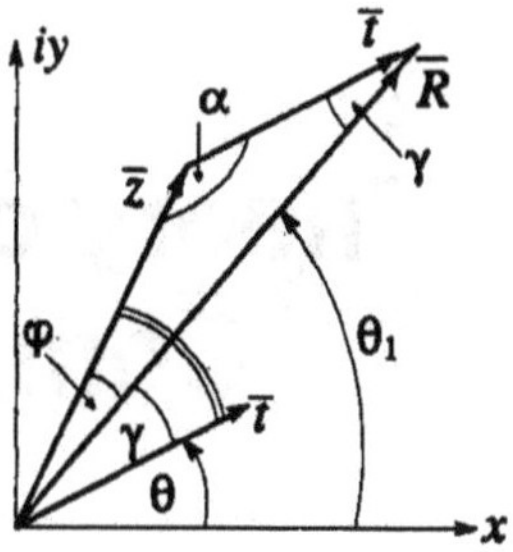

Fig. 6-2

$$\overline{RR}^* = (\bar{z} + \bar{t})(\bar{z}^* + \bar{t}^*)$$

$$= |\bar{z}|^2 + |\bar{t}|^2 + \bar{t}\bar{z}^* + \bar{z}\bar{t}^*$$

$$= |\bar{z}|^2 + |\bar{t}|^2 + |\bar{t}||\bar{z}|\left(e^{i(\theta-\varphi-\theta_1)} + e^{-i(\theta-\varphi-\theta_1)}\right)$$

$$= |\bar{z}|^2 + |\bar{t}|^2 + 2|\bar{t}||\bar{z}|\cos(180 - \alpha) = |\bar{z}|^2 + |\bar{t}|^2 - 2|\bar{t}||\bar{z}|\cos\alpha.$$

Hence, $|\bar{R}|^2 = \overline{RR}^*$.

6.1.4 Functions of a Complex Variable

$f(\bar{z})$, a function of the complex variable $\bar{z}$, can be considered a vector field in that one can associate a particular $f(\bar{z})$ with every $\bar{z}$ in the $\bar{z}$ plane. Since $f(\bar{z})$ is also a complex function, it can be expressed as:

$$f(\bar{z}) = \varphi(xy) + i\psi(xy) \rightarrow f^*(\bar{z}) = \varphi(xy) - i\psi(xy).$$

Similarly, in polar (or vector) form, $f(\bar{z}) = |f(\bar{z})|e^{i\Omega}$,

where $|f(\bar{z})| = \sqrt{\varphi^2(xy) + \psi^2(xy)}$ and $\Omega = \tan^{-1}\dfrac{\psi(xy)}{\varphi(xy)}$.

6.1.5 Calculus of Functions of Complex Variables

We deal first with single-valued functions.
Consider $f(\bar{z}) = \bar{z} = \rho e^{i\theta}$. From Fig. 6-3 we note that $f(\bar{z})$ is unchanged by going around $2\pi n$ times.

Hence, $f(\bar{z}) = \bar{z} = \rho e^{i(\theta+2\pi n)}$, where n is integer-valued, is single-valued.

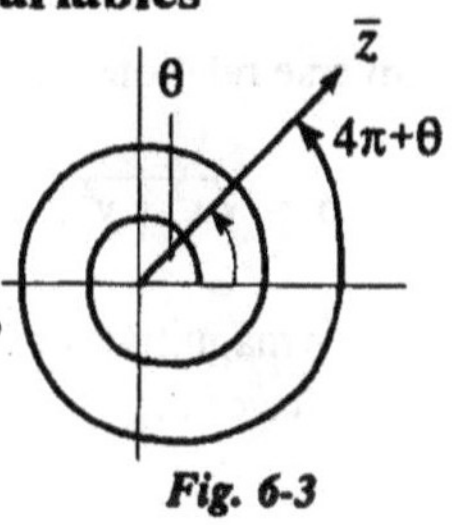

Fig. 6-3

Consider the function

$$f(\bar{z}) = \ln \bar{z} = \ln\left(\rho e^{i(\theta+2n\pi)}\right) = \ln \rho + i(\theta + 2n\pi).$$

Here we note that $f(\bar{z})$ is multiple-valued, in fact infinitely valued. In order to work with $f(\bar{z})$, we must make it single-valued (otherwise it would not be an analytic function — analyticity will be discussed later). We do this by introducing a cut, so that we limit values of θ to $0 < \theta < 2\pi$ (Fig. 6-4). Since one is not permitted to cross the cut, n is restricted to be zero-valued and we can now speak of a single-valued function $f(\bar{z})$ in this cut plane.

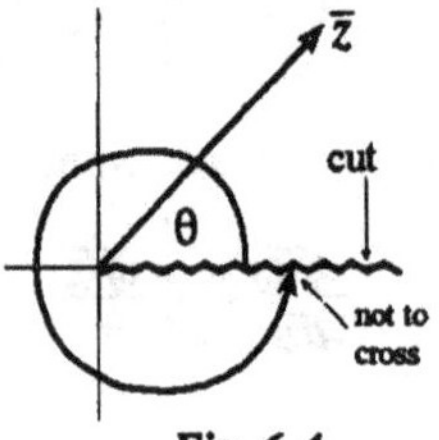

Fig. 6-4

Derivatives:

$$\frac{df(\bar{z})}{d\bar{z}} = \lim_{\Delta\bar{z}\to 0} \frac{f(\bar{z}+\Delta\bar{z}) - f(\bar{z})}{\Delta\bar{z}}$$

independent of how $\Delta\bar{z} \to 0$. The independence condition is necessary, since there are many possible directions for $\Delta\bar{z} \to 0$ (Fig. 6-5).

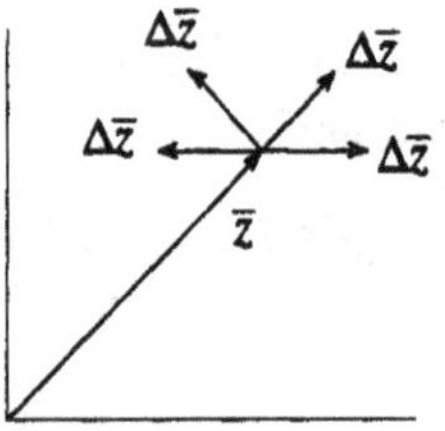

Fig. 6-5

6.2 Analytic Functions

A function $f(\bar{z})$ is analytic at a point $\bar{z}$ if and only if the derivative exists at that point. A function $f(\bar{z})$ is analytic over a region R if $f(\bar{z})$ is analytic at every point $\bar{z}$ in the region. [This region may turn out to be quite odd, consisting of loops, rings, etc.]

Theorem:

If $f(\bar{z})$ is analytic, and if it can be expressed in the form $f(\bar{z}) = \varphi(xy) + i\psi(xy)$:

$$\frac{\partial\varphi(xy)}{\partial x} = \frac{\partial\psi(xy)}{\partial y}, \qquad \frac{\partial\varphi(xy)}{\partial y} = -\frac{\partial\psi(xy)}{\partial x}.$$

These expressions are known as the Cauchy-Riemann conditions. Henceforth we will drop the variables x and y in φ and ψ with the understanding that they are neverthess implied.

Proof: Since $f(\bar{z})$ is analytic, $\Delta\bar{z} \to 0$ in any direction. For the case where $\Delta\bar{z} = \Delta x$ (Fig. 6-6),

$$\frac{df(\bar{z})}{d\bar{z}} = \frac{\partial f(\bar{z})}{\partial x} = \frac{\partial\varphi}{\partial x} + i\frac{\partial\psi}{\partial x},$$

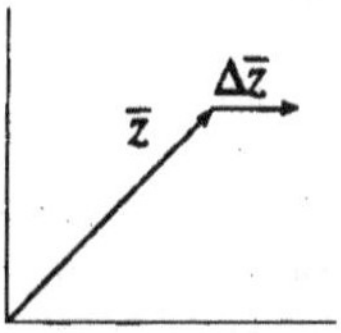

Fig. 6-6

while for the case where $\Delta\bar{z} = i\,\Delta y$ (Fig. 6-7),

$$\frac{df(\bar{z})}{d\bar{z}} = \frac{1}{i}\frac{\partial f(\bar{z})}{\partial y} = -i\left\{\frac{\partial\varphi}{\partial y} + i\frac{\partial\psi}{\partial y}\right\} = \frac{\partial\psi}{\partial y} - i\frac{\partial\varphi}{\partial y}.$$

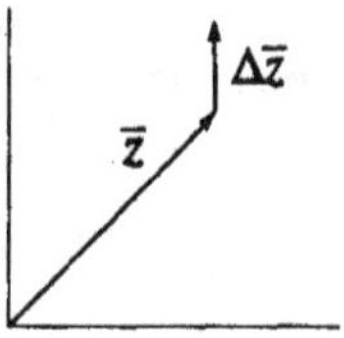

Fig. 6-7

Since $df(\bar{z})/d\bar{z}$ exists and is independent of how $\Delta\bar{z} \to 0$, these two expressions should be equal; hence

$$\frac{\partial\varphi}{\partial x} + i\frac{\partial\psi}{\partial x} = \frac{\partial\psi}{\partial y} - i\frac{\partial\varphi}{\partial y},$$

which, according to our definition of equality of complex numbers, implies that

$$\frac{\partial\varphi(xy)}{\partial x} = \frac{\partial\psi(xy)}{\partial y}. \qquad \frac{\partial\varphi(xy)}{\partial y} = -\frac{\partial\psi(xy)}{\partial x}.$$

Theorem: (Converse of above) If $f(\bar{z})$ can be expressed as $\varphi + i\psi$, and if φ and ψ obey the Cauchy-Riemann conditions, then:

(a) $f(\bar{z})$ is a function of $\bar{z}$ only (i.e. not $\bar{z}^*$).
(b) $f(\bar{z})$ is analytic.

Proof: $\bar{z} = x + iy$. $\bar{z}^* = x - iy$.

$$\frac{\partial f}{\partial x} = \frac{\partial\varphi}{\partial x} + i\frac{\partial\psi}{\partial x}. \qquad -i\frac{\partial f}{\partial y} = \frac{\partial\psi}{\partial y} - i\frac{\partial\varphi}{\partial y}.$$

For $f = f(\bar{z}, \bar{z}^*)$:

$$\frac{\partial f}{\partial x} = \frac{\partial f}{\partial\bar{z}}\frac{\partial\bar{z}}{\partial x} + \frac{\partial f}{\partial\bar{z}^*}\frac{\partial\bar{z}^*}{\partial x} = \frac{\partial f}{\partial\bar{z}} + \frac{\partial f}{\partial\bar{z}^*}.$$

$$\Downarrow$$

(1) $$\frac{\partial f}{\partial\bar{z}} + \frac{\partial f}{\partial\bar{z}^*} = \frac{\partial\varphi}{\partial x} + i\frac{\partial\psi}{\partial x}.$$

$$\frac{\partial f}{\partial y} = \frac{\partial f}{\partial \bar{z}}\frac{\partial \bar{z}}{\partial y} + \frac{\partial f}{\partial \bar{z}^*}\frac{\partial \bar{z}^*}{\partial y} = i\frac{\partial f}{\partial \bar{z}} + (-i)\frac{\partial f}{\partial \bar{z}^*}.$$

$$\Downarrow$$

(2) $$\frac{\partial f}{\partial \bar{z}} - \frac{\partial f}{\partial \bar{z}^*} = \frac{\partial \psi}{\partial y} - i\frac{\partial \varphi}{\partial y}.$$

Subtracting equation (2) from equation (1), we have:

$$2\frac{\partial f}{\partial \bar{z}^*} = \left(\frac{\partial \varphi}{\partial x} - \frac{\partial \psi}{\partial y}\right) + i\left(\frac{\partial \psi}{\partial x} + \frac{\partial \varphi}{\partial y}\right) = 0,$$

if the Cauchy-Riemann conditions are satisfied.

Hence, $\dfrac{\partial f}{\partial \bar{z}^*} = 0$ and $f = f(\bar{z})$ only. Adding equations (1) and (2), we have:

$$2\frac{\partial f}{\partial \bar{z}} = \left(\frac{\partial \varphi}{\partial x} + \frac{\partial \psi}{\partial y}\right) + i\left(\frac{\partial \psi}{\partial x} - \frac{\partial \varphi}{\partial y}\right).$$

Applying the Cauchy-Riemann conditions:

(3) $$\frac{\partial f}{\partial \bar{z}} = \frac{\partial \varphi}{\partial x} + i\frac{\partial \psi}{\partial x}.$$

(4) $$\frac{\partial f}{\partial \bar{z}} = \frac{\partial \psi}{\partial y} - i\frac{\partial \varphi}{\partial y}.$$

Since all the derivatives on the right side exist, $\partial f/\partial \bar{z}$ exists and $f(\bar{z})$ is analytic. This completes the proof, since we have shown that $\partial f/\partial \bar{z}$ exists as we approach from the x-axis and from the y-axis, i.e. independent of how $\Delta \bar{z} \to 0$.

6.3 Applications of Analytic Functions

6.3.1 Electrostatics

Consider a charge-free region ($\rho = 0$) in which the potentials on conducting surfaces are specified. (We are dealing with a static case for which $\nabla \times \mathbf{E} = 0$, since $\mathbf{H} = 0$.) We seek $\mathbf{E}$ and φ in the neighborhood of the conductor.

Under the static conditions, Maxwell's equations for **E** are $\nabla\cdot\mathbf{E} = 0$ and $\nabla\times\mathbf{E} = 0$. We have shown (Methods in Mathematical Phycics I) that if the curl of a vector vanishes in a certain region, then this vector can be expressed as the gradient of a scalar. Hence, $\mathbf{E} = -\nabla\varphi$ (the negative sign is used to produce certain desired end results). Thus, $\nabla\cdot\nabla\varphi = 0 \rightarrow \nabla^2\varphi = 0$. Since $\nabla\varphi\cdot d\mathbf{r} = d\varphi$, then along φ = const., $d\varphi = 0$ and $\nabla\varphi$ is perpendicular to the surface. Hence the gradient of a function forms a set of lines perpendicular to the set of lines φ = const. (Fig. 6-8).

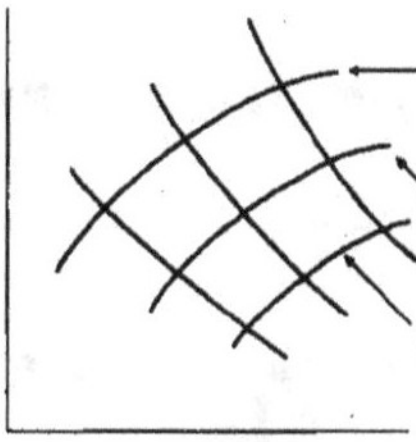

Fig. 6-8

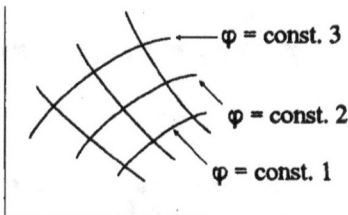

Consider now the analytic function $f(z) = \varphi + i\psi$, where φ is the electrostatic potential ($\mathbf{E} = -\nabla\varphi$). Then,

$$\mathbf{E}\cdot\nabla\psi = -\left(\frac{\partial\varphi}{\partial x}\frac{\partial\psi}{\partial x} + \frac{\partial\varphi}{\partial y}\frac{\partial\psi}{\partial y}\right)$$

$$= \frac{\partial\psi}{\partial y}\frac{\partial\psi}{\partial x} - \frac{\partial\psi}{\partial x}\frac{\partial\psi}{\partial y} = 0,$$

since φ and ψ obey the Cauchy-Riemann conditions for an analytic function $f(z)$.

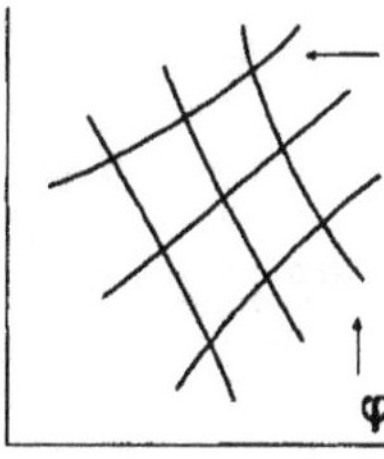

Fig. 6-9

Thus $\mathbf{E}\cdot\nabla\psi = 0$, which implies (see Fig. 6-9):

Lines of **E** are perpendicular to lines of $\nabla\psi$.
Lines of **E** are parallel to lines of ψ = const.
Lines of **E** are perpendicular to lines of φ = const.
Hence, lines of ψ = const. are perpendicular to lines of φ = const.

In the Problems it is shown that if $f(z)$ is analytic and can be expressed as $f(z) = \varphi + i\psi$, then $\nabla^2\varphi = 0$ and $\nabla^2\psi = 0$. Furthermore, since **E** is everywhere on the surface perpendicular to the surface, and since φ = const. are lines perpendicular to **E**, then φ = const. on the surface. Hence if you specify an analytic function $f(z) = \varphi + i\psi$, this is sufficient to determine φ (and from this $-\nabla\varphi$), since φ satisfies $\nabla^2\varphi = 0$ and φ = const. on the surface of the conductor.

[There is a method, known as the Schwartz-Christoffel method (not to be studied in this course) for generating any desired analytic function, given the conducting surface.]

Special cases:

Case I: We state that $f(\bar{z}) = \frac{A}{i}\bar{z}^n$ is a solution to the problem of Fig. 6-10.

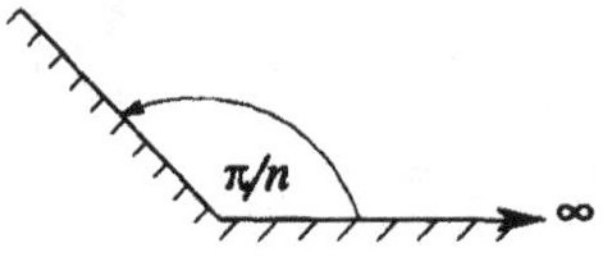

Fig. 6-10

Proof: Note first that $f(\bar{z})$ is analytic. Then express it in the form $\varphi + i\psi$.

$$f(\bar{z}) = \frac{A}{i}\rho^n e^{in\theta} = \frac{A}{i}\rho^n(\cos n\theta + i\sin n\theta)$$

$$= A\rho^n(\sin n\theta - i\cos n\theta),$$

so that $\varphi = A\rho^n \sin n\theta$ and $\psi = -A\rho^n\cos n\theta$. Note that along $\theta = 0, \varphi = 0$ while along $\theta = \pi/n, \varphi = 0$. Hence, $\varphi = 0$ along the length of the conducting surface. Remembering that lines of φ are perpendicular to lines of ψ, we can map the lines of φ and **E** as shown in Fig. 6-11.

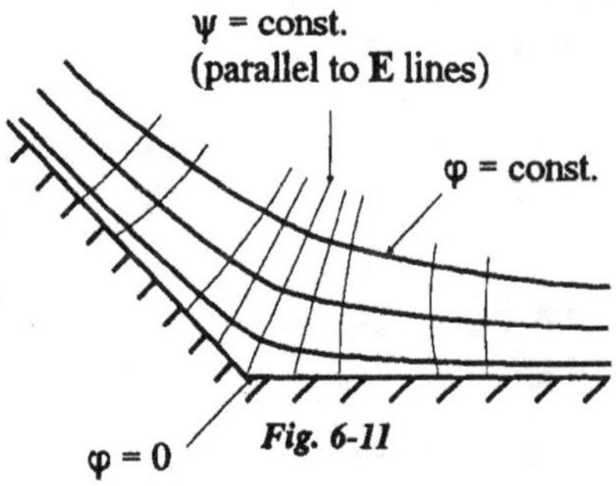

Fig. 6-11

Case II: As a special case of the above, set $n = 1$ (Fig. 6-12).

$$f(\bar{z}) = \frac{A}{i}\bar{z} = \frac{A}{i}\rho e^{i\theta} = A\rho\sin\theta - iA\rho\cos\theta.$$

$$\varphi = A\rho\sin\theta = Ay.$$

$$\psi = -A\rho\cos\theta = -Ax.$$

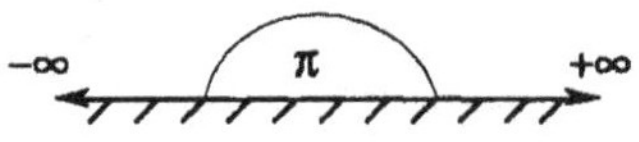

Fig. 6-12

Hence, lines of φ = const. are represented by y = const. while lines of ψ = const. are represented by x = const. (Fig. 6-13).

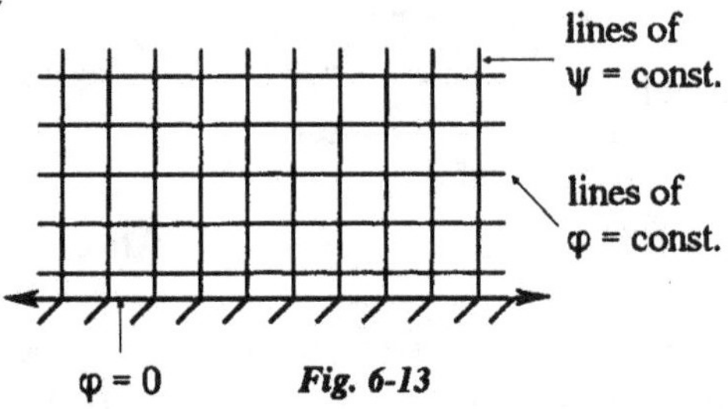

Fig. 6-13

Furthermore,

$$\mathbf{E} = -\nabla\varphi$$

$$= -\nabla Ay = -Ay_o$$

Case III: As a special case, set $n = 2$.

$$f(\bar{z}) = \frac{A}{i}(\bar{z})^2 = \frac{A}{i}\rho^2(\cos 2\theta + i\sin 2\theta).$$

$$\varphi = A\rho^2 \sin 2\theta.$$

$$\psi = -A\rho^2 \cos 2\theta.$$

Lines of φ = const. are represented by $A\rho^2\sin 2\theta = A\rho^2 2\sin\theta\cos\theta = 2Axy$ = const. (hyperbolas). Lines of constant ψ are represented by $-A\rho^2\cos 2\theta = -A\rho^2(\cos^2\theta - \sin^2\theta) = -A(x^2 - y^2)$ = const., or $x^2 = y^2$ + const. (equilateral hyperbolas) (Fig. 6-14).

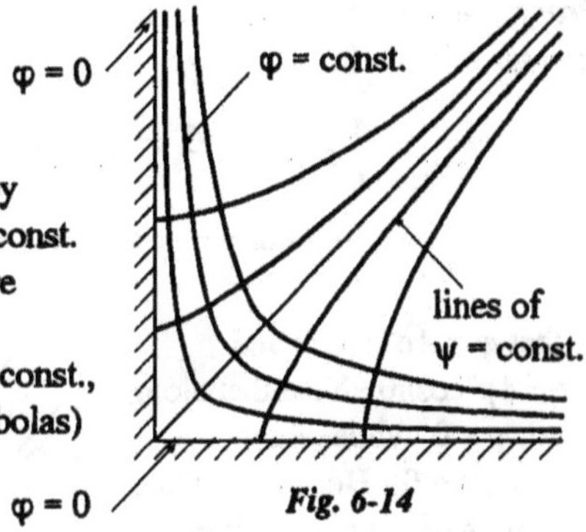

Fig. 6-14

Case IV: For the case $n = 1/2$, $\pi/n = 2\pi$ and the problem reduces to finding the solution for a knife edge (Fig. 6-15).

$$\varphi = A\rho^{1/2}\sin(\theta/2).$$

$$\psi = -A\rho^{1/2}\cos(\theta/2).$$

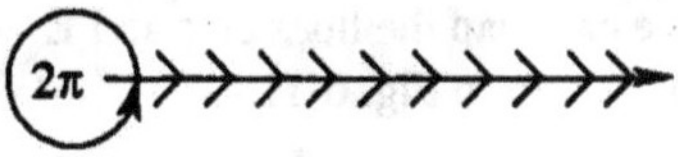

Fig. 6-15

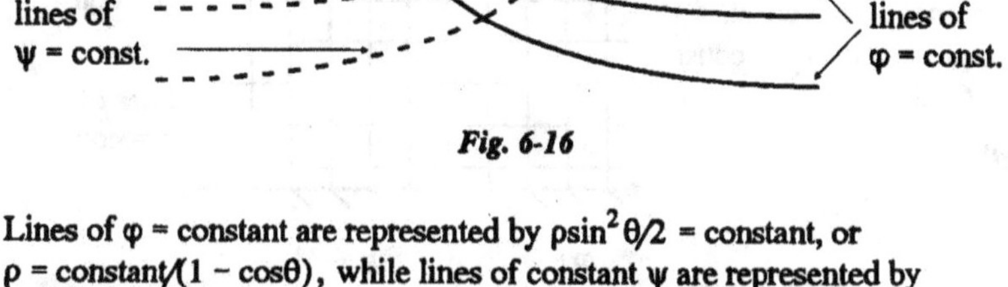

Fig. 6-16

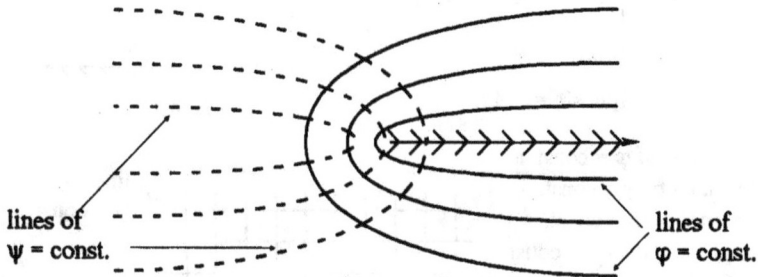

Lines of φ = constant are represented by $\rho\sin^2\theta/2$ = constant, or ρ = constant/$(1 - \cos\theta)$, while lines of constant ψ are represented by $\rho\cos^2\theta/2$ = constant, or ρ = constant/$(1 + \cos\theta)$ (Fig. 6-16).

6.3.2 Hydrodynamics

This is the case in which there exists a body about which there flows an incompressible fluid. One further specifies that there be irrotational motion (no whirls). Incompressibility implies that $d\rho/dt = 0$, or

$$\frac{\partial\rho}{\partial t} + \frac{\partial\rho}{\partial x}\frac{dx}{dt} + \frac{\partial\rho}{\partial y}\frac{dy}{dt} + \frac{\partial\rho}{\partial z}\frac{dz}{dt} = \frac{\partial\rho}{\partial t} + \nabla\rho\cdot\mathbf{v}.$$

Since the equation of continuity demands that

$$\partial\rho/\partial t + \nabla\cdot\rho\mathbf{v} = 0 \quad \rightarrow \quad \partial\rho/\partial t + \nabla\rho\cdot\mathbf{v} + \rho\nabla\cdot\mathbf{v} = 0,$$

it follows that the incompressibility condition says that $\nabla\cdot\mathbf{v} = 0$.

For irrotational motion,

$$\nabla\times\mathbf{v} = 0 \quad \rightarrow \quad \mathbf{v} = -\nabla\varphi,$$

which from $\nabla\cdot\mathbf{v} = 0$ implies $\nabla^2\varphi = 0$.

Note here the important difference from the electrostatic case. Whereas **E** was perpendicular to the conducting surface and φ = constant on the surface, for the hydrodynamic case **v** is parallel to the boundary and $|\mathbf{v}|$ = constant on the boundary surface.

Since **v** is parallel to lines of ψ = constant and perpendicular to lines of φ = constant, we again have lines of φ = constant perpendicular to lines of ψ = constant.

Mapping of lines of **v** and φ:

To solve the hydrodynamic problem we seek a φ satisfying $\nabla^2\varphi = 0$ with lines of ψ = constant parallel to the surface and lines of φ = constant perpendicular to the surface, in addition to $\mathbf{v} = -\nabla\varphi$, such that $|\mathbf{v}|$ = constant on the surface. (Fig. 6-17).

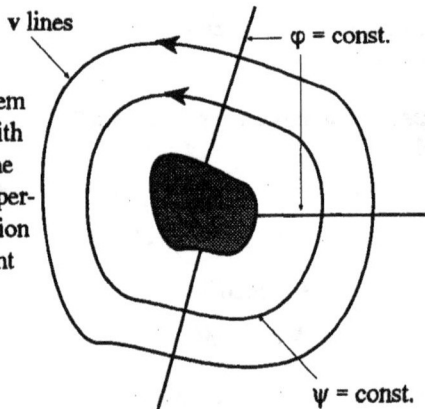

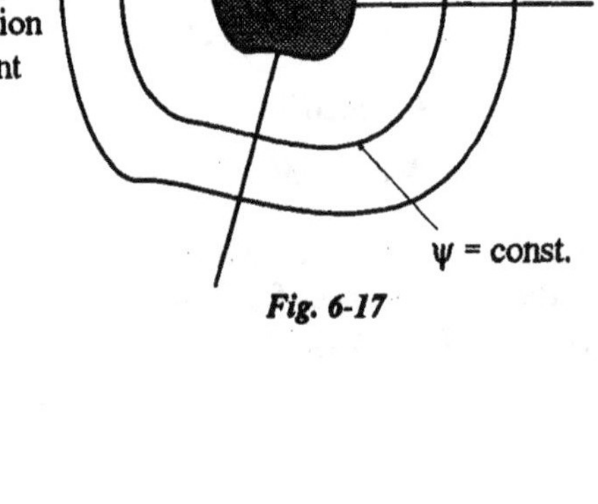

Fig. 6-17

We claim the solution to the hydrodynamic problem illustrated in Fig. 6-18 to be

$$f(z) = Az^n = A\rho^n\cos n\theta + iA\rho^n\sin n\theta.$$

$$\varphi = A\rho^n\cos n\theta. \quad \psi = A\rho^n\sin n\theta.$$

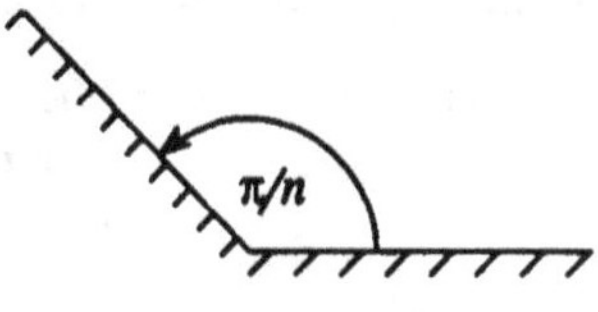

Fig. 6-18

Note that along the boundary surfaces $\psi = 0$, while for other regions of space, lines of constant ψ are as shown in Fig. 6-19. Lines of **v** are parallel to lines of constant ψ.

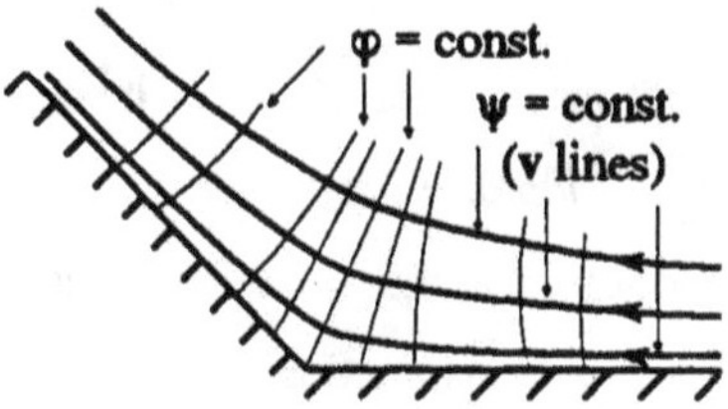

Fig. 6-19

Lines are seen to be exactly reversed as compared to the electrostatic case.

Special Cases

Case I: For case of $n = 1$,

$$f(z) = A\rho\cos\theta + iA\rho\sin\theta.$$

$$\varphi = A\rho\cos\theta = Ax.$$

$$\psi = A\rho\sin\theta = Ay.$$

$$\mathbf{v} = -\nabla\varphi = -A\mathbf{i}_o.$$

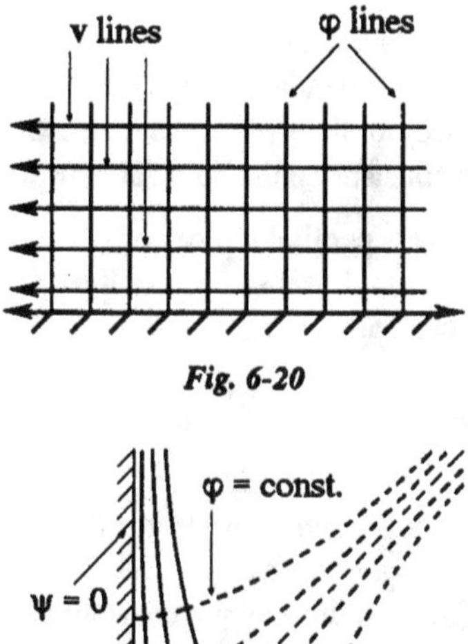

Fig. 6-20

Mapping of lines of φ and **v** is shown in Fig. 6-20.

Case II: For the case $n = 2$, $\pi/n = \pi/2$, and

$$f(z) = A\rho^2\cos 2\theta + iA\rho^2\sin 2\theta.$$

$$\varphi = A\rho^2\cos 2\theta = A(x^2 - y^2).$$

$$\psi = A\rho^2\sin 2\theta = 2Axy.$$

$$\mathbf{v} = -\nabla\varphi = -A(2x\mathbf{i}_o - 2y\mathbf{j}_o).$$

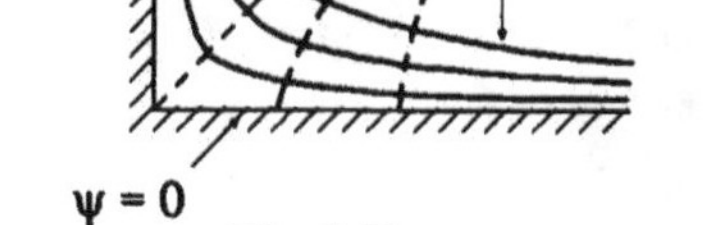

Fig. 6-21

Lines of φ = const. give equilateral hyperbolas while lines of ψ = const. give hyperbolas (Fig. 6-21).

Case III: For $n = 1/2$,

$$f(\bar{z}) = A\bar{z}^{1/2} = A\rho^{1/2}\cos\theta/2 + iA\rho^{1/2}\sin\theta/2.$$

$$\varphi = A\rho^{1/2}\cos\theta/2. \quad \psi = A\rho^{1/2}\sin\theta/2.$$

Lines of φ = constant = lines of $\rho(1 + \cos\theta)$ = constant, while lines of ψ = constant = lines of $\rho(1 - \cos\theta)$ = constant (Fig. 6-22).

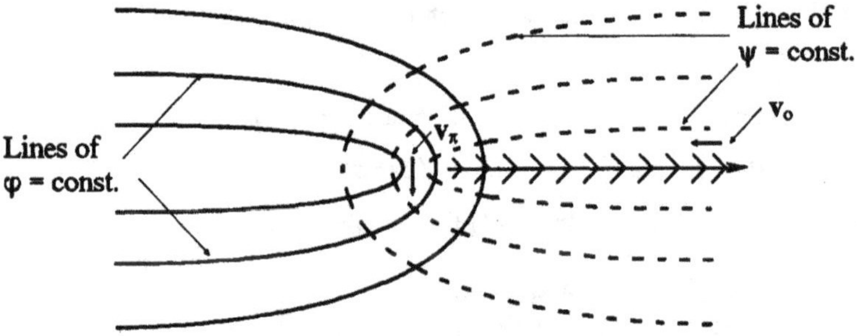

Fig. 6-22

$$\mathbf{v} = -\nabla\varphi = -\rho_o\left(\frac{\partial\varphi}{\partial\rho}\right) - \frac{\theta_o}{\rho}\frac{\partial\varphi}{\partial\theta} = -\frac{A\cos\theta/2}{2\rho^{1/2}}\rho_o + \frac{A\sin\theta/2}{2\rho^{1/2}}\theta_o$$

$$= -\frac{A}{2\rho^{1/2}}(\rho_o\cos\theta/2 - \theta_o\sin\theta/2).$$

$$|\mathbf{v}| = |\nabla\varphi| = \frac{A}{2\rho^{1/2}} \underset{\rho\to 0}{\longrightarrow} \infty.$$

At $\theta = 0$ and π, $\mathbf{v}$ is in the $-\rho_o$ and $+\theta_o$ direction, respectively (i.e. the fluid goes whipping around the sharp corner).

Consider a function $f(\bar{z}) = \ln\bar{z}$ of the type previously discussed, which at first does not appear to be analytic, that is, its derivative does not exist everywhere, i.e. $df(\bar{z})/d\bar{z} = 1/\bar{z}$, which does not exist at the origin.

We seek to evaluate

$$\frac{df(\bar{z})}{d\bar{z}} = \lim_{\Delta\bar{z}\to 0}\frac{f(\bar{z}+\Delta\bar{z}) - f(\bar{z})}{\Delta\bar{z}}.$$

Considering Fig. 6-23:

$$\bar{z} + \Delta\bar{z} = \rho e^{i\theta}.$$

$$f(\bar{z} + \Delta\bar{z}) = \ln(\rho e^{i\theta}) = \ln\rho + i\theta.$$

$$f(\bar{z}) = \ln\left[\rho' e^{i(2\pi-\varphi)}\right] = \ln\rho' + i(2\pi - \varphi).$$

$$\text{(a)} \qquad f(\bar{z} + \Delta\bar{z}) - f(\bar{z}) = \ln(\rho/\rho') + i(\theta + \varphi - 2\pi)\,.$$

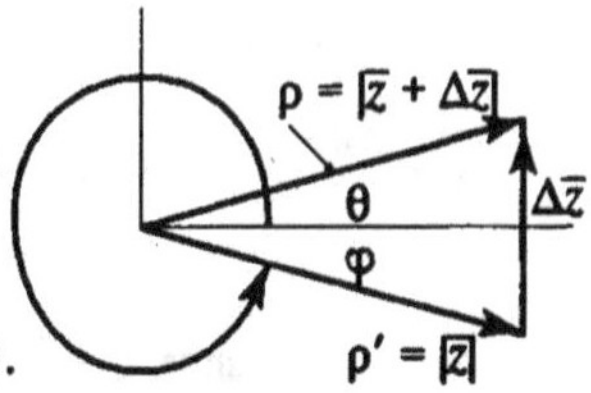

Fig. 6-23

Introducing a cut at $y = 0$ so that $\Delta\bar{z}$ cannot cross the x-axis, we have from Fig. 6-24:

$$f(\bar{z} + \Delta\bar{z}) = \ln(\rho'' e^{i\omega}) = \ln\rho'' + i\omega.$$

$$f(\bar{z}) = \ln\left[\rho' e^{i(2\pi-\varphi)}\right] = \ln\rho' + i(2\pi - \varphi).$$

$$\text{(b)} \qquad f(\bar{z} + \Delta\bar{z}) - f(\bar{z}) = \ln(\rho''/\rho') + i(\omega + \varphi - 2\pi)\,.$$

In the limit as $\Delta\bar{z} \to 0$, (a) and (b) yield different values for $df(\bar{z})/d\bar{z}$, so that the function is not analytic.

By introducing the cut at the x-axis ($y = 0$), we limit $f(\bar{z})$ to be single-valued and, in addition, we make $f(\bar{z}) = \ln\bar{z}$ analytic in the cut plane $0 < \theta < 2\pi$. Some examples will now clarify this discussion.

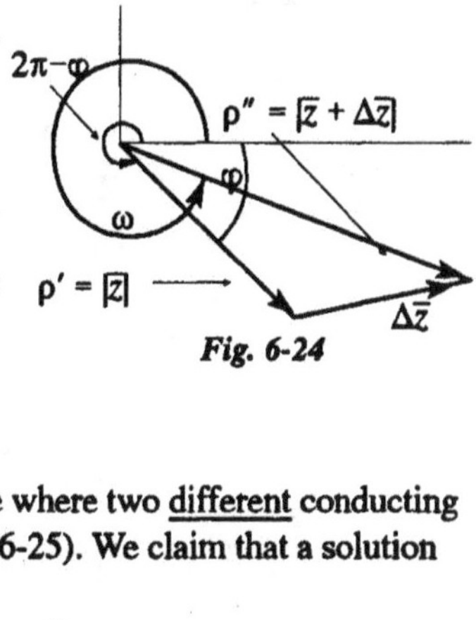

Fig. 6-24

Case IV (Electrostatics): Consider the case where two different conducting surfaces make angles with each other (Fig. 6-25). We claim that a solution to the problem is given by:

$$f(\bar{z}) = \frac{A}{i}\ln\bar{z} = \frac{A}{i}\ln(\rho e^{i\theta})$$

$$= \frac{A}{i}\ln\rho + A\theta.$$

$$\varphi = A\theta, \quad \psi = -A\ln\rho.$$

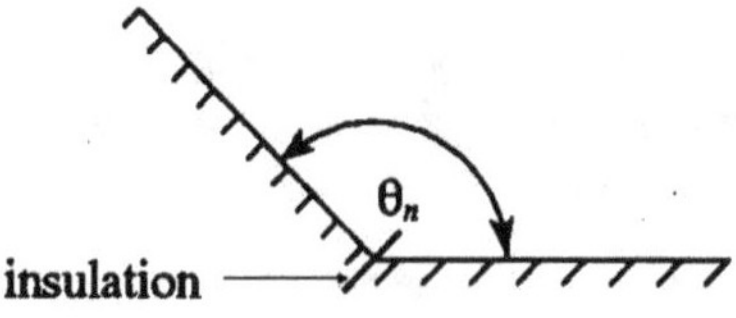

Fig. 6-25

Lines of φ = const. are given by θ = const. while lines of ψ = const. are given by ρ = const.

At $\theta = 0$, $\varphi = 0$, and at $\theta = \theta_n$, $\varphi = A\theta_n$.

Also,

$$\mathbf{E} = -\nabla\varphi = \frac{-\theta_o}{\rho}\frac{\partial\varphi}{\partial\theta} = \frac{-A\theta_o}{\rho}.$$

Mapping of lines of **E** and φ is shown in Fig. 6-26.

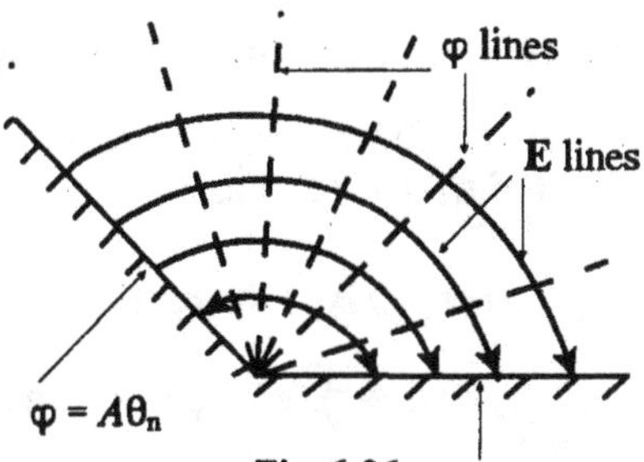

Fig. 6-26

Case V (Hydrodynamics):

The corresponding hydrodynamical case for which $f(\bar{z}) = (A/i)\ln\bar{z}$ represents a solution is given by observing that one wants ψ = constant to yield lines parallel to the surface boundary, while one wants lines of φ = constant to be perpendicular to the surface boundary. Observing these lines in the *Electrostatic* case above, it is obvious that the corresponding physical problem is as shown in Fig. 6-27.

At $\theta = 0$, $\varphi = 0$ while at $\theta = 2\pi$, $\varphi = 2\pi A$.

Note that although there is a discontinuity in φ at $\theta = 0$, $\mathbf{v} = -\nabla\varphi = -(A/\rho)\theta_o$, and is continuous at all points.

The cut at $\theta = 0$ is necessary so as to be able to utilize $f(\bar{z}) = (A/i)\ln\bar{z}$.

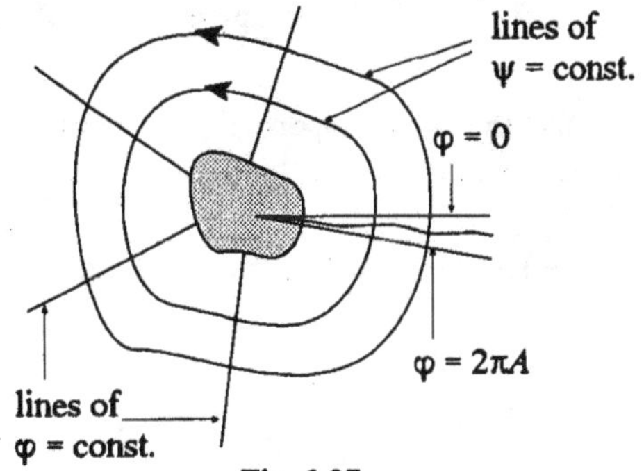

Fig. 6-27

6.4 Integral Calculus of Complex Variables

$\int_\Gamma f(\bar{z})d\bar{z}$ is defined as the line integral of a function of a complex variable over a path Γ in the complex plane (Fig. 6-28). $\oint_C f(\bar{z})d\bar{z}$ is designated as the contour integral over the path C in the direction shown (Fig. 6-29).

Fig. 6-29

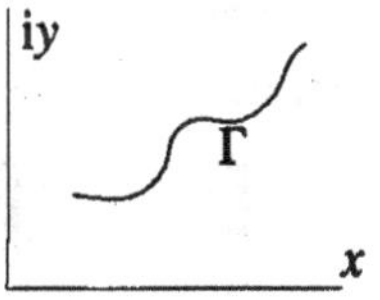

Fig. 6-28

6.4.1 Cauchy Integral Theorem

Theorem: If $f(\bar{z})$ is analytic on and within the contour C, then the line integral of $f(\bar{z})$ over C is zero, i.e. $\oint_C f(\bar{z})d\bar{z} = 0$.

Proof: Since $f(\bar{z})$ is analytic, derivatives of φ and ψ exist.

$$\oint_C f(\bar{z})d\bar{z} = \oint_C (\varphi + i\psi)(dx + idy) = \oint_C (\varphi dx - \psi dy) + i\oint_C (\psi dx + \varphi dy).$$

The general expression for Stokes' theorem (*Math Physics I*) is given by

$$\sum_{ijk}\int_S \varepsilon_{ijk}\frac{\partial}{\partial x_j}T_{kl_1l_2\ldots l_n}\,dS_i = \sum_i\int_s T_{il_1l_2\ldots l_n}\,ds_i\,.$$

For T a scalar, this reduces to

$$\sum_{ij}\int_S \varepsilon_{ijk}\frac{\partial\varphi}{\partial x_j}dS_i = \int_s \varphi\, ds_k \quad\rightarrow\quad \int_S d\mathbf{S}\times\nabla\varphi = \int_s \varphi d\mathbf{s}.$$

Note that the only component of $d\mathbf{S}$ that exists is the z component, i.e. $d\mathbf{S} = dxdy\mathbf{k}_0$. It follows that

$$\int_s \varphi dx = -\int_S dxdy\frac{\partial\varphi}{\partial y}. \qquad \int_s \psi dy = \int_S dxdy\frac{\partial\psi}{\partial x}.$$

Hence,

$$\oint_C f(\bar{z})d\bar{z} = \iint\left[-\frac{\partial\varphi}{\partial y} - \frac{\partial\psi}{\partial x}\right]dxdy + i\iint\left[-\frac{\partial\psi}{\partial y} + \frac{\partial\varphi}{\partial x}\right]dxdy = 0,$$

since φ and ψ satisfy the C-R conditions.

Corollary 1: If $f(\bar{z})$ is analytic on and within a closed curve C, and if $\bar{z}_1$ and $\bar{z}_2$ are two points lying within the region R bounded by C, then $\int_{\bar{z}_1}^{\bar{z}_2} f(\bar{z})d\bar{z}$ is independent of the path.

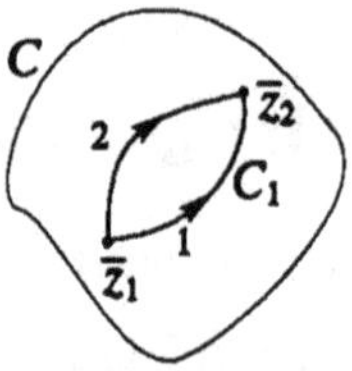

Fig. 6-30

Proof: Referring to Fig. 6-30, by Cauchy's theorem:

$$\oint_{C_1} f(\bar{z})d\bar{z} = \int_1 f(\bar{z})d\bar{z} - \int_2 f(\bar{z})d\bar{z} = 0 \quad\rightarrow\quad \int_1 f(\bar{z})d\bar{z} = \int_2 f(\bar{z})d\bar{z},$$

i.e. independent of the path.

Corollary 2: If a simple closed curve C_2 lies entirely in the region bounded by another closed curve C_1, and if $f(\bar{z})$ is analytic on C_1 and C_2 and in the region between C_1 and C_2, then

$$\int_{C_1} f(\bar{z})d\bar{z} = \int_{C_2} f(\bar{z})d\bar{z}.$$

Proof: Referring to Fig. 6-31, introduce a cut Γ such that the two (dotted) lines are infinitesimally close to one another. Then, by Cauchy's theorem:

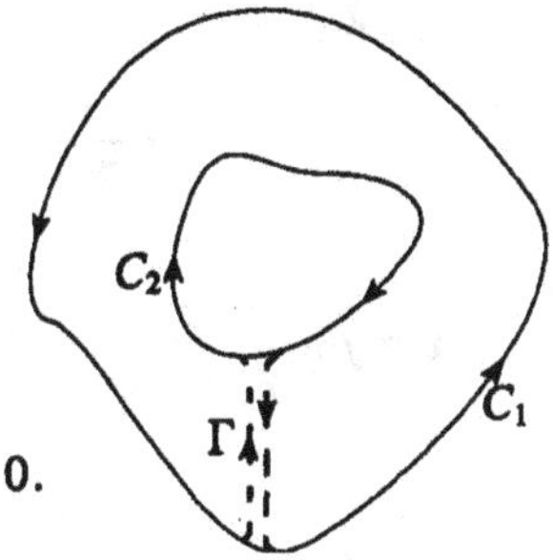

Fig. 6-31

$$\oint_{C_1} f(\bar{z})d\bar{z} + \int_{\Gamma} f(\bar{z})d\bar{z} + \oint_{-C_2} f(\bar{z})d\bar{z} - \int_{\Gamma} f(\bar{z})d\bar{z} = 0.$$

Hence,

$$\oint_{C_1} f(\bar{z})d\bar{z} = \oint_{C_2} f(\bar{z})d\bar{z}.$$

(Both C_1 and C_2 are taken in the counter-clockwise direction.)

6.4.2 Residue Theorem

Theorem: If $f(\bar{z})$ is analytic on and within a closed curve C, and if $\bar{a}$ is a point lying entirely within the region bounded by C, then

$$f(\bar{a}) = \frac{1}{2\pi i}\oint_C \frac{f(\bar{z})d\bar{z}}{(\bar{z}-\bar{a})}.$$

Proof: From Corollary 2 above, $\oint_C = \oint_{C_1}$. Define $\varepsilon(\bar{z})$ to be

$$\varepsilon(\bar{z}) = \frac{f(\bar{z}) - f(\bar{a})}{(\bar{z}-\bar{a})} - f'(\bar{a}),$$

a function of $\bar{z}$ on the circle C_1 of radius δ centered about point $\bar{a}$ (Fig. 6-32).

$$\lim_{\delta\to 0}\varepsilon(\bar{z}) = \lim_{\delta\to 0}\frac{f(\bar{z}) - f(\bar{a})}{(\bar{z}-\bar{a})} - f'(\bar{a}) = 0.$$

Hence, $\varepsilon(\bar{z})\to 0$ as $\delta\to 0$.

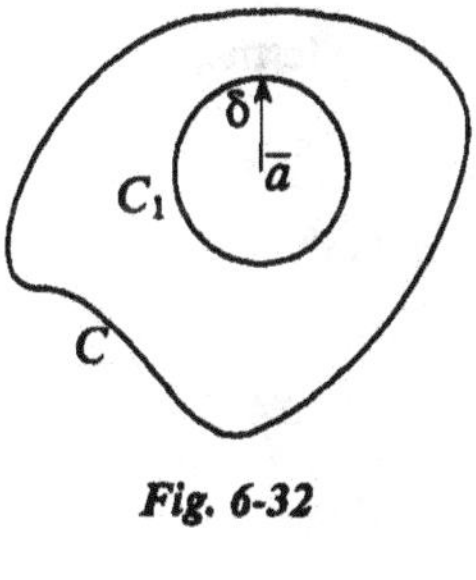

Fig. 6-32

Substituting for $f(\bar{z})$:

$$\frac{1}{2\pi i}\oint_{C_1}\frac{f(\bar{z})d\bar{z}}{(\bar{z}-\bar{a})} = \frac{1}{2\pi i}\oint_{C_1}\left\{\frac{f(\bar{a})+\varepsilon(\bar{z})(\bar{z}-\bar{a})+f'(\bar{a})(\bar{z}-\bar{a})}{(\bar{z}-\bar{a})}\right\}$$

$$= \frac{1}{2\pi i}\left[\mathrm{I}+\mathrm{II}+\mathrm{III}\right],$$

where

$$\mathrm{I} = f(\bar{a})\oint_{C_1}\frac{d\bar{z}}{(\bar{z}-\bar{a})}. \quad \mathrm{II} = \oint_{C_1}\varepsilon(\bar{z})d\bar{z}. \quad \mathrm{III} = f'(\bar{a})\oint_{C_1}d\bar{z}.$$

Term II: Since $\varepsilon(\bar{z})$ is analytic on C_1 and on C and in the region between C and C_1, $\mathrm{II} = \oint_{C_1}\varepsilon(\bar{z})d\bar{z} = \oint_{C}\varepsilon(\bar{z})d\bar{z}$. Since $\varepsilon(\bar{z})\to 0$ as $\delta\to 0$, $\varepsilon(\bar{z})$ is analytic over the entire region within C. Hence, $\mathrm{II} = 0$.

Term I: I can be evaluated using vector substitution (refer to Fig. 6-33):

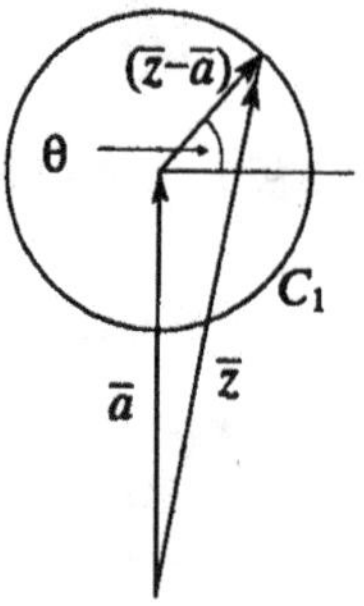

Fig. 6-33

$$(\bar{z}-\bar{a}) = \delta e^{i\theta}. \quad d\bar{z} = \delta i e^{i\theta}d\theta.$$

$$\frac{d\bar{z}}{(\bar{z}-\bar{a})} = i d\theta.$$

$$\mathrm{I} = f(\bar{a})\oint_{C_1}\frac{d\bar{z}}{(\bar{z}-\bar{a})} = i f(\bar{a})\oint_{C_1}d\theta = 2\pi i f(\bar{a}).$$

Term III: Similarly,

$$\mathrm{III} = f'(\bar{a})\oint_{C_1}d\bar{z} = f'(\bar{a})\delta i\oint_{C_1}e^{i\theta}d\theta = f'(\bar{a})\delta\left[e^{i\theta}\right]_0^{2\pi} = 0.$$

Collecting terms,

$$\oint_{C_1}\frac{f(\bar{z})d\bar{z}}{(\bar{z}-\bar{a})} = f(\bar{a}) = \oint_{C}\frac{f(\bar{z})d\bar{z}}{(\bar{z}-\bar{a})}.$$

Theorem: If $f(\bar{z})$ is analytic on and within a closed curve C, and if $\bar{a}$ is a point lying entirely within the region bounded by C, then

$$f'(\bar{a}) = \frac{1}{2\pi i}\oint_C \frac{f(\bar{z})d\bar{z}}{(\bar{z}-\bar{a})^2}.$$

Proof: Define a function

$$g(\bar{z}) \equiv \rho(\bar{z})(\bar{z}-\bar{a}) = \frac{f(\bar{z})-f(\bar{a})}{(\bar{z}-\bar{a})} - f'(\bar{a}),$$

defined on the circle C_1. In addition, $\lim_{\delta\to 0} g(\bar{z}) = 0.$

Substituting for $f(\bar{z})$:

$$\oint_{C_1} \frac{f(\bar{z})d\bar{z}}{(\bar{z}-\bar{a})^2} = \oint_{C_1} \left\{\frac{f(\bar{a}) + \rho(\bar{z})(\bar{z}-\bar{a})^2 + f'(\bar{a})(\bar{z}-\bar{a})}{(\bar{z}-\bar{a})^2}\right\}$$

$$= \left[\mathrm{I} + \mathrm{II} + \mathrm{III}\right],$$

where

$$\mathrm{I} = \oint_{C_1} \rho(\bar{z})d\bar{z}. \quad \mathrm{II} = f'(\bar{a})\oint_{C_1} \frac{d\bar{z}}{(\bar{z}-\bar{a})}. \quad \mathrm{III} = f(\bar{a})\oint_{C_1} \frac{d\bar{z}}{(\bar{z}-\bar{a})^2}.$$

<u>Term I:</u> As $\delta\to 0$,

$$\oint_{C_1} \rho(\bar{z})d\bar{z} = \oint_{C_1} \frac{g(\bar{z})d\bar{z}}{(\bar{z}-\bar{a})} = \oint_{C_1} g(\bar{z})id\theta \to 0,$$

since $g(\bar{z}) \to 0$ as $\delta\to 0$. (Also, $\rho(\bar{z})$ is analytic within C_1.)

<u>Term III:</u>

$$\oint_{C_1} \frac{d\bar{z}}{(\bar{z}-\bar{a})^2} = \oint_{C_1} \frac{i\delta e^{i\theta}}{\delta^2 e^{2i\theta}}d\theta = \frac{i}{\delta}\oint_{C_1} e^{-i\theta}d\theta = 0.$$

Collecting terms,

$$\frac{1}{2\pi i}\oint_{C_1} \frac{f(\bar{z})d\bar{z}}{(\bar{z}-\bar{a})^2} = \frac{f'(\bar{a})}{2\pi i}\oint_{C_1} \frac{d\bar{z}}{(\bar{z}-\bar{a})} \Rightarrow f'(\bar{a}) = \oint_C \frac{f(\bar{z})d\bar{z}}{(\bar{z}-\bar{a})^2}.$$

Theorem: If $f(\bar{z})$ is analytic on and within a closed curve C, then $f'(\bar{z})$ is analytic within C.

[Note that one usually states conditions for the contour and the region enclosed, and ends up with a result that holds only within the region.]

Proof: When one states that $f'(\bar{z})$ is analytic one means that

$$f''(\bar{a}) = \lim_{\bar{h}\to 0} \frac{f'(\bar{a}+\bar{h}) - f'(\bar{a})}{\bar{h}}$$

exists and is independent of how $\bar{h} \to 0$ (Fig. 6-34).

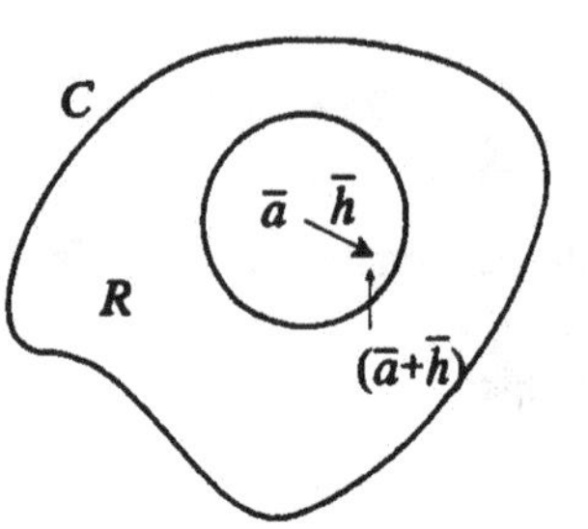

Fig. 6-34

From the theorem proved above:

$$f'(\bar{a}+\bar{h}) = \frac{1}{2\pi i}\oint_C \frac{f(\bar{z})d\bar{z}}{(\bar{z}-\bar{a}-\bar{h})^2}.$$

$$f'(\bar{a}) = \frac{1}{2\pi i}\oint_C \frac{f(\bar{z})d\bar{z}}{(\bar{z}-\bar{a})^2}.$$

$$f''(\bar{a}) = \lim_{\bar{h}\to 0} \frac{1}{(2\pi i\bar{h})}\oint_C f(\bar{z})d\bar{z}\left[\frac{1}{(\bar{z}-\bar{a}-\bar{h})^2} - \frac{1}{(\bar{z}-\bar{a})^2}\right]$$

$$= \lim_{\bar{h}\to 0} \frac{1}{(2\pi i)}\oint_C \frac{f(\bar{z})[2(\bar{z}-\bar{a})-\bar{h}]}{(\bar{z}-\bar{a}-\bar{h})^2(\bar{z}-\bar{a})^2}\,d\bar{z}$$

$$= \lim_{\bar{h}\to 0} \frac{1}{(2\pi i)}\left\{2\oint_C \frac{f(\bar{z})d\bar{z}}{(\bar{z}-\bar{a}-\bar{h})^2(\bar{z}-\bar{a})} - \bar{h}\oint_C \frac{f(\bar{z})d\bar{z}}{(\bar{z}-\bar{a}-\bar{h})^2(\bar{z}-\bar{a})^2}\right\}.$$

Since $\bar{a}$ and $(\bar{a}+\bar{h})$ are within the contour C, and since $\bar{z}$ is evaluated on C, $(\bar{z}-\bar{a}-\bar{h})^2$ and $(\bar{z}-\bar{a})^2$ are finite. Hence, $\oint_C \frac{f(\bar{z})d\bar{z}}{(\bar{z}-\bar{a}-\bar{h})^2(\bar{z}-\bar{a})^2}$ is finite and has some maximum value. Therefore, in the limit as $\bar{h} \to 0$, this term vanishes, and we are left with

$$f''(\bar{a}) = \frac{1}{(\pi i)}\oint_C \frac{f(\bar{z})d\bar{z}}{(\bar{z}-\bar{a})^3},$$

independent of $\bar{h}$.

Theorem: If $f(\bar{z})$ is analytic on and within a contour C surrounding a region R, then all of its derivatives exist in R and are analytic.

$$f^n(\bar{a}) = \frac{n!}{2\pi i}\oint_C \frac{f(\bar{z})d\bar{z}}{(\bar{z}-\bar{a})^{n+1}}.$$

Proof: The proof follows from the previous theorem in like manner.

6.5 Taylor's Theorem

Theorem: If C is a circle of radius R with $\bar{a}$ as center (Fig. 6-35), and if $f(\bar{z})$ is analytic on and within C, then

$$f(\bar{z}) = f(\bar{a}) + (\bar{z}-\bar{a})f'(\bar{a}) + (\bar{z}-\bar{a})^2\frac{f''(\bar{a})}{2!} + \cdots = \sum_{n=0}^{\infty}\frac{f^n(\bar{a})}{n!}(\bar{z}-\bar{a})^n,$$

for all $|(\bar{z}-\bar{a})| < R$.

Proof: The residue theorem states that

$$(1) \qquad f(\bar{z}) = f(\bar{a}+\bar{h}) = \frac{1}{2\pi i}\oint_C \frac{f(\bar{z})d\bar{z}}{(\bar{z}-\bar{a}-\bar{h})}.$$

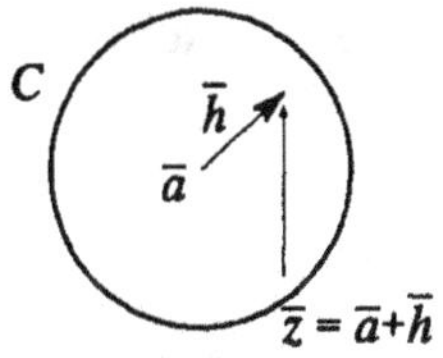

Fig. 6-35

Consider the following expansion:

$$\left[\frac{1}{(\bar{z}-\bar{a})} + \frac{\bar{h}}{(\bar{z}-\bar{a})^2} + \frac{\bar{h}^2}{(\bar{z}-\bar{a})^3} + \cdots + \frac{\bar{h}^N}{(\bar{z}-\bar{a})^{N+1}} + \frac{\bar{h}^{N+1}}{(\bar{z}-\bar{a})^{N+1}(\bar{z}-\bar{a}-\bar{h})}\right]$$

$$= \frac{1}{(\bar{z}-\bar{a}-\bar{h})}\left[\frac{(\bar{z}-\bar{a}-\bar{h})}{(\bar{z}-\bar{a})} + \frac{\bar{h}(\bar{z}-\bar{a}-\bar{h})}{(\bar{z}-\bar{a})^2} + \cdots + \frac{\bar{h}^{N+1}}{(\bar{z}-\bar{a})^{N+1}}\right]$$

$$= \frac{1}{(\bar{z}-\bar{a}-\bar{h})}\left[1 - \frac{\bar{h}}{(\bar{z}-\bar{a})} + \frac{\bar{h}}{(\bar{z}-\bar{a})} - \frac{\bar{h}^2}{(\bar{z}-\bar{a})^2} + \frac{\bar{h}^2}{(\bar{z}-\bar{a})^2} + \cdots +\right]$$

$$+ \cdots + \frac{1}{(\bar{z}-\bar{a}-\bar{h})}\left[-\frac{\bar{h}^{N+1}}{(\bar{z}-\bar{a})^{N+1}} + \frac{\bar{h}^{N+1}}{(\bar{z}-\bar{a})^{N+1}}\right]$$

$$= \frac{1}{(\bar{z}-\bar{a}-\bar{h})}.$$

Substituting the expansion for $\dfrac{1}{(\bar{z}-\bar{a}-\bar{h})}$ into expression (1) for $f(\bar{a}+\bar{h})$:

$$f(\bar{a}+\bar{h}) = \frac{1}{2\pi i}\oint_C f(\bar{z})\left[\frac{1}{(\bar{z}-\bar{a})}+\frac{\bar{h}}{(\bar{z}-\bar{a})^2}+\cdots+\frac{\bar{h}^{N+1}}{(\bar{z}-\bar{a})^{N+1}(\bar{z}-\bar{a}-\bar{h})}\right]d\bar{z}$$

$$= f(\bar{a}) + \bar{h}\frac{f'(\bar{a})}{1!} + \bar{h}^2\frac{f''(\bar{a})}{2!} + \cdots + \bar{h}^N\frac{f^N(\bar{a})}{N!} + \cdots$$

$$+ \frac{\bar{h}^{N+1}}{2\pi i}\oint_C \frac{f(\bar{z})d\bar{z}}{(\bar{z}-\bar{a})^{N+1}(\bar{z}-\bar{a}-\bar{h})}.$$

We would like now to show that as $N \to \infty$:

$$\lim_{N\to\infty}\frac{\bar{h}^{N+1}}{2\pi i}\oint_C \frac{f(\bar{z})d\bar{z}}{(\bar{z}-\bar{a})^{N+1}(\bar{z}-\bar{a}-\bar{h})} = 0.$$

Consider, therefore:

$$\left|f(\bar{a}+\bar{h}) - \sum_{n=0}^{N}\frac{f^n(\bar{a})\bar{h}^n}{n!}\right| = \left|\frac{\bar{h}^{N+1}}{2\pi i}\oint_C \frac{f(\bar{z})d\bar{z}}{(\bar{z}-\bar{a})^{N+1}(\bar{z}-\bar{a}-\bar{h})}\right|.$$

$f(\bar{z})$ is bounded since it is evaluated on C, and has some maximum bound, say M. On C, then, $|f(\bar{z})| < M$. Also, $(\bar{z}-\bar{a}-\bar{h})$ is bound, since $(\bar{a}+\bar{h})$ must be within C. Hence, $\left|\dfrac{f(\bar{z})}{(\bar{z}-\bar{a}-\bar{h})}\right| < W$, and, since $h < R$,

$$\left|\frac{\bar{h}^{N+1}}{2\pi i}\oint_C \frac{f(\bar{z})d\bar{z}}{(\bar{z}-\bar{a})^{N+1}(\bar{z}-\bar{a}-\bar{h})}\right| < \left|\frac{\bar{h}^{N+1}}{R^{N+1}}\frac{2\pi RW}{2\pi}\right| = \left|\frac{W\bar{h}^{N+1}}{R^N}\right| \underset{N\to\infty}{\to} 0.$$

Consider the function $f(\bar{z}) = (1-\bar{z})^{-1}$. For a circle of radius $R = 1-\varepsilon$ $(\varepsilon > 0)$ about the origin (Fig. 6-36), $f(\bar{z})$ is analytic on and within C. Thus, $f(\bar{z}) = (1-\bar{z})^{-1}$ will be expandable in a power series

$$\frac{1}{(1-\bar{z})} = \sum_{n=0}^{\infty}\frac{f^n(0)}{n!}(\bar{z})^n = \sum_{n=0}^{\infty}(\bar{z})^n.$$

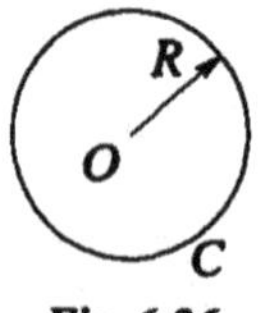

Fig. 6-36

6.6 Laurent Theorem

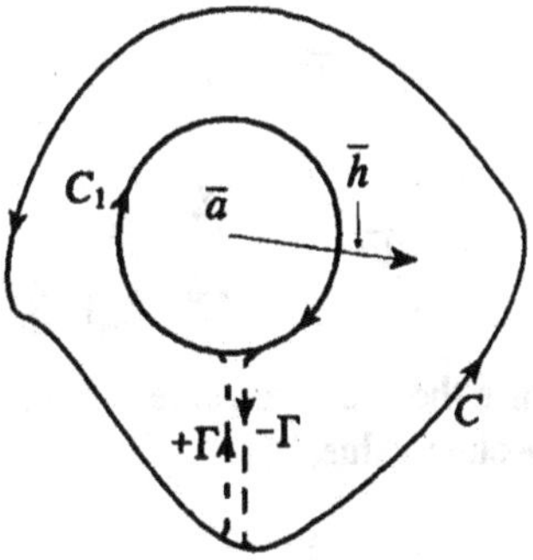

Fig. 6-37

Theorem: If $f(\bar{z})$ is analytic on C and on C_1 and in the region between them (Fig. 6-37), then for $\bar{z} = (\bar{a} + \bar{h})$ ($\bar{z}$ lying in the region between C and C_1):

$$f(\bar{z}) = f(\bar{a} + \bar{h})$$

$$= \sum_{n=0}^{\infty} c_n(\bar{z} - \bar{a})^n + \sum_{m=1}^{\infty} \frac{b_m}{(\bar{z} - \bar{a})^m},$$

where

$$c_n = \frac{1}{2\pi i}\oint_C \frac{f(\bar{t})d\bar{t}}{(\bar{t} - \bar{a})^{n+1}}, \qquad b_m = \frac{1}{2\pi i}\oint_{C_1} (\bar{t} - \bar{a})^{m-1} f(\bar{t})d\bar{t}$$

Note that if $f(\bar{z})$ is analytic <u>within</u> C_1 as well, then by Cauchy's theorem, $b_m = 0$ since $(\bar{t} - \bar{a})$ is finite while $f(\bar{z})$ is analytic. Note also that in this case the c_n reduce to $f^n(\bar{a})/n!$, so that

$$f(\bar{a} + \bar{h}) = \sum_{n=0}^{\infty} \frac{f^n(\bar{a})}{n!}(\bar{z} - \bar{a})^n,$$

which is identical to Taylor's theorem.

Proof: Introduce a cut Γ (Fig. 6-37), so that $f(\bar{z})$ is analytic over the entire curve $(C+\Gamma-C_1-\Gamma)$. Since $f(\bar{z})$ is now analytic over the contour and within the contour, we have

$$f(\bar{a} + \bar{h}) = \frac{1}{2\pi i}\oint_C \frac{f(\bar{t})d\bar{t}}{(\bar{t} - \bar{a} - \bar{h})} - \frac{1}{2\pi i}\oint_{C_1} \frac{f(\bar{t})d\bar{t}}{(\bar{t} - \bar{a} - \bar{h})}.$$

We now apply the expansion we used before in treating Taylor's theorem. For the first integral $(\bar{t} - \bar{a} > \bar{h})$:

$$\frac{1}{(\bar{t} - \bar{a} - \bar{h})} = \sum_{n=0}^{N} \frac{\bar{h}^n}{(\bar{t} - \bar{a})^{n+1}} + \frac{\bar{h}^{N+1}}{(\bar{t} - \bar{a})^{N+1}(\bar{t} - \bar{a} - \bar{h})}.$$

$$f(\bar{a}+\bar{h}) = \sum_{n=0}^{N} c_n\bar{h}^n + \frac{\bar{h}^{N+1}}{2\pi i}\oint_C \frac{f(\bar{t})d\bar{t}}{(\bar{t}-\bar{a})^{N+1}(\bar{t}-\bar{a}-\bar{h})}.$$

$$\left|f(\bar{a}+\bar{h}) - \sum_{n=0}^{N} c_n\bar{h}^n\right| = \left|\frac{\bar{h}^{N+1}}{2\pi i}\oint_C \frac{f(\bar{t})d\bar{t}}{(\bar{t}-\bar{a})^{N+1}(\bar{t}-\bar{a}-\bar{h})}\right|.$$

Since the absolute value of a sum is always less than the sum of the absolute value,

$$\left|\frac{\bar{h}^{N+1}}{2\pi i}\oint_C \frac{f(\bar{t})d\bar{t}}{(\bar{t}-\bar{a})^{N+1}(\bar{t}-\bar{a}-\bar{h})}\right| < \frac{|\bar{h}^{N+1}|}{2\pi}\oint_C \left|\frac{f(\bar{t})d\bar{t}}{(\bar{t}-\bar{a})^{N+1}(\bar{t}-\bar{a}-\bar{h})}\right|.$$

Since $(\bar{t}-\bar{a}-\bar{h})$ remains finite no matter how close to the boundary $(\bar{a}=\bar{h})$ gets, and since $f(\bar{t})$ is analytic and bounded on C, we have that $\left|\frac{f(\bar{t})d\bar{t}}{(\bar{t}-\bar{a}-\bar{h})}\right| < M$. Thus, since $\bar{h} < R$,

$$\left|\frac{\bar{h}^{N+1}}{2\pi i}\oint_C \frac{f(\bar{t})d\bar{t}}{(\bar{t}-\bar{a})^{N+1}(\bar{t}-\bar{a}-\bar{h})}\right| < \frac{|\bar{h}^{N+1}|M}{2\pi R^{N+1}}\oint_C |d\bar{t}| = MR\frac{|\bar{h}^{N+1}|}{R^{N+1}} \underset{N\to\infty}{\to} 0.$$

For the second integral $(\bar{h} > \bar{t}-\bar{a})$: $\dfrac{1}{(\bar{t}-\bar{a}-\bar{h})} = -\dfrac{1}{\bar{h}-(\bar{t}-\bar{a})}$

$$= -\left[\frac{1}{\bar{h}} + \frac{(\bar{t}-\bar{a})}{\bar{h}^2} + \frac{(\bar{t}-\bar{a})^2}{\bar{h}^3} + \dots + \frac{(\bar{t}-\bar{a})^{N+1}}{\bar{h}^{N+1}[\bar{h}-(\bar{t}-\bar{a})]}\right].$$

$$f(\bar{a}+\bar{h}) = \frac{1}{2\pi i}\oint_{C_1} f(\bar{t})d\bar{t}\left\{\sum_{n=1}^{N+1}\frac{(\bar{t}-\bar{a})^{n-1}}{\bar{h}^n} - \frac{(\bar{t}-\bar{a})^{N+1}}{\bar{h}^{N+1}[\bar{t}-\bar{a}-\bar{h})]}\right\}.$$

$$\left|f(\bar{a}+\bar{h}) - \frac{1}{2\pi i}\oint_{C_1} f(\bar{t})d\bar{t}\sum_{n=1}^{N+1}\frac{(\bar{t}-\bar{a})^{n-1}}{\bar{h}^n}\right| = \left|\frac{1}{2\pi i\bar{h}^{N+1}}\oint_{C_1}\frac{f(\bar{t})d\bar{t}\,(\bar{t}-\bar{a})^{N+1}}{(\bar{t}-\bar{a}-\bar{h})}\right|.$$

As above, on $C_1, (\bar{t}-\bar{a})^{N+1} = R_{C_1}^{N+1} = R_0^{N+1}$, while $\left|\frac{f(\bar{t})d\bar{t}}{(\bar{t}-\bar{a}-\bar{h})}\right| < W$ and $|d\bar{t}| = R_0 d\theta$.

Hence, since $R_o < \bar{h}$,

$$\left| \frac{1}{2\pi i \bar{h}^{N+1}} \oint_{C_1} \frac{f(\bar{t})d\bar{t}\,(\bar{t}-\bar{a})^{N+1}}{(\bar{t}-\bar{a}-\bar{h})} \right| < \frac{2\pi W R_o^{N+1}}{2\pi |\bar{h}^{N+1}|} R_o \underset{N\to\infty}{\to} 0.$$

Thus we have shown that as $N \to \infty$,

$$f(\bar{z}) = f(\bar{a}+\bar{h}) = \sum_{n=0}^{\infty} c_n(\bar{z}-\bar{a})^n + \sum_{m=1}^{\infty} \frac{b_m}{(\bar{z}-\bar{a})^m}.$$

6.7 Singularities

The most important property of a function $f(\bar{z})$ is its singularities. (We want to consider only single-valued functions.)

If $f(\bar{z})$ is analytic except over certain isolated points, then these points are called isolated singularities.

Let point $\bar{a}$ be an isolated singularity of $f(\bar{z})$ (Fig. 6-38). Then since it is isolated, one can always find a finite region R surrounding it and within which $f(\bar{z})$ is analytic. Then, over this region R we can expand $f(\bar{z})$ in a Laurent expansion.

$$f(\bar{z}) = \sum_{n=0}^{\infty} c_n(\bar{z}-\bar{a})^n + \sum_{m=1}^{\infty} \frac{b_m}{(\bar{z}-\bar{a})^m}.$$

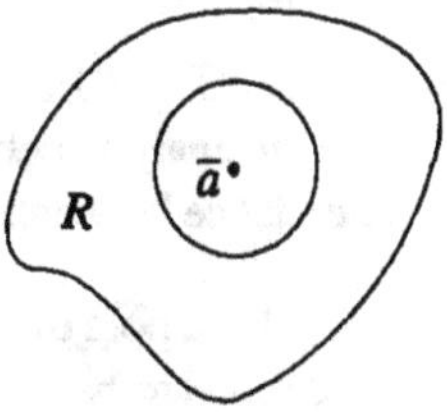

Fig. 6-38

Types of Singularities

(a) If $b_N \neq 0$ and $b_m = 0$ for all $m > N$, then point $\bar{a}$ is a *pole* of order N. A "simple" pole is one of order 1, in which case we have only $b_1/(\bar{z}-\bar{a})$.

(b) If $f(\bar{z})$ is single-valued and point $\bar{a}$ is a singular point, and if the series $b_m/(\bar{z}-\bar{a})^m$ is infinite (i.e. one cannot find an N such that (a) above holds), then point $\bar{a}$ is called an *essential* singularity of $f(\bar{z})$.

(c) If $f(\bar{z})$ can be expanded in a Laurent series about a point $\bar{a}$, and if $c_n = 0$ for all $n > N$ and $c_N \neq 0$, then point $\bar{a}$ is defined as a *zero* of order N.

The behavior of $f(\bar{z})$ near an essential singularity can be quite different from the behavior near a pole of finite order, as $\bar{z} \to \bar{a}$.

For a pole of finite order N as $\bar{z} \to \bar{a}$: $f(\bar{z}) \to b_N/(\bar{z} - \bar{a})^N$.

For an essential singularity, one might get the results 0, 1, ∞, etc. Consider, for example, the function

$$f(\bar{z}) = e^{1/\bar{z}} = \sum_{m=0}^{\infty} \frac{1}{n!} \frac{1}{(\bar{z})^m},$$

which we note has an essential singularity at the origin (none of the b_m vanish). Approaching the origin from 3 directions (Fig. 6-39):

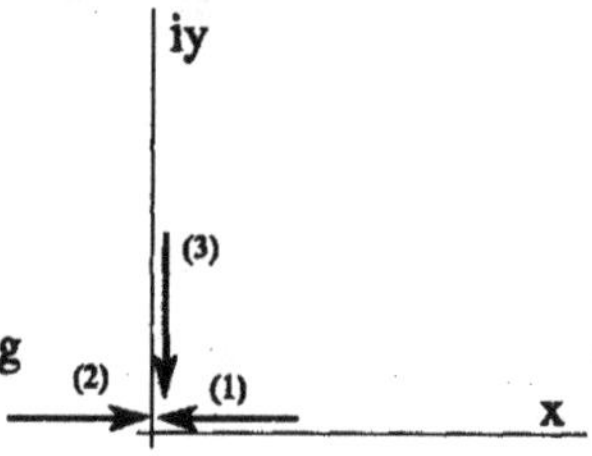

Fig. 6-39

(1) At $y = 0$, $\bar{z} = x$: $e^{1/\bar{z}} \to e^{1/x} \underset{x \to 0+}{\to} \infty$.

(2) At $y = 0$, $\bar{z} = x$: $e^{1/\bar{z}} \to e^{1/x} \underset{x \to 0-}{\to} 0$.

(3) At $x = 0$, $\bar{z} = iy$: $\left|e^{1/\bar{z}}\right| \to \left|e^{-i/y}\right| \underset{y \to 0+}{\to} 1$.

Consider the point infinity as a singular point. If one makes the transformation $\bar{z} \to 1/\bar{z}'$, then the singularity at infinity is transformed to one of finite consideration.

If we define $f(\bar{z}) = f(1/\bar{z}') = g(\bar{z}')$, then the analytic behavior of $g(\bar{z}')$ at $\bar{z}' = 0$ will give the analytic behavior of $f(\bar{z})$ at $\bar{z} = \infty$, thus yielding a direct correspondence between infinity and the origin.

(a) If $\bar{z}' = 0$ is a pole of order N for $g(\bar{z}')$, i.e. all b_m vanish for $m > N$, then $\bar{z} = \infty$ is defined to be a pole of order N for $f(\bar{z})$, where

$$f(\bar{z}) = \sum_{m=1}^{\infty} b_m \bar{z}^m + \sum_{n=0}^{\infty} \frac{c_n}{\bar{z}^n},$$

and the principal part of $f(\bar{z})$ for a pole of order N is given by $\sum_{m=1}^{\infty} b_m \bar{z}^m$ (as $\bar{z}' \to 0, f(\bar{z}) \to b_N \bar{z}^N$).

(b) If $\bar{z}' = 0$ is an essential singularity for $g(\bar{z}')$, then $\bar{z} = \infty$ is defined to be an essential singularity for $f(\bar{z})$.

(c) If $\bar{z}' = 0$ is a zero of order N for $g(\bar{z}')$, then $\bar{z} = \infty$ is defined to be a zero of order N for $f(\bar{z})$.

6.8 Liouville Theorem

Theorem: If $f(\bar{z})$ has no singularities anywhere, including ∞ (i.e. $|f(\bar{z})| < K$ everywhere), then $f(\bar{z})$ = constant.

[This is equivalent to stating that the potential is finite and bounded in a charge-free region, in which case V is constant everywhere.]

Referring to Fig. 6-40:

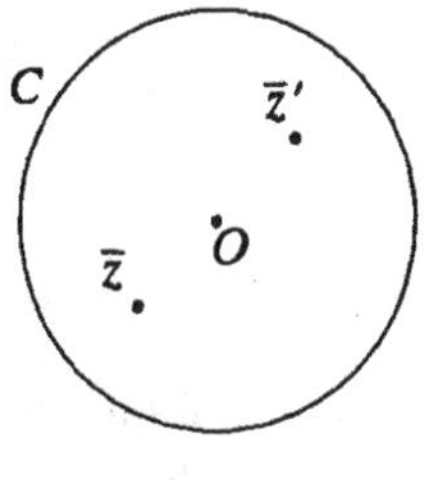

Fig. 6-40

$$f(\bar{z}) = \frac{1}{2\pi i}\oint_C \frac{f(\bar{\xi})d\bar{\xi}}{(\bar{\xi} - \bar{z})}.$$

$$f(\bar{z}') = \frac{1}{2\pi i}\oint_C \frac{f(\bar{\xi})d\bar{\xi}}{(\bar{\xi} - \bar{z}')}.$$

In the above, $\bar{\xi}$ is evaluated on the circle C.
We would like to show that $f(\bar{z}) = f(\bar{z}')$.

Proof: Consider $f(\bar{z}) - f(\bar{z}') = \dfrac{1}{2\pi i}\displaystyle\oint_C \frac{f(\bar{\xi})(\bar{z} - \bar{z}')d\bar{\xi}}{(\bar{\xi} - \bar{z}')(\bar{\xi} - \bar{z})}$.

Since by hypothesis $f(\bar{\xi})$ is analytic everywhere, choose some convenient origin O and expand the circle such that, if R is the radius, then

$|(\bar{\xi} - \bar{z})|$ and $|(\bar{\xi} - \bar{z}')| > \dfrac{R}{2}$. It follows that, as $R \to \infty$,

$$|f(\bar{z}) - f(\bar{z}')| \leq \frac{R}{2\pi}\oint_C \left|\frac{f(\bar{\xi})(\bar{z} - \bar{z}')d\theta}{(\bar{\xi} - \bar{z})}\right| < \frac{2\pi KR[\bar{z} - \bar{z}']}{2\pi R^2/4} = \frac{4K[\bar{z} - \bar{z}']}{R} \to 0.$$

Corollary: If $f(\bar{z})$ has only a finite number of poles, including ∞, then $f(\bar{z})$ is a rational function (i.e. a finite fraction of finite polynomials).

Proof: Let $\bar{a}_1\, \bar{a}_2 \ldots \bar{a}_N$ and ∞ be poles of order $N_1\, N_2 \ldots N_N$ and N_∞. Consider the principal part of $f(\bar{z})$ (PP_s) at one of the poles, say $\bar{a}_s$ ($s = 1, 2 \ldots n$).

$$\text{PP}_s = \sum_{t=1}^{N_s} \frac{b_{s,t}}{(\bar{z} - \bar{a}_s)^t}.$$

Similarly for the principal part of $f(\overline{z})$ at infinity:

$$\mathrm{PP}_\infty = \sum_{t=1}^{N_\infty} b_{\infty,t}\overline{z}^t.$$

Define the function

$$g(\overline{z}) = \left[f(\overline{z}) - \sum_{s=1}^{n}\sum_{t=1}^{N_s} \frac{b_{s,t}}{(\overline{z}-\overline{a}_s)^t} - \sum_{t=1}^{N_\infty} b_{\infty,t}\overline{z}^t \right].$$

Note that there are no poles of $g(\overline{z})$, even at infinity, for at any pole of $f(\overline{z})$, the principal part of $f(\overline{z})$ will be canceled by the corresponding $\sum_{t=1}^{N_i} \frac{b_{i,t}}{(\overline{z}-\overline{a}_i)^t}$, and similarly for the pole at infinity. Furthermore, since the summations do not extend to infinity, we have no essential singularity either. In fact, the function $g(\overline{z})$ has no singularity everywhere and thus by Liouville's theorem, $g(\overline{z})$ = constant. Hence,

$$f(\overline{z}) = C + \sum_{s=1}^{n}\sum_{t=1}^{N_s} \frac{b_{s,t}}{(\overline{z}-\overline{a}_s)^t} - \sum_{t=1}^{N_\infty} b_{\infty,t}\overline{z}^t,$$

and is therefore a rational function.

Converse: Any rational function $f(\overline{z})$, expressible as

$$f(\overline{z}) = \frac{\sum_{n=0}^{N} d_n\overline{z}^n}{\sum_{n=0}^{N} e_n\overline{z}^n},$$

has the property that for $N \neq 0$, the numerator has a pole of order N at infinity, while the zeros of the denominator yield poles of order N.

Hence, any rational function has a finite number of poles (including infinity) or else is a constant. If $f(\overline{z})$ is not rational, then it must have an essential singularity, and vice versa.

6.9 Multiple-Valued Functions

Consider $f(z) = \ln z = \ln(\rho e^{i\theta}) = \ln\rho + i\theta$, and refer to Fig. 6-41:

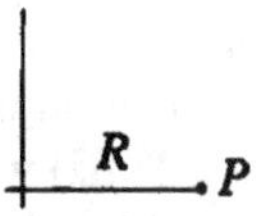

At point P, $\theta = 0 \rightarrow f(z) = \ln R$.

At point P, $\theta = 2\pi \rightarrow f(z) = \ln R + 2\pi i$.

Fig. 6-41

Thus, point P is seen to be multiple-valued for $f(z) = \ln R + 2\pi ik$, where $k = 0, 1, \ldots \infty$.

One can reduce the multiple-valued function to a single-valued one by the introduction of a cut extending from $z = 0$ to infinity in such a manner that θ not be allowed to rotate more than 360°. Then, considering $f(z)$ at any point P (Fig. 6-42), $f(z) = \ln R + i\theta_1$ and is unique, since we cannot cross the cut in performing the rotation. The sacrifice one makes in reducing a multiple-valued function to a single-valued one is that one loses continuity across the cut.

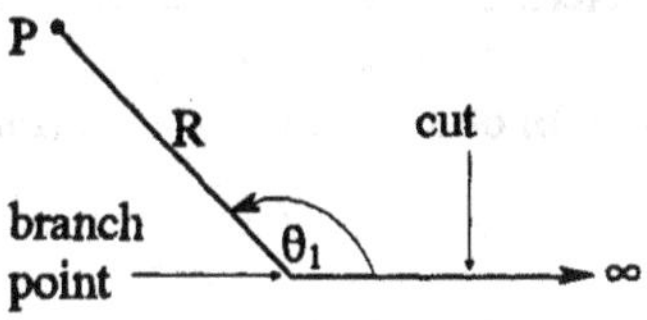

Fig. 6-42

Hence, for the function $f(z) = \ln z$, with a cut extending from $z = 0$ to infinity, one now has a single-valued function, but one which is not analytic along the cut. In the remaining region, all properties of the single-valued function $f(z)$ are retained. If analytic in the remaining region, then one applies those previous theorems dealing with analytic functions. On the other hand, if certain poles exist, one can apply those theorems relating to poles, etc. The point $z = 0$ is called a *branch* point. It represents a peculiar type of singularity in the domain of multiple-valued functions.

Consider the function $f(z) = z^{\alpha}$, where α is not a positive integer (else $f(z)$ would be single-valued at $\theta = 0, 2\pi, 4\pi$, etc.). For $z = Re^{i\theta}$, $f(z) = R^{\alpha}e^{i\alpha\theta}$. Note that once again one has a multiple-valued function, since

At point P, $\theta = 0 \rightarrow f(z) = R^{\alpha}$.

At point P, $\theta = 2\pi \rightarrow f(z) = R^{\alpha}e^{2\pi i\alpha}$.

Hence, a cut is again required, with the same considerations applying as above.

Riemann Surfaces

Assume that continuity is demanded, i.e. one allows multiple values. For the case $\alpha = 1/2$, $f(z) = z^{\alpha} = R^{1/2}e^{i\theta/2}$. Considering points (a) and (b) of Fig. 6-43:

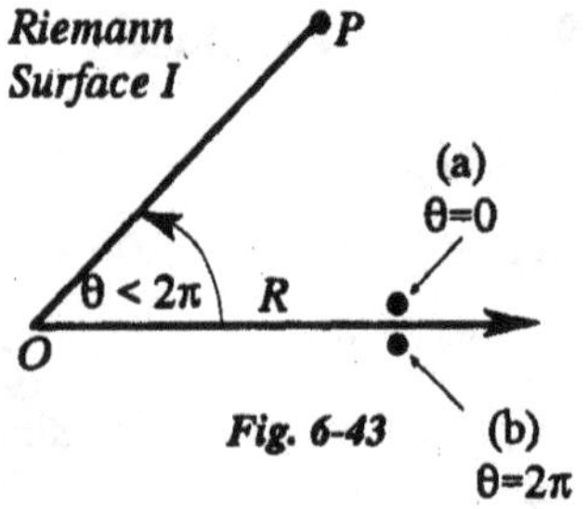

Fig. 6-43

$$f(z)_{(a)} = R^{1/2}.$$

$$f(z)_{(b)} = R^{1/2}e^{i\pi} = -R^{1/2}.$$

$$f(z)_P = R^{1/2}e^{i\theta/2}.$$

In order for $f(z)$ to remain continuous, we must define so-called Riemann Surfaces, such that when θ reaches point (b), we move on to a new surface, Riemann Surface II, i.e. (b)′ is the continuation on Riemann Surface II of point (b) of Riemann Surface I (Fig. 6-44).

$$f(z)_{(b)'} = -R^{1/2}.$$

$$f(z)_{P'} = R^{1/2}e^{i(\pi+\theta/2)} = -R^{1/2}e^{i\theta/2} = -f(z)_P.$$

$$f(z)_{(c)} = R^{1/2}e^{2\pi i} = R^{1/2} = f(z)_{(a)}.$$

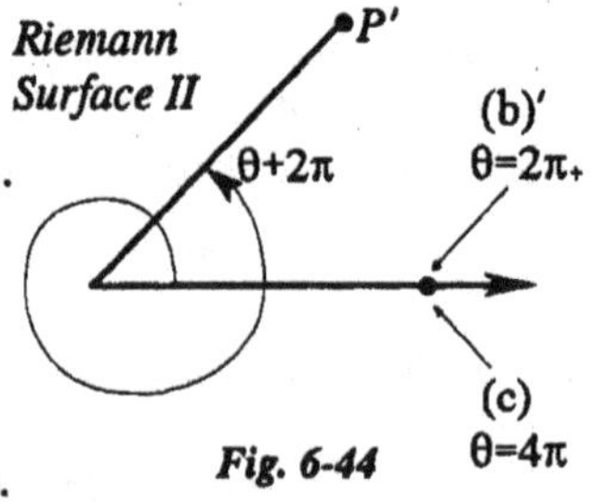

Fig. 6-44

After reaching point (b)′, one continues around to point (c), then returns to Riemann Surface I. In this way $f(z)$ remains continuous. The combination of the two Riemannian sheets yields a function $f(z) = z^{1/2}$ which is single-valued on any one sheet, and is a continuously varying function.

For $\alpha = 1/n$, $f(z) = z^{1/n}$ $(n > 0)$, we will require n Riemannian sheets, since we will have to revolve n times until the same function returns, i.e.

$$\text{At } (R, \theta)\ f(z) = R^{1/n}e^{i\theta/n}.$$

$$\text{At } \theta = 0,\ f(z) = R^{1/n}.$$

$$\text{At } \theta = 2\pi n,\ f(z) = R^{1/n}e^{i/n(2\pi n)} = R^{1/n}.$$

If α is an irrational number, we would need an infinite number of Riemannian sheets. The ln function similarly requires an infinite number of sheets, since $f(z) = \ln R + i\theta$, and no matter how many revolutions one makes, one can never revert back to $f(z) = \ln R$, corresponding to $\theta = 0$.

6.10 Theory of Residues

Definition: If a function $f(\bar{z})$ has a pole $\bar{z}_1$ of order n, and if

$$f(\bar{z}) = \varphi(\bar{z}) + \frac{b_1}{(\bar{z}-\bar{z}_1)} + \frac{b_2}{(\bar{z}-\bar{z}_1)^2} + \ldots + \frac{b_{n-1}}{(\bar{z}-\bar{z}_1)^{n-1}} + \frac{b_n}{(\bar{z}-\bar{z}_1)^n},$$

where $\varphi(\bar{z})$ is analytic at $\bar{z} = \bar{z}_1$, then b_1 (the coefficient of $1/(\bar{z}-\bar{z}_1)$), is defined to be the *residue* of $f(\bar{z})$ at $\bar{z} = \bar{z}_1$.

Theorem: If $f(\bar{z})$ has, as singularities, only poles $\bar{z}_1, \bar{z}_2, \ldots \bar{z}_n$ inside a closed curve C, then

$$\oint_C f(\bar{z})d\bar{z} = 2\pi i \sum_{j=1}^{n} R_j,$$

where R_j is the residue of $f(\bar{z})$ at $\bar{z} = \bar{z}_j$ ($j = 1,2,\ldots n$).

Proof: Referring to Fig. 6-45:

$$\oint_{C'} = \oint_C - \sum_{j=1}^{n} \oint_{C_j} = 0,$$

since $f(\bar{z})$ is analytic over the region between C and $\sum_j C_j$. Hence,

$$\oint_C = \sum_{j=1}^{n} \oint_{C_j}.$$

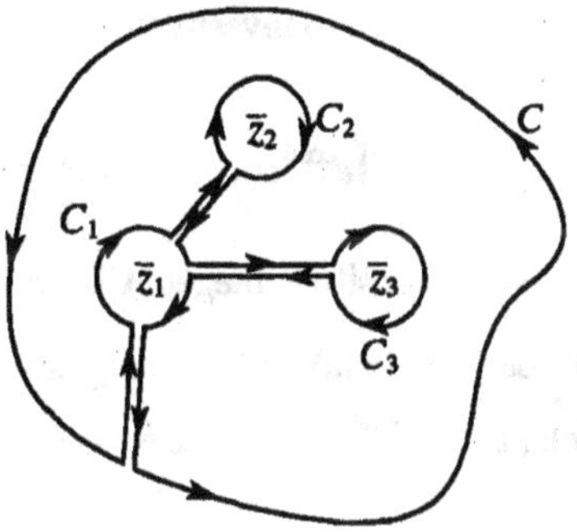

Fig. 6-45

Considering one of the C_j, say C_1:

$$\oint_{C_1} f(\bar{z})d\bar{z} = \oint_{C_1} \varphi(\bar{z})d\bar{z} + \sum_{m=1}^{n} \oint_{C_1} \frac{b_m d\bar{z}}{(\bar{z}-\bar{z}_1)^m}.$$

Since $\varphi(\bar{z})$ is analytic at $\bar{z} = \bar{z}_1$, the first integral on the right vanishes, by Cauchy's theorem. Setting $\bar{z} - \bar{z}_1 = \varepsilon e^{i\theta}$, $d\bar{z} = i\varepsilon e^{i\theta} d\theta$, so that

$$\oint_{C_1} f(\bar{z})d\bar{z} = \oint_{C_1} \frac{b_m d\bar{z}}{(\bar{z}-\bar{z}_1)^m} = \int_0^{2\pi} \frac{i\varepsilon b_m e^{i\theta} d\theta}{\varepsilon^m e^{im\theta}} = \frac{i b_m}{\varepsilon^{m-1}} \int_0^{2\pi} e^{-i(m-1)\theta} d\theta$$

$$= \begin{cases} 0 \text{ for } m \neq 1. \\ 2\pi i b_1 = 2\pi i R_1 \text{ for } m = 1. \end{cases} \quad \rightarrow \quad \oint_C f(\bar{z})d\bar{z} = 2\pi i \sum_{j=1}^{n} R_j.$$

6.10.1 Complex Integration

Type I: $\int_0^{2\pi} F(\cos\theta, \sin\theta)d\theta$

Under what conditions can we use the theorem of residues? We want to convert this type of integral to a contour integration over the unit circle. Over the unit circle, $|z| = 1$, $z = e^{i\theta}$, $dz = ie^{i\theta}d\theta = iz\,d\theta$. We also have:

$\cos\theta = (e^{i\theta} + e^{-i\theta})/2 = (z + 1/z)/2$; $\sin\theta = (e^{i\theta} - e^{-i\theta})/2i = (z - 1/z)/2i$.

Thus, $$\int_0^{2\pi} F(\cos\theta, \sin\theta)d\theta = \frac{1}{i}\oint_{\substack{\text{unit}\\ \text{circle}}} F\left[\frac{(z+1/z)}{2}, \frac{(z-1/z)}{2i}\right]\frac{dz}{z} = \frac{1}{i}\oint_{\substack{\text{unit}\\ \text{circle}}} \frac{f(z)dz}{z},$$

where

$$f(z) = F\left[\cos\theta = \frac{1}{2}(z + 1/z), \sin\theta = \frac{1}{2i}(z - 1/z)\right].$$

We then proceed to investigate $\frac{f(z)}{z}$, looking for the poles within the unit circle.

<u>Example</u> (a) $\int_0^{2\pi} e^{\cos\theta}\cos(n\theta - \sin\theta)d\theta$ or (b) $\int_0^{2\pi} e^{\cos\theta}\sin(n\theta - \sin\theta)d\theta$,

where n is a positive integer. Observe that $\int_0^{2\pi} e^{\cos\theta}e^{-i(n\theta - \sin\theta)}d\theta$ has a real part identical to (a) and an imaginary part identical to (b). Over the unit circle, $z^{-n} = e^{-in\theta}$, $-nz^{-n-1}dz = -ine^{-in\theta}d\theta \to e^{-in\theta}d\theta = \frac{1}{i}\frac{dz}{z^{n+1}}$. Thus,

$$\int_0^{2\pi} e^{\cos\theta}e^{-i(n\theta - \sin\theta)}d\theta = \frac{1}{i}\oint_{\substack{\text{unit}\\ \text{circle}}} e^{\cos\theta + i\sin\theta}\frac{dz}{z^{n+1}} = \frac{1}{i}\oint_{\substack{\text{unit}\\ \text{circle}}} \frac{e^{z}dz}{z^{n+1}} = \frac{1}{i}\oint_{\substack{\text{unit}\\ \text{circle}}} f(z)dz.$$

Note that $f(z)$ has a pole at $z = 0$. We seek the residue at this pole. The procedure is to expand $f(z)$ and select the coefficient of the $1/z$ term.

$$\frac{e^{z}}{z^{n+1}} = \frac{1}{z^{n+1}}\sum_{m=0}^{\infty}\frac{z^m}{m!} \to \text{Residue} = \frac{1}{n!}.$$

Thus

$$\int_0^{2\pi} e^{\cos\theta}\cos(n\theta - \sin\theta)d\theta = \frac{2\pi}{n!}; \quad \int_0^{2\pi} e^{\cos\theta}\sin(n\theta - \sin\theta)d\theta = 0.$$

Type II: $\int_{-\infty}^{+\infty} e^{ikx}F(x)dx.$

This type of integration occurs quite often in the use of Fourier integrals:

$$f(x) = \frac{1}{(2\pi)^{1/2}}\int_{-\infty}^{+\infty} g(k)e^{ikx}dk; \qquad g(k) = \frac{1}{(2\pi)^{1/2}}\int_{-\infty}^{+\infty} f(x)e^{-ikx}dx.$$

We seek the requirements for transforming the given Type II integral into one involving contour integration.

Jordan's Lemma: Let C be a contour along which $z = Re^{i\theta}$, where R is a constant and where θ runs from 0 to π. If $|F(z)| < |1/z^m|$ for all $|z| > R_o$ (where $m > 0$, even fractional), and $k > 0$, then

$$\lim_{R\to\infty}\int_C e^{ikz}F(z)dz = 0.$$

Proof: Over path C, $z = R\cos\theta + iR\sin\theta$, so that

$$\left|e^{ikz}\right| = \left|e^{ikR\cos\theta - kR\sin\theta}\right| = \left|e^{ikR\cos\theta}\right|e^{-kR\sin\theta} = e^{-kR\sin\theta}.$$

For $R > R_o$,

$$\left|\int_C e^{ikz}F(z)dz\right| \le \int_C \left|e^{ikz}\right||F(z)||dz| \le \int_0^{\pi}|F(z)|e^{-kR\sin\theta}Rd\theta < 2\int_0^{\pi/2}\frac{e^{-kR\sin\theta}Rd\theta}{R^m}.$$

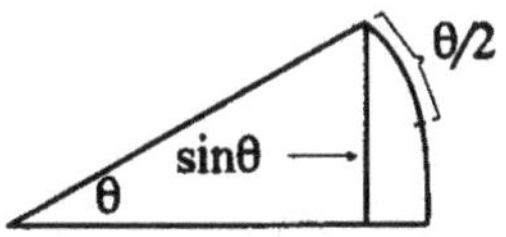

Fig. 6-46

Since $\sin\theta > \theta/2$ for $0 \le \theta \le \pi/2$ (Fig. 6-46),

$$\left|\int_C e^{ikz}F(z)dz\right| < 2\int_0^{\pi/2}\frac{e^{-kR\theta/2}Rd\theta}{R^m}$$

$$= -\frac{4}{kR^m}\left|e^{-kR\theta/2}\right|_0^{\pi/2} = \frac{4}{kR^m}\left[1 - e^{-kR\pi/4}\right].$$

Case (a): For $k > 0$ and finite, $\lim_{R\to\infty} e^{-kR\pi/4} = 0$, so that $\left|\int_C e^{ikz}F(z)dz\right| \to 0$ as $R \to \infty$. Thus, we choose the contour for which θ runs from 0 to π (Fig. 6-47).

Fig. 6-47

Case (b): For $k = 0$, $\left|\int_C e^{ik\bar{z}} F(\bar{z}) d\bar{z}\right| = \left|\int_C F(\bar{z}) d\bar{z}\right|$, and if we require that $|F(\bar{z})| < \left|1/\bar{z}^{\varepsilon}\right|$ for all $R > R_0$,

$$\left|\int_C e^{ik\bar{z}} F(\bar{z}) d\bar{z}\right| < 2\int_0^{\pi/2} \frac{R d\theta}{R^{\varepsilon}} \rightarrow 0$$

for $\varepsilon > 1$.

[A fractional value for $\varepsilon < 1$ means that $\left|\int_C F(\bar{z}) d\bar{z}\right| \rightarrow \infty$ as $R \rightarrow \infty$.] Thus, for $k = 0$, we can use the same contour, i.e. from $\theta = 0$ to π.

Case (c): For $k < 0$ (say $k = -\alpha$),

$$\left|e^{ik\bar{z}}\right| = \left|e^{-i\alpha\bar{z}}\right| = \left|e^{-i\alpha R\cos\theta} e^{\alpha R\sin\theta}\right| = e^{\alpha R\sin\theta}.$$

$$\left|\int_C e^{ik\bar{z}} F(\bar{z}) d\bar{z}\right| \le \int_C e^{\alpha R\sin\theta} |F(\bar{z})||d\bar{z}| < 2\int_0^{\pi/2} \frac{e^{\alpha R\sin\theta} R d\theta}{R^m}$$

$$< \frac{2}{R^m}\int_0^{\pi/2} e^{\alpha R\theta/2} R d\theta = \frac{4}{\alpha R^m}\left[e^{\alpha R\theta/2}\right]_0^{\pi/2}.$$

Note that if we take the contour integral from $\theta = 0$ to $\pi/2$, we get

$$\frac{4}{\alpha R^m}\left[e^{\alpha R\pi/4} - 1\right] \rightarrow \infty \text{ as } R \rightarrow \infty.$$

However, if we take the contour integral from $\theta = 0$ to $-\pi/2$, we get

$$\frac{4}{\alpha R^m}\left[e^{-\alpha R\pi/4} - 1\right] \rightarrow 0 \text{ as } R \rightarrow \infty.$$

Hence for $k < 0$, choose the contour in the lower half-plane (Fig. 6-48).

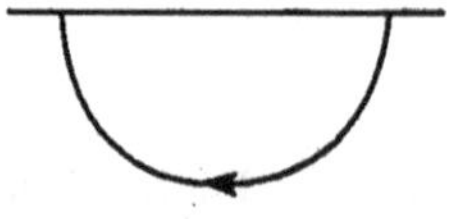

Fig. 6-48

Example (i)

$$\int_0^{\infty} \frac{\sin x}{x}\, dx = \frac{1}{2}\int_{-\infty}^{+\infty} \frac{\sin x}{x}\, dx.$$

The last step follows since the sin function is even, i.e. replacing x by $-x$:

$$\int_{x=0}^{x=-\infty} \frac{\sin(-x)}{(-x)}\, d(-x) = -\int_0^{-\infty} \frac{\sin x}{x}\, dx = \int_{-\infty}^{0} \frac{\sin x}{x}\, dx.$$

Referring to Fig. 6-49:

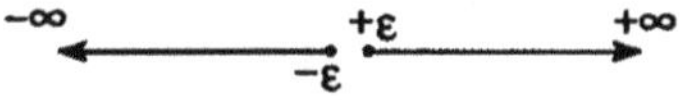

Fig. 6-49

$$\frac{1}{2}\int_{-\infty}^{+\infty} = \frac{1}{2}\left\{\int_{-\infty}^{-\varepsilon} + \int_{-\varepsilon}^{+\varepsilon} + \int_{+\varepsilon}^{+\infty}\right\},$$

where

$$\lim_{\varepsilon \to 0} \int_{-\varepsilon}^{+\varepsilon} \frac{\sin x}{x}\, dx = \lim_{\varepsilon \to 0} 2\varepsilon \to 0.$$

Hence, as $\varepsilon \to 0$,

$$\frac{1}{2}\int_{-\infty}^{+\infty} \frac{\sin x}{x}\, dx = \frac{1}{2}\left\{\int_{-\infty}^{-\varepsilon} \frac{\sin x}{x}\, dx + \int_{+\varepsilon}^{+\infty} \frac{\sin x}{x}\, dx\right\}.$$

Note that the imaginary part of

$$\frac{1}{2}\left\{\int_{-\infty}^{-\varepsilon} \frac{e^{ikx}}{x}\, dx + \int_{+\varepsilon}^{+\infty} \frac{e^{ikx}}{x}\, dx\right\} = \frac{1}{2}\int_{-\infty}^{+\infty} \frac{\sin x}{x}\, dx = \int_0^{\infty} \frac{\sin x}{x}\, dx,$$

which is the integral we wish to evaluate. Therefore, let us consider $\frac{1}{2}\oint_C \frac{e^{iz}}{z}\, dz$ over the contour C (Fig. 6-50).

Letting $R \to \infty$, and noting that $\frac{e^{iz}}{z}$ has no singularities in the region,

$$\frac{1}{2}\oint_C \frac{e^{iz}}{z}\, dz = 0.$$

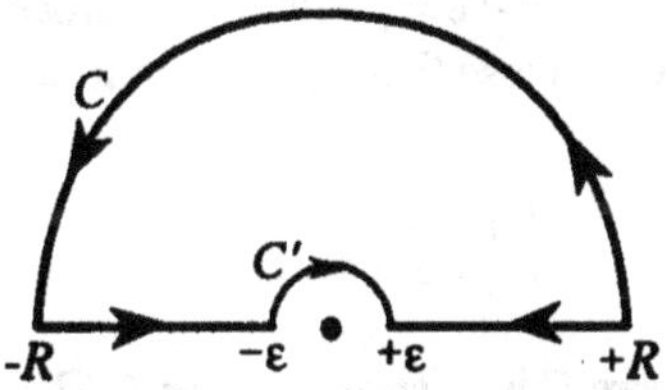

Fig. 6-50

Furthermore, since in this case $k = 1$ and $F(\bar{z}) = 1/\bar{z}$, so that the Lemma is satisfied, we have that

$$\frac{1}{2}\int_{\substack{\text{arc}\\ R\to\infty}} \frac{e^{i\bar{z}}d\bar{z}}{\bar{z}} = 0 \quad \rightarrow \quad \frac{1}{2}\int_{-\infty}^{+\infty} = \frac{1}{2}\left[\int_{-\infty}^{-\varepsilon} + \int_{C'} + \int_{+\varepsilon}^{+\infty}\right] = 0,$$

so that (in what follows $\mathrm{I_m}$ means the imaginary part of)

$$\int_0^{\infty} \frac{\sin x}{x}\,dx = \frac{1}{2}\int_{-\infty}^{+\infty} \frac{\sin x dx}{x} = \frac{\mathrm{I_m}}{4}\oint_{C} \frac{e^{i\bar{z}}d\bar{z}}{\bar{z}} = \frac{\mathrm{I_m}}{4}\,2\pi i \cdot \text{Residue at } (\bar{z} = 0)$$

$$= \frac{\mathrm{I_m}}{4}\,2\pi i = \frac{\pi}{2}.$$

<u>Example (ii)</u>

An atom starts to radiate at time $t = 0$ (its vector potential at that time is $\mathbf{A}_o$). We seek the expression for the vector potential $A(\mathbf{r}t)$ at a later time t at a distance r from the the atom. The solution is given by the expression:

$$A(\mathbf{r}t) = \mathbf{A}_o \int_{-\infty}^{+\infty} \frac{e^{-i\omega(t - r/c)}}{r(\omega - \omega_o + i\gamma)}\,d\omega,$$

where

ω_o = natural frequency of the atom;
$1/\gamma$ = lifetime of the atom (necessarily > 0);
$\mathbf{A}_o$ = the vector potential at time $t = 0$ (constant vector).

We consider the ω plane and note that $1/r(\omega - \omega_o + i\gamma)$ has a singularity at $\omega = \omega_o - i\gamma$.

(a) For $t < r/c$ (i.e. $r >$ distance covered by the speed of light in a time t), $e^{-i\omega(t - r/c)} = e^{ib\omega}$, with $b > 0$. Since $F(\omega) \approx 1/\omega$, the Lemma is satisfied, and if we choose the upper contour, then

$$\frac{1}{r}\int_{C} \frac{e^{-i\omega b}d\omega}{(\omega - \omega_o + i\gamma)} = 0,$$

where C is as shown in Fig. 6-51.

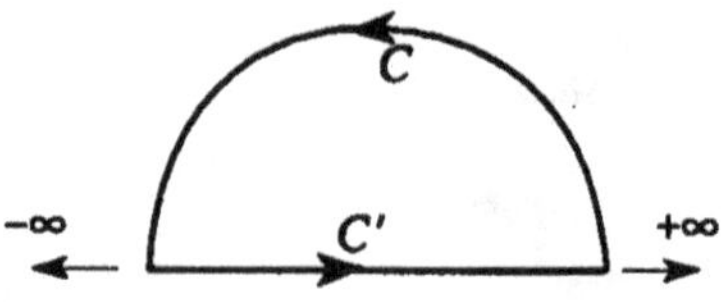

Fig. 6-51

$$A(rt) = A_0 \oint_{C+C'} \frac{e^{-i\omega(t - r/c)}}{(\omega - \omega_0 + i\gamma)r} d\omega = 2\pi i \sum_i R_i = 0,$$

since there are no poles within the contour. This result implies that wave fronts cannot travel faster than the speed of light.

(b) For $t > r/c$, we choose the lower half-plane, θ running from 0 to $-\pi/2$: again, from the Lemma, the integral over C vanishes, so that

$$A(rt) = A_0 \oint_{C+C'} \frac{e^{-i\omega(t - r/c)}}{(\omega - \omega_0 + i\gamma)r} d\omega = \frac{2\pi i A_0}{r} e^{-i\omega_0(t-r/c) - \gamma(t-r/c)},$$

which we recognize as the expression for a travelling plane wave with an amplitude $2\pi i A_0/r$ and with an exponential damping factor, i.e. for $t > r/c$, and $\gamma > 0$, we have a damping factor $e^{-\gamma(t-r/c)}$.

Type III: $\int_0^{+\infty} x^{\alpha-1} F(x) dx$,

where α is not an integer, $F(z)$ is analytic on the positive real axis, and where $z^\alpha F(z) \to 0$ as $z \to 0$ and $|z| \to \infty$.

[Note that since z^α is multiple-valued ($z^\alpha = R^\alpha e^{i\alpha\theta}$ and has different values at $\theta = 0$ and $\theta = 2\pi$ provided α is not an integer), $z^{\alpha-1}$ is multiple-valued as well.]

If $F(z)$ has a pole or an essential singularity on the real axis, then $\int_0^\infty$ has no meaning. It can still have meaning, though, if $F(z)$ is not analytic on the real axis. Consider the z plane. Note that $z^{(\alpha-1)} F(z)$ has a branch point at $z = 0$ because z^α has a branch point at $z = 0$. Introducing a cut from $z = 0$ to $z = \infty$ (Fig. 6-52):

$$\oint_C z^{(\alpha-1)} F(z) dz$$

$$= \left\{ \int_A^B + \int_B^{B'} + \int_{B'}^{A'} + \int_{A'}^{A} \right\} z^{(\alpha-1)} F(z) dz .$$

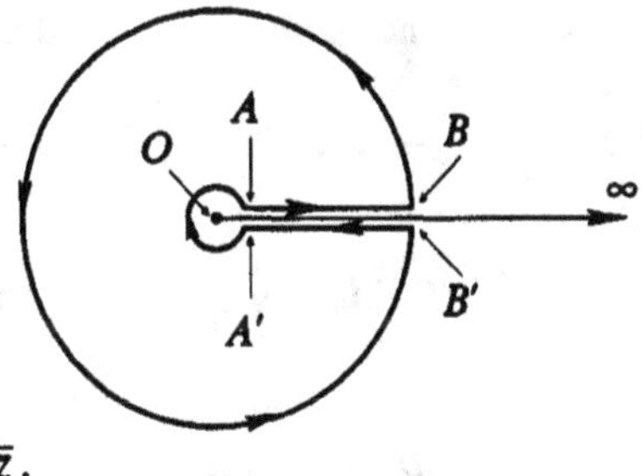

Fig. 6-52

$$\int_{B}^{B'} \bar{z}^{\alpha} F(\bar{z}) \frac{d\bar{z}}{\bar{z}} = \int_{B}^{B'} \bar{z}^{\alpha} F(\bar{z}) \frac{iRe^{i\theta} d\theta}{Re^{i\theta}} = \int_{B}^{B'} \left[\bar{z}^{\alpha} F(\bar{z})\right]_{\text{at } \bar{z}=Re^{i\theta}} i d\theta \to 0,$$

as $|\bar{z}| \to \infty$, since by hypothesis, $\left[\bar{z}^{\alpha} F(\bar{z})\right] \to 0$ as $|\bar{z}| \to \infty$.

$$\int_{A'}^{A} \bar{z}^{\alpha} F(\bar{z}) \frac{d\bar{z}}{\bar{z}} = \int_{A'}^{A} \bar{z}^{\alpha} F(\bar{z}) \frac{iRe^{i\theta} d\theta}{Re^{i\theta}} = \int_{B}^{B'} \left[\bar{z}^{\alpha} F(\bar{z})\right]_{\text{at } \bar{z}=Re^{i\theta}} i d\theta \to 0,$$

as $|\bar{z}| \to \infty$, since by hypothesis, $\left[\bar{z}^{\alpha} F(\bar{z})\right] \to 0$ as $|\bar{z}| \to \infty$.

There remains

$$\oint_{C} \bar{z}^{(\alpha-1)} F(\bar{z}) d\bar{z} = \int_{A}^{B} \bar{z}^{(\alpha-1)} F(\bar{z}) d\bar{z} - \int_{A'}^{B'} \bar{z}^{(\alpha-1)} F(\bar{z}) d\bar{z}.$$

Along AB: $\theta = 0$, $\bar{z}^{(\alpha-1)} F(\bar{z}) = x^{\alpha-1} F(x)$.

Along $A'B'$: $\theta = 2\pi$, $\bar{z}^{(\alpha-1)} F(\bar{z}) = x^{\alpha-1} e^{2\pi\alpha i} F(x)$.

Thus,

$$\oint_{C} \bar{z}^{(\alpha-1)} F(\bar{z}) d\bar{z} = \int_{0}^{\infty} x^{\alpha-1} F(x) dx - \int_{0}^{\infty} x^{\alpha-1} e^{2\pi\alpha i} F(x) dx,$$

and

$$\int_{0}^{\infty} x^{\alpha-1} F(x) dx = \frac{1}{1 - e^{2\pi\alpha i}} \oint_{C} \bar{z}^{(\alpha-1)} F(\bar{z}) d\bar{z}.$$

<u>Example (i)</u> $\int_{0}^{\infty} x^{\alpha-1} \frac{dx}{1+x}$, where $0 < \alpha < 1$. Note that $\frac{\bar{z}^{(\alpha-1)}}{1+\bar{z}}$ has a pole at $\bar{z} = -1$ and a branch point at $\bar{z} = 0$, and that

$\bar{z}^{\alpha} F(\bar{z}) = \frac{\bar{z}^{\alpha}}{1+\bar{z}} \to 0$ as $\left\{ \begin{matrix} \bar{z} \to 0 \\ \bar{z} \to \infty \end{matrix} \right\}$, i.e. we have a Type III integral. Thus,

$$\oint_{C} \bar{z}^{(\alpha-1)} F(\bar{z}) d\bar{z} = \oint_{C} \frac{\bar{z}^{(\alpha)} d\bar{z}}{\bar{z}(1+\bar{z})} = 2\pi i(-1)^{\alpha-1} = 2\pi i e^{i\pi(\alpha-1)} = -2\pi i e^{i\alpha\pi},$$

and

$$\int_{0}^{\infty} x^{\alpha-1} \frac{dx}{1+x} = \frac{-2\pi i e^{i\alpha\pi}}{1 - e^{2\pi\alpha i}} = \frac{2\pi i}{e^{i\alpha\pi} - e^{-i\alpha\pi}} = \frac{\pi}{\sin\alpha\pi}.$$

6.11 Analytic Continuation

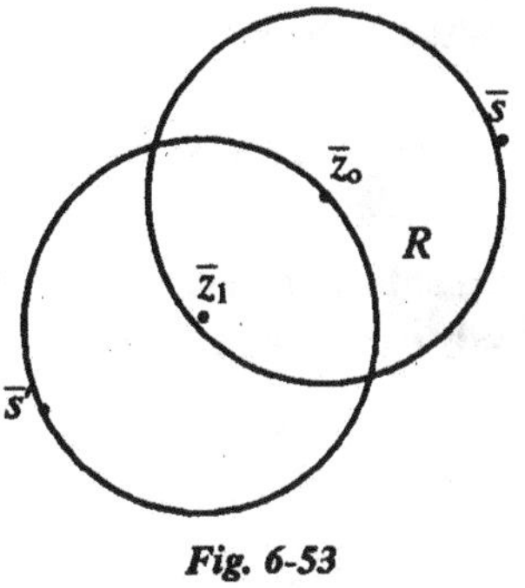

Fig. 6-53

Suppose a function $f(\bar{z})$ is analytic at $\bar{z} = \bar{z}_o$ (Fig. 6-53). Expanding $f(\bar{z})$ about point $\bar{z}_o$,

$$f(\bar{z}) = \sum_{n=o}^{\infty} a_n(\bar{z} - \bar{z}_o)^n.$$

If point $\bar{s}$ is the nearest singularity, then the series has a radius of convergence equal to $(\bar{s} - \bar{z}_o)$.

The question to be considered is the following: Can one find a function which will be analytic over a region greater than the radius of convergence of $f(\bar{z})$?

Consider a point $\bar{z}_1$ lying within region R (Fig. 6-53). According to Taylor's theorem, one can expand $f(\bar{z})$ in a Taylor series about point $\bar{z}_1$. The radius of convergence will now be $(\bar{s}' - \bar{z}_1)$, where $\bar{s}'$ is the nearest singularity to $\bar{z}_1$.

Thus, for a function having an isolated singularity, one can cover the entire domain, except for this isolated singularity, by the method of analytic continuation.

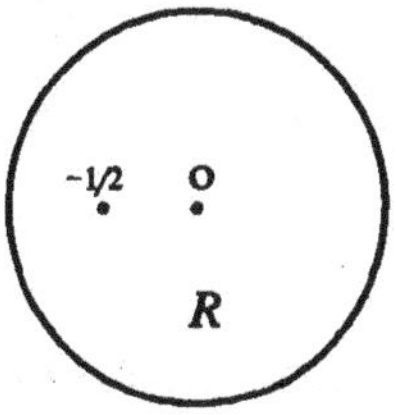

Fig. 6-54

For $f(\bar{z}) = 1/(1 - \bar{z})$ and analytic at $\bar{z} = 0$,

$$f(\bar{z}) = \frac{1}{1 - \bar{z}} = 1 + \bar{z} + \bar{z}^2 + \bar{z}^3 + \ldots = \sum_{n=o}^{\infty} \bar{z}^n,$$

valid only for $|\bar{z}| < 1$.

Let R (Fig. 6-54) be the region of convergence $|\bar{z}| < 1$. Expanding $f(\bar{z})$ about the point $x = -1/2$ ($\rho = 1, \bar{z} = e^{i\theta}$ on the circle):

$$f(\bar{z}) = f(-1/2) + f'(-1/2)(\bar{z} + 1/2) + \frac{f''(-1/2)(\bar{z} + 1/2)^2}{2} + \ldots$$

$$= \sum_{n=o}^{\infty} \frac{f^n(-1/2)(\bar{z} + 1/2)^n}{n!}.$$

Evaluating term by term:

$$f(z) = \frac{2}{3} + \frac{4}{9}(z + 1/2) + \frac{16}{2\cdot 27}(z + 1/2)^2 + \ldots$$

[Note that the very same result could have been obtained from the following perfectly calculable quantities:

$$f(z) = \sum_{n=0}^{\infty} z^n. \qquad f(-1/2) = \sum_{n=0}^{\infty} (-1/2)^n.$$

$$f'(z) = \sum_{n=1}^{\infty} n z^{(n-1)}. \qquad f'(-1/2) = \sum_{n=1}^{\infty} n(-1/2)^{(n-1)}.$$

$$f''(z) = \sum_{n=2}^{\infty} n(n-1) z^{(n-2)}. \qquad f''(-1/2) = \sum_{n=2}^{\infty} n(n-1)(-1/2)^{(n-2)}.$$

etc.]

Within the circle of convergence $|z| < 1$, and rewriting:

$$\frac{1}{1 - z} = \frac{1}{\frac{3}{2} - (z + 1/2)} = \frac{1}{\frac{3}{2}\left[1 - \frac{(z + 1/2)}{3/2}\right]} = \frac{2}{3}\left[1 - \frac{(z + 1/2)}{3/2}\right]^{-1}.$$

$$= \frac{2}{3}\left[1 + \frac{(z + 1/2)}{3/2} + \frac{(z + 1/2)^2}{(3/2)^2} + \frac{(z + 1/2)^3}{(3/2)^3} + \ldots\right],$$

which is seen to be the same series expansion as the above, namely,

$$f(z) = \sum_{n=0}^{\infty} \frac{f^n(-1/2)(z + 1/2)^n}{n!},$$

and which converges for $\left|\frac{(z + 1/2)}{3/2}\right| < 1$, or for $|z + 1/2| < 3/2$.

Hence, the new radius of convergence = 3/2 (Fig. 6-55).

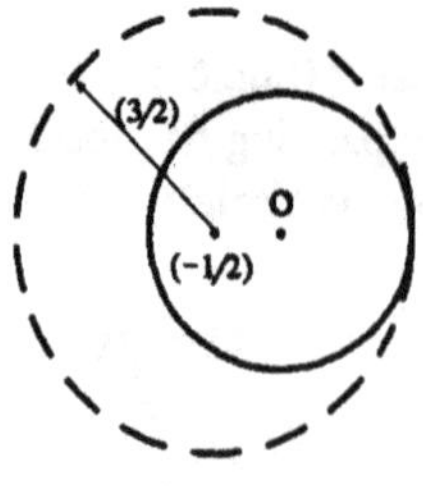

Fig. 6-55

Theorem: If F_1 and F_2 are analytic in a connected region R, and if $F_1 = F_2$ in region R_1, which is included in R, then $F_1 = F_2$ everywhere.

Proof: Define $F_1 - F_2 = f(\underline{z})$ to be analytic in region R, with $f(\underline{z}) = 0$ in R_1 (Fig. 6-56). Choose a point $\bar{a}$ within R_1 and draw a circle tangent to S_R at any point, say B. Since $f(\underline{z})$ is analytic on and within curve C,

we have that $f(\underline{z}) = \sum_{n=0}^{\infty} a_n(\underline{z} - \bar{a})^n$,

where $a_n = f^n(\bar{a})/n!$

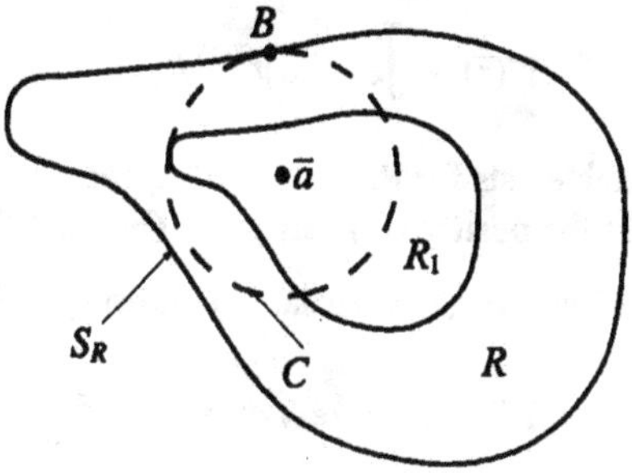

Fig. 6-56

Since point $\bar{a}$ is in region R_1 and $f(\underline{z}) = 0$ in R_1 by hypothesis, all $a_n = 0$, so that $f(\underline{z}) = 0$ in C, or $F_1 = F_2$ within C. By selecting some new point $\bar{b}$ in region C, we extend by analytic continuation and find, in this manner, that $f(\underline{z}) = 0$ in R, or that $F_1 = F_2$ in R.

What does this mean? Suppose that the analytic continuation of the function

$$f(\underline{z}) = \frac{1}{1-\underline{z}} = 1 + \underline{z} + \underline{z}^2 + \underline{z}^3 + \ldots + \underline{z}^n + \ldots = \sum_{n=0}^{\infty} \underline{z}^n$$

beyond $|\underline{z}| < 1$ yielded a particular function $G(\underline{z})$. In a connected region that excludes $\underline{z} = 1$, $G(\underline{z})$ and $1/(1 - \underline{z})$ are analytic, yet are equal to each other within a restricted region $|\underline{z}| < 1$. The above theorem shows that this $G(\underline{z})$ must equal $1/(1 - \underline{z})$ over the entire connected region. In fact, any method of analytic continuation will always yield $1/(1 - \underline{z})$.

6.11.1 Gamma Function (Γ)

The Gamma function $\Gamma(\underline{z})$ is one that will be shown to behave like $n!$ at certain isolated points. $\Gamma(\underline{z})$ is defined as:

$$\Gamma(\underline{z}) = \int_0^{\infty} e^{-t} t^{(\underline{z}-1)} dt.$$

$\Gamma(\underline{z})$ converges for $\mathrm{Re}(\underline{z}) > 0$, and is of the Type III integral discussed above.

For $\mathrm{Re}(\bar{z}) > \varepsilon > 0$, the integrand is uniformly convergent, and we can differentiate under the integral sign:

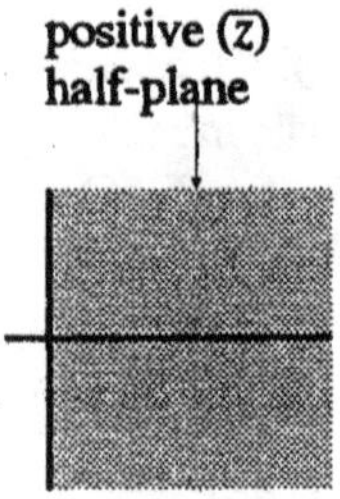

Fig. 6-57

$$\Gamma'(\bar{z}) = \int_0^\infty e^{-t}(\ln t)t^{(\bar{z}-1)}dt,$$

which exists for $\mathrm{Re}(\bar{z}) > \varepsilon > 0$. Hence, $\Gamma(\bar{z})$ is analytic over the positive $(\bar{z})$ half-plane (Fig. 6-57).

For $\mathrm{Re}(\bar{z}) > \varepsilon > 0$, partial integration yields:

$$\Gamma(\bar{z}+1) = \int_0^\infty e^{-t}t^{\bar{z}}dt = \left|-e^{-t}t^{\bar{z}}\right|_0^\infty + \bar{z}\int_0^\infty e^{-t}t^{\bar{z}-1}dt.$$

But for $\mathrm{Re}(\bar{z}) > 0$, $\left|e^{-t}t^{\bar{z}}\right|_0^\infty \to 0$, so that

$$\Gamma(\bar{z}+1) = \bar{z}\,\Gamma(\bar{z}).$$

Evaluating:

$$\Gamma(1) = \int_0^\infty e^{-t}dt = \left|e^{-t}\right|_\infty^0 = 1.$$

$$\Gamma(2) = 1\cdot\Gamma(1) = 1.$$

$$\Gamma(3) = \Gamma(2+1) = 2\cdot\Gamma(1) = 2\cdot 1 = 2!$$

$$\Gamma(4) = \Gamma(3+1) = 3\cdot\Gamma(2) = 3\cdot 2\cdot 1 = 3!$$

.

.

.

$$\Gamma(n) = (n-1)!$$

We now proceed to the problem of extending $\Gamma(\bar{z})$ into the $\mathrm{Re}(\bar{z}) < 0$ plane.

Consider a point $\bar{z}$ located in region I, the strip between $\mathrm{Re}(\bar{z}) = 0$ and $\mathrm{Re}(\bar{z}) = -1$ (Fig. 6-58). $(\bar{z}+1)$ then locates a point in the $\mathrm{Re}(\bar{z}) > \varepsilon$ region. But since $\Gamma(\bar{z}+1)$ is defined and analytic in this region, $\Gamma(\bar{z})$ is analytic in shaded region I (excluding poinr $\bar{z} = 0$).

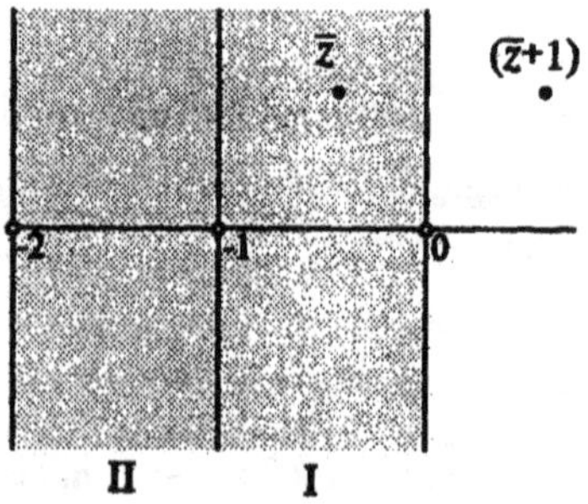

Fig. 6-58

Repeating the process, locating a point in the second shaded region II,

$$\Gamma(\bar{z}) = \frac{1}{\bar{z}}\Gamma(\bar{z}+1) = \frac{1}{\bar{z}(\bar{z}+1)}\Gamma(\bar{z}+2),$$

so that for $\bar{z} \neq 0, \bar{z} \neq -1$ and for $\Gamma(\bar{z})$ analytic, $\Gamma(\bar{z})$ will be analytic. By this method we succeed in expanding the region in which $\Gamma(\bar{z})$ is analytic to all but the points $\bar{z} = 0, -1, -2, \ldots -n \ldots$, this analytic function $\Gamma(\bar{z})$ reducing, in the region $\text{Re}(\bar{z}) > 0$, to $\int_0^\infty e^{-t} t^{(\bar{z}-1)} dt$.

To determine the behavior of $\Gamma(\bar{z})$ at points $-n$, let $\bar{z} = -n + \bar{\delta}$ where $\bar{\delta}$ is very small (Fig. 6-59). Note that the point $(-n + \bar{\delta})$ is related to a corresponding point in the region where $\Gamma(\bar{z})$ is analytic.

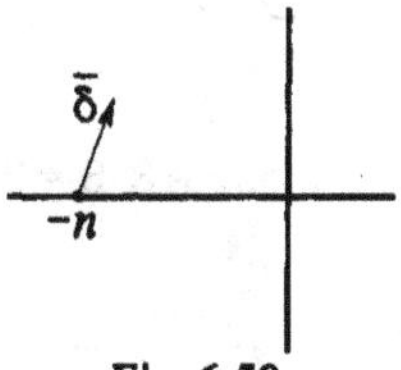

Fig. 6-59

$$\Gamma(\bar{z}) = \frac{1}{\bar{z}}\Gamma(\bar{z}+1) = \frac{1}{\bar{z}(\bar{z}+1)}\Gamma(\bar{z}+2) = \ldots$$

$$= \frac{1}{\bar{z}(\bar{z}+1)(\bar{z}+2)\ldots(\bar{z}+n)}\Gamma(\bar{z}+n+1).$$

$$\lim_{\delta\to 0}\Gamma(-n+\delta) = \lim_{\delta\to 0}\frac{\Gamma(\delta+1)}{(-n+\delta)(-n+\delta+1)\ldots(-1+\delta)\delta}$$

$$= \frac{\Gamma(1)}{(-n)(-n+1)\ldots(-1)\delta} = \frac{(-1)^n\,\Gamma(1)}{n!\delta} = \frac{(-1)^n}{n!\delta}.$$

Hence, at points $-n$,

$$\Gamma(-n+\delta) \to \frac{(-1)^n}{n!\delta}.$$

We seek an expression for $\Gamma(\bar{z})$ at $\text{Re}(\bar{z}) < 0$, one that is not obtained by the use of the extended expression $\Gamma(\bar{z}) = \Gamma(\bar{z}+1)/\bar{z}$.

For $\text{Re}(\bar{z}) > 0$, we have $\Gamma(\bar{z}) = \int_0^\infty e^{-t} t^{(\bar{z}-1)} dt$. We consider the complex $\bar{t}$ plane, noting the fact that $e^{-t} t^{(\bar{z}-1)}$ is not analytic in the plane.

We therefore introduce a cut extending from 0 to infinity (Fig. 6-60).

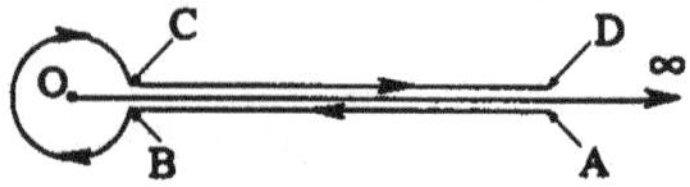

Fig. 6-60

(a) $$\int_C^D e^{-\bar t}\bar t^{(\bar z-1)}d\bar t = \int_0^\infty e^{-|\bar t|}\bar t^{(\bar z-1)}d\bar t = \Gamma(\bar z),$$

since along $CD, \theta = 0$ and $\bar t = |\bar t|e^{i\theta} = |\bar t|$.

(b) $$\int_A^B e^{-\bar t}\bar t^{(\bar z-1)}d\bar t = -\int_B^\infty e^{-\bar t}\bar t^{(\bar z-1)}d\bar t = -\int_0^\infty e^{-|\bar t|}|\bar t|e^{2\pi i(\bar z-1)}d\bar t = -e^{2\pi i\bar z}\Gamma(\bar z),$$

since along $AB, \theta = 2\pi$ and $\bar t^{(\bar z-1)} = |\bar t|^{\bar z-1}e^{2\pi i(\bar z-1)}$.

(c) $$\int_{\text{arc } B}^{C} e^{-\bar t}\bar t^{(\bar z-1)}d\bar t = \int_0^{2\pi} e^{-\bar t}e^{i\theta(\bar z-1)}|\bar t|^{\bar z}i\,d\theta = i|\bar t|^{\bar z}\int_0^{2\pi}\exp(-|\bar t|e^{i\theta})e^{i\theta(\bar z-1)}d\theta \to 0$$

as $|\bar t| \to 0$ for $\mathrm{Re}(\bar z) > 0$. Hence,

$$\int_{\substack{\text{open contour}\\ ABCD}} e^{-\bar t}\bar t^{(\bar z-1)}d\bar t = (1 - e^{2\pi i\bar z})\Gamma(\bar z),$$

or

(d) $$\Gamma(\bar z) = \frac{1}{1 - e^{2\pi i\bar z}}\int_{ABCD} e^{-\bar t}\bar t^{(\bar z-1)}d\bar t.$$

We see then that $\Gamma(\bar z)$ is defined for all $\mathrm{Re}(\bar z) > 0$ except for integral values of $\bar z$ (i.e. when $e^{2\pi i\bar z} = 1$). Since we want $\Gamma(\bar z)$ for $\mathrm{Re}(\bar z) < 0$ as well, let us consider the deformed path (*DP*) shown in Fig. 6-61. For this region,

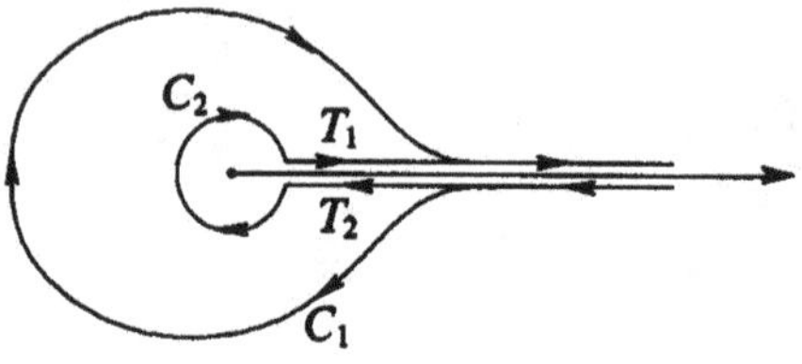

Fig. 6-61

$$\oint_{C_1-T_1-C_2-T_2} = 0 \quad \Rightarrow \quad \int_{C_1} = \int_{C_2} + \int_{T_1+T_2}.$$

Thus, equation (d) above holds for non-integral values of $\bar z$ and for all $\mathrm{Re}(\bar z)$. Since the integrand is uniformly convergent, one can differentiate under the integral sign.

$$\Gamma'(\bar{z}) = \frac{1}{1 - e^{2\pi i\bar{z}}} \int_{\bar{D}P} e^{-\bar{t}}\bar{t}^{(\bar{z}-1)} \ln\bar{t}\, d\bar{t} + \frac{2\pi i e^{2\pi i\bar{z}}}{(1 - e^{2\pi i\bar{z}})^2} \int_{\bar{D}P} e^{-\bar{t}}\bar{t}^{(\bar{z}-1)} d\bar{t}$$

$$= \frac{1}{1 - e^{2\pi i\bar{z}}} \left\{ \int_{\bar{D}P} e^{-\bar{t}}\bar{t}^{(\bar{z}-1)} \ln\bar{t}\, d\bar{t} + \frac{2\pi i e^{2\pi i\bar{z}}}{(1 - e^{2\pi i\bar{z}})} \int_{\bar{D}P} e^{-\bar{t}}\bar{t}^{(\bar{z}-1)} d\bar{t} \right\},$$

and exists; hence $\Gamma(\bar{z})$ is analytic over all $\mathrm{Re}(\bar{z})$ except for integral $\bar{z}$ values.

Chapter 7

Theory of Ordinary Differential Equations

7.1 Ordinary Differential Equations in Physics

7.1.1 Vibrating String

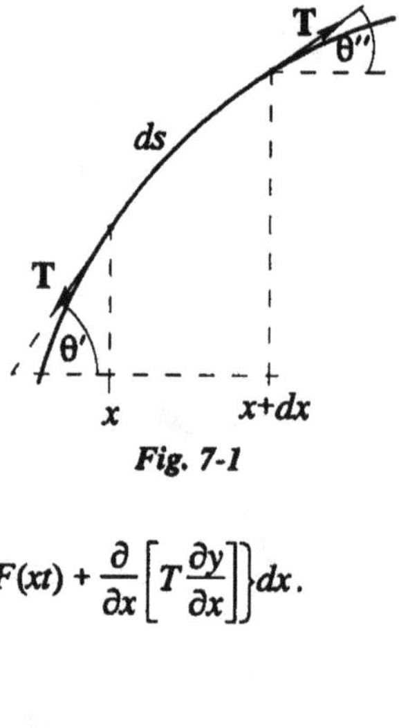

Fig. 7-1

Consider a fragment of string of length $ds = dx$, where we assume the slope of the string to be fairly small (Fig. 7-1).

If $\rho(x)$ = linear density, $T(x)$ = tension in the string, and $F(xt)dx$ = external force in the y-direction acting on dx, then the total force acting on length dx is given by

$$F(xt)dx + \left[T_{x+dx}\sin\theta'' - T_x \sin\theta'\right]$$

$$= F(xt)dx + \left[T\frac{\partial y}{\partial x}\right]_{x+dx} - \left[T\frac{\partial y}{\partial x}\right]_x = \left\{F(xt) + \frac{\partial}{\partial x}\left[T\frac{\partial y}{\partial x}\right]\right\}dx.$$

The equation of motion then becomes

$$\rho(x)dx\frac{\partial^2 y}{\partial t^2} = \left\{F(xt) + \frac{\partial}{\partial x}\left[T\frac{\partial y}{\partial x}\right]\right\}dx$$

or

$$\left\{\rho(x)\frac{\partial^2}{\partial t^2} - \frac{\partial}{\partial x}T\frac{\partial}{\partial x}\right\}y(xt) = F(xt).$$

If we define the operator $\mathcal{L} \equiv \left\{\rho(x)\frac{\partial^2}{\partial t^2} - \frac{\partial}{\partial x}T\frac{\partial}{\partial x}\right\}$, the equation of motion can be written as

(1) $\mathcal{L}y(xt) = F(xt)$.

In order to reduce the above partial differential equation to one of "ordinary" form, we make use of the Fourier transform, as follows:

$$F(xt) = \frac{1}{\sqrt{2\pi}}\int_{-\infty}^{+\infty} G_\omega(x)e^{i\omega t}d\omega. \qquad y(xt) = \frac{1}{\sqrt{2\pi}}\int_{-\infty}^{+\infty} f_\omega(x)e^{i\omega t}d\omega.$$

$$\frac{\partial y(xt)}{\partial x} = \frac{1}{\sqrt{2\pi}}\int_{-\infty}^{+\infty}\frac{\partial f_\omega(x)}{\partial x}e^{i\omega t}d\omega. \qquad \frac{\partial^2 y(xt)}{\partial t^2} = \frac{1}{\sqrt{2\pi}}\int_{-\infty}^{+\infty} -\omega^2 f_\omega(x)e^{i\omega t}d\omega.$$

Substituting into partial differential equation (1) above:

$$\int_{-\infty}^{+\infty}\left\{-\omega^2\rho(x)f_\omega(x) - \frac{\partial}{\partial x}T\frac{\partial f_\omega(x)}{\partial x} - G_\omega(x)\right\}e^{i\omega t}d\omega = 0.$$

Multiplying by $e^{-i\omega' t}$ and integrating over t from $\omega' t = 0$ to 2π:

$$\int_{-\infty}^{+\infty}\int_{0}^{2\pi}\left\{-\omega^2\rho(x)f_\omega(x) - \frac{\partial}{\partial x}T\frac{\partial f_\omega(x)}{\partial x} - G_\omega(x)\right\}e^{i\omega t}e^{-i\omega' t}d\omega dt$$

$$= \int_{-\infty}^{+\infty}\left\{-\omega^2\rho(x)f_\omega(x) - \frac{\partial}{\partial x}T\frac{\partial f_\omega(x)}{\partial x} - G_\omega(x)\right\}\delta_{\omega\omega'}\,d\omega = 0.$$

$$\Downarrow$$

(2) $\left[-\omega^2\rho(x) - \frac{\partial}{\partial x}T\frac{\partial}{\partial x}\right]f_\omega(x) = G_\omega(x)$.

Since derivatives with respect to ω do not appear in equation (2), the partial derivatives become total derivatives, with ω regarded as parameter, i.e.

(3) $\left[-\omega^2\rho(x) - \frac{d}{dx}T\frac{d}{dx}\right]f(x) = G(x)$.

Defining $p(x) = \frac{dT(x)}{dx}/T(x)$ and $g(x) = \frac{\omega^2\rho(x)f(x)}{T(x)}$, equation (3) becomes

$$(4) \qquad \frac{d^2f(x)}{dx^2} + p(x)\frac{df(x)}{dx} + g(x) = -\frac{G(x)}{T(x)},$$

a known function of x.

7.1.2 General Properties of the Linear Operator $\mathcal{L}$

(where $\mathcal{L}$ may contain partial derivatives of several variables)

$$\mathcal{L}\Psi = F \text{ (known function)}. \qquad \mathcal{L}(\Psi_1 + \Psi_2) = \mathcal{L}\Psi_1 + \mathcal{L}\Psi_2 .$$

(a) If Ψ_1 and Ψ_2 both satisfy the *inhomogeneous* equation $\mathcal{L}\Psi = F$, then the difference $\Psi_1 - \Psi_2$ must satisfy the *homogeneous* equation $\mathcal{L}(\Psi_1 - \Psi_2) = 0$.

(b) If Ψ_o satisfies the homogeneous equation $\mathcal{L}\Psi_o = 0$ and Ψ_1 satisfies the equation $\mathcal{L}\Psi_1 = F$, then the sum $\Psi_o + \Psi_1$ must satisfy $\mathcal{L}(\Psi_o + \Psi_1) = F$.

With the aid of these above seemingly trivial statements, we seek (i) the most general solution of $\mathcal{L}\Psi_o = 0$; (ii) a particular solution Ψ_1 satisfying $\mathcal{L}\Psi_1 = F$. The most general solution of the inhomogeneous equation $\mathcal{L}\Psi = F$ will then be given by $\Psi = \Psi_o + \Psi_1$.

[Note that the above properties hold in general for any linear operator, not necessarily one only involving differential operations.]

Define a Green's function $G(x_1x_2\ldots x_nx_1'x_2'\ldots x_n')$, such that

$$\mathcal{L}_{x_1x_2\ldots x_n}G(x_1x_2\ldots x_nx_1'x_2'\ldots x_n') = \delta(x_1 - x_1')\delta(x_2 - x_2')\ldots\,\delta(x_n - x_n').$$

This definition suggests a method for finding a general solution to $\mathcal{L}\Psi = F$, by first finding a particular solution Ψ_1.

Theorem: A particular solution for

$$\mathcal{L}_{x_1x_2\ldots x_n}\Psi_1(x_1x_2\ldots x_n) = F(x_1x_2\ldots x_n)$$

is given by

$$\Psi_1(x_1x_2\ldots x_n) = \int G(x_1x_2\ldots x_nx_1'x_2'\ldots x_n')F(x_1'x_2'\ldots x_n')dx_1'dx_2'\ldots dx_n'.$$

Proof: $\mathcal{L}$ is a linear operator, and thus can be taken inside the integral, so that

$$\mathcal{L}_{x_1x_2\ldots x_n}\Psi_1(x_1x_2\ldots x_n)$$

$$= \int \delta(x_1 - x_1')\ldots\, \delta(x_n - x_n')F(x_1'x_2'\ldots x_n')dx_1'dx_2'\ldots dx_n'$$

$$= F(x_1x_2\ldots x_n).$$

Note the following:

(1) Finding the general solution of $\mathcal{L}\Psi_0 = 0$ does not depend upon F.

(2) Finding G such that $\mathcal{L}G = \delta^n(x_i - x_i')$ also does not depend upon F.

Thus, the Green's function method will always yield a particular function corresponding to a particular type of operator. This function can then be used with any F one pleases. For example, for the case of $\nabla^2\Psi = F(xyz)$, $G \sim 1/|\mathbf{r} - \mathbf{r}'|$.

7.1.3 The 3-Dimensional Schrödinger Equation

For the case of central forces, $V(r)$ is a function of r only. The total energy is expressed by $H = \frac{p^2}{2m} + V(r)$, where $\mathbf{p} = m\mathbf{v}$. The equation of motion for particles is obtained by replacing $\mathbf{p}$ by $\frac{\hbar}{i}\nabla$ and H by $-\frac{\hbar}{i}\frac{\partial}{\partial t}$. Thus,

$$\text{(1)} \qquad -\frac{\hbar}{i}\frac{\partial\Psi}{\partial t} = \left[-\frac{\hbar^2}{2m}\nabla^2 + V(r)\right]\Psi(\mathbf{r}t)$$

[Compare equation (1) with the wave equation for electromagnetic waves, $\nabla^2\varphi - \frac{1}{c^2}\frac{\partial^2\varphi}{\partial t^2} = 0$ and with the heat flow/diffusion equation, $\nabla^2 T - \lambda\frac{\partial T}{\partial t} = 0$. Note the resemblance between the latter equation and the Schrödinger equation when we set $V(r) = 0$ and convert the real t axis into an imaginary it axis. The characteristics of the solution of the heat flow equation will then apply directly to the Schrödinger equation.]

The program now is to transform the Schrödinger equation, a partial differential equation, into an ordinary differential equation, and we do this by making use of the Fourier transform again.

$$\Psi(\mathbf{r}t) = \frac{1}{\sqrt{2\pi}}\int_{-\infty}^{+\infty}\Phi_\omega(\mathbf{r})e^{i\omega t}d\omega \quad \rightarrow \quad \dot{\Psi}(\mathbf{r}t) = \frac{1}{\sqrt{2\pi}}\int_{-\infty}^{+\infty}i\omega\Phi_\omega(\mathbf{r})e^{i\omega t}d\omega.$$

Substituting into equation (1):

$$-\frac{\hbar}{\sqrt{2\pi}}\int_{-\infty}^{+\infty}\omega\Phi_\omega(\mathbf{r})e^{i\omega t}d\omega = \frac{1}{\sqrt{2\pi}}\int_{-\infty}^{+\infty}\left\{-\frac{\hbar^2}{2m}\nabla^2 + V(r)\right]\Phi_\omega(\mathbf{r})e^{i\omega t}d\omega$$

Multiplying by $e^{-i\omega' t}$, integrating from $\omega' t = 0$ to 2π, and using the orthonormality relations for $e^{i\theta}$, we have:

$$E\Phi_\omega(\mathbf{r}) = \left\{-\frac{\hbar^2}{2m}\nabla^2 + V(r)\right]\Phi_\omega(\mathbf{r}),$$

where we have set $E = -\hbar\omega$.

Defining the operator $\mathcal{L} \equiv -\dfrac{\hbar^2}{2m}\nabla^2 + V(r) - E$, we can write $\mathcal{L}\Phi_\omega(\mathbf{r}) = 0$, where $\mathcal{L} = \mathcal{L}_{r\theta\varphi}$.

[Since there are no operations involving the parameter ω, we will henceforth drop the subscript ω.]

Thus we have transformed the Schrödinger partial differential equation (1) involving time to a three-dimensional, time-independent partial differential equation.

7.1.4 Separation of Variables

In spherical coordinates,

$$\nabla^2 = \frac{1}{r^2}\frac{\partial}{\partial r}\left[r^2\frac{\partial}{\partial r}\right] + \frac{1}{r^2\sin\theta}\frac{\partial}{\partial\theta}\sin\theta\frac{\partial}{\partial\theta} + \frac{1}{r^2\sin^2\theta}\frac{\partial^2}{\partial\varphi^2},$$

so that $\mathcal{L}\Phi(\mathbf{r}) = 0$ becomes

(1) $$-\frac{\hbar^2}{2m}\left[\frac{1}{r^2}\frac{\partial}{\partial r}\left(r^2\frac{\partial}{\partial r}\right) + V(r) - E\right]\Phi(\mathbf{r})$$

$$-\frac{\hbar^2}{2m}\left[\frac{1}{r^2\sin\theta}\frac{\partial}{\partial\theta}\sin\theta\frac{\partial}{\partial\theta} + \frac{1}{r^2\sin^2\theta}\frac{\partial^2}{\partial\varphi^2}\right]\Phi(\mathbf{r}) = 0.$$

Multiplying through by r^2, we see that the first operator,

$$-\frac{\hbar^2}{2m} r^2 \left[\frac{1}{r^2}\frac{\partial}{\partial r}\left(r^2 \frac{\partial}{\partial r}\right) + V(r) - E\right],$$

is a function of r only, while the second operator,

$$-\frac{\hbar^2}{2m}\left[\frac{1}{\sin\theta}\frac{\partial}{\partial\theta}\sin\theta\frac{\partial}{\partial\theta} + \frac{1}{\sin^2\theta}\frac{\partial^2}{\partial\varphi^2}\right],$$

is a function of θ and φ only. Hence, we can write $\mathcal{L}_{r\theta\varphi} = \mathcal{L}_r + \mathcal{L}_{\theta\varphi}$.

In general, if $\mathcal{L}_{x_1x_2}\Psi = 0$ and if $\mathcal{L}_{x_1x_2} = \mathcal{L}_{x_1} + \mathcal{L}_{x_2}$, where $\mathcal{L}_{x_1}$ is a function of x_1 only and $\mathcal{L}_{x_2}$ is a function of x_2 only, then $\Psi = F_1(x_1)\cdot F_2(x_2)$ will be a solution. [Later we will show it to be the most general solution.]

Assuming such a solution $\Psi = F_1(x_1)\cdot F_2(x_2)$,

$$\mathcal{L}_{x_1x_2}\Psi = \mathcal{L}_{x_1}\Psi + \mathcal{L}_{x_2}\Psi = F_2\mathcal{L}_{x_1}F_1 + F_1\mathcal{L}_{x_2}F_2 = 0.$$

Dividing by $F_1\cdot F_2$,

$$\frac{\mathcal{L}_{x_1x_2}\Psi}{F_1\cdot F_2} = \frac{\mathcal{L}_{x_1}F_1}{F_1} + \frac{\mathcal{L}_{x_2}F_2}{F_2} = 0,$$

which can be true only if

$$\frac{\mathcal{L}_{x_1}F_1}{F_1} = \lambda; \quad \frac{\mathcal{L}_{x_2}F_2}{F_2} = -\lambda. \quad \Rightarrow \quad \mathcal{L}_{x_1}F_1 = \lambda F_1; \quad \mathcal{L}_{x_2}F_2 = -\lambda F_2.$$

Let us therefore attempt a solution $\Phi(\mathbf{r}) = R(r)Y(\theta\varphi)$. Substituting into (1):

$$-\frac{\hbar^2}{2m}\left[\frac{1}{r^2}\frac{\partial}{\partial r}\left(r^2\frac{\partial}{\partial r}\right) + \frac{1}{r^2\sin\theta}\frac{\partial}{\partial\theta}\sin\theta\frac{\partial}{\partial\theta} + \frac{1}{r^2\sin^2\theta}\frac{\partial^2}{\partial\varphi^2}\right]\Phi(\mathbf{r})$$

$$+ V(r)\Phi(\mathbf{r}) - E\,\Phi(\mathbf{r}) = 0.$$

We multiply by r^2 and, following the prescription outlined above, divide by $R(r)Y(\theta\varphi)$.

$$\frac{r^2}{R(r)}\left\{-\frac{\hbar^2}{2m}\left[\frac{1}{r^2}\frac{\partial}{\partial r}r^2\frac{\partial}{\partial r}\right]+V(r)-E\right\}R(r)$$

$$-\frac{\hbar^2}{2m}\frac{1}{Y(\theta\varphi)}\left[\frac{1}{\sin\theta}\frac{\partial}{\partial\theta}\sin\theta\frac{\partial}{\partial\theta}+\frac{1}{\sin^2\theta}\frac{\partial^2}{\partial\varphi^2}\right]Y(\theta\varphi) = 0.$$

What we have here essentially is $F(r) + F(\theta\varphi) = 0$, which can be true only if $F(r) = -\lambda'$ and $F(\theta\varphi) = \lambda'$. Thus,

$$-\frac{\hbar^2}{2m}\left[\frac{1}{\sin\theta}\frac{\partial}{\partial\theta}\sin\theta\frac{\partial}{\partial\theta}+\frac{1}{\sin^2\theta}\frac{\partial^2}{\partial\varphi^2}\right]Y(\theta\varphi) = \lambda' Y(\theta\varphi).$$

We set $\lambda' = -\hbar^2\lambda/2m$, so that

$$\left[\frac{1}{\sin\theta}\frac{\partial}{\partial\theta}\sin\theta\frac{\partial}{\partial\theta}+\frac{1}{\sin^2\theta}\frac{\partial^2}{\partial\varphi^2}\right]Y(\theta\varphi) = \lambda Y(\theta\varphi).$$

$\lambda = -l(l+1)$ are the eigenvalues, and

$$Y_{lm_l}(\theta\varphi) = \left(\frac{(2l+1)(l-m)!}{4\pi(l+m)!}\right)^{1/2} P_l^m(\cos\theta)e^{im\varphi}$$

are the eigenfunctions of the Laplacian operator in θ and φ.

Since $F(r) = -\lambda' = \lambda\hbar^2/2m = -(l)(l+1)\hbar^2/2m$, we have

$$\frac{r^2}{R(r)}\left\{-\frac{\hbar^2}{2m}\left[\frac{1}{r^2}\frac{\partial}{\partial r}r^2\frac{\partial}{\partial r}\right]+V(r)-E\right\}R(r) = -(l)(l+1)\hbar^2/2m,$$

or,

$$(2) \qquad r^2\left[\frac{1}{r^2}\frac{\partial}{\partial r}r^2\frac{\partial}{\partial r}+\frac{2m}{\hbar^2}\left[E-V(r)\right]-(l)(l+1)\right]R(r) = 0.$$

We now seek to show that $\Phi(\mathbf{r}) = R(r)Y(\theta\varphi)$ represents the most general solution of

$$\left\{-\frac{\hbar^2}{2m}\nabla^2+V(r)-E\right\}\Phi(\mathbf{r}) = 0.$$

Since the $Y_{lm_l}(\theta\varphi)$ form a complete and orthonormal set, any function of θ and φ can be expanded in a series of spherical harmonics, i.e.

$$\Phi(r\theta\varphi) = \sum_{lm_l} R_{lm_l}(r) Y_{lm_l}(\theta\varphi).$$

Substituting into the 3-dimensional, time-independent partial differential equation (1), and multiplying through by r^2:

$$\sum_{lm_l} Y_{lm_l}(\theta\varphi) r^2 \left[-\frac{\hbar^2}{2m}\left(\frac{1}{r^2}\frac{\partial}{\partial r} r^2 \frac{\partial}{\partial r}\right) + V(r) - E \right] R_{lm_l}(r)$$

$$+ \sum_{lm_l} -\frac{\hbar^2}{2m} R_{lm_l}(r) \left[\frac{1}{\sin\theta}\frac{\partial}{\partial\theta}\sin\theta\frac{\partial}{\partial\theta} + \frac{1}{\sin^2\theta}\frac{\partial^2}{\partial\varphi^2} \right] Y_{lm_l}(\theta\varphi) = 0.$$

Since the $Y_{lm_l}(\theta\varphi)$ are eigensolutions of the second part of the partial differential equation above with eigenvalues $-l(l+1)$, we have

$$\sum_{lm_l} Y_{lm_l}(\theta\varphi) \left[-\frac{\hbar^2}{2m}\left(\frac{1}{r^2}\frac{\partial}{\partial r} r^2 \frac{\partial}{\partial r}\right) + V(r) - E \right] R_{lm_l}(r)$$

$$+ \sum_{lm_l} -\frac{\hbar^2}{2m} R_{lm_l}(r)(-l)(l+1) Y_{lm_l}(\theta\varphi) = 0.$$

Factoring,

$$\sum_{lm_l} Y_{lm_l}(\theta\varphi) \left\{ r^2 \left[-\frac{\hbar^2}{2m}\left(\frac{1}{r^2}\frac{\partial}{\partial r} r^2 \frac{\partial}{\partial r}\right) + V(r) - E \right] + \frac{l(l+1)\hbar^2}{2m} \right\} R_{lm_l}(r) = 0.$$

Until now nothing has been said about $V(r)$ in the proof of a general solution. If $V(r)$ is a function of r only, however, then the entire bracketed expression $\{\ \}$ above is also a function of r only. Furthermore, since the $Y_{lm_l}(\theta\varphi)$ form an orthonormal set,

$$\int_0^{2\pi} d\varphi \int_0^{\pi} Y^*_{l'm'_l}(\theta\varphi) Y_{lm_l}(\theta\varphi) \sin\theta d\theta = \delta_{ll'}\, \delta_{m_l m'_l}.$$

Thus,

$$\int_0^{2\pi} d\varphi \int_0^{\pi} \sin\theta d\theta Y^*_{l'm_{l'}}(\theta\varphi) \sum_{lm_l} Y_{lm_l}(\theta\varphi) \left\{ r^2 \left[\frac{-\hbar^2}{2m} \left(\frac{1}{r^2} \frac{\partial}{\partial r} r^2 \frac{\partial}{\partial r} \right) + V(r) - E \right] R_{lm_l}(r) \right.$$

$$+ \int_0^{2\pi} d\varphi \int_0^{\pi} \sin\theta d\theta Y^*_{l'm_{l'}}(\theta\varphi) \sum_{lm_l} Y_{lm_l}(\theta\varphi) \frac{l(l+1)\hbar^2}{2m} R_{lm_l}(r)$$

$$= \sum_{lm_l} \delta_{ll'} \delta_{m_l m_{l'}} \left\{ r^2 \left[\frac{-\hbar^2}{2m} \left(\frac{1}{r^2} \frac{\partial}{\partial r} r^2 \frac{\partial}{\partial r} \right) + V(r) - E \right] + \frac{l(l+1)\hbar^2}{2m} \right\} R_{lm_l}(r) = 0.$$

$$\Downarrow$$

$$\left\{ r^2 \left[\frac{-\hbar^2}{2m} \left(\frac{1}{r^2} \frac{\partial}{\partial r} r^2 \frac{\partial}{\partial r} \right) + V(r) - E \right] + \frac{l'(l'+1)\hbar^2}{2m} \right\} R_{l'm_{l'}}(r) = 0.$$

Since only derivatives of r are involved, we are left with ordinary and not partial derivatives, so that

$$\frac{1}{r^2} \frac{d}{dr} \left(r^2 \frac{dR(r)}{dr} \right) - \frac{2m}{\hbar^2} \left[V(r) - E + \frac{l(l+1)\hbar^2}{2mr^2} \right] R(r) = 0.$$

Rewriting,

$$\frac{1}{r^2} \left[r^2 \frac{d^2R}{dr^2} + 2r \frac{dR}{dr} \right] - \frac{2m}{\hbar^2} \left[V(r) - E + \frac{l(l+1)\hbar^2}{2m} \right] R(r) = 0.$$

If we define $p(r) \equiv \frac{2}{r}$ and $q(r) \equiv -\frac{2m}{\hbar^2} \left[V(r) - E + \frac{l(l+1)\hbar^2}{2m} \right]$, we have

(3) $$\frac{d^2R(r)}{dr^2} + p(r) \frac{dR(r)}{dr} + q(r)R(r) = 0.$$

In what follows we shall discuss more fully this second-order differential equation. But before doing so, consider the first-order differential equation: $\frac{dg(x)}{dx} + u(x)g(x) = \upsilon(x)$, where $u(x)$ and $\upsilon(x)$ are well-known functions.

We make the following transformations:

$$g(x) \equiv G(x)\exp\left[-\int_0^x u(x')dx'\right];$$

$$\upsilon(x) \equiv V(x)\exp\left[-\int_0^x u(x')dx'\right],$$

where $V(x)$ is known since $u(x)$ and $\upsilon(x)$ are known. We then have:

$$\frac{dg(x)}{dx} = \frac{dG(x)}{dx}\exp\left[-\int_0^x u(x')dx'\right] - G(x)u(x)\exp\left[-\int_0^x u(x')dx'\right].$$

$$\frac{dg(x)}{dx} + u(x)g(x) = \frac{dG(x)}{dx}\exp\left[-\int_0^x u(x')dx'\right] = V(x)\exp\left[-\int_0^x u(x')dx'\right].$$

Hence, $\frac{dG(x)}{dx} = V(x)$, a known function, and $G(x) = \int_0^x V(x')dx'$ + constant.

7.2 Ordinary Points and Singular Points

Consider the differential equation

$$(1) \qquad \frac{d^2f(z)}{dz^2} + p(z)\frac{df(z)}{dz} + q(z)f(z) = 0.$$

If we regard z as the complex variable $\bar{z}$ and study $f(\bar{z})$ in the complex plane, we will know the behavior of $f(\bar{z})$ along the real axis.

Definition: A point in the $\bar{z}$ plane which is a singularity of either $p(\bar{z})$ or $q(\bar{z})$ (or both) is called a singular point of the differential equation (1). Any point which is not a singular point is called an ordinary point.

Program: We shall attempt to find solutions for ordinary points, and then study the analytic behavior of $f(\bar{z})$ near singular points.

If $\bar{z}_0$ is an ordinary point, one can always find a point $\bar{z}$ near $\bar{z}_0$ such that $p(\bar{z})$ is analytic over the path from $\bar{z}$ to $\bar{z}_0$ (Fig. 7-2).

Define $f(\bar{z}) = \Phi(\bar{z})\exp\left[-\frac{1}{2}\int_{\bar{z}_0}^{\bar{z}} p(\bar{z}')d\bar{z}'\right]$, where $p(\bar{z})$ is analytic over the path. If $f(\bar{z})$ is analytic, then $\Phi(\bar{z})$ is analytic.

Setting $s(\bar{z}) = -\frac{1}{2}\int_{\bar{z}_0}^{\bar{z}} p(\bar{z}')d\bar{z}'$, we can write $f(\bar{z}) = \Phi(\bar{z})e^{s(\bar{z})}$.

It then follows that

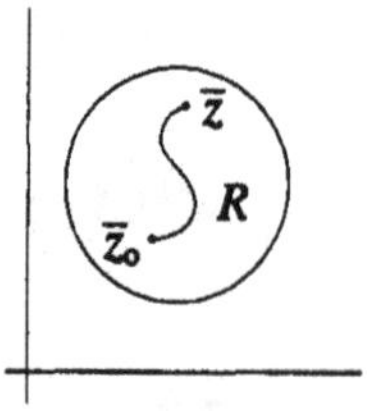

Fig. 7-2

$$\frac{df(\bar{z})}{d\bar{z}} = \frac{d\Phi(\bar{z})}{d\bar{z}}e^{s(\bar{z})} - \frac{p(\bar{z})}{2}\Phi(\bar{z})e^{s(\bar{z})} = \frac{d\Phi(\bar{z})}{d\bar{z}}e^{s(\bar{z})} - \frac{p(\bar{z})f(\bar{z})}{2}.$$

$$\frac{d^2f(\bar{z})}{d\bar{z}^2} = \frac{d^2\Phi(\bar{z})}{d\bar{z}^2}e^{s(\bar{z})} - \frac{p(\bar{z})}{2}\frac{d\Phi(\bar{z})}{d\bar{z}}e^{s(\bar{z})} - \frac{p(\bar{z})}{2}\frac{df(\bar{z})}{d\bar{z}} - \frac{f(\bar{z})}{2}\frac{dp(\bar{z})}{d\bar{z}}$$

$$= \frac{d^2\Phi(\bar{z})}{d\bar{z}^2}e^{s(\bar{z})} - \frac{p(\bar{z})}{2}\left[\frac{df(\bar{z})}{d\bar{z}} + \frac{d\Phi(\bar{z})}{d\bar{z}}e^{s(\bar{z})}\right] - \frac{f(\bar{z})}{2}\frac{dp(\bar{z})}{d\bar{z}}$$

$$= \frac{d^2\Phi(\bar{z})}{d\bar{z}^2}e^{s(\bar{z})} - \frac{p(\bar{z})}{2}\left[2\frac{df(\bar{z})}{d\bar{z}} + \frac{f(\bar{z})p(\bar{z})}{2}\right] - \frac{f(\bar{z})}{2}\frac{dp(\bar{z})}{d\bar{z}}.$$

Collecting terms and substituting back into differential equation (1):

$$0 = \frac{d^2f(\bar{z})}{d\bar{z}^2} + p(\bar{z})\frac{df(\bar{z})}{d\bar{z}} + q(\bar{z})f(\bar{z})$$

$$= e^{s(\bar{z})}\left[\frac{d^2\Phi(\bar{z})}{d\bar{z}^2} - \frac{f(\bar{z})p^2(\bar{z})}{4e^{s(\bar{z})}} - \frac{f(\bar{z})}{2e^{s(\bar{z})}}\frac{dp(\bar{z})}{d\bar{z}} + \frac{f(\bar{z})q(\bar{z})}{e^{s(\bar{z})}}\right]$$

$$= e^{s(\bar{z})}\left[\frac{d^2\Phi(\bar{z})}{d\bar{z}^2} - \frac{\Phi(\bar{z})p^2(\bar{z})}{4} - \frac{\Phi(\bar{z})}{2}\frac{dp(\bar{z})}{d\bar{z}} + \Phi(\bar{z})q(\bar{z})\right]$$

or,

(2) $$0 = \frac{d^2\Phi(\bar{z})}{d\bar{z}^2} + \Phi(\bar{z})J(\bar{z}),$$

where we define $J(\bar{z}) = q(\bar{z}) - \frac{p^2(\bar{z})}{4} - \frac{1}{2}\frac{dp(\bar{z})}{d\bar{z}}$.

Note that all ordinary points of the differential equation (1) are also ordinary points of equation (2), since $J(z)$ is analytic if $p(z)$ and $q(z)$ are analytic. We now proceed to investigate

$$(2)\qquad \frac{d^2\Phi(z)}{dz^2} + \Phi(z)J(z) = 0.$$

7.2.1 Iteration Method

Consider the differential equation

$$(1)\qquad \frac{d^2\Phi^\lambda(z)}{dz^2} + \lambda\Phi^\lambda(z)J(z) = 0,$$

which differs from the equation under investigation in the appearance of λ, whose meaning is such that $\left|\Phi^\lambda(z)\right|_{\lambda=1} = \Phi(z)$.

Assuming (later to be shown to be true) that $\Phi^\lambda(z)$ can be expanded in a power series of λ:

$$\Phi^\lambda(z) = \Phi_o(z) + \lambda\Phi_1(z) + \lambda^2\Phi_2(z) + \ldots = \sum_{n=o}^{\infty}\lambda^n\,\Phi_n(z).$$

Assuming also that one can differentiate within the summation:

$$\frac{d^2\Phi_o(z)}{dz^2} + \lambda\frac{d^2\Phi_1(z)}{dz^2} + \lambda^2\frac{d^2\Phi_2(z)}{dz^2} + \ldots + \lambda^n\frac{d^2\Phi_n(z)}{dz^2} + \ldots$$

$$+ \lambda J(z)\left[\Phi_o(z) + \lambda\Phi_1(z) + \lambda^2\Phi_2(z) + \ldots + \lambda^n\Phi_n(z) + \ldots\right] = 0$$

Collecting terms in powers of λ:

$$\frac{d^2\Phi_o(z)}{dz^2} + \lambda\left[\frac{d^2\Phi_1(z)}{dz^2} + J(z)\Phi_o(z)\right] + \lambda^2\left[\frac{d^2\Phi_2(z)}{dz^2} + J(z)\Phi_1(z)\right]$$

$$+ \lambda^3\left[\frac{d^2\Phi_3(z)}{dz^3} + J(z)\Phi_2(z)\right] + \ldots + \lambda^n\left[\frac{d^2\Phi_n(z)}{dz^n} + J(z)\Phi_{n-1}(z)\right] + \ldots = 0,$$

which can be satisfied only if the coefficients of each $\lambda^i = 0$.

$$\frac{d^2\Phi_o(\overline{z})}{d\overline{z}^2} = 0. \quad \frac{d^2\Phi_1(\overline{z})}{d\overline{z}^2} + J(\overline{z})\Phi_o(\overline{z}) = 0. \quad \frac{d^2\Phi_2(\overline{z})}{d\overline{z}^2} + J(\overline{z})\Phi_1(\overline{z}) = 0.$$

In general,

$$(2) \quad \begin{cases} \dfrac{d^2\Phi_n(\overline{z})}{d\overline{z}^2} + J(\overline{z})\Phi_{n-1}(\overline{z}) = 0 \text{ for } n \geq 1. \\ \\ \dfrac{d^2\Phi_o(\overline{z})}{d\overline{z}^2} = 0 \text{ for } n = 0. \end{cases}$$

Note here the iteration method:

- Start with $\dfrac{d^2\Phi_o(\overline{z})}{d\overline{z}^2} = 0.$
- Solve for $\Phi_o(\overline{z})$.
- Insert $\Phi_o(\overline{z})$ into $\dfrac{d^2\Phi_1(\overline{z})}{d\overline{z}^2} = -J(\overline{z})\Phi_o(\overline{z})$.
- Solve for $\Phi_1(\overline{z})$, $\Phi_2(\overline{z})$, etc., up to $\Phi_n(\overline{z})$.
- Knowing all $\Phi_n(\overline{z})$ (as $n \to \infty$), you now know $\Phi^\lambda(\overline{z})$ from the power series expansion.

Consider again equation (1): We must find $\Phi^\lambda(\overline{z})$ which satisfies

$$\left|\Phi^\lambda(\overline{z})\right|_{\overline{z}_o} = A, \quad \left|\frac{d\Phi^\lambda(\overline{z})}{d\overline{z}}\right|_{\overline{z}_o} = B,$$

where A and B are given boundary conditions, and also satisfies

$$\frac{d^2\Phi^\lambda(\overline{z})}{d\overline{z}^2} + \lambda\Phi^\lambda(\overline{z})J(\overline{z}) = 0.$$

Note that $\Phi_o(\overline{z}) = A + B(\overline{z} - \overline{z}_o)$ satisfies $\dfrac{d^2\Phi_o(\overline{z})}{d\overline{z}^2} = 0$ for $n = 0$. For $n \geq 1$ it follows from $\dfrac{d^2\Phi_n(\overline{z})}{d\overline{z}^2} = -J(\overline{z})\Phi_{n-1}(\overline{z})$ that $\dfrac{d\Phi_n(\overline{z})}{d\overline{z}} = -\int_{\overline{z}_o}^{\overline{z}} J(\overline{z}')\Phi_{n-1}(\overline{z}')d\overline{z}'$.

This latter expression is satisfied by

$$\Phi_n(\bar{z}) = -\int_{\bar{z}_o}^{\bar{z}} (\bar{z} - \bar{z}')J(\bar{z}')\Phi_{n-1}(\bar{z}')d\bar{z}',$$

since

$$\frac{d\Phi_n(\bar{z})}{d\bar{z}} = -\left[(\bar{z} - \bar{z}')J(\bar{z}')\Phi_{n-1}(\bar{z}')\right]_{\bar{z}'=\bar{z}} - \int_{\bar{z}_o}^{\bar{z}} \frac{d}{d\bar{z}}(\bar{z} - \bar{z}')J(\bar{z}')\Phi_{n-1}(\bar{z}')d\bar{z}'$$

$$= 0 - \int_{\bar{z}_o}^{\bar{z}} J(\bar{z}')\Phi_{n-1}(\bar{z}')d\bar{z}'.$$

We have $\Phi^\lambda(\bar{z}) = \Phi_o(\bar{z}) + \lambda\Phi_1(\bar{z}) + \lambda^2\Phi_2(\bar{z}) + \ldots$, from which

$$\left|\Phi^\lambda(\bar{z})\right|_{\bar{z}_o} = \left|\Phi_o(\bar{z})\right|_{\bar{z}_o} \text{ only,}$$

and

$$\left|\frac{d\Phi^\lambda(\bar{z})}{d\bar{z}}\right|_{\bar{z}_o} = \left|\frac{d\Phi_o(\bar{z})}{d\bar{z}}\right|_{\bar{z}_o} \text{ only,}$$

since, for $n \geq 1$, $\Phi_n(\bar{z}_o) = 0$ and $\left|\frac{d\Phi_n(\bar{z})}{d\bar{z}}\right|_{\bar{z}_o} = 0$.

Hence, boundary conditions $\left|\Phi^\lambda(\bar{z})\right|_{\bar{z}_o} = A$ and $\left|\frac{d\Phi^\lambda(\bar{z})}{d\bar{z}}\right|_{\bar{z}_o} = B$ are satisfied by

$$\Phi_o(\bar{z}) = A + B(\bar{z} - \bar{z}_o).$$

(3)

$$\Phi_n(\bar{z}) = -\int_{\bar{z}_o}^{\bar{z}} (\bar{z} - \bar{z}')J(\bar{z}')\Phi_{n-1}(\bar{z}')d\bar{z}'.$$

We have thus found a $\Phi^\lambda(\bar{z})$ satisfying $\frac{d^2\Phi^\lambda(\bar{z})}{d\bar{z}^2} + \lambda\Phi^\lambda(\bar{z})J(\bar{z}) = 0$ and also satisfying the proper boundary conditions.

Let us now proceed to investigate the series $\Phi^\lambda(\bar{z}) = \sum_{n=0}^{\infty} \lambda^n\Phi_n(\bar{z})$.

Let the upper limit of $J(\bar{z})$ and $\Phi_o(\bar{z})$ along the path $\bar{z} \to \bar{z}_o$ be U and V respectively, such that $|J(\bar{z})| \leq U$ and $|\Phi_o(\bar{z})| \leq V$.

Theorem: $|\Phi_n(\bar{z})| \le \dfrac{VU^n}{n!}|\bar{z}-\bar{z}_o|^{2n}$.

Proof: The method of proof will be by induction, i.e. we assume it to be true for $n-1$ and show it to be true for n. Then if true for $n=0$, the proof is complete.

From (3) above,

$$|\Phi_n(\bar{z})| < -\int_{\bar{z}_o}^{\bar{z}} |(\bar{z}-\bar{z}')||J(\bar{z}')||\Phi_{n-1}(\bar{z}')||d\bar{z}'|.$$

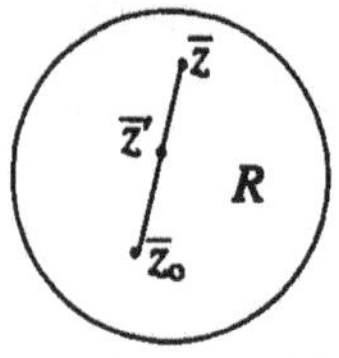

Fig. 7-3

Since $\Phi_n(\bar{z})$ is analytic in R, the path is immaterial. Hence we choose the straight path from $\bar{z}_o$ to $\bar{z}$ (Fig. 7-3). Then, since $\bar{z}'$ is a point on the path, $|\bar{z}-\bar{z}'| \le |\bar{z}-\bar{z}_o|$.

Assuming the theorem true for $n-1$: $|\Phi_{n-1}(\bar{z}')| \le V\dfrac{U^{n-1}}{(n-1)!}|\bar{z}'-\bar{z}_o|^{2n-2}$.

$$|\Phi_n(\bar{z})| \le |(\bar{z}-\bar{z}_o)|\int_{\bar{z}_o}^{\bar{z}}|J(\bar{z}')||\Phi_{n-1}(\bar{z}')||d\bar{z}'|$$

$$= \frac{U|(\bar{z}-\bar{z}_o)|VU^{n-1}}{(n-1)!}\int_{\bar{z}_o}^{\bar{z}}|(\bar{z}'-\bar{z}_o)|^{2n-2}|d\bar{z}'|$$

Setting $t = |\bar{z}'-\bar{z}_o|$, so that $dt = |d\bar{z}'|$, we have

$$|\Phi_n(\bar{z})| \le \frac{U^nV|(\bar{z}-\bar{z}_o)|}{(n-1)!}\int_0^{|\bar{z}-\bar{z}_o|}t^{2n-2}dt = \frac{VU^n|(\bar{z}-\bar{z}_o)|^{2n}}{(n-1)!(2n-1)}.$$

But for $n \ge 1$, $2n-1 \ge n$, so that

$$|\Phi_n(\bar{z})| \le \frac{VU^n|(\bar{z}-\bar{z}_o)|^{2n}}{n!},$$

and the theorem is true for n. For $n=0$, $|\Phi_o(\bar{z})| \le V$. Hence the theorem is true for $n=0$, thus completing the proof.

To show that $\Phi^\lambda(\bar{z})$ can indeed be expanded in a power series in λ, consider the power series $e^{\lambda\xi} = \sum_0^\infty \dfrac{\lambda^n\xi^n}{n!}$, which converges for all values of ξ, with a radius of convergence of $\lambda = \infty$.

Setting $\xi = U(\bar{z} - \bar{z}_o)^2$, we have from the theorem above,

$$|\Phi_n(\bar{z})| \le \frac{VU^n|(\bar{z} - \bar{z}_o)|^{2n}}{n!} = \frac{V\xi^n}{n!} \rightarrow |\lambda^n \Phi_n(\bar{z})| \le \frac{\lambda^n V\xi^n}{n!}.$$

Thus, $\sum_{n=o}^{\infty} \lambda^n \Phi_n(\bar{z})$ converges uniformly for any finite λ. Furthermore, $\Phi^{\lambda}(\bar{z}) = \sum_{n=o}^{\infty} \lambda^n \Phi_n(\bar{z})$ defines an analytic function of $\bar{z}$ with a radius of convergence equal to infinity.

For every ordinary second-order differential equation, there is a solution $\Phi(\bar{z})$, an analytic function, at points near $\bar{z}_o$, the ordinary point, such that $\Phi(\bar{z})$ has two arbitrary constants, namely the integration constants, and these constants may be fixed by fixing $\Phi(\bar{z}_o)$ and $\Phi'(\bar{z}_o) \equiv \left|\frac{d\Phi(\bar{z})}{d\bar{z}}\right|_{\bar{z}_o}$. In addition, this solution is unique, as we now show.

Theorem: If the values of $\varphi(\bar{z}_o)$ and $\varphi'(\bar{z}_o)$ are fixed, then $\varphi(\bar{z})$ is unique.

Let $\varphi_1(\bar{z})$ and $\varphi_2(\bar{z})$ both be solutions, so that $\varphi_1(\bar{z}_o) = \varphi_2(\bar{z}_o)$ and $\varphi_1'(\bar{z}_o) = \varphi_2'(\bar{z}_o)$. We will prove that $\varphi_1(\bar{z}) = \varphi_2(\bar{z})$ not only at $\bar{z}_o$ but at points near $\bar{z}_o$.

Proof: Define $\Phi(\bar{z}) \equiv \varphi_1(\bar{z}) - \varphi_2(\bar{z})$, so that $\Phi(\bar{z}_o) \equiv \varphi_1(\bar{z}_o) - \varphi_2(\bar{z}_o) = 0$ and $\Phi'(\bar{z}_o) \equiv \varphi_1'(\bar{z}_o) - \varphi_2'(\bar{z}_o) = 0$. Since $\varphi_1(\bar{z})$ and $\varphi_2(\bar{z})$ are solutions, $\Phi(\bar{z})$ is also a solution, i.e.

$$\text{(4)} \qquad \frac{d^2\Phi(\bar{z})}{d\bar{z}^2} + p(\bar{z})\frac{d\Phi(\bar{z})}{d\bar{z}} + q(\bar{z})\Phi(\bar{z}) = 0.$$

At $\bar{z} = \bar{z}_o$, $\left|\frac{d^2\Phi(\bar{z})}{d\bar{z}^2}\right|_{\bar{z}_o} = -\left[p(\bar{z})\frac{d\Phi(\bar{z})}{d\bar{z}} + q(\bar{z})\Phi(\bar{z})\right]_{\bar{z}_o} = 0$. This follows since $\frac{d\Phi(\bar{z})}{d\bar{z}} = \Phi(\bar{z}) = 0$ at $\bar{z} = \bar{z}_o$. Taking derivatives of (4) and evaluating at $\bar{z}_o$,

$$\left|\frac{d^3\Phi(\bar{z})}{d\bar{z}^3}\right|_{\bar{z}_o} = -\left[p'(\bar{z})\frac{d\Phi(\bar{z})}{d\bar{z}} + p(\bar{z})\frac{d^2\Phi(\bar{z})}{d\bar{z}^2} + q'(\bar{z})\Phi(\bar{z}) + q(\bar{z})\frac{d\Phi(\bar{z})}{d\bar{z}}\right]_{\bar{z}_o} = 0,$$

since $\frac{d^2\Phi(\bar{z})}{d\bar{z}^2} = \frac{d\Phi(\bar{z})}{d\bar{z}} = \Phi(\bar{z}) = 0$ at $\bar{z} = \bar{z}_o$.

Thus, $\left|\dfrac{d^3\Phi(\bar{z})}{d\bar{z}^3}\right|_{\bar{z}_o} = 0$. Since Φ is analytic and can be expanded in a Taylor series about $\bar{z}_o$,

$$\Phi(\bar{z}) = \Phi(\bar{z}_o) + (\bar{z} - \bar{z}_o)\Phi'(\bar{z}_o) + \frac{(\bar{z} - \bar{z}_o)^2}{2!}\Phi''(\bar{z}_o) + \ldots$$

$$= \sum_{n=0}^{\infty} \frac{\Phi^n(\bar{z}_o)(\bar{z} - \bar{z}_o)^n}{n!} = 0.$$

By directly observing $p(\bar{z})$ and $q(\bar{z})$, one can determine at once the minimum value of the radius of convergence of the solution $\Phi(\bar{z})$. An ordinary point of the differential equation is defined as a regular point of the solution $\Phi(\bar{z})$.

7.2.2 Power Series Method

Concerning equation

$$(1) \qquad \frac{d^2f(\bar{z})}{d\bar{z}^2} + p(\bar{z})\frac{df(\bar{z})}{d\bar{z}} + q(\bar{z})f(\bar{z}) = 0,$$

let $\xi = \bar{z} - \bar{z}_o$ and perform the expansion $f(\bar{z}) = \sum_{n=0}^{\infty} a_n\xi^n$, where the radius of convergence is not greater than the distance of $\bar{z}_o$ to the nearest singularity. Since both $p(\bar{z})$ and $q(\bar{z})$ are analytic at $\bar{z}_o$ and in the region near $\bar{z}_o$, we can write

$$p(\bar{z}) = \sum_{m=0}^{\infty} p_m\xi^m. \qquad q(\bar{z}) = \sum_{m=0}^{\infty} q_m\xi^m.$$

$$\frac{df(\bar{z})}{d\bar{z}} = \frac{df(\bar{z})}{d\xi} = \sum_{n=0}^{\infty} na_n\xi^{n-1}. \qquad \frac{d^2f(\bar{z})}{d\bar{z}^2} = \frac{d^2f(\bar{z})}{d\xi^2} = \sum_{n=0}^{\infty} n(n-1)a_n\xi^{n-2}.$$

$$q(\bar{z})f(\bar{z}) = \sum_{m=0}^{\infty} q_m\xi^m \sum_{n=0}^{\infty} a_n\xi^n. \qquad p(\bar{z})\frac{df(\bar{z})}{d\bar{z}} = \sum_{m=0}^{\infty} p_m\xi^m \sum_{n=0}^{\infty} na_n\xi^{n-1}.$$

Substituting back into equation (1) and collecting terms involving ξ^n,

$$\sum_{n=0}^{\infty} \xi^n \left\{(n+2)(n+1)a_{n+2} + \sum_{m=0}^{n}(n-m+1)p_m a_{n-m+1} + \sum_{m=0}^{n} q_m a_{n-m}\right\} = 0.$$

This can be true only if all the coefficients separately vanish. Thus,

$$(2) \qquad (n+2)(n+1)a_{n+2} + \sum_{m=0}^{n}(n-m+1)p_m a_{n-m+1} + \sum_{m=0}^{n} q_m a_{n-m} = 0.$$

Equations (2) are called *difference equations*, with parameter n running from 0 to infinity.

For $n = 0$: $2a_2 + a_1 p_o + a_o q_o = 0.$

For $n = 1$: $6a_3 + 2a_2 p_o + a_1 p_1 + a_1 q_o + a_o q_1 = 0.$

etc.

We ask: What advantages do these difference equations offer over the differential equations? By fixing a_o and a_1 arbitrarily, case $n = 0$ determines a_2; case $n = 1$ determines a_3, etc. Thus, $f(\overline{z}_o) = a_o, f'(\overline{z}_o) = a_1, f''(\overline{z}_o) = 2a_2$, etc, thus determining $f(\overline{z}) = \sum_{n=o}^{\infty} a_n \xi^n$.

(a) For $a_o = 1$ and $a_1 = 0$, solution $f_1(\overline{z})$ satisfies $f_1(\overline{z}_o) = 1$ and $f_1'(\overline{z}_o) = 0$.

(b) For $a_1 = 1$ and $a_o = 0$, solution $f_2(\overline{z})$ satisfies $f_2(\overline{z}_o) = 0$ and $f_2'(\overline{z}_o) = 1$.

Thus, if boundary conditions require that $f(\overline{z}_o) = A$ and $f'(\overline{z}_o) = B$, then we can write $f(\overline{z}) = Af_1(\overline{z}) + Bf_2(\overline{z})$, since $f(\overline{z}_o) = Af_1(\overline{z}_o) + Bf_2(\overline{z}_o) = A$ and $f'(\overline{z}_o) = Af_1'(\overline{z}_o) + Bf_2'(\overline{z}_o) = B$.

7.3 Hermite Polynomials

The Hermite equation, given by

$$(1) \qquad \frac{d^2 f(\overline{z})}{d\overline{z}^2} - 2\overline{z}\frac{df(\overline{z})}{d\overline{z}} + 2\lambda f(\overline{z}) = 0,$$

where $p(\overline{z}) = -2\overline{z}$ and $q(\overline{z}) = 2\lambda$ are both analytic everywhere except at $\overline{z} = \infty$ for $p(\overline{z})$, will be studied by the power series method of the previous section.

The solution $f(\overline{z})$ is regular at all points of space except at $\overline{z} = \infty$, and is therefore an entire function, i.e. analytic everywhere except at infinity, with an infinite radius of convergence. If $f(\overline{z})$ is a polynomial, it has a pole at infinity; if it is not a polynomial, it has an essential singularity at infinity.

Since $f(\bar{z})$ is an entire function, it can be expanded about the point $\bar{z}_o$ in a Taylor series.

$$f(\bar{z}) = \sum_{n=0}^{\infty} a_n \bar{z}^n. \qquad \frac{df(\bar{z})}{d\bar{z}} = \sum_{n=1}^{\infty} n a_n \bar{z}^{n-1}.$$

$$\bar{z}\frac{df(\bar{z})}{d\bar{z}} = \sum_{n=1}^{\infty} n a_n \bar{z}^n. \qquad \frac{d^2 f(\bar{z})}{d\bar{z}^2} = \sum_{n=2}^{\infty} n(n-1) a_n \bar{z}^{n-2}.$$

Substituting into (1) and collecting powers of $\bar{z}^n$:

$$\sum_{n=0}^{\infty} \xi^n \left[(n+2)(n+1)a_{n+2} - 2na_n + 2\lambda a_n\right] = 0,$$

which holds only if $(n+2)(n+1)a_{n+2} - 2na_n + 2\lambda a_n = 0$ $[n = 0, 1 \dots \infty]$, or

$$\text{(2)} \qquad a_{n+2} = \frac{2(n-\lambda)a_n}{(n+2)(n+1)}.$$

For $n = 0$: $a_2 = \dfrac{-2\lambda a_o}{2!}$.

For $n = 1$: $a_3 = \dfrac{2(1-\lambda)a_1}{3!}$.

For $n = 2$: $a_4 = \dfrac{2(2-\lambda)a_2}{4\cdot 3} = \dfrac{-2\cdot 2(2-\lambda)\lambda a_o}{4!}$.

For $n = 3$: $a_5 = \dfrac{2(3-\lambda)a_3}{5\cdot 4} = \dfrac{2\cdot 2(3-\lambda)(1-\lambda)a_1}{5!}$.

etc.

We see that even and odd indices form two separate groups:

(i) For $a_o = 1$, $a_1 = 0$, $f_1(\bar{z})$ satisfies $f_1(0) = a_o = 1$ and $f_1'(0) = a_1 = 0$, so that $f_1(\bar{z})$ is an even function in $\bar{z}$ and all terms in odd powers of $\bar{z}$ vanish.

(ii) For $a_1 = 1$, $a_o = 0$, $f_2(\bar{z})$ satisfies $f_2(0) = a_o = 0$ and $f_2'(0) = a_1 = 1$, so that $f_2(\bar{z})$ is an odd function in $\bar{z}$ and all terms in even powers of $\bar{z}$ vanish.

We now ask: What values of λ will make $f(\bar{z})$ a polynomial? (These λ values will be seen to correspond to the eigenvalues of a harmonic oscillator.) We do not seek those values of λ making $f(\bar{z})$ a non-polynomial because these values are rejected on physical grounds. This point will be discussed later.

If the polynomial is to be of order n, then we want $a_{n+1} = a_{n+2} = \ldots = 0$. [Note from equation (2) above that if $\lambda = n$ (positive integer), then $a_{n+2} = a_{n+4} = a_{n+6} = \ldots = 0$.]

(A) For n = even integer:

Choosing $a_o = a_o$, $a_1 = 0$, we have $a_3 = a_5 = a_7 = \ldots = 0$, and our solution will be a polynomial of even powers in $\bar{z}$ up to $a_n\bar{z}^n$. Thus,

$$f(\bar{z}) = a_o + a_2\bar{z}^2 + a_4\bar{z}^4 + \ldots + a_n\bar{z}^n,$$

where

$$a_{n+2} = \frac{2(n-\lambda)a_n}{(n+2)(n+1)} \quad \rightarrow \quad a_n = \frac{2(n-2-\lambda)a_{n-2}}{n(n-1)}.$$

Thus,

$$a_2 = \frac{-2\lambda a_o}{2!}.$$

$$a_4 = \frac{-2(\lambda-2)a_2}{4\cdot 3} = \frac{(-2)^2\lambda(\lambda-2)a_o}{4!}.$$

$$a_6 = \frac{-2(\lambda-4)a_4}{4\cdot 3} = \frac{(-2)^3\lambda(\lambda-2)(\lambda-4)a_o}{6!}.$$

etc.

Choose $a_o = \dfrac{(-1)^{n/2}n!}{(n/2)!}$, so that for $\lambda = n$ (even integer):

$$a_n = \frac{(-2)^{n/2}\lambda(\lambda-2)(\lambda-4)\ldots(\lambda-n+2)(-1)^{n/2}n!}{n!(n/2)!} \quad \rightarrow \quad 2^n$$

To evaluate next lower coefficients, i.e. a_{n-2}, a_{n-4}, etc., we make use of:

$$a_{n-2} = \frac{n(n-1)a_n}{2(n-2-\lambda)}.$$

$$a_{n-2} = \frac{n(n-1)2^n}{2(n-2-\lambda)} = \frac{n(n-1)2^n}{2(-2)} = -n(n-1)2^{n-2}.$$

$$a_{n-4} = a_{n-2}\frac{(n-2)(n-3)}{2(n-4-\lambda)} = \frac{-2^{n-2}n(n-1)(n-2)(n-3)}{-2^3}$$

$$= \frac{2^{n-4}n(n-1)(n-2)(n-3)}{2!}.$$

Thus

$$f_1(\bar{z}) = a_n\bar{z}^n + a_{n-2}\bar{z}^{n-2} + a_{n-4}\bar{z}^{n-4} + \ldots$$

$$= (2\bar{z})^n - \frac{n(n-1)(2\bar{z})^{n-2}}{1!} + \frac{n(n-1)(n-2)(n-3)(2\bar{z})^{n-4}}{2!} - \ldots$$

The above expression defines the *normalized Hermite polynomials*, i.e.

(3) $$H_n(\bar{z}) = (2\bar{z})^n - \frac{n(n-1)(2\bar{z})^{n-2}}{1!} + \frac{n(n-1)(n-2)(n-3)(2\bar{z})^{n-4}}{2!} - \ldots$$

$[H_0(\bar{z}) = 1;\ H_1(\bar{z}) = 2\bar{z};\ H_2(\bar{z}) = 4\bar{z}^2 - 2;$ etc.]

(B) For n = odd integer:

Choosing $a_0 = 0,\ a_1 \neq 0$, we have $a_2 = a_4 = a_6 = \ldots = 0$, and our solution will be a polynomial of odd powers in $\bar{z}$ up to $a_n\bar{z}^n$. Thus,

$$f(\bar{z}) = a_1\bar{z} + a_3\bar{z}^3 + a_5\bar{z}^5 + \ldots + a_n\bar{z}^n,$$

where

$$a_3 = \frac{-2(\lambda-1)a_1}{3!}. \qquad a_5 = \frac{(-2)^2(\lambda-1)(\lambda-3)a_1}{5!}.$$

$$a_7 = \frac{(-2)^3(\lambda-1)(\lambda-3)(\lambda-5)a_1}{7!}.$$

$$\vdots$$

$$a_n = \frac{(-2)^{(n-1)/2}(\lambda-1)(\lambda-3)\ldots(\lambda-n+2)a_1}{n!}.$$

7.3.1 Hermite Polynomials and the Harmonic Oscillator

Consider the 1-dimensional harmonic oscillator, where the potential (V) and total energy (E) are expressed, respectively, by:

$$V(x) = \frac{kx^2}{2}. \qquad E(x) = \frac{p^2}{2m} + V(x).$$

Replacing p by $\frac{\hbar}{i}\frac{\partial}{\partial x}$ and the E by $i\hbar\frac{\partial}{\partial t}$, we have the Schrödinger equation:

$$(1) \qquad i\hbar\frac{\partial\Psi}{\partial t} = -\frac{\hbar^2}{2m}\frac{\partial^2\Psi}{\partial x^2} + \frac{kx^2\Psi}{2} \quad \Rightarrow \quad E\Psi = \left(-\frac{\hbar^2}{2m}\nabla^2 + V(x)\right)\Psi.$$

In Quantum Mechanics one specifies $\int_{-\infty}^{+\infty}\Psi^2 dx$ to be finite, implying that there is some finite probability of finding the particle. This is enough to cast out certain solutions for Ψ. Classically the frequency of oscillation is given by $\omega^2 = k/m$ and the particle can be situated anyplace on the energy curve (Fig. 7-4).

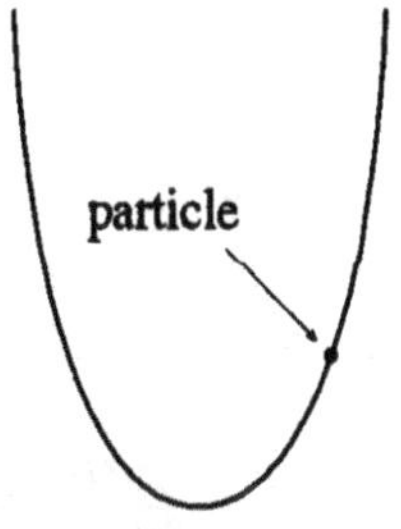

Fig. 7-4

Multiplying the Schrödinger equation by $2/\hbar\omega$ and setting $\omega^2 = k/m$:

$$\frac{2}{\hbar\omega}E\Psi = \frac{2}{\hbar\omega}\left(-\frac{\hbar^2}{2m}\frac{d^2\Psi}{dx^2} + \frac{kx^2\Psi}{2}\right) = -\frac{\hbar}{(mk)^{1/2}}\frac{d^2\Psi}{dx^2} + \frac{(mk)^{1/2}}{\hbar}x^2\Psi.$$

If we now let $\bar{z}^2 = \frac{(mk)^{1/2}}{\hbar}x^2$, we have

$$(2) \qquad \frac{d^2\Psi}{d\bar{z}^2} - \bar{z}^2\Psi = -\frac{2E}{\hbar\omega}\Psi.$$

Note the similarity between Schrödinger's equation (2) and the Hermite equation $\frac{d^2f(\bar{z})}{d\bar{z}^2} - 2\bar{z}\frac{df(\bar{z})}{d\bar{z}} + 2\lambda f(\bar{z}) = 0$. In section 7.2 we have seen how to remove the $\frac{df(\bar{z})}{d\bar{z}}$ term from a 2nd-order differential equation, and we proceed to do so now.

Setting $f(\bar{z}) = \Psi(\bar{z})e^{\bar{z}^2/2}$,

$$\frac{df(\bar{z})}{d\bar{z}} = \frac{d\Psi(\bar{z})}{d\bar{z}}e^{\bar{z}^2/2} + \bar{z}\Psi(\bar{z})e^{\bar{z}^2/2} = \bar{z}f(\bar{z}) + \frac{d\Psi(\bar{z})}{d\bar{z}}e^{\bar{z}^2/2}.$$

$$\frac{d^2f(\bar{z})}{d\bar{z}^2} = f(\bar{z}) + \bar{z}\frac{df(\bar{z})}{d\bar{z}} + \frac{d^2\Psi(\bar{z})}{d\bar{z}^2}e^{\bar{z}^2/2} + \bar{z}e^{\bar{z}^2/2}\frac{d\Psi(\bar{z})}{d\bar{z}}$$

$$= f(\bar{z}) + \bar{z}\left(\bar{z}f(\bar{z}) + e^{\bar{z}^2/2}\frac{d\Psi(\bar{z})}{d\bar{z}}\right) + \frac{d^2\Psi(\bar{z})}{d\bar{z}^2}e^{\bar{z}^2/2} + \bar{z}e^{\bar{z}^2/2}\frac{d\Psi(\bar{z})}{d\bar{z}}$$

$$= f(\bar{z}) + \bar{z}^2f(\bar{z}) + 2\bar{z}e^{\bar{z}^2/2}\frac{d\Psi(\bar{z})}{d\bar{z}} + \frac{d^2\Psi(\bar{z})}{d\bar{z}^2}e^{\bar{z}^2/2}.$$

$$-2\bar{z}\frac{df(\bar{z})}{d\bar{z}} = -2\bar{z}\left(\bar{z}f(\bar{z}) + \frac{d\Psi(\bar{z})}{d\bar{z}}e^{\bar{z}^2/2}\right) = -2\bar{z}^2f(\bar{z}) - 2\bar{z}\frac{d\Psi(\bar{z})}{d\bar{z}}e^{\bar{z}^2/2}.$$

Collecting:

$$0 = \frac{d^2f(\bar{z})}{d\bar{z}^2} - 2\bar{z}\frac{df(\bar{z})}{d\bar{z}} + 2\lambda f(\bar{z})$$

$$= f(\bar{z}) + \bar{z}^2f(\bar{z}) + 2\bar{z}e^{\bar{z}^2/2}\frac{d\Psi(\bar{z})}{d\bar{z}} + \frac{d^2\Psi(\bar{z})}{d\bar{z}^2}e^{\bar{z}^2/2}$$

$$- 2\bar{z}^2f(\bar{z}) - 2\bar{z}\frac{d\Psi(\bar{z})}{d\bar{z}}e^{\bar{z}^2/2} + 2\lambda f(\bar{z})$$

$$= f(\bar{z}) - \bar{z}^2f(\bar{z}) + 2\lambda f(\bar{z}) + \frac{d^2\Psi(\bar{z})}{d\bar{z}^2}e^{\bar{z}^2/2}$$

$$= e^{\bar{z}^2/2}\left(\Psi(\bar{z}) - \bar{z}^2\Psi(\bar{z}) + 2\lambda\Psi(\bar{z}) + \frac{d^2\Psi(\bar{z})}{d\bar{z}^2}\right).$$

Thus,

$$\text{(3)} \qquad \frac{d^2\Psi(\bar{z})}{d\bar{z}^2} - \bar{z}^2\Psi(\bar{z}) + \Psi(\bar{z})(2\lambda + 1) = 0.$$

Comparing equations (3) and (2), we see that the Hermite equation (3) is the Schrödinger equation (2) for the 1-dimensional harmonic oscillator, where $2E/\hbar\omega = 2\lambda + 1$. For $\lambda = n$ (positive integer), $2E/\hbar\omega = 2n + 1$, and normalized solutions are given by $\Psi_n(\bar{z}) = H_n(\bar{z})e^{-\bar{z}^2/2}$ (Fig. 7-5).

[We want the probability integral $\int_{-\infty}^{+\infty} |\Psi(\bar{z})|^2 d\bar{z}$ to be a finite quantity, and it will be so only if $\lambda = n$. If one uses any solution other than the polynomial one for λ, and there is an essential singularity at infinity, the integral will blow up.]

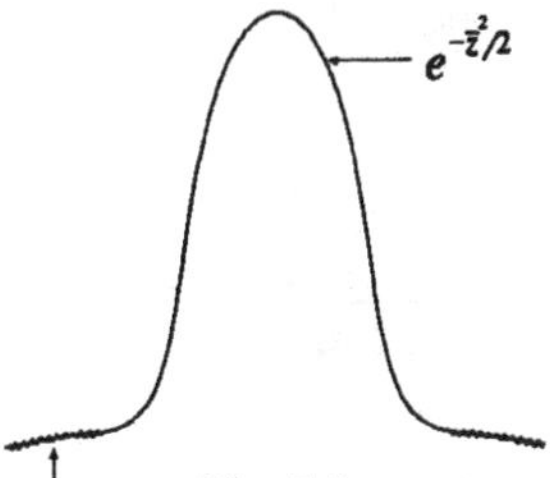

Fig. 7-5

$H_n(\bar{z})$ (determines the wiggles)

Since $E_n = \hbar\omega(n + 1/2)$, where $n \to 0$ to ∞, only specified values of E are allowed, and the possible occupied states for the particle are as shown in Fig. 7-6.

Why do we normalize $H_n(\bar{z})$?
Because if $H_n(\bar{z})$ is expressed as

$$(2\bar{z})^2 - \frac{n(n-1)(2\bar{z})^{n-2}}{1!} + \frac{n(n-1)(n-2)(n-3)(2\bar{z})^{n-4}}{2!} - \ldots,$$

then it can be regarded as a contour integral, as we now proceed to demonstrate.

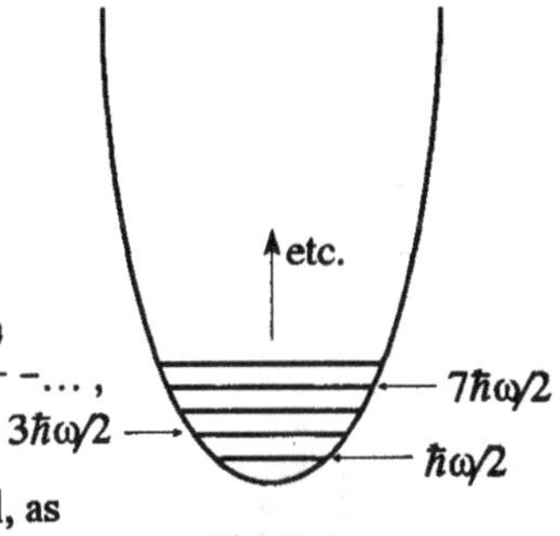

Fig. 7-6

7.3.2 Hermite Polynomial Theorems

(In what follows, we will drop the complex conjugate sign ($\bar{\ }$). It will be obvious from the content whether the variable is to be regarded as complex or not.)

Theorem:

$$(1) \qquad H_n(z) = \frac{n!}{2\pi i}\oint \frac{e^{-t^2+2zt}}{t^{n+1}}\,dt,$$

where $\oint$ is taken about a circle surrounding $t = 0$ in the complex t plane.

Proof: For the Hermite equation $\frac{d^2f(z)}{dz^2} - 2z\frac{df(z)}{dz} + 2\lambda f(z) = 0$, where $\lambda = n$ (positive integer), solutions are given by

$$f(z) = H_n(z) = (2z)^2 - \frac{n(n-1)(2z)^{n-2}}{1!} + \frac{n(n-1)(n-2)(n-3)(2z)^{n-4}}{2!} - \cdots.$$

From $f(z) = \frac{1}{2\pi i}\oint \frac{e^{-t^2+2zt}}{t^{n+1}}dt$, we have $\frac{df(z)}{dz} = \frac{1}{2\pi i}\oint \frac{2e^{-t^2+2zt}}{t^n}dt$.

$$\frac{d^2f(z)}{dz^2} = \frac{1}{2\pi i}\oint \frac{4e^{-t^2+2zt}}{t^{n-1}}dt. \qquad z\frac{df(z)}{dz} = \frac{1}{2\pi i}\oint \frac{2ze^{-t^2+2zt}}{t^n}dt.$$

For $\lambda = n$,

$$\frac{d^2f(z)}{dz^2} - 2z\frac{df(z)}{dz} + 2nf(z) = \frac{1}{2\pi i}\oint \frac{(2t^2 - 4zt + 2n)e^{-t^2+2zt}}{t^{n+1}}dt$$

$$= -\frac{2}{2\pi i}\oint \left[\frac{-n}{t^{n+1}} + \frac{2zt - 2t^2}{t^{n+1}}\right]e^{-t^2+2zt}dt$$

$$= -\frac{2}{2\pi i}\oint \frac{\partial}{\partial t}\left[\frac{e^{-t^2+2zt}}{t^n}\right]dt = 0.$$

The expression vanishes because the integrand is a perfect differential and the contour integral of a perfect differential is zero.

Hence, $\frac{1}{2\pi i}\oint \frac{e^{-t^2+2zt}}{t^{n+1}}dt$ is a solution of the Hermite equation, and since it is a polynomial in z, it is identical with $H_n(z)$ except for a constant multiplier. To determine this multiplier, we evaluate $H_n(0)$ and $f_n(0)$ and compare.

$$H_n(0) = a_0 = \frac{(-1)^{n/2}n!}{(n/2)!}. \qquad f(0) = \frac{1}{2\pi i}\oint \frac{e^{-t^2}}{t^{n+1}}dt = \frac{1}{2\pi i}2\pi i\frac{(-1)^{n/2}}{(n/2)!}.$$

Thus the constant multiplier is $n!$ so that $H_n(z) = \frac{n!}{2\pi i}\oint \frac{e^{-t^2+2zt}}{t^{n+1}}dt.$

Theorem:

$$(2) \qquad e^{-t^2+2zt} = \sum_{n=0}^{\infty} \frac{t^n}{n!} H_n(z),$$

and defines the *generating function* of the Hermite polynomials.

Proof: From theorem (1), $\frac{2\pi i H_n(z)}{n!} = \oint \frac{e^{-t^2+2zt}}{t^{n+1}} dt$, and from the theorem of residues, it is obvious that $H_n(z)/n!$ must be the coefficient of t^n in the power series expansion of e^{-t^2+2zt}. Hence, $e^{-t^2+2zt} = \sum_{n=0}^{\infty} \frac{t^n H_n(z)}{n!}$.

Theorem:

$$(3) \qquad \int_{-\infty}^{+\infty} e^{-z^2} H_n(z) H_m(z) dz = \delta_{nm} 2^n n! \sqrt{\pi}.$$

Proof: From the generating function (2) above,

$$e^{-t^2+2zt} = \sum_{n=0}^{\infty} \frac{t^n H_n(z)}{n!} \quad \text{and} \quad e^{-t^2+2zt} = \sum_{m=0}^{\infty} \frac{t^m H_m(z)}{m!},$$

we construct the following:

$$\sum_{n,m=0}^{\infty} \frac{s^m t^n}{n! m!} \int_{-\infty}^{+\infty} e^{-z^2} H_n(z) H_m(z) dz = \int_{-\infty}^{+\infty} e^{-t^2-s^2+2tz+2sz-z^2} dz$$

$$= e^{2st} \int_{-\infty}^{+\infty} e^{-(z-s-t)^2} dz = \sqrt{\pi} \sum_{n=0}^{\infty} \frac{2st^n}{n!}.$$

The right-hand side of the equation tells us that powers of s and t are equal. Two power series must agree term by term if they are to be equal. Hence, comparing coefficients of $(st)^n$ on both sides of the above equation, we have

$$\int_{-\infty}^{+\infty} e^{-z^2} H_n(z) H_m(z) dz = n! \sqrt{\pi} 2^n \delta_{nm}.$$

The reason for the *clumsy* e^{-z^2} term inside the integral derives from the nature of the solution of the Schrödinger equation for the 1-dimensional harmonic oscillator. By virtue of theorem (3) above, the normalized solutions of the 1-dimensional harmonic oscillator will be

$$(4) \qquad \Psi_n(z) = \left[\frac{1}{2^n n!\sqrt{\pi}}\right]^{1/2} e^{-z^2/2} H_n(z).$$

In other words,

$$\int_{-\infty}^{+\infty} \Psi_n(z)^2 dz = \int_{-\infty}^{+\infty} \frac{e^{-z^2} H_n^2(z)}{2^n n!\sqrt{\pi}} dz = 1 \quad \rightarrow \quad \int_{-\infty}^{+\infty} \Psi_n(z)\Psi_m(z) dz = \delta_{nm}.$$

The Ψ_n of equation (4) above are the eigenfunctions of the Schrödinger equation for the 1-dimensional harmonic oscillator, i.e.

$$\left(\frac{d^2}{dz^2} - z^2\right)\Psi_n(z) = -\frac{2E_n}{\hbar\omega}\Psi_n(z),$$

with eigenvalues $E_n = (n + 1/2)\hbar\omega$.

Theorem:

$$(5) \qquad \frac{dH_n(z)}{dz} = 2nH_{n-1}(z),$$

which enables one to obtain lower terms of $H_n(z)$ by differentiating higher ones.

Proof: From the contour integral definition of $H_n(z)$, equation (1), we have:

$$\frac{dH_n(z)}{dz} = \frac{n!}{2\pi i}\oint \frac{2e^{-t^2+2zt}}{t^n} dt = \frac{2n(n-1)!}{2\pi i}\oint \frac{e^{-t^2+2zt}}{t^n} dt = 2nH_{n-1}(z).$$

[Note the ease with which the $H_n(z)$ can be manipulated once it can be expressed in the form of a contour integral.]

Theorem:

$$(6) \qquad H_{n+1}(z) - 2zH_n(z) + 2nH_{n-1} = 0.$$

Proof:

$$H_{n+1}(z) - 2zH_n(z) + 2nH_{n-1} = \frac{n!}{2\pi i}\oint \left[\frac{n+1}{t^{n+2}} - \frac{2z}{t^{n+1}} + \frac{2}{t^n}\right] e^{-t^2+2zt} dt.$$

The integrand represents a perfect differential, so that

$$\frac{\partial}{\partial t}\left[\frac{e^{-t^2+2zt}}{t^{n+1}}\right] = -\left[\frac{(n+1)}{t^{n+2}} - \frac{2z}{t^{n+1}} + \frac{2}{t^n}\right]e^{-t^2+2zt} \Rightarrow \oint \frac{\partial}{\partial t}\left[\frac{e^{-t^2+2zt}}{t^{n+1}}\right]dt = 0.$$

7.4 Behavior of Solutions Near Singular Points

Consider the equation

$$(1) \qquad \frac{d^2f(z)}{dz^2} + p(z)\frac{df(z)}{dz} + q(z)f(z) = 0,$$

where $p(z)$ and $q(z)$ possess isolated singularities. Let z_o be an isolated singular point. Do the solutions become singular, and if so, what *kind* of singularity do they then have? Or are the solutions analytic?

Assume that at $z = z_o$, $p(z)$ has a pole of order m and $q(z)$ has a pole of order n, where m and n cannot both be zero, otherwise z_o would then be an ordinary point.

Let $\xi = z - z_o$. By the definition of poles of order m and n, respectively,

$$p(z) = \frac{p_{-m}}{\xi^m} + \frac{p_{-m+1}}{\xi^{m-1}} + \ldots \qquad q(z) = \frac{q_{-n}}{\xi^n} + \frac{q_{-n+1}}{\xi^{n-1}} + \ldots$$

Assume that near z_o, $f(z) = \xi^s \sum_{k=0}^{\infty} a_k \xi^k$, where s is any number.

Let p have at most a simple pole at $z = z_o$ (i.e. let $m \le 1$), and let $q(z)$ have at most a pole of second order at $z = z_o$ (i.e. let $n \le 2$). Then

$$p(z) = \frac{p_{-1}(z)}{\xi} + p_o(z) + p_1(z)\xi + \ldots = \sum_{m=0}^{\infty} p_{m-1}(z)\xi^{m-1}.$$

$$q(z) = \frac{q_{-2}(z)}{\xi^2} + \frac{q_{-1}(z)}{\xi} + + q_o(z) + q_1(z)\xi + \ldots = \sum_{m=0}^{\infty} q_{m-2}(z)\xi^{m-2}.$$

From $f(z) = \sum_{k=0}^{\infty} a_k \xi^{s+k} \Rightarrow \frac{df(z)}{dz} = \sum_{k=0}^{\infty} (s+k)a_k\xi^{s+k-1}$.

$$\frac{d^2f(z)}{dz^2} = \sum_{k=0}^{\infty}(s+k)(s+k-1)a_k\xi^{s+k-2}.$$

$$p(z)\frac{df(z)}{dz} = \sum_{k=0}^{\infty}(s+k)a_k\xi^{s+k-1}\sum_{m=0}^{\infty}p_{m-1}\xi^{m-1}$$

$$= \sum_{k=0}^{\infty}(s+k)a_k\sum_{m=0}^{\infty}p_{m-1}\xi^{s+k+m-2}$$

$$= \sum_{k=0}^{\infty}(s+k-m)a_{k-m}\sum_{m=0}^{k}p_{m-1}\xi^{s+k-2}.$$

$$q(z)f(z) = \sum_{k=0}^{\infty}a_k\xi^{s+k}\sum_{m=0}^{\infty}q_{m-2}\xi^{m-2} = \sum_{k=0}^{\infty}a_{k-m}\sum_{m=0}^{k}q_{m-2}\xi^{s+k-2}.$$

Substituting into equation (1):

$$\sum_{k=0}^{\infty}\xi^{s+k-2}\left\{a_k(s+k)(s+k-1)+\sum_{m=0}^{k}\left[(s+k-m)a_{k-m}p_{m-1}+q_{m-2}a_{k-m}\right]\right\} = 0,$$

from which it follows that

$$a_k(s+k)(s+k-1)+\sum_{m=0}^{k}(s+k-m)a_{k-m}p_{m-1}+\sum_{m=0}^{k}q_{m-2}a_{k-m} = 0.$$

For $k = 0$, we have the indicial equation:

(2) $\quad s(s-1)+sp_{-1}+q_{-2} = 0.$

For $k \geq 1$:

(3) $$a_k\left[(s+k)(s+k-1)+(s+k)p_{-1}+q_{-2}\right]$$

$$= -\sum_{m=1}^{k}a_{k-m}\left[p_{m-1}(s+k-m)+q_{m-2}\right].$$

Given an arbitrary a_o, one can solve for all the a_i, yielding $f(z) = \xi^s \sum_{n=o}^{\infty} a_n \xi^n$, i.e. for each value of s corresponding to solutions of the indicial equation, we get $a_o, a_1, a_2, \ldots$etc, and therefore $f_1(z)$ and $f_2(z)$.

For the indicial equation, $F(s) \equiv s(s-1) + sp_{-1} + q_{-2} = 0$, having roots s_1 and s_2, $f_1(s) \equiv F(s_1) = 0$ and $f_2(s) \equiv F(s_2) = 0$.

Consider $F(s+n) \equiv (s+n)(s+n-1) + (s+n)p_{-1} + q_{-2}$. If $F(s) = 0$, enabling us to solve for s_1 and s_2, while $F(s+n) \neq 0$, then given an arbitrary a_o, we can solve for the a_n. If, however, $F(s+n) = 0$, which will be true when $s_1 - s_2 = t$ (a positive integer), then for $s = s_1$, you get $f_1(z)$ but for $s = s_2$, you also get $f_1(z)$. We seek the second solution $f_2(z)$ in an explicit form, when $s_1 - s_2 = t$ (a positive integer).

Let $f(z) = f_1(z)\Phi(z)$, where $f_1(z)$ corresponds to s_1, the larger of the two roots.

$$\frac{df(z)}{dz} = \Phi(z)\frac{df_1(z)}{dz} + f_1(z)\frac{d\Phi(z)}{dz}.$$

$$\frac{d^2f(z)}{dz^2} = 2\frac{d\Phi(z)}{dz}\frac{df_1(z)}{dz} + \Phi(z)\frac{d^2f_1(z)}{dz^2} + f_1(z)\frac{d^2\Phi(z)}{dz^2}.$$

$$p(z)\frac{df(z)}{dz} = p(z)\Phi(z)\frac{df_1(z)}{dz} + p(z)f_1(z)\frac{d\Phi(z)}{dz}.$$

$$q(z)f(z) = q(z)f_1(z)\Phi(z).$$

Substituting into equation (1):

$$\Phi(z)\frac{d^2f_1(z)}{dz^2} + 2\frac{d\Phi(z)}{dz}\frac{df_1(z)}{dz} + f_1(z)\frac{d^2\Phi(z)}{dz^2} + p(z)\Phi(z)\frac{df_1(z)}{dz}$$

$$+ pf_1(z)\frac{d\Phi(z)}{dz} + q(z)f_1(z)\Phi(z) = 0.$$

$$\Phi(z)\left[\frac{d^2f_1(z)}{dz^2} + p(z)\frac{df_1(z)}{dz} + q(z)f_1(z)\right] + 2\frac{d\Phi(z)}{dz}\frac{df_1(z)}{dz}$$

$$+ f_1(z)\frac{d^2\Phi(z)}{dz^2} + p(z)f_1(z)\frac{d\Phi(z)}{dz} = 0.$$

Since the bracketed expression satisfies equation (1) for $f_1(z)$, we have

$$\frac{d^2\Phi(z)}{dz^2} + \left[\frac{2}{f_1(z)}\frac{df_1(z)}{dz} + p(z)\right]\frac{d\Phi(z)}{dz} = 0.$$

$$\Downarrow$$

$$\frac{\dfrac{d^2\Phi(z)}{dz^2}}{\dfrac{d\Phi(z)}{dz}} + \frac{2}{f_1(z)}\frac{df_1(z)}{dz} + p(z) = 0.$$

This last equation can be written as

$$\frac{d}{dz}\ln\frac{d\Phi(z)}{dz} + 2\frac{d}{dz}\ln f_1(z) + \frac{d}{dz}\int^z p(z')dz' = 0,$$

where, since $p(z)$ is analytic everywhere except at point z_o, we can use any path of integration that does not include z_o. Rewriting and integrating:

$$\ln\frac{d\Phi(z)}{dz} + 2\ln f_1(z) + \int^z p(z')dz' = \ln A,$$

$$\Downarrow$$

$$\frac{d\Phi(z)}{dz} = \frac{A}{f_1(z)^2}\exp\left\{-\int^z p(z')dz'\right\}.$$

Substituting $f_1(z) = (z - z_o)^{s_1}\sum_{n=0}^{\infty} a_n(z - z_o)^n$ and $p(z) = \sum_{m=0}^{\infty} p_{m-1}(z)(z - z_o)^{m-1}$ into the expression for $d\Phi(z)/dz$:

$$\frac{d\Phi(z)}{dz} = \frac{A\exp\left[-p_{-1}\ln(z - z_o) - \sum_{n=0}^{\infty} b_n(z - z_o)^n\right]}{(z - z_o)^{2s_1}\left[\sum_{n=0}^{\infty} a_n(z - z_o)^n\right]^2}.$$

Defining

$$h(z) = \frac{A \exp\left[-\sum_{n=0}^{\infty} b_n(z-z_o)^n\right]}{\left[\sum_{n=0}^{\infty} a_n(z-z_o)^n\right]^2},$$

we have

$$\frac{d\Phi(z)}{dz} = \frac{h(z)}{(z-z_o)^{2s_1+p_{-1}}}.$$

Since $h(z)$ is completely analytic at $z = z_o$, let us consider the term $2s_1 + p_{-1}$. The indicial equation (2) states that $F(s_1) = 0$ and $F(s_2) = 0$, or,

$$s_1(s_1 - 1) + p_{-1}s_1 + q_{-2} = 0 \text{ and } s_2(s_2 - 1) + p_{-1}s_2 + q_{-2} = 0.$$

Subtracting, $s_1^2 - s_2^2 + (p_{-1} - 1)(s_1 - s_2) = 0$. For $s_1 \neq s_2$, and factoring out $s_1 - s_2$, we have $[s_1 + s_2 + p_{-1} = 0]$. Substituting $s_1 = s_2 + t$ (positive integer), we are left with $[2s_1 + p_{-1} = 1 + t$ (positive integer)$]$. Thus,

$$\frac{d\Phi(z)}{dz} = \frac{h(z)}{(z-z_o)^{1+t}}.$$

(Since $p(z)$ is known, the b_n are known, and hence $h(z)$ is known.) Since $h(z)$ is completely regular, it can be expanded in a Taylor series, i.e.

$$h(z) = \sum_{n=0}^{\infty} h_n(z-z_o)^n,$$

so that

$$\frac{d\Phi(z)}{dz} = \frac{\sum_{n=0}^{\infty} h_n(z-z_o)^n}{(z-z_o)^{1+t}}$$

$$= \frac{h_o + h_1(z-z_o) + h_2(z-z_o)^2 \ldots + h_t(z-z_o)^t + h_{t+1}(z-z_o)^{t+1} + \ldots}{(z-z_o)^{1+t}}.$$

$$= \frac{h_o}{(z-z_o)^{1+t}} + \frac{h_1}{(z-z_o)^t} + \ldots + \frac{h_t}{(z-z_o)} + h_{t+1} + h_{t+2}(z-z_o) + \ldots$$

$$f_1(z) = \sum_{n=o}^{\infty} a_n(z-z_o)^{s_1+n} \text{ and } f(z) = f_1(z)\Phi(z) = f_1(z)\left[\Phi_o + \int^{z} \frac{d\Phi(z')}{dz'}dz'\right],$$

so that

$$f(z) = f_1(z)\left\{\Phi_o - \frac{h_o}{t(z-z_o)^t} - \frac{h_1}{(t-1)(z-z_o)^{t-1}} + \ldots\right\}$$

$$+ f_1(z)\left\{h_t \ln(z-z_o) + h_{t+1}(z-z_o) + h_{t+2}(z-z_o)^2 + \ldots\right\}$$

$$= f_1(z)\left\{\Phi_o - \frac{h_o}{t(z-z_o)^t} - \frac{h_1}{(t-1)(z-z_o)^{t-1}} + \ldots\right\}$$

$$+ f_1(z)h_t \ln(z-z_o) + f_1(z)\left\{h_{t+1}(z-z_o) + h_{t+2}(z-z_o)^2 + \ldots\right\}$$

$$\equiv A + B + C.$$

$$A = \sum_{n=o}^{\infty} a_n(z-z_o)^{s_1+n}\left\{\Phi_o - \frac{h_o}{t(z-z_o)^t} - \frac{h_1}{(t-1)(z-z_o)^{t-1}} + \ldots\right\}$$

$$= a_o(z-z_o)^{s_1}\Phi_o - \frac{a_o h_o(z-z_o)^{s_1-t}}{t} - \frac{a_o h_1(z-z_o)^{s_1-t+1}}{t-1} + \ldots$$

$$+ a_1(z-z_o)^{s_1+1}\Phi_o - \frac{a_1 h_o(z-z_o)^{s_1-t+1}}{t} - \frac{a_1 h_1(z-z_o)^{s_1-t+2}}{t-1} + \ldots$$

Setting $s_2 = s_1 - t$:

$$A = \sum_{n=o}^{\infty} a_n\Phi_o(z-z_o)^{s_1+n} + \sum_{n=o}^{\infty} d_n(z-z_o)^{s_2+n}.$$

$$C = \sum_{n=o}^{\infty} a_n(z-z_o)^{s_1+n}\left\{h_{t+1}(z-z_o) + h_{t+2}(z-z_o)^2 + \ldots\right\}.$$

$$= \sum_{n=o}^{\infty} \omega_n(z-z_o)^{s_1+n+1},$$

but represents no new contribution to $f(z)$.

Combining,

$$f_2(z) = f_1(z)\Phi(z) = \sum_{n=0}^{\infty} k_n(z-z_o)^{s_2+n} + h_t f_1(z)\ln(z-z_o).$$

If

$h_t \neq 0$, there is a branch point at $z = z_o$.

$h_t = 0$ and s_2 is integer-valued, $f_2(z)$ *could* be analytic.

$h_t = 0$ and both s_1 and s_2 are positive integers, $f_2(z)$ *is* analytic.

Summarizing:

If $p(z) = \dfrac{p_{-1}}{(z-z_o)} + p_o + \ldots$ and $q(z) = \dfrac{p_{-2}}{(z-z_o)^2} + \dfrac{q_{-1}}{(z-z_o)} + q_o + \ldots$, then z_o is a regular singular point, and the functions $f_1(z)$ and $f_2(z)$ have at most a branch point, since $s(s-1) + s(p_{-1}) + q_{-2} = 0$ with roots s_1 and s_2.

(a) For $s_1 \neq s_2$ (positive integer), then $f(z) = (z-z_o)^s \sum_{n=0}^{\infty} a_n(z-z_o)^n$ yields two distinct solutions $f_1(z)$ and $f_2(z)$.

(b) For $s_1 - s_2 = t$ (positive integer), we have $f_1(z) = (z-z_o)^s \sum_{n=0}^{\infty} a_n(z-z_o)^n$ for s_1 the higher of the two roots, and $f_2(z) = (z-z_o)^{s_2} \sum_{n=0}^{\infty} k_n(z-z_o)^n + h_t f_1(z)\ln(z-z_o)$, where if $h_t = 0$ and s_1 and s_2 are positive integers, then $f_1(z)$ and $f_2(z)$ are both analytic.

Behavior of solutions at infinity: Consider again

(1) $$\frac{d^2f(z)}{dz^2} + p(z)\frac{df(z)}{dz} + q(z)f(z) = 0.$$

Setting $z = 1/z'$ we study the behavior of $f(z)$ at infinity by studying the behavior of $f(1/z')$ at $z' = 0$.

If $f(z) = a + \dfrac{b}{z} + \dfrac{c}{z^2} \ldots \;\rightarrow\; f(1/z') = a + bz' + cz'^2 + \ldots$, and at $z' = 0$, $f(1/z')$ is regular; hence $f(z)$ is regular at $z = \infty$.

If $f(z) = az + bz^2 \;\rightarrow\; f(1/z') = \dfrac{a}{z'} + \dfrac{b}{(z')^2}$, and $f(1/z')$ has a pole of order 2 at $z' = 0$; hence $f(z)$ has a pole of order 2 at $z = \infty$.

We write differential equation (1) as a function of z', setting $z = 1/z'$, noting that $\frac{d}{dz} = \frac{d}{dz'}\frac{dz'}{dz} = -\frac{1}{z^2}\frac{d}{dz'} = -z'^2\frac{d}{dz'}$. Thus,

$$\frac{d^2f(z)}{dz^2} = -z'^2\frac{d^2f(z)(-1)}{dz'^2\ z^2} + \frac{df(z)2z'}{dz'\ z^2} = z'^4\frac{d^2f(z)}{dz'^2} + 2z'^3\frac{df(z)}{dz'}.$$

$$p(z)\frac{df(z)}{dz} = -p(z)z'^2\frac{df(z)}{dz'}.$$

Collecting terms and substituting into (1):

$$z'^4\frac{d^2f(z)}{dz'^2} + 2z'^3\frac{df(z)}{dz'} - p(z)z'^2\frac{df(z)}{dz'} + q(z)f(z) = 0.$$

$$\Downarrow$$

$$\frac{d^2f(z)}{dz'^2} + \left[\frac{2}{z'} - \frac{p(z)}{z'^2}\right]\frac{df(z)}{dz'} + \frac{q(z)f(z)}{z'^4} = 0.$$

The terms that we now have to investigate are

$$p'(z') = \frac{2}{z'} - \frac{p(z)}{z'^2} = 2z - p(z)z^2.$$

$$q'(z') = \frac{q(z)}{z'^4} = q(z)z^4.$$

Case (a): If $2z - p(z)z^2$ and $q(z)z^4$ are analytic at $z = \infty$, then $f(z)$ is analytic at infinity, and infinity is an ordinary point.

Case (b): If $\frac{2}{z'} - \frac{p(z)}{z'^2}$ has a pole of order 1 at $z' = 0$ (i.e. if $-p(z)/z'$ is analytic at $z' = 0$) and $q(z)/z'^4$ has a pole of order 2 at $z' = 0$ (i.e. if $q(z)/z'^2$ is analytic at $z' = 0$), then $f(z)$ is analytic at infinity (or has at most a branch point), and infinity is a regular singular point.

If case (b) is not true, then infinity is an irregular singular point, and we have an essential singularity at infinity.

7.5 Bessel Functions

Bessel's equation is given by

(1) $$\frac{d^2f(z)}{dz^2} + \frac{1}{z}\frac{df(z)}{dz} + \left(1 - \frac{\lambda^2}{z^2}\right)f(z) = 0.$$

Note that $p(z) = 1/z$ has a simple pole at $z = 0$, while $q(z) = 1 - (\lambda/z)^2$ has a pole of second order at $z = 0$. Hence, $z = 0$ is a regular singular point for $f(z)$.

Note also that $p(z)z$ is analytic at $z = \infty$, but that $q(z)z^2$ is *not* analytic at $z = \infty$. Hence, $z = \infty$ is an irregular singular point, and both solutions have an essential singularity at $z = \infty$.

Behavior near z = 0:

From $p(z) = \frac{p_{-1}(z)}{z} + p_o + p_1 z + \ldots$ and $q(z) = \frac{q_{-2}(z)}{z^2} + \frac{q_{-1}(z)}{z} + q_o + \ldots$, we see that $p_{-1} = 1$ and $q_{-2} = -\lambda^2$. The indicial equation becomes

$$s(s-1) + s - \lambda^2 = 0 \quad \rightarrow \quad s^2 = \lambda^2 \quad \rightarrow \quad S = \pm\lambda.$$

Choosing $\lambda > 0$, $s_1 = +\lambda$ and $s_2 = -\lambda$, so that $s_1 - s_2 = 2\lambda$.

1) For $2\lambda \neq$ integer (i.e. λ is not an integer),

$$f(z) \rightarrow \begin{Bmatrix} z^{\lambda} \sum_{n=0}^{\infty} a_n z^n \\ z^{-\lambda} \sum_{n=0}^{\infty} b_n z^n \end{Bmatrix} \rightarrow \text{both solutions have branch points.}$$

2) For $\lambda =$ integer (i.e. 2λ is an integer),

$$f(z) \rightarrow \begin{Bmatrix} z^{\lambda} \sum_{n=0}^{\infty} a_n z^n \text{ (analytic)} \\ z^{-\lambda} \sum_{n=0}^{\infty} b_n z^n + h_t z^{\lambda} \sum_{n=0}^{\infty} a_n z^n \ln z \end{Bmatrix} \rightarrow \text{branch point if } h_t \neq 0.$$

3) For $\lambda = 1/2$ integer (i.e. 2λ is an integer),

$$f(z) \to \left\{ \begin{array}{l} z^{\lambda} \sum_{n=0}^{\infty} a_n z^n \text{ (branch point)} \\ z^{-\lambda} \sum_{n=0}^{\infty} b_n z^n + h_t z^{\lambda} \sum_{n=0}^{\infty} a_n z^n \ln z \end{array} \right\} \to \text{branch point even if } h_t \neq 0.$$

The Bessel equation occurs quite often in physics due to its intimate relationship with the Laplacian operator. Consider the wave equation

$$\nabla^2 \Phi(\mathbf{r}t) - \frac{\ddot{\Phi}(\mathbf{r}t)}{c^2} = 0.$$

Applying a Fourier transformation with time as parameter,

$$\Phi_t(\mathbf{r}) = \int_{-\infty}^{+\infty} \Psi_\omega(\mathbf{r}) e^{i\omega t} d\omega \quad \to \quad \Psi_\omega(\mathbf{r}) = \frac{1}{2\pi} \int_{-\infty}^{+\infty} \Phi_t(\mathbf{r}) e^{-i\omega t} dt.$$

$$\frac{\partial^2 \Phi_t(\mathbf{r})}{\partial t^2} = \int_{-\infty}^{+\infty} -\omega^2 \Psi_\omega(\mathbf{r}) e^{i\omega t} d\omega. \qquad \nabla_r^2 \Phi_t(\mathbf{r}) = \int_{-\infty}^{+\infty} \nabla_r^2 \Psi_\omega(\mathbf{r}) e^{i\omega t} d\omega.$$

Substituting into the wave equation:

$$\int_{-\infty}^{+\infty} e^{i\omega t} \left[\nabla_r^2 \Psi_\omega(\mathbf{r}) + \frac{\omega^2}{c^2} \Psi_\omega(\mathbf{r}) \right] d\omega = 0.$$

Since the $e^{i\omega t}$ represents an orthonormal set,

(2) $$\nabla^2 \Psi_\omega(\mathbf{r}) + k_o^2 \Psi_\omega(\mathbf{r}) = 0,$$

where $k_o^2 = \omega^2/c^2$. For $k_o = 0$, the wave equation reduces to Laplace's equation, while for $k_o \neq 0$, one has the time-independent Schrödinger equation.

(A) Make the following cylindrical coordinate transformation:

$$\Psi(\rho\varphi z) = \sum_{-\infty}^{+\infty} \int_{-\infty}^{+\infty} f_{m,k}(\rho) e^{im\varphi} e^{ikz} dk.$$

$$\nabla^2\Psi(\rho\varphi z) = \frac{1}{\rho}\frac{\partial}{\partial\rho}\left(\rho\frac{\partial(\Psi(\rho\varphi z))}{\partial\rho}\right) + \frac{1}{\rho^2}\frac{\partial^2\Psi(\rho\varphi z)}{\partial\varphi^2} + \frac{\partial^2\Psi(\rho\varphi z)}{\partial z^2}$$

$$= \sum_{-\infty}^{+\infty}\int_{-\infty}^{+\infty}\left[\frac{1}{\rho}\frac{\partial}{\partial\rho}\left(\rho\frac{\partial f(\rho)}{\partial\rho}\right) - \frac{m^2 f(\rho)}{\rho^2} - k^2 f(\rho)\right]e^{im\varphi}e^{ikz}dk.$$

$$k_o^2\Psi(\rho\varphi z) = \sum_{-\infty}^{+\infty}\int_{-\infty}^{+\infty} f(\rho)k_o^2 e^{im\varphi}e^{ikz}dk.$$

Combining and substituting into equation (2):

$$\sum_{-\infty}^{+\infty}\int_{-\infty}^{+\infty}\left[\frac{1}{\rho}\frac{d}{d\rho}\left(\rho\frac{df(\rho)}{d\rho}\right) - \frac{m^2 f(\rho)}{\rho^2} + (k_o^2 - k^2)f(\rho)\right]e^{im\varphi}e^{ikz}dk = 0.$$

Employing the orthonormality properties of the $e^{im\varphi}$, we have

$$\frac{1}{\rho}\frac{d}{d\rho}\left(\rho\frac{df(\rho)}{d\rho}\right) - \frac{m^2 f(\rho)}{\rho^2} + \kappa^2 f(\rho) = 0,$$

where $\kappa^2 \equiv k_o^2 - k^2$. Expanding this equation,

$$\frac{d^2f(\rho)}{d\rho^2} + \frac{1}{\rho}\frac{df(\rho)}{d\rho} + \left(\kappa^2 - \frac{m^2}{\rho^2}\right)f(\rho) = 0.$$

Setting $\kappa\rho \equiv z$, $\frac{df(\rho)}{d\rho} = \kappa\frac{df(\rho)}{dz}$ and $\frac{d^2f(\rho)}{d\rho^2} = \kappa^2\frac{d^2f(\rho)}{dz^2}$, so that

$$\kappa^2\frac{d^2f(\rho)}{dz^2} + \frac{\kappa^2}{z}\frac{df(\rho)}{dz} + \left(\kappa^2 - \frac{\kappa^2 m^2}{z^2}\right)f(\rho) = 0.$$

$$\Downarrow$$

(3) $$\frac{d^2f(\rho)}{dz^2} + \frac{1}{z}\frac{df(\rho)}{dz} + \left(1 - \frac{m^2}{z^2}\right)f(\rho) = 0,$$

which we immediately recognize as Bessel's equation with $\lambda = m$ (integer).

Thus,

$$f(z) \to \begin{cases} z^{\lambda} \sum_{n=0}^{\infty} a_n z^n \text{ (analytic)} \\ z^{-\lambda} \sum_{n=0}^{\infty} b_n z^n + h_t z^{\lambda} \sum_{n=0}^{\infty} a_n z^n \ln z \text{ (branch point)} \end{cases}$$

(B) Make the following spherical coordinate transformation:

$$\Psi(r\theta\varphi) = \sum_{lm_l} R_{lm_l}(r) Y_{lm_l}(\theta\varphi).$$

$$\nabla^2\Psi(r\theta\varphi) = \frac{1}{r^2}\frac{\partial}{\partial r}\left(r^2\frac{\partial\Psi(r\theta\varphi)}{\partial r}\right) + \frac{1}{r^2\sin\theta}\frac{\partial}{\partial\theta}\left(\sin\theta\frac{\partial\Psi(r\theta\varphi)}{\partial\theta}\right)$$

$$+ \frac{1}{r^2\sin^2\theta}\frac{\partial^2\Psi(r\theta\varphi)}{\partial\varphi^2}$$

$$= \sum_{lm_l}\left[\frac{1}{r^2}\frac{d}{dr}\left(r^2\frac{dR_{lm_l}(r)}{dr}\right)\right] Y_{lm_l}(\theta\varphi)$$

$$+ \sum_{lm_l}\left[\frac{1}{r^2\sin\theta}\frac{\partial}{\partial\theta}\left(\sin\theta\frac{\partial Y(\theta\varphi)}{\partial\theta}\right) + \frac{1}{r^2\sin^2\theta}\frac{\partial^2 Y(\theta\varphi)}{\partial\varphi^2}\right] R_{lm_l}(r).$$

$$k_o^2\Psi(r\theta\varphi) = \sum_{lm_l} k_o^2 R_{lm_l}(r) Y_{lm_l}(\theta\varphi).$$

Combining and subsituting into equation (2):

$$\sum_{lm_l} r^2\left[\frac{1}{r^2}\frac{d}{dr}\left(r^2\frac{dR_{lm_l}(r)}{dr}\right)\right] Y_{lm_l}(\theta\varphi)$$

$$+ \sum_{lm_l}\left[\frac{1}{\sin\theta}\frac{\partial}{\partial\theta}\left(\sin\theta\frac{\partial Y(\theta\varphi)}{\partial\theta}\right) + \frac{1}{\sin^2\theta}\frac{\partial^2 Y(\theta\varphi)}{\partial\varphi^2}\right] R_{lm_l}(r)$$

$$+ \sum_{lm_l} k_o^2 R_{lm_l}(r) Y_{lm_l}(\theta\varphi) = 0.$$

We know from Part I, Section 5.3.8, that

$$\left[\frac{1}{\sin\theta}\frac{\partial}{\partial\theta}\left(\sin\theta\frac{\partial}{\partial\theta}\right)+\frac{1}{\sin^2\theta}\frac{\partial^2}{\partial\varphi^2}\right]Y_{lm_l}(\theta\varphi) = -l(l+1)Y_{lm_l}(\theta\varphi).$$

$$\sum_{lm_l}\left[\frac{1}{r^2}\frac{d}{dr}\left(r^2\frac{dR(r)}{dr}\right)+k_o^2R(r)-\frac{l(l+1)}{r^2}R(r)\right]Y_{lm_l}(\theta\varphi).$$

Employing the orthonormality relationships for the $Y_{lm_l}(\theta\varphi)$:

(4) $$\frac{1}{r^2}\frac{d}{dr}\left(r^2\frac{dR(r)}{dr}\right)+k_o^2R(r)-\frac{l(l+1)}{r^2}R(r) = 0.$$

Setting $R(r)=\frac{f(r)}{\sqrt{r}}$,

$$\frac{dR(r)}{dr} = -\frac{f(r)}{2r^{3/2}}+\frac{1}{\sqrt{r}}\frac{df(r)}{dr}. \qquad r^2\frac{dR(r)}{dr} = -\frac{r^{1/2}f(r)}{2}+r^{3/2}\frac{df(r)}{dr}.$$

$$\frac{d}{dr}\left(r^2\frac{dR(r)}{dr}\right) = -\frac{f(r)}{4r^{1/2}}-\frac{r^{1/2}}{2}\frac{df(r)}{dr}+r^{3/2}\frac{d^2f(r)}{dr^2}+\frac{3r^{1/2}}{2}\frac{df(r)}{dr}$$

$$= -\frac{f(r)}{4r^{1/2}}+r^{3/2}\frac{d^2f(r)}{dr^2}+r^{1/2}\frac{df(r)}{dr}.$$

Collecting and substituting into equation (4):

$$-\frac{f(r)}{4r^{5/2}}+\frac{1}{r^{1/2}}\frac{d^2f(r)}{dr^2}+\frac{1}{r^{3/2}}\frac{df(r)}{dr}+\frac{k_o^2f(r)}{r^{1/2}}-\frac{l(l+1)f(r)}{r^{5/2}} = 0.$$

$$\Downarrow$$

(5) $$\frac{d^2f(r)}{dr^2}+\frac{1}{r}\frac{df(r)}{dr}+\left[k_o^2-\frac{(l+1/2)^2}{r^2}\right]f(r) = 0.$$

Setting $k_or = z$,

$$\frac{df(r)}{dr} = k_o\frac{df(z)}{dz} \text{ and } \frac{d^2f(r)}{dr^2} = k_o^2\frac{d^2f(z)}{dz^2}.$$

Substituting into equation (5):

$$(6)\qquad \frac{d^2f(z)}{dz^2}+\frac{1}{z}\frac{df(z)}{dz}+\left[1-\frac{(l+1/2)^2}{z^2}\right]f(z) = 0,$$

which we immediately recognize as Bessel's equation with $\lambda = l + 1/2$, i.e. half-integral λ. Thus,

$$f_1(z) = z^{\lambda}\sum_{n=0}^{\infty} a_n z^n \text{ (branch point)}$$

$$f_2(z) = z^{\lambda}\sum_{n=0}^{\infty} b_n z^n + \ldots \text{ (branch point)}$$

Summarizing:

The Fourier transform was used to remove the time dependence from the wave equation. Then, by the technique of separation of variables, one arrives at Bessel's equation.

(a) In cylindrical coordinates, integer values of λ arise from the m in $e^{im\varphi}$.
(b) In spherical coordinates, half-integer values of λ arise from the eigenvalues $(l + 1/2)$.

7.5.1 Bessel's Equation — Singularities

Since $p(z)$ has a simple pole at $z = 0$ and $q(z)$ has a pole of second order at $z = 0$, $z = 0$ is a regular singular point and $f_1(z)$ and $f_2(z)$ have at most branch points. In addition, since $q(z)z^2$ is not analytic at $z = \infty$, $z = \infty$ is an irregular singular point and both solutions have an essential singularity at $z = \infty$.

Expanding solutions of Bessel's equation about the point $z = 0$:

$$(1)\qquad f(z) = z^{\lambda}\sum_{n=0}^{\infty} a_n z^n,$$

where $\lambda = \pm|\lambda|$.

$$\frac{d^2f(z)}{dz^2} = \sum_{n=0}^{\infty}(\lambda+n)(\lambda+n-1)a_n z^{n+\lambda-2}. \qquad \frac{1}{z}\frac{df(z)}{dz} = \sum_{n=0}^{\infty}(n+\lambda)a_n z^{n+\lambda-2}.$$

$$\frac{\lambda^2 f(z)}{z^2} = \sum_{n=o}^{\infty} a_n \lambda^2 z^{n+\lambda-2}.$$

Substituting into Bessel's equation and collecting in powers of $(n + \lambda - 2)$:

$$\sum_{n=o}^{\infty} z^{n+\lambda-2}\left(a_n\left[(\lambda + n)(\lambda + n - 1) + (n + \lambda) - \lambda^2\right] + a_{n-2}\right) = 0.$$

Since this expression is analytic for all $z \neq 0$:

$$a_n\left[(\lambda + n)(\lambda + n - 1 + 1) - \lambda^2\right] + a_{n-2} = 0.$$

$$\Downarrow$$

(2) $$a_n = -\frac{a_{n-2}}{n(2\lambda + n)}.$$

Setting $a_1 = 0$ and a_o (arbitrary) $= a_o$ and substituting into (1):

$$f(z) = z^\lambda\left(a_o + a_2 z^2 + a_4 z^4 \ldots\right)$$

$$= z^\lambda\left(a_o - \frac{a_o z^2}{2(2\lambda+2)} + \frac{a_o z^4}{2\cdot 4(2\lambda+2)(2\lambda+4)} - \ldots\right)$$

$$= a_o z^\lambda \sum_{m=o}^{\infty} \frac{(-1)^m z^{2m}}{2\cdot 4\cdots 2m(2\lambda+2)(2\lambda+4)\cdots(2\lambda+2m)},$$

with the added specification that $f(z) = a_o z^\lambda$ for $m = 0$. Note that $2\cdot 4\cdots 2m = 2^m m!$, while

$$(2\lambda+2)(2\lambda+4)\cdots(2\lambda+2m) = 2^m(\lambda+1)(\lambda+2)\cdots(\lambda+m)$$

$$= \frac{2^m(\lambda + m)!}{\lambda!} = 2^m \frac{\Gamma(\lambda+m+1)}{\Gamma(\lambda+1)}.$$

Thus,

$$f(z) = a_o z^\lambda \sum_{m=o}^{\infty} \frac{(-1)^m (z/2)^{2m}\Gamma(\lambda+1)}{m!\Gamma(\lambda+m+1)}.$$

We choose $a_0 = \dfrac{1}{2^\lambda \Gamma(\lambda+1)}$, so that

$$f(z) = \sum_{m=0}^{\infty} \frac{(-1)^m (z/2)^{2m+\lambda}}{m!\,\Gamma(\lambda+m+1)}.$$

This, then, is what we define to be the Bessel function $J_\lambda(z)$.

$$(3) \qquad J_\lambda(z) = \sum_{m=0}^{\infty} (z/2)^{2m+\lambda} \frac{(-1)^m}{m!\,\Gamma(\lambda+m+1)}.$$

If λ is neither a half integer nor a full integer, so that $2\lambda \neq$ integer, we have two independent solutions, i.e. $J_{+|\lambda|} \neq J_{-|\lambda|}$, where both series start out with different powers of z.

If, however, 2λ *is* integer-valued, but the ln term is absent (i.e. $h_t = 0$):

(a) For $\lambda = \pm n$ (n a positive integer)

$$(4) \qquad J_{+n}(z) = \sum_{m=0}^{\infty} (z/2)^{2m+n} \frac{(-1)^m}{m!(m+n)!},$$

and the solution is analytic everywhere except at infinity. As for $J_{-n}(z)$, it exists only when $\Gamma(\lambda+m+1)$ is at least equal to $\Gamma(1)$, i.e. $\lambda+m+1 > 0$ because $\Gamma(0)$ represents a pole.

Hence,

$$(5) \qquad J_{-n}(z) = \sum_{m=n}^{\infty} (z/2)^{2m-n} \frac{(-1)^m}{m!\,\Gamma(m-n+1)}.$$

If we set $m - n = s$, so that $m = s + n$, then

$$J_{-n}(z) = \sum_{s=0}^{\infty} (z/2)^{2s+n} \frac{(-1)^n(-1)^s}{(s+n)!s!} = (-1)^n \sum_{s=0}^{\infty} (z/2)^{2s+n} \frac{(-1)^s}{(s+n)!s!},$$

and except for the factor $(-1)^n$, $J_{-n}(z)$ is identical to $J_{+n}(z)$. Hence, for λ integer-valued, we have only one solution.

(b) For λ non-integral valued, the general solution of Bessel's equation is given by $f(z) = AJ_{+|\lambda|} + BJ_{-|\lambda|}$.

Question: How do we find the second solution $f_2(z)$ for integral-valued λ? We could use the general method, namely, $f_2(z) = f_1(z)\Phi(z)$. Instead, we note that $h_t \neq 0$ and thereby expect a ln term, i.e. a branch point. Consider the expression $J_{+|\lambda|}(z) - (-1)^n J_{-|\lambda|}(z)$, which $\to 0$ as $\lambda \to n$ (positive integer). For all λ, both $J_{+|\lambda|}(z)$ and $J_{-|\lambda|}(z)$ are solutions of Bessel's equation. We therefore ask the following:

Is

$$(6) \qquad Y_n(z) \equiv \lim_{\lambda \to n} \frac{J_{+|\lambda|}(z) - (-1)^n J_{-|\lambda|}(z)}{\lambda - n}$$

different from $J_n(z)$? If it is, then we have found our second solution, since (6) satisfies Bessel's equation and is different from $J_n(z)$.

Evaluating for $n = 0$:

$$Y_o(z) = \lim_{\lambda \to o} \frac{J_{+|\lambda|}(z) - J_{-|\lambda|}(z)}{\lambda} = 2\left[\frac{\partial J_\lambda(z)}{\partial \lambda}\right]_{\lambda = o}.$$

From equation (3),

$$\frac{\partial J_\lambda(z)}{\partial \lambda} = \sum_{m=o}^{\infty} \frac{(z/2)^{2m+\lambda}(-1)^m \ln(z/2)}{m!\Gamma(\lambda+m+1)}$$

$$- \sum_{m=o}^{\infty} \frac{(z/2)^{2m+\lambda}(-1)^m}{m!}\left[\frac{1}{\Gamma(\lambda+m+1)^2}\frac{\partial \Gamma(\lambda+m+1)}{\partial \lambda}\right].$$

The bracketed expression can be written as

$$\frac{1}{\Gamma(\lambda+m+1)^2}\frac{\partial \Gamma(\lambda+m+1)}{\partial \lambda} = \frac{1}{\Gamma(\lambda+m+1)}\frac{\partial}{\partial \lambda}\ln\Gamma(\lambda+m+1).$$

Defining $\Psi(\lambda+m+1) \equiv \dfrac{\partial}{\partial \lambda}\ln\Gamma(\lambda+m+1)$:

$$Y_o(z) = 2\sum_{m=o}^{\infty} \frac{(z/2)^{2m}(-1)^m \ln(z/2)}{m!\Gamma(m+1)} + 2\sum_{m=o}^{\infty} \frac{(z/2)^{2m}(-1)^m}{m!}\left[-\frac{\Psi(m+1)}{\Gamma(m+1)}\right].$$

Rewriting:

$$(7) \qquad Y_o(z) = 2J_o(z)\ln(z/2) - 2\sum_{m=o}^{\infty}\frac{(z/2)^{2m}(-1)^m\,\Psi(m+1)}{m!^2}.$$

We see that $Y_o(z)$ is certainly different from $J_o(z)$; in fact it contains the expected ln term plus an analytic term.

7.5.2 Bessel Function — Theorems

The Bessel function

$$(1) \qquad J_\lambda(z) = \sum_{m=o}^{\infty}(z/2)^{2m+\lambda}\frac{(-1)^m}{m!\Gamma(\lambda+m+1)},$$

is a solution of Bessel's equation

$$(2) \qquad \frac{d^2J_\lambda(z)}{dz^2} + \frac{1}{z}\frac{dJ_\lambda(z)}{dz} + \left(1 - \frac{\lambda^2}{z^2}\right)J_\lambda(z) = 0,$$

which we saw occurs quite often in physics because of its relationship to the Laplace operator.

We seek to relate $J_\lambda(z)$ to an integral in the complex domain, similar to what was done for Hermite polynomials; we do this because the integral representation furnishes us with much more information.

Theorem: (for λ = integer–valued n)

$$(3) \qquad J_n(z) = \frac{(z/2)^n}{2\pi i}\oint\frac{1}{t^{n+1}}\exp\left[t - \frac{z^2}{4t}\right]dt.$$

The integral is taken about a circle surrounding $t = 0$ in the complex t plane.

Proof:

Method (A). Show that equation (3) satisfies equation (2).

$$\frac{dJ_n(z)}{dz} = \frac{n}{z}J_n(z) - \frac{(z/2)^n}{2\pi i}\oint\frac{z}{2t}\frac{1}{t^{n+1}}\exp\left(t - \frac{z^2}{4t}\right)dt.$$

$$\frac{d^2J_n(z)}{dz^2} = -\frac{n}{z^2}+\frac{n^2}{z^2}J_n(z) - \frac{n}{z}\frac{(z/2)^n}{2\pi i}\oint \frac{z}{2t}\frac{1}{t^{n+1}}\exp\left(t-\frac{z^2}{4t}\right)dt$$

$$+\frac{(z/2)^n}{2\pi i}\oint\left[\frac{z^2}{4t^2}-\frac{n+1}{2t}\right]\frac{1}{t^{n+1}}\exp\left(t-\frac{z^2}{4t}\right)dt.$$

$$\frac{1}{z}\frac{dJ_n(z)}{dz} = \frac{n}{z^2}J_n(z) - \frac{(z/2)^n}{2\pi i}\oint\frac{1}{2t}\frac{1}{t^{n+1}}\exp\left(t-\frac{z^2}{4t}\right)dt.$$

Substituting into Bessel's equation (2) for $\lambda = n$:

$$\frac{(z/2)^n}{2\pi i}\oint\left[\frac{z^2}{4t^2}-\frac{2n+2}{2t}+1\right]\frac{1}{t^{n+1}}\exp\left(t-\frac{z^2}{4t}\right)dt$$

$$= \frac{(z/2)^n}{2\pi i}\oint\frac{\partial}{\partial t}\frac{\exp\left(t-z^2/4t\right)}{t^{n+1}}dt = 0.$$

Hence, $J_\lambda(z)$ as given by the integral representation (3) satisfies Bessel's equation (2).

Method (B). Show that the expression for $J_\lambda(z)$ as given in equation (1) is represented by the contour integral expression (3).

Consider $e^{-z^2/4t} = \sum_{m=0}^{\infty}\left(\frac{-z^2}{4t}\right)^m\frac{1}{m!}$. For $t \neq 0$, this series converges uniformly and absolutely, i.e. has *good* properties. Substituting into equation (3):

$$J_n(z) = \sum_{m=0}^{\infty}\frac{(z/2)^n}{2\pi i}\oint\frac{(-1)^m}{m!}\frac{e^t}{t^{n+m+1}}(z/2)^{2m}dt,$$

where the summation has been taken outside of the integral since the series is uniformly convergent.

From the theorem of residues,

$$\frac{1}{2\pi i}\oint\frac{e^t dt}{t^{n+m+1}} = \frac{1}{(n+m)!} \rightarrow J_n(z) = \sum_{m=0}^{\infty}\frac{(z/2)^{n+2m}(-1)^m}{m!(n+m)}dt \underset{(\lambda=n)}{\equiv} J_\lambda(z)$$

For λ not integer-valued, we know that our solutions will contain branch points at $t = 0$, and branch cuts must be introduced. In such a case, the integral $\oint \frac{\partial}{\partial t} \frac{\exp\left(t - z^2/4t\right)}{t^{\lambda+1}} dt$ is not taken over a closed contour and hence does not vanish. Instead we choose a particular path of integration C shown in Fig. 7-7.

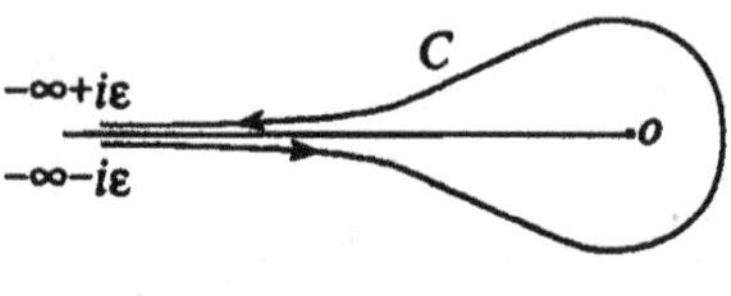

Fig. 7-7

For this path,

$$\frac{(z/2)^\lambda}{2\pi i} \int_C \frac{\partial}{\partial t} \frac{\exp\left(t - z^2/4t\right)}{t^{\lambda+1}} dt = \frac{(z/2)^\lambda}{2\pi i} \left[\frac{\exp\left(t - z^2/4t\right)}{t^{\lambda+1}} \right]_{-\infty - i\varepsilon}^{\infty + i\varepsilon} = 0.$$

Thus, we have here found one of many possible contours, such that even for λ not an integer, the integral representation for $J_n(z)$ still satisfies (2).

<u>Question:</u> Is

$$\frac{1}{\Gamma(\lambda+m+1)} = \frac{1}{2\pi i} \int_C \frac{e^t dt}{t^{\lambda+m+1}} ?$$

If it is true, then from method (B) above,

$$J_\lambda(z) = \frac{(z/2)^\lambda}{2\pi i} \int_C \frac{\exp\left(t - z^2/4t\right)}{t^{\lambda+1}} dt = \sum_{m=0}^{\infty} \frac{(z/2)^\lambda}{2\pi i} \int_C \frac{(-1)^m}{m!} \frac{e^t (z/2)^{2m}}{t^{\lambda+m+1}} dt$$

$$= \sum_{m=0}^{\infty} \frac{(z/2)^{2m+\lambda} (-1)^m}{\Gamma(m+1)\Gamma(m+\lambda+1)},$$

and the integral representation would be identical with the Bessel function expansion. We seek then to prove that

$$\frac{1}{\Gamma(\lambda+m+1)} = \frac{1}{2\pi i} \int_C \frac{e^t dt}{t^{\lambda+m+1}},$$

where the path C is as shown in Fig. 7-7.

Theorem: $\Gamma(z)\Gamma(1-z) = \dfrac{\pi}{\sin\pi z}$, where Re(z) > 0 but < 1.

Proof: Consider the shaded strip (Fig. 7-8) in the z-plane. Both $\Gamma(z)$ and $\Gamma(1-z)$ are analytic in the strip, as are π and $\sin\pi z$. Then, by the theorem proved in Section 6.11.1, if $\Gamma(z)\Gamma(1-z) = \pi/\sin\pi z$ in region I, it will be true for all regions of analytic continuation, i.e. if $F_1(z) \equiv \Gamma(z)\Gamma(1-z)$, which can be analytically continued, and is analytic over all z except at certain isolated points, and if $F_2(z) \equiv \pi/\sin\pi z$, which is analytic at all values of z except for certain isolated points $z = 1, 2, \ldots n$, then by showing that $F_1(z) = F_2(z)$ in the shaded region, $F_1(z)$ will equal $F_2(z)$ over the entire region.

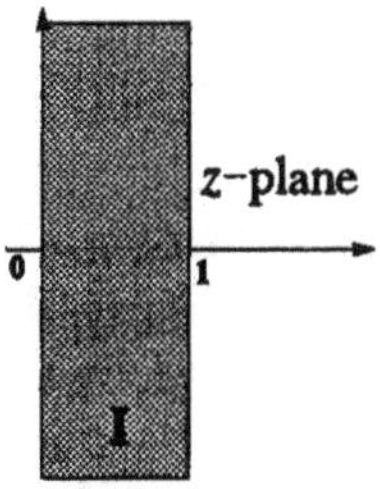

Fig. 7-8

Let $z' = z+1$, so that

$$\Gamma(z')\Gamma(1-z') = \Gamma(z+1)\Gamma(-z) = z\,\Gamma(z)\frac{1}{-z}\Gamma(1-z) = -\Gamma(z)\Gamma(1-z),$$

thus relating points at $z' = z+1$ to points in the shaded region, where the expression is analytic.

The Γ-function was defined (Section 6.11.1) as $\Gamma(z) \equiv \int_0^\infty e^{-t}t^{z-1}dt$ for Re(z) > 0. Setting $t = x^2$:

$$\Gamma(z) = \int_0^\infty e^{-x^2}x^{2z-2}2x\,dx = 2\int_0^\infty e^{-x^2}x^{2z-1}dx.$$

$$\Gamma(1-z) = 2\int_0^\infty e^{-x^2}x^{1-2z}dx = 2\int_0^\infty e^{-y^2}y^{-(2z-1)}dy.$$

$$\Gamma(z)\Gamma(1-z) = 4\int_0^\infty\int_0^\infty e^{-(y^2+x^2)}(x/y)^{(2z-1)}dx\,dy.$$

Making the transformation $x = r\cos\theta$, $y = r\sin\theta$, we have $x^2 + y^2 = r^2$, $(x/y)^{2z-1} = \cot\theta^{(2z-1)}$ and $\iint dx\,dy = \iint r\,dr\,d\theta$. Hence,

$$\Gamma(z)\Gamma(1-z) = 4\int_0^\infty\int_0^{\pi/2} e^{-r^2}r\cot\theta^{(2z-1)}dr\,d\theta = 4\int_0^{\pi/2}\left[\frac{-e^{-r^2}}{2}\right]_0^\infty \cot\theta^{2z-1}d\theta$$

$$= 2\int_0^{\pi/2}\cot\theta^{2z-1}d\theta.$$

Letting $\xi = \cot\theta$, $d\xi = -\csc^2\theta d\theta \;\rightarrow\; d\theta = -\dfrac{d\xi}{1+\xi^2}$. Substituting once more:

$$\Gamma(z)\Gamma(1-z) = 2\int_{\infty}^{0} \frac{-\xi^{2z-1}d\xi}{1+\xi^2} = 2\int_{0}^{\infty} \frac{\xi^{2z-1}}{1+\xi^2}\,d\xi.$$

Note that for Re(z) between 0 and 1, $2z$ is not an integer, so that $\int_0^\infty \frac{\xi^{2z-1}}{1+\xi^2}\,d\xi$ falls into the type III integral discussed in Sect. 6.10.1. To test this, we must see whether $\xi^{2z}/(1+\xi^2) \rightarrow 0$ as $\xi \rightarrow 0$ and ∞; and it does, by observation. Note also that $\xi^{2z-1}/(1+\xi^2)$ has a branch point at $\xi = 0$. Hence, let us consider the contour shown in Fig. 7-9. In Section 6.10.1 we showed that

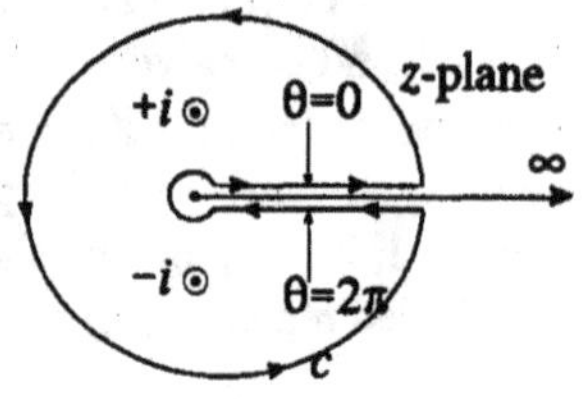

Fig. 7-9

$$\int_0^\infty \frac{\xi^{2z-1}}{1+\xi^2}\,d\xi = \frac{1}{1-e^{4\pi i z}}\oint \frac{\xi^{2z-1}}{1+\xi^2}\,d\xi.$$

To show that this holds for the contour of Fig. 7-9, note that over the large outer circle, $\xi^{2z}/(1+\xi^2) \rightarrow 0$ as $\xi \rightarrow \infty$, while over the small circle, $\xi^{2z}/(1+\xi^2) \rightarrow 0$ as $\xi \rightarrow 0$; hence,

$$\oint \frac{\xi^{2z-1}}{1+\xi^2}\,d\xi = \int_0^\infty \frac{\xi^{2z-1}}{1+\xi^2}\,d\xi - \int_0^\infty e^{i4\pi z}\frac{\xi^{2z-1}}{1+\xi^2}\,d\xi = \left[1-e^{i4\pi z}\right]\oint \frac{\xi^{2z-1}}{1+\xi^2}\,d\xi.$$

This last step follows since the first integral on the right is evaluated at $\theta = 0$ while the second integral is evaluated at $\theta = \pi/2$. Thus,

$$2\int_0^\infty \frac{\xi^{2z-1}}{1+\xi^2}\,d\xi = \frac{2}{1-e^{i4\pi z}}\oint \frac{\xi^{2z-1}}{1+\xi^2}\,d\xi.$$

Evaluating residues:

$$\text{Res}\oint \frac{\xi^{2z-1}}{1+\xi^2}\,d\xi = \left|\frac{\xi^{2z-1}}{2i}\right|_{\xi=+i} - \left|\frac{\xi^{2z-1}}{2i}\right|_{\xi=-i}.$$

We must be careful here since $i = e^{i\pi/2}, e^{i5\pi/2}, e^{i9\pi/2}$, etc., while $-i = e^{-i\pi/2}, e^{-i5\pi/2}, e^{-i9\pi/2}$, etc. Within the limits of our integration, however, $+i = e^{i\pi/2}$, and $-i = e^{i3\pi/2}$.

$$\text{Residue} \oint \frac{\xi^{2z-1}}{1+\xi^2}\,d\xi = \frac{e^{i\pi(2z-1)/2}}{2i} - \frac{e^{i3\pi(2z-1)/2}}{2i} = \frac{e^{i\pi z}e^{-i\pi/2}}{2i} - \frac{e^{i3\pi z}e^{-i3\pi/2}}{2i}$$

$$= -\frac{e^{i\pi z}}{2} - \frac{e^{i3\pi z}}{2} = -\frac{1}{2}\left(e^{i\pi z} + e^{i3\pi z}\right).$$

Hence,

$$2\int_0^\infty \frac{\xi^{2z-1}}{1+\xi^2}\,d\xi = \frac{-2\pi i}{1-e^{i4\pi z}}\left(e^{i\pi z} + e^{i3\pi z}\right) = \frac{-2\pi i e^{i\pi z}(1+e^{i2\pi z})}{(1+e^{i2\pi z})(1-e^{i2\pi z})}$$

$$= \frac{2\pi i}{e^{i\pi z} - e^{-i\pi z}} = \frac{\pi}{\sin\pi z}.$$

We have thus shown that

$$\Gamma(z)\Gamma(1-z) = 2\int_0^\infty \frac{\xi^{2z-1}}{1+\xi^2}\,d\xi = \frac{\pi}{\sin\pi z}.$$

We return now to the problem of showing that

$$\frac{1}{\Gamma(\lambda+m+1)} = \frac{1}{2\pi i}\int_C \frac{e^t dt}{t^{\lambda+m+1}}.$$

In Section 6.11.1 we showed that

$$\Gamma(z) = \frac{1}{1-e^{i2\pi z}}\int_C e^{-s}s^{z-1}ds,$$

where the contour C is as shown in Fig. 7-10. From the expression $\Gamma(z)\Gamma(1-z) = -2\pi i e^{i\pi z}/\left(1-e^{i2\pi z}\right)$, we have

$$\Gamma(z)\left(1-e^{i2\pi z}\right) = \frac{-2\pi i e^{i\pi z}}{\Gamma(1-z)} = \int_C e^{-s}s^{z-1}ds$$

Setting $\xi = 1-z$,

$$\frac{1}{\Gamma(\xi)} = \frac{-1}{2\pi i e^{i\pi(1-\xi)}}\int_C \frac{e^{-s}}{s^\xi}\,ds = \frac{+1}{2\pi i e^{-i\pi\xi}}\int_C \frac{e^{-s}}{s^\xi}\,ds.$$

The contour C in the t-plane (Fig. 7-7) is to be compared with the contour C' in the s plane (Fig. 7-10). We perform the inversion $t = -s$ so that contour $C' \to$ contour $-C$ and

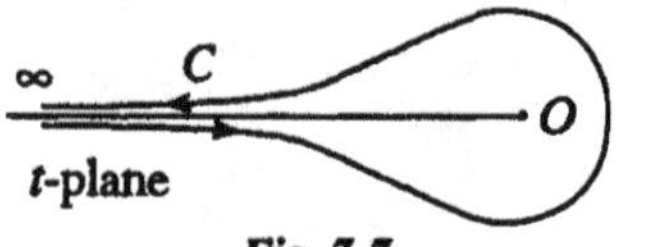

Fig. 7-7

$$\frac{1}{\Gamma(\xi)} = \frac{(e^{i\pi})^{\xi}}{2\pi i}\int_{-C}\frac{-e^t dt}{(-1)^{\xi}t^{\xi}} = \frac{1}{2\pi i}\int_C \frac{e^t dt}{t^{\xi}},$$

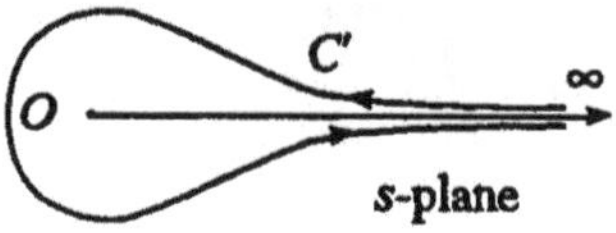

Fig. 7-10

and by setting $\xi = (\lambda+m+1)$:

$$\frac{1}{\Gamma(\lambda+m+1)} = \frac{1}{2\pi i}\int_C \frac{e^t dt}{t^{\lambda+m+1}},$$

thus completing the proof of the integral representation theorem.

<u>Recursion relations</u>

Theorem:

(4) $$J_{\lambda-1}(z) + J_{\lambda+1}(z) = \frac{2\lambda}{z}J_{\lambda}(z).$$

Proof: Using the integral representation (3):

$$J_{\lambda-1}(z) + J_{\lambda+1}(z) - \frac{2\lambda}{z}J_{\lambda}(z) = (z/2)^{\lambda-1}\frac{1}{2\pi i}\int_C \frac{e^{t-z^2/4t}}{t^{\lambda}}dt$$

$$+ (z/2)^{\lambda+1}\frac{1}{2\pi i}\int_C \frac{e^{t-z^2/4t}}{t^{\lambda+2}}dt - \lambda(z/2)^{\lambda-1}\frac{1}{2\pi i}\int_C \frac{e^{t-z^2/4t}}{t^{\lambda+1}}dt$$

$$= (z/2)^{\lambda-1}\frac{1}{2\pi i}\int_C \frac{e^{t-z^2/4t}}{t^{\lambda}}\left[1 + \frac{z^2}{4t^2} - \frac{\lambda}{t}\right]dt$$

$$= (z/2)^{\lambda-1}\frac{1}{2\pi i}\int_C \frac{\partial}{\partial t}\left[\frac{e^{t-z^2/4t}}{t^{\lambda}}\right]dt$$

$$= \frac{(z/2)^{\lambda-1}}{2\pi i}\left[\frac{e^{t-z^2/4t}}{t^{\lambda}}\right]_{-\infty-i\varepsilon}^{-\infty+i\varepsilon} = 0.$$

Theorem:

(5) $$\frac{dJ_\lambda(z)}{dz} - \frac{\lambda}{z}J_\lambda(z) = -J_{\lambda+1}(z) \quad \rightarrow \quad 2\frac{dJ_\lambda(z)}{dz} = J_{\lambda-1}(z) - J_{\lambda+1}(z)$$

Proof:

$$\frac{dJ_\lambda(z)}{dz} = \frac{\lambda}{z}J_\lambda(z) - \frac{(z/2)^\lambda}{2\pi i}\int_C \frac{z}{2t}\frac{e^{t-z^2/4t}}{t^{\lambda+1}}\,dt = \frac{\lambda}{z}J_\lambda(z) - J_{\lambda+1}(z).$$

From Theorem (4),

$$\frac{2\lambda}{z}J_\lambda(z) = J_{\lambda-1}(z) + J_{\lambda+1}(z).$$

$$\Downarrow$$

$$\frac{dJ_\lambda(z)}{dz} = \frac{\lambda}{z}J_\lambda(z) - J_{\lambda+1}(z) = \frac{1}{2}\left[J_{\lambda-1}(z) - J_{\lambda+1}(z)\right].$$

$$\Downarrow$$

$$2\frac{dJ_\lambda(z)}{dz} = J_{\lambda-1}(z) - J_{\lambda+1}(z).$$

[Note that Theorem (4) could have been proven using a power series method, but it is too clumsy. The integral representation demonstrates again its great usefulness.]

7.5.3 Bessel Functions vis-à-vis Sine and Cosine Functions

(1) $$J_{1/2}(z) = (z/2)^{1/2}\sum_{m=0}^{\infty}\frac{(-1)^m(z/2)^{2m}}{m!\Gamma(m+3/2)},$$

where $\Gamma(m+3/2) = (m+1/2)\Gamma(m+1/2) = (m+1/2)(m-1/2)\cdots(1/2)\Gamma(1/2)$.
Setting $y = t^{1/2}$ in $\Gamma(1/2) = \int_0^\infty e^{-t}t^{-1/2}dt$:

$$\Gamma(1/2) = \int_0^\infty e^{-t}t^{-1/2}dt = 2\int_0^\infty e^{-t}d(t^{1/2}) \quad \rightarrow \quad 2\int_0^\infty e^{-y^2}dy = \sqrt{\pi}.$$

Hence,

$$\Gamma(m+3/2) = \frac{(2m+1)(2m-1)(2m-3)\cdots 1\cdot\sqrt{\pi}}{2^{m+1}}\,\frac{2m(2m-2)(2m-4)\cdots 2}{2^m m!}$$

$$= \frac{(2m+1)!\sqrt{\pi}}{2^{2m+1}m!}.$$

Substituting into (1):

$$J_{1/2}(z) = (z/2)^{1/2}\sum_{m=0}^{\infty}\frac{(-1)^m(z/2)^{2m}2^{2m+1}m!}{m!(2m+1)!\sqrt{\pi}} = (2/\pi z)^{1/2}\sum_{m=0}^{\infty}\frac{(-1)^m z^{2m+1}}{(2m+1)!}$$

$$= (2/\pi z)^{1/2}\left\{z - \frac{z^3}{3!} + \frac{z^5}{5!} - \cdots\right\}.$$

Thus,

(2) $$J_{1/2}(z) = (2/\pi)^{1/2}\frac{\sin z}{\sqrt{z}},$$

and is plotted in Fig. 7-11.

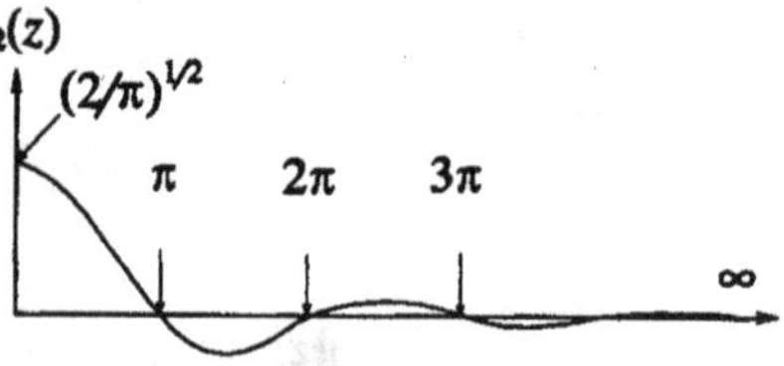

Fig. 7-11

To evaluate $J_{3/2}(z)$, we make use of recursion relation (5) for $\lambda = 1/2$, i.e.

$$J_{\lambda+1}(z) = \frac{\lambda}{z}J_\lambda(z) - \frac{dJ_\lambda(z)}{dz} \quad\rightarrow\quad J_{3/2}(z) = \frac{1}{2z}J_{1/2}(z) - \frac{dJ_{1/2}(z)}{dz}$$

$$= \frac{1}{2z}\sqrt{2/\pi z}\,\sin z - \sqrt{2/\pi}\left\{\frac{\cos z}{\sqrt{z}} - \frac{\sin z}{2z\sqrt{z}}\right\}$$

$$= \sqrt{2/\pi z}\left\{\frac{\sin z}{z} - \cos z\right\}.$$

Thus,

(3) $$J_{3/2}(z) = \sqrt{2/\pi z}\left\{\frac{\sin z}{z} - \cos z\right\}.$$

Next higher order half-integral Bessel functions are obtained by continuing to use recursion relation (5).

$$J_{5/2}(z) = \frac{3}{2z}J_{3/2}(z) - \frac{dJ_{3/2}(z)}{dz} = \frac{3}{2z}\left[\sqrt{2/\pi z}\left(\frac{\sin z}{z} - \cos z\right)\right]$$

$$-\left[\sqrt{2/\pi z}\left(\frac{z\cos z - \sin z}{z^2} + \sin z\right) - \frac{1}{\sqrt{2\pi}z^{3/2}}\left(\frac{\sin z}{z} - \cos z\right)\right]$$

$$= \sqrt{2/\pi z}\left[\frac{2}{z}\left(\frac{\sin z}{z} - \cos z\right) - \left(\frac{z\cos z - \sin z + z^2\sin z}{z^2}\right)\right]$$

$$= \sqrt{2/\pi z}\left[\frac{-3\cos z}{z} + \frac{3\sin z}{z^2} - \sin z\right].$$

Thus,

$$(4) \qquad J_{5/2}(z) = \sqrt{2/\pi z}\left[\sin z\left(\frac{3}{z^2} - 1\right) - \frac{3\cos z}{z}\right].$$

For λ half-integer valued, we seek the asymptotic behavior of $J_\lambda(z)$ as $z \to \infty$. The general behavior of $J_\lambda(z)$ for $\lambda = 1/2$ integer can be obtained using recursion relation (5), $\frac{dJ_\lambda(z)}{dz} - \frac{\lambda}{z}J_\lambda(z) = -J_{\lambda+1}(z)$. If one neglects $\frac{\lambda}{z}J_\lambda(z)$ as $z \to \infty$, $J_{\lambda+1}(z) \to -\frac{dJ_\lambda(z)}{dz}$. Then, since $J_{1/2}(z) = (2/\pi)^{1/2}\sin z/\sqrt{z}$,

$$J_{3/2}(z) \to -(2/\pi)^{1/2}\frac{\cos z}{\sqrt{z}}.$$

$$J_{5/2}(z) \to -(2/\pi)^{1/2}\frac{\sin z}{\sqrt{z}} = (-1)^2(2/\pi z)^{1/2}\frac{d^2(\sin z)}{dz^2} = -(2/\pi z)^{1/2}\sin z.$$

$$J_{7/2}(z) \to -\frac{dJ_{5/2}(z)}{dz} = (-1)^3(2/\pi z)^{1/2}\frac{d^3(\sin z)}{dz^3} = (2/\pi z)^{1/2}\cos z.$$

$$J_{9/2}(z) \to -\frac{dJ_{7/2}(z)}{dz} = (2/\pi z)^{1/2}\sin z.$$

From this point on, the cycle repeats, i.e. $J_{9/2}(z)$ behaves similarly to $J_{1/2}(z)$.

The behavior of $J_\lambda(z)$ as $z \to \infty$ is seen to be similar to the behavior of sines and cosines in the 4 quadrants, i.e.

$$J_{1/2} \to +\sin z\,;\; J_{3/2} \to -\cos z\,;\; J_{5/2} \to -\sin z\,;\; J_{7/2} \to +\cos z\,;$$

$$J_{9/2} \to +\sin z\,;\; \text{etc.}$$

Thus, for λ half-integer valued, as $z \to \infty$,

$$(5) \qquad J_\lambda(z) \;\to\; (2/\pi z)^{1/2} \cos\left[z - \frac{\lambda\pi}{2} - \frac{\pi}{4}\right].$$

We will now show that equation (5) holds for *any* λ. Consider Bessel's equation

$$\frac{d^2J_\lambda(z)}{dz^2} + \frac{1}{z}\frac{dJ_\lambda(z)}{dz} + \left(1 - \frac{\lambda^2}{z^2}\right)J_\lambda(z) = 0.$$

To aid in the investigation of $J_\lambda(z)$ as $z \to \infty$, one removes the $1/z$ term by the following transformation. Setting $J_\lambda(z) = g(z)/\sqrt{z}$, we have:

$$\frac{dJ_\lambda(z)}{dz} = \frac{1}{\sqrt{z}}\frac{dg(z)}{dz} - \frac{g(z)}{2z^{3/2}}. \qquad \frac{1}{z}\frac{dJ_\lambda(z)}{dz} = \frac{1}{z^{3/2}}\frac{dg(z)}{dz} - \frac{g(z)}{2z^{5/2}}.$$

$$\frac{d^2J_\lambda(z)}{dz^2} = \frac{1}{\sqrt{z}}\frac{d^2g(z)}{dz^2} - \frac{1}{2z^{3/2}}\frac{dg(z)}{dz} - \frac{1}{2z^{3/2}}\frac{dg(z)}{dz} + \frac{3g(z)}{4z^{5/2}}.$$

Substituting into Bessel's equation:

$$\frac{1}{\sqrt{z}}\frac{d^2g(z)}{dz^2} - \frac{1}{z^{3/2}}\frac{dg(z)}{dz} + \frac{3g(z)}{4z^{5/2}} + \frac{1}{z^{3/2}}\frac{dg(z)}{dz} - \frac{g(z)}{2z^{5/2}} + \frac{g(z)}{z^{1/2}} - \frac{\lambda^2 g(z)}{z^{5/2}} = 0.$$

$$\Downarrow$$

$$(6) \qquad \frac{d^2g(z)}{dz^2} + \left(1 - \frac{(\lambda^2 - 1/4)}{z^2}\right)g(z) \;\to\; \frac{d^2g(z)}{dz^2} + g(z) = 0.$$

The last step follows if one neglects the term $\frac{(\lambda^2 - 1/4)}{z^2}$ for large z.

Solutions of (6) are given by $g(z) = a_\lambda \cos(z + b_\lambda) \rightarrow J_\lambda(z) = \dfrac{a_\lambda \cos(z + b_\lambda)}{\sqrt{z}}$.
We wish to show that $a_\lambda = (2/\pi)^{1/2}$ (independent of λ) and $b_\lambda = -\dfrac{\lambda\pi}{2} - \dfrac{\pi}{4}$.
From recursion relation (5), neglecting $\dfrac{\lambda J_\lambda(z)}{z}$ for large z, we have

$$J_{\lambda+1}(z) = -\frac{dJ_\lambda(z)}{dz} = \frac{a_{\lambda+1}\cos(z + b_{\lambda+1})}{\sqrt{z}}.$$

Thus, for large z,

$$\frac{a_{\lambda+1}\cos(z + b_{\lambda+1})}{\sqrt{z}} = -\frac{a_\lambda}{\sqrt{z}}\frac{d}{dz}\cos(z + b_\lambda) = \frac{a_\lambda \sin(z + b_\lambda)}{\sqrt{z}}.$$

Hence,

$$a_\lambda = a_{\lambda+1} = a_{\lambda+2} \ldots = \alpha \text{ (independent of } \lambda\text{), and}$$

$$b_\lambda = -\frac{\lambda\pi}{2} + \beta \text{ (independent of } \lambda\text{).}$$

This last follows since

$$\cos(z + b_{\lambda+1}) = \cos\left(z - \frac{\lambda\pi}{2} + \beta - \frac{\pi}{2}\right) = \sin\left(z - \frac{\lambda\pi}{2} + \beta\right)$$

$$= \sin(z + b_\lambda).$$

We have shown, therefore, that for <u>any</u> λ,

$$(7) \qquad J_\lambda(z) \rightarrow \frac{\alpha\cos(z - \lambda\pi/2 + \beta)}{\sqrt{z}}.$$

Comparing (7) with (5), the behavior of $J_\lambda(z)$ for half-integer λ, we would like to say that $\alpha = (2/\pi)^{1/2}$ and $\beta = -\pi/4$. However, the result (5) has been shown to be true only for values of λ succeeding by 1. In other words, for $\lambda = 1/3, 4/3, \ldots 101\tfrac{1}{3}$, etc., values of α and β do not change. We do not know whether these α and β values are $(2/\pi)^{1/2}$ and $-\pi/4$, respectively for $\lambda = 1/2$. This result can be found in *Modern Analysis* (Whittaker and Watson, Cambridge University Press, 1952).

7.5.4 Hankel Functions

(1)
$$H_\lambda^1(z) = \frac{+i}{\sin\lambda\pi}\left[e^{-i\lambda\pi}J_\lambda(z) - J_{-\lambda}(z)\right].$$

$$H_\lambda^2(z) = \frac{-i}{\sin\lambda\pi}\left[e^{+i\lambda\pi}J_\lambda(z) - J_{-\lambda}(z)\right].$$

Hankel functions $H_\lambda^{1,2}(z)$ are obviously solutions of Bessel's equation since $J_\lambda(z)$ and $J_{-\lambda}(z)$ are solutions. For λ integer valued, one understands $H_n^{1,2}(z)$ to be $\lim_{\lambda\to n} H_\lambda^{1,2}(z)$.

For z real, $H_\lambda^2(z)$ is clearly the complex conjugate of $H_\lambda^1(z)$. We saw in Section 7.5.1 that since $g(z)z^2$ is not analytic at z = infinity, infinity is an irregular singular point, where both solutions $J_\lambda(z)$ and $J_{-\lambda}(z)$, and therefore $H_\lambda^{1,2}(z)$, have an essential singularity at z = infinity.

Let $\lambda = n + \varepsilon$, where n is an integer. Then, as $\varepsilon\to 0$ (or as $\lambda\to n$),

$$\sin\lambda\pi = \sin(n\pi+\varepsilon\pi) = \varepsilon\pi\cos n\pi = (-1)^n\varepsilon\pi, \text{ and}$$

$$e^{\pm i\lambda\pi} = e^{\pm(n+\varepsilon)i\pi} = (-1)^n \text{ as } \varepsilon\to 0.$$

$$\lim_{\lambda\to n} H_\lambda^{1,2}(z) = \frac{\pm i}{(-1)^n\pi}\lim_{\varepsilon\to 0}\left[\frac{(-1)^n J_{n+\varepsilon}(z) - J_{-n-\varepsilon}(z)}{\varepsilon}\right].$$

(2)
$$\lim_{\lambda\to n} H_\lambda^{1,2}(z) = \frac{\pm i}{\pi}\lim_{\varepsilon\to 0}\left[\frac{J_{n+\varepsilon}(z) - (-1)^n J_{-n-\varepsilon}(z)}{\varepsilon}\right].$$

We saw in equation (6), Section 7.5.1, that $\lim_{\lambda\to n}\left[\frac{J_\lambda(z) - (-1)^n J_{-\lambda}(z)}{\lambda - n}\right]$ was exactly the second solution of Bessel's equation for integral λ. Hence, the Hankel functions are second solutions of Bessel's equation for integral λ.

Using the asymptotic behavior of $J_\lambda(z)$ from Section 7.5.3, equation (5):

(3)
$$2H_\lambda^1(z) = \frac{2ie^{-i\pi\lambda/2}}{\sin\lambda\pi}(2/\pi z)^{1/2}\left[e^{-i\pi\lambda/2}\cos(z - \lambda\pi/2 - \pi/4)\right]$$

$$- \frac{2ie^{-i\pi\lambda/2}}{\sin\lambda\pi}(2/\pi z)^{1/2}\left[e^{+i\pi\lambda/2}\cos(z + \lambda\pi/2 - \pi/4)\right].$$

Substituting $2\cos(z - \lambda\pi/2 - \pi/4) = e^{iz-i\lambda\pi/2-i\pi/4} + e^{-iz+i\lambda\pi/2+i\pi/4}$, into (3):

$$2H_\lambda^1(z) = \frac{ie^{-i\pi\lambda/2}}{\sin\lambda\pi}(2/\pi z)^{1/2}\left[e^{-i\pi\lambda/2}\left(e^{iz-i\lambda\pi/2-i\pi/4} + e^{-iz+i\lambda\pi/2+i\pi/4}\right)\right]$$

$$-\frac{ie^{-i\pi\lambda/2}}{\sin\lambda\pi}(2/\pi z)^{1/2}\left[e^{+i\pi\lambda/2}\left(e^{iz+i\lambda\pi/2-i\pi/4} + e^{-iz-i\lambda\pi/2+i\pi/4}\right)\right]$$

$$= \frac{ie^{-i\pi\lambda/2}}{\sin\lambda\pi}(2/\pi z)^{1/2}\left[e^{iz-i\pi/4}\left(e^{-i\lambda\pi} - e^{+i\lambda\pi}\right) + e^{-iz+i\pi/4}(0)\right]$$

$$= \frac{-ie^{-i\pi\lambda/2}}{\sin\lambda\pi}(2/\pi z)^{1/2}\left[e^{iz-i\pi/4}\sin\lambda\pi\right]\cdot 2i$$

$$= 2e^{-i\pi\lambda/2}(2/\pi z)^{1/2}e^{iz-i\pi/4}.$$

Thus,

$$H_\lambda^1(z) \rightarrow (2/\pi z)^{1/2}e^{i(z-\pi\lambda/2-\pi/4)}.$$

(4)

$$H_\lambda^2(z) \rightarrow (2/\pi z)^{1/2}e^{-i(z-\pi\lambda/2-\pi/4)}.$$

$H_\lambda^2(z)$ is obtained by replacing i by $-i$ in $H_\lambda^1(z)$.

Chapter 8

Theory of Partial Differential Equations

8.1 Examples of Field Equations in Physics

Poisson's equation

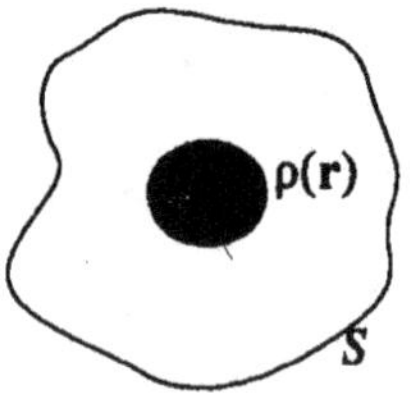

Fig. 8-1

(1) $\nabla^2\Phi(\mathbf{r}) = -4\pi\rho(\mathbf{r}),$

where the charge distribution $\rho(\mathbf{r})$ is given. For $\rho(\mathbf{r}) = 0$, equation (1) reduces to Laplace's equation $\nabla^2\Phi(\mathbf{r}) = 0$.

Boundary Conditions (Fig. 8-1)

Either $\Phi(\mathbf{r})$ or $\partial\Phi(\mathbf{r})/\partial n$ on S is known. These boundary conditions are sufficient to determine $\Phi(\mathbf{r})$ uniquely everywhere.

Wave Equation

(2) $\nabla^2\Phi(\mathbf{r}t) - \dfrac{1}{c^2}\dfrac{\partial^2\Phi(\mathbf{r}t)}{\partial t^2} = 0.$

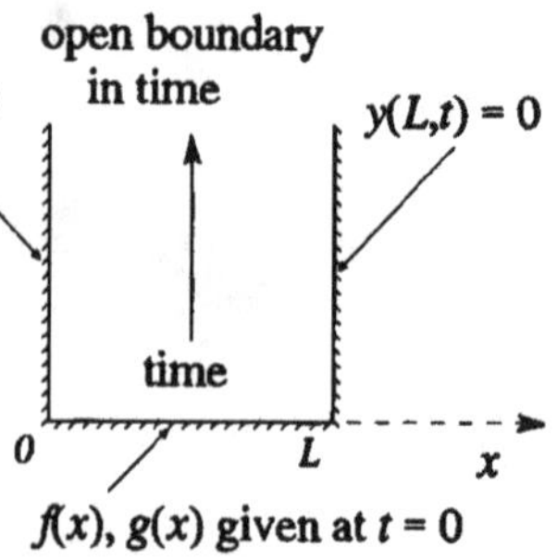

Fig. 8-2

An example is the vibration of a string of constant density under constant tension. The equation for such a string in 1 dimension is

$$\frac{\partial^2 y}{\partial x^2} - \frac{1}{\upsilon^2}\frac{\partial^2 y}{\partial t^2} = 0.$$

<u>Boundary Conditions</u> (Fig. 8-2)

At $x = 0$ and $x = L$, $y = 0$ for all time.

At time $t = 0$, $y = f(x)$ and $y' = g(x)$ are given.

These boundary conditions then determine a unique solution to the wave equation.

Heat Conduction

$$(3) \qquad \nabla^2 T(\mathbf{r}t) = \lambda \frac{\partial T(\mathbf{r}t)}{\partial t},$$

where $\lambda > 0$. For a 1-dimensional rod, equation (3) $\rightarrow \dfrac{\partial^2 T(xt)}{\partial x^2} = \lambda \dfrac{\partial T(xt)}{\partial t}$.

<u>Boundary Conditions</u> (Fig. 8-3)

At $t = 0$, $T = T(x,0)$ is given.

At $x = 0$, $T = T(0,t)$ is given, and

At $x = L$, $T = T(L,t)$ is given

for all time.

The solution of T for all times and positions along the 1-dimensional rod is then uniquely determined.

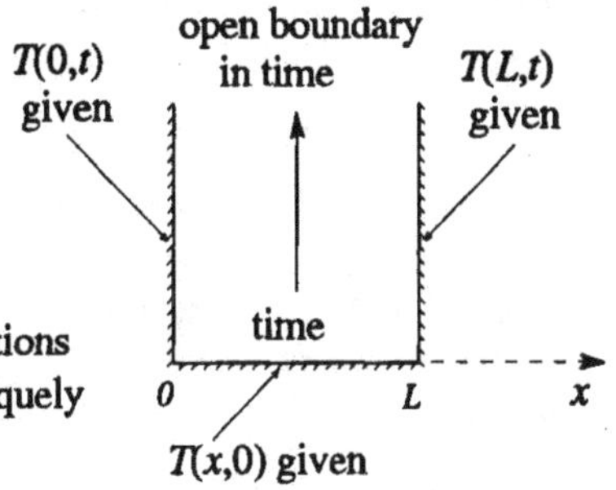

Fig. 8-3

In general, for second-order *ordinary* differential equations, solutions are given by:

$$f(x) = Af_1(x) + Bf_2(x),$$

where $f_1(x)$ and $f_2(x)$ are each solutions of the differential equation, and the two arbitrary constants A and B are determined by specifying:

(a) $f(x{=}0)$ and $f'(x{=}0)$ as given. This corresponds to the boundary conditions of an open-boundary problem.

(b) $f(x{=}0)$ and $f(x{=}L)$ as given, or $f(x{=}0)$ and $f'(x{=}L)$ as given. This corresponds to the boundary conditions of a closed-boundary problem in one dimension, $x = 0$ to $x = L$.

For Poisson's equation (1): Since time is not involved, one needs the value of $f(\mathbf{r})$ at all points on the boundary.

For the wave equation (2): Since the boundary is closed in 1 dimension, one needs $f(x=0)$ and $f(x=L)$. On the other hand, since time is an open boundary (of second order), one needs $y(x,t=0)$ and $y'(x,t=0)$.

For the heat conduction equation (3): Since the boundary is closed in 1 dimension, one needs $T(x=0,t)$ and $T(x=L,t)$. Time, however, is open and of first order, so that the value of $T(x,t=0)$ will suffice.

The Schrödinger equation

$$\left(-\frac{\hbar^2}{2m}\nabla^2 + V\right)\Psi(\mathbf{r}t) = -\frac{\hbar}{i}\frac{\partial}{\partial t}\Psi(\mathbf{r}t)$$

resembles the heat conduction equation in that the equation is of second order in space and first order in time.

8.2 Theory of Characteristics

Consider any surface S on which Φ and $\left[\nabla\Phi\right]_n$ are known. Is the solution completely specified? Consider the following (restricting ourselves to 2 dimensions):

The most general second-order partial differential equation is given as

(1)
$$A\frac{\partial^2\Phi(xy)}{\partial x^2} + 2B\frac{\partial^2\Phi(xy)}{\partial x\partial y} + C\frac{\partial^2\Phi(xy)}{\partial y^2} + D\frac{\partial\Phi(xy)}{\partial x} + E\frac{\partial\Phi(xy)}{\partial y} + F\Phi(xy) = 0.$$

Problem: Given Φ and $\left[\nabla\Phi\right]_n$ along PQ, the segment of a curve (Fig. 8-4). Consider a point $O(xy)$ on PQ. Can one find $\Phi(xy)$ in the neighborhood of point O? In other words, can one find all the derivatives of $\Phi(xy)$?

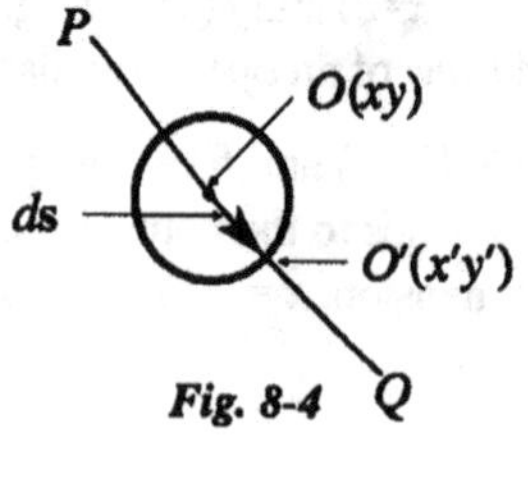

Fig. 8-4

Solution: Given the initial conditions, i.e. that Φ and $\left[\nabla\Phi\right]_n$ are known, $\left[\nabla\Phi\right]_{tang}$ is known. Performing a rotation, one can then get $\left[\nabla\Phi\right]_x = \dfrac{\partial\Phi}{\partial x}$ and $\left[\nabla\Phi\right]_y = \dfrac{\partial\Phi}{\partial y}$.

Choose $ds = (dx, dy)$ along PQ. Since $\dfrac{\partial\Phi}{\partial x}$ and $\dfrac{\partial\Phi}{\partial y}$ are known at $O(x,y)$ and at $O'(x',y')$, $d\left(\dfrac{\partial\Phi}{\partial x}\right)_{PQ}$ and $d\left(\dfrac{\partial\Phi}{\partial y}\right)_{PQ}$ are known We then have

$$d\left(\frac{\partial\Phi}{\partial x}\right)_{PQ} = \left(\frac{\partial\Phi}{\partial x}\right)_{O'} - \left(\frac{\partial\Phi}{\partial x}\right)_{O} = \frac{\partial^2\Phi}{\partial x^2}dx + \frac{\partial^2\Phi}{\partial x\partial y}dy.$$

$$d\left(\frac{\partial\Phi}{\partial y}\right)_{PQ} = \left(\frac{\partial\Phi}{\partial y}\right)_{O'} - \left(\frac{\partial\Phi}{\partial y}\right)_{O} = \frac{\partial^2\Phi}{\partial y\partial x}dx + \frac{\partial^2\Phi}{\partial y^2}dy.$$

From equation (1):

$$\left[-D\frac{\partial\Phi}{\partial x} - E\frac{\partial\Phi}{\partial y} - F\Phi\right]_{PQ} = A\frac{\partial^2\Phi}{\partial x^2} + 2B\frac{\partial^2\Phi}{\partial x\partial y} + C\frac{\partial^2\Phi}{\partial y^2}.$$

The left-hand sides of the above 3 equations are known by the arguments above. Rewriting as a set of simultaneous equations:

$$\frac{\partial^2\Phi}{\partial x^2}dx + \frac{\partial^2\Phi}{\partial x\partial y}dy + 0 = d\left(\frac{\partial\Phi}{\partial x}\right)_{PQ}.$$

$$0 + \frac{\partial^2\Phi}{\partial x\partial y}dx + \frac{\partial^2\Phi}{\partial y^2}dy = d\left(\frac{\partial\Phi}{\partial y}\right)_{PQ}.$$

$$\frac{\partial^2\Phi}{\partial x^2}A + \frac{\partial^2\Phi}{\partial x\partial y}2B + \frac{\partial^2\Phi}{\partial y^2}C = \left[-D\frac{\partial\Phi}{\partial x} - E\frac{\partial\Phi}{\partial y} - F\Phi\right]_{PQ}.$$

We get solutions when the determinant does not vanish, i.e. when

$$\det\begin{vmatrix} dx & dy & 0 \\ 0 & dx & dy \\ A & 2B & C \end{vmatrix} = C\,dx^2 + A\,dy^2 - 2B\,dx dy \neq 0.$$

Thus, the requirement for knowing the values of $\frac{\partial^2\Phi}{\partial x^2}$, $\frac{\partial^2\Phi}{\partial x\partial y}$ and $\frac{\partial^2\Phi}{\partial y^2}$ is:

(2) $\quad C\,dx^2 + A\,dy^2 - 2B\,dxdy \neq 0.$

Now that second-order derivatives are known, can one find third-order derivatives? Differentiating equation (1) (of which Φ is a solution) with respect to x:

$$A\frac{\partial^3\Phi}{\partial x^3} + 2B\frac{\partial^3\Phi}{\partial x^2\partial y} + C\frac{\partial^3\Phi}{\partial x\partial y^2} = -D\frac{\partial^2\Phi}{\partial x^2} - E\frac{\partial^2\Phi}{\partial x\partial y} - F\frac{\partial\Phi}{\partial x}.$$

As before:

$$d\left(\frac{\partial^2\Phi}{\partial x^2}\right)_{PQ} = \left(\frac{\partial^2\Phi}{\partial x^2}\right)_{O'} - \left(\frac{\partial^2\Phi}{\partial x^2}\right)_{O} = \frac{\partial^3\Phi}{\partial x^3}dx + \frac{\partial^3\Phi}{\partial x^2\partial y}dy.$$

$$d\left(\frac{\partial^2\Phi}{\partial x\partial y}\right)_{PQ} = \left(\frac{\partial^2\Phi}{\partial x\partial y}\right)_{O'} - \left(\frac{\partial^2\Phi}{\partial x\partial y}\right)_{O} = \frac{\partial^3\Phi}{\partial x^2\partial y}dx + \frac{\partial^3\Phi}{\partial x\partial y^2}dy.$$

Setting up the simultaneous equations again:

$$A\frac{\partial^3\Phi}{\partial x^3} + 2B\frac{\partial^3\Phi}{\partial x^2\partial y} + C\frac{\partial^3\Phi}{\partial x\partial y^2} = \text{known}.$$

$$dx\frac{\partial^3\Phi}{\partial x^3} + dy\frac{\partial^3\Phi}{\partial x^2\partial y} + 0 = \text{known}.$$

$$0 + dx\frac{\partial^3\Phi}{\partial x^2\partial y} + dy\frac{\partial^3\Phi}{\partial x\partial y^2} + 0 = \text{known}.$$

We get solutions when the determinant does not vanish, i.e. when

$$\det\begin{vmatrix} A & 2B & C \\ dx & dy & 0 \\ 0 & dx & dy \end{vmatrix} = A\,dy^2 - 2B\,dxdy + C\,dx^2 \neq 0.$$

By repeated processes, i.e. taking derivatives with respect to x and y, one can find $\partial^{n+m}\Phi/\partial x^n\partial y^m$ and, assuming the Taylor series to be convergent, one can find Φ in the neighborhood of the point $O(x,y)$ on PQ.

8.2.1 The Characteristic Curve

We seek the curve expressed by the equation

(1) $A\,dy^2 + C\,dx^2 - 2B\,dxdy = 0.$

Rewriting:

$$A\left(\frac{dy}{dx}\right)^2 + C - 2B\left(\frac{dy}{dx}\right) = 0 \quad \rightarrow \quad Ay'^2 + C - 2By' = 0.$$

Solving for y:

$$y = \int_x \frac{dy}{dx}\,dx = \int_x y'dx = \int_x \frac{2B \pm \sqrt{4B^2 - 4AC}}{2A}\,dx,$$

yielding, in general, two families of curves, where the two integration constants are the parameters of the corresponding families.

Special Cases

(i) *Elliptic* type of partial differential equation, for which $B^2 < AC$. In this case, dy/dx is complex, yielding two *complex* families, complex conjugates of each other.

Example: Laplace's equation $\frac{\partial^2\Phi}{\partial x^2} + \frac{\partial^2\Phi}{\partial y^2} = 0$, where $A = 1$, $C = 1$, $B = 0$, and $B^2 < AC$.

The boundary for the electrostatic problem will never be a characteristic curve, since for elliptic types the characteristic curves are complex. If Φ is specified on the surface, it will be known everywhere inside.

(ii) *Hyperbolic* type of partial differential equation, for which $B^2 > AC$. In this case we have two *real* families of curves.

Example: Wave equation $\frac{\partial^2\Phi}{\partial x^2} - \frac{1}{c^2}\frac{\partial^2\Phi}{\partial t^2} = 0$, where $A = 1$, $B = 0$, $C = -1$ (setting $y = ct$) and $B^2 > AC$.

Equation (1) becomes

$$dy^2 - dx^2 = 0 \rightarrow dx/dy = \pm 1 \rightarrow dx/cdt = \pm 1 \rightarrow x = \pm ct + \text{constant}.$$

The two families of curves, given by $x + ct = \mu$ and $x - ct = \lambda$, are shown in Fig. 8-5. Note that the curves must intersect at right angles since the slopes are ± 1, respectively.

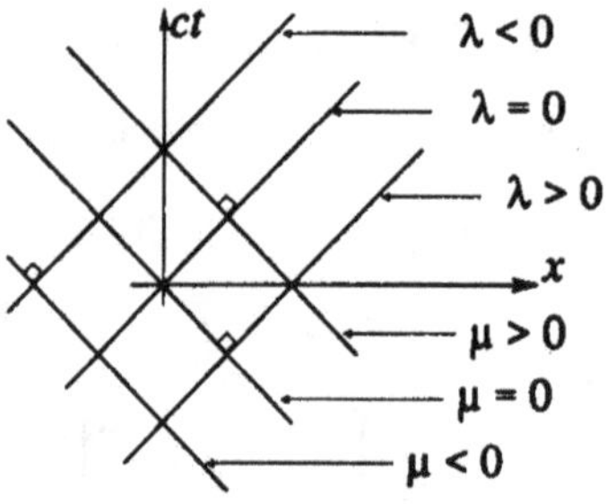

Fig. 8-5

(iii) *Parabolic* type of partial differential equation, for which $B^2 = AC$. In this case we have one family of curves $dy/dx = B/A$.

Example: Heat conduction equation $\dfrac{\partial^2 \Phi}{\partial x^2} - \lambda \dfrac{\partial \Phi}{\partial t} = 0$, where $A = 1, B = 0, C = 0$.

The characteristic curves are determined from $Ady^2 = 0 \rightarrow dt^2 = 0$, or $t =$ constant (Fig. 8-6). Once Φ is assigned along the characteristic curve, $\dfrac{\partial \Phi}{\partial x}$ is known, so that $\dfrac{\partial \Phi}{\partial t} = \dfrac{1}{\lambda}\dfrac{\partial^2 \Phi}{\partial x^2}$ will be determined. Hence, one can assign only Φ (not both Φ and $\partial\Phi/\partial n = \partial\Phi/\partial t$); otherwise one may overdetermine the solution.

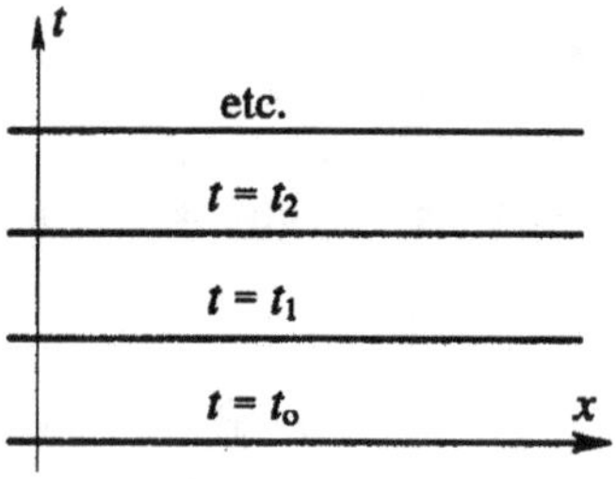

Fig. 8-6

8.2.2 The One-Dimensional Wave Equation

In the previous section we showed that for the hyperbolic type differential equation the characteristic curves are given by $x + ct = \mu$ and $x - ct = \lambda$, or $x = (\mu + \lambda)/2$ and $ct = (\mu - \lambda)/2$, so that

$$\frac{\partial}{\partial x} = 2\left(\frac{\partial}{\partial \mu} + \frac{\partial}{\partial \lambda}\right) \rightarrow \frac{\partial^2}{\partial x^2} = 4\left(\frac{\partial}{\partial \mu} + \frac{\partial}{\partial \lambda}\right)\left(\frac{\partial}{\partial \mu} + \frac{\partial}{\partial \lambda}\right)$$

$$= 4\left(\frac{\partial^2}{\partial \mu^2} + 2\frac{\partial^2}{\partial \lambda \partial \mu} + \frac{\partial^2}{\partial \lambda^2}\right).$$

$$\frac{\partial}{\partial (ct)} = 2\left(\frac{\partial}{\partial \mu} - \frac{\partial}{\partial \lambda}\right) \rightarrow \frac{\partial^2}{\partial (ct)^2} = 4\left(\frac{\partial}{\partial \mu} - \frac{\partial}{\partial \lambda}\right)\left(\frac{\partial}{\partial \mu} - \frac{\partial}{\partial \lambda}\right)$$

$$= 4\left(\frac{\partial^2}{\partial \mu^2} - 2\frac{\partial^2}{\partial \lambda \partial \mu} + \frac{\partial^2}{\partial \lambda^2}\right).$$

Substituting into the wave equation:

$$\frac{\partial^2 \Phi}{\partial x^2} - \frac{1}{c^2}\frac{\partial^2 \Phi}{\partial t^2} = 16\frac{\partial^2 \Phi}{\partial \mu \partial \lambda} = 0.$$

$$\Downarrow$$

$$\frac{\partial}{\partial \mu}\frac{\partial \Phi}{\partial \lambda} = 0 \Rightarrow \frac{\partial \Phi}{\partial \lambda} = f(\lambda) \text{ only} \Rightarrow \Phi = f_1(\lambda) + f_2(\mu).$$

$$\frac{\partial}{\partial \lambda}\frac{\partial \Phi}{\partial \mu} = 0 \Rightarrow \frac{\partial \Phi}{\partial \mu} = g(\mu) \text{ only} \Rightarrow \Phi = f_3(\lambda) + f_4(\mu).$$

Combining,

(1) $\Phi(\mu,\lambda) = F(\lambda) + G(\mu)$.

The D'Alembert solution of the 1-dimensional wave equation is given by

(2) $\Phi(x+ct, x-ct) = F(x-ct) + G(x+ct)$,

where $x + ct$ represents a wave travelling in the -x direction and $x - ct$ represents a wave travelling in the $+x$ direction (c = wave velocity). For x from 0 to L, the characteristic curves are as shown in Fig. 8-7.

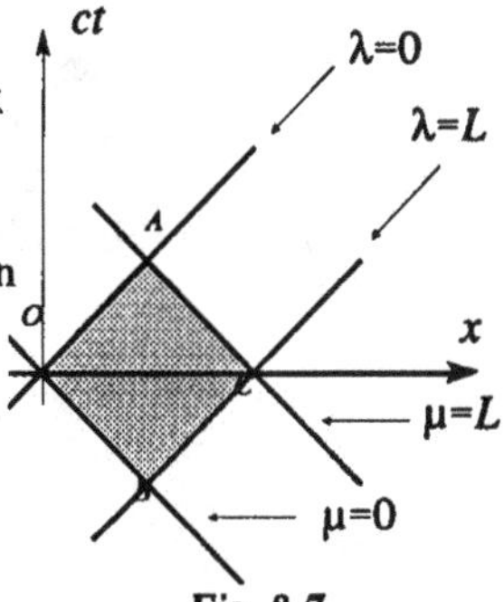

Fig. 8-7

Question: Under what conditions can $F(x-ct)$ and $G(x-ct)$ be found, i.e. under what conditions can one find Φ?

(A) Suppose Φ is known along OA and OB. Can one arbitrarily assign $\partial\Phi/\partial n$ and thereby find Φ in the shaded domain? Consider the following:

Along OA: $\Phi|_{OA} = F(0) + G(\mu)$, where μ runs from $\mu = 0$ to $\mu = L$, is *known*.

Along OB: $\Phi|_{OB} = F(\lambda) + G(0)$, where λ runs from $\lambda = 0$ to $\lambda = L$, is *known*.

Hence, in the shaded domain $OACB$, wherein $0 \le \lambda \le L$ and $0 \le \mu \le L$, we now know $F(0) + G(0) + F(\lambda) + G(\mu)$. Therefore, $\Phi = F(\lambda) + G(\mu)$ is known, where λ and μ take on their respective limits.

Since Φ is now known in the shaded domain, one cannot arbitrarily assign normal derivatives at the characteristic curves (at least inside this domain), because the solution is already known, and one may, as mentioned before, overdetermine the solution.

(B) Suppose $\Phi(x,t{=}0)$ and $\dot{\Phi}(x,t{=}0)$ are known for $0 \le x \le L$. We define $\Phi(x,t{=}0) \equiv u(x)$ and $\dot{\Phi}(x,t{=}0)/c \equiv \upsilon(x)$, where $u(x)$ and $\upsilon(x)$ are known within the region $0 \le x \le L$. With these conditions specified, in what regions will the solutions be known?

If Φ satisfies the wave equation, then from equation (2) at $t = 0$:

(3) $$\Phi(x,t{=}0) \;=\; F(x) + G(x) \;=\; u(x).$$

(4) $$\dot{\Phi}(x,t{=}0)/c \;=\; -F'(x) + G'(x) \;=\; \upsilon(x).$$

Integrating equation (4):

$$\int F'(x')dx' - \int G'(x')dx' \;=\; -\int^{x} \upsilon(x')dx' \;\rightarrow\; F(x) - G(x) \;=\; -\int^{x} \upsilon(x')dx'.$$

Combining with equation (3) and solving:

$$F(x) \;=\; \frac{1}{2}\left[u(x) - \int^{x} \upsilon(x')dx'\right]. \quad G(x) \;=\; \frac{1}{2}\left[u(x) + \int^{x} \upsilon(x')dx'\right].$$

Hence, in the interval $0 \le \lambda \le L$ and $0 \le \mu \le L$,

$$F(\lambda) \;=\; \frac{1}{2}\left[u(\lambda) - \int^{\lambda} \upsilon(x')dx'\right] \text{ and } G(\mu) \;=\; \frac{1}{2}\left[u(\mu) + \int^{\mu} \upsilon(x')dx'\right],$$

so that

(5) $$\Phi(\mu,\lambda) \;=\; F(\lambda) + G(\mu) \;=\; \frac{u(\lambda) + u(\mu)}{2} - \frac{1}{2}\int_{\mu}^{\lambda} \upsilon(x')dx'.$$

Thus, given $u(x)$ and $\upsilon(x)$ in the domain $0 \le \lambda \le L$ and $0 \le \mu \le L$, Φ is known in the shaded region. Once again, one cannot specify $\partial\Phi/\partial n$ within the region for fear of overdetermining the solution.

(C) If now, *in addition*, one specifies $\Phi(x=0,t)$ and $\Phi(x=L,t)$ for $t > 0$, then on physical grounds one would expect the solution Φ for all time from $t = 0$ to $t = \infty$ and for all x from $x = 0$ to $x = L$ to be known. To show this mathematically, consider the following:

In the desired domain, $\lambda = x - ct$ runs from $\lambda = L$ to $\lambda = -\infty$, while $\mu = x + ct$ runs from $\mu = 0$ to $\mu = +\infty$. From discussion (B), $F(\lambda)$ and $G(\mu)$ are known in the domain $0 \leq \mu/\lambda \leq L$. Hence, by hypothesis,

(6) $F(-ct) + G(ct)$ *is known* for $t \geq 0$.

(7) $F(L-ct) + G(L+ct)$ *is known* for $t \geq 0$.

From (6), since $G(\mu)$ is known for argument μ from 0 to L, $F(-ct)$ is known from argument $ct = 0$ to $ct = L$, or $F(\lambda)$ from $\lambda = 0$ to $\lambda = -L$.

From (7), since $F(L-ct)$ is known from $L-ct = 0$ to $L-ct = L$ (or from $ct = L$ to $ct = 0$), then $G(L+ct) = G(2L+ct-L)$ is known from argumental values of L to $2L$, i.e. $G(\mu)$ from $\mu = L$ to $\mu = 2L$ is known.

Mapping procedure (Fig. 8-8)

From (B), Φ in region A is known.

Knowing $G(\mu)$ from $\mu = 0$ to L, $F(\lambda)$ from $\lambda = 0$ to $-L$ is known.
Knowing $F(\lambda)$ from $\lambda = 0$ to L, $G(\mu)$ from $\mu = L$ to $\mu = 2L$ is known.

Thus, Φ is known in region B.

Knowing $G(\mu)$ from $\mu = L$ to $2L$, $F(\lambda)$ from $\lambda = -L$ to $-2L$ is known.
Knowing $F(\lambda)$ from $\lambda = 0$ to $-L$, $G(\mu)$ from $\mu = 2L$ to $\mu = 3L$ is known.

Thus, Φ is known in region C.

Continuing in the forward direction, regions shown in Fig. 8-8 will be mapped out. Proceeding in the backward direction will map out the remaining regions.

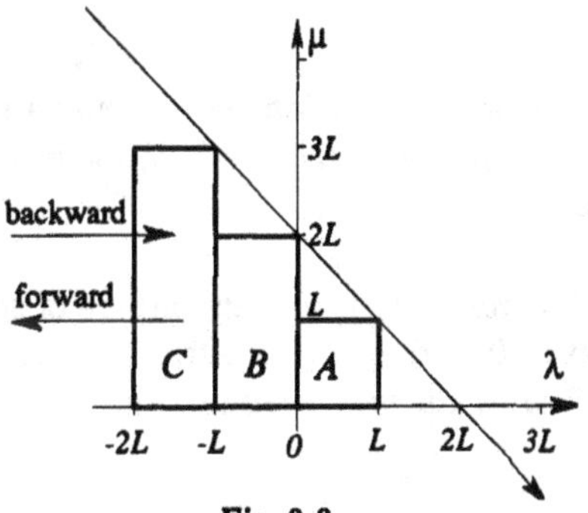

Fig. 8-8

Chapter 9

Heat Conduction

9.1 Fundamental Equations

If κ = thermal conductivity, E = energy content per unit volume, S = energy supplied per unit volume per unit time by some external agency, and c = specific heat per unit volume, then $\mathbf{Q} = -\kappa\nabla T$, where $\mathbf{Q}$ is the heat flux, and $E = cT$. From the first law of thermodynamics,

$$\frac{\partial E}{\partial t} = -\nabla\cdot\mathbf{Q} + S,$$

where $-\nabla\cdot\mathbf{Q}$ represents the heat influx. Substituting for E and $\mathbf{Q}$, we have:

$$c\frac{\partial T}{\partial t} = \kappa\nabla^2 T + S.$$

Defining $s = \frac{S}{c}$ and $\lambda = \frac{\kappa}{c}$, we get

$$\text{(1)} \qquad \frac{\partial T}{\partial t} = \lambda\nabla^2 T + s.$$

Equation (1) is an inhomogeneous partial differential equation of the parabolic type (see Section 8.2.1), and is of first-order in time and second-order in space.

Theorem: If s is a given function of space and time, and $T(\mathbf{r},t{=}0)$ is given at $t = 0$, and if for $t \geq 0$ either $T(\mathbf{r},t)$ or $(\nabla T)_n$ at the boundary is given, then $T(\mathbf{r},t)$ is *uniquely* determined for all $t > 0$.

Proof: Define $\mathcal{F}(\mathbf{r},t) \equiv T_1(\mathbf{r},t) - T_2(\mathbf{r},t)$, where $T_1(\mathbf{r},t)$ and $T_2(\mathbf{r},t)$ both satisfy the same conditions above. From (1), $\dfrac{\partial T_1(\mathbf{r},t)}{\partial t} - \lambda\nabla^2 T_1(\mathbf{r},t) = s$ and $\dfrac{\partial T_2(\mathbf{r},t)}{\partial t} - \lambda\nabla^2 T_2(\mathbf{r},t) = s$. Subtracting, we have $\dfrac{\partial \mathcal{F}}{\partial t} - \lambda\nabla^2\mathcal{F} = 0$, and after multiplying by $\mathcal{F}$, we get

$$\mathcal{F}\frac{\partial \mathcal{F}}{\partial t} = \lambda\mathcal{F}\nabla^2\mathcal{F} = \lambda\left\{\nabla\cdot[\mathcal{F}\nabla\mathcal{F}] - (\nabla\mathcal{F})^2\right\}.$$

Integrating over all space,

$$\frac{\partial}{\partial t}\int_V \frac{\mathcal{F}^2}{2}d\tau = \lambda\int_S \mathcal{F}\nabla\mathcal{F}\cdot d\mathbf{S} - \lambda\int_V (\nabla\mathcal{F})^2 d\tau .$$

By hypothesis, either $\mathcal{F} = 0$ or $\nabla\mathcal{F}\cdot\boldsymbol{n} = 0$ on the boundary. Integrating over time:

$$\int_{t=0}^{t>0} dt\,\frac{\partial}{\partial t}\int_V \frac{\mathcal{F}^2}{2}d\tau = \int_V \left|\frac{\mathcal{F}^2}{2}d\tau\right|_{t=0}^{t>0} = -\lambda\int_{t=0}^{t>0} dt\int_V (\nabla\mathcal{F})^2 d\tau .$$

Since $\mathcal{F} = 0$ at $t = 0$, we are left with $\displaystyle\int_V \left|\frac{\mathcal{F}^2}{2}d\tau\right|_{t>0}$ = negative quantity. Since the integral on the left side is a positive quantity, this result is clearly impossible unless $\mathcal{F} \equiv 0$.

9.2 Infinite Medium

(A) HOMOGENEOUS SOLUTION ($s = 0$)

Consider the homogeneous equation

$$(1) \qquad \frac{\partial T}{\partial t} - \lambda\,\nabla^2 T = 0.$$

Performing a Fourier analysis,

$$(2) \qquad T(\mathbf{r},t) = \frac{1}{(2\pi)^{3/2}}\int_{-\infty}^{+\infty} F(\mathbf{k},t)e^{i\mathbf{k}\cdot\mathbf{r}}d^3\mathbf{k}.$$

Substituting into (1):

$$\frac{1}{(2\pi)^{3/2}}\int_{-\infty}^{+\infty} d^3\mathbf{k}\left\{\frac{\partial F(\mathbf{k},t)}{\partial t} + \lambda k^2 F(\mathbf{k},t)\right\}e^{i\mathbf{k}\cdot\mathbf{r}} = 0.$$

Multiplying by $e^{-i\mathbf{k}'\cdot\mathbf{r}}$ for $\mathbf{k}'$ an arbitrary vector in $\mathbf{k}$ space and integrating over all space:

$$\int_{-\infty}^{+\infty} e^{-i\mathbf{k}'\cdot\mathbf{r}} d^3\mathbf{r}\int_{-\infty}^{+\infty} d^3\mathbf{k}\left\{\frac{\partial F(\mathbf{k},t)}{\partial t} + \lambda k^2 F(\mathbf{k},t)\right\}e^{i\mathbf{k}\cdot\mathbf{r}} = 0.$$

$$\Downarrow$$

$$\int_{-\infty}^{+\infty} d^3\mathbf{k}\left\{\frac{\partial F(\mathbf{k},t)}{\partial t} + \lambda k^2 F(\mathbf{k},t)\right\}\delta^3(\mathbf{k}-\mathbf{k}') = 0.$$

$$\Downarrow$$

$$\frac{\partial F(\mathbf{k}',t)}{\partial t} + \lambda k'^2 F(\mathbf{k}',t) = 0 \quad \Rightarrow \quad \frac{\partial F(\mathbf{k},t)}{\partial t} = -\lambda k^2 F(\mathbf{k},t).$$

The last step follows since $\mathbf{k}'$ is arbitrary. Expressing $F(\mathbf{k},t) = f(\mathbf{k})e^{-k^2\lambda t}$, we have from (2):

$$T(\mathbf{r},t) = \frac{1}{(2\pi)^{3/2}}\int_{-\infty}^{+\infty} f(\mathbf{k})e^{i\mathbf{k}\cdot\mathbf{r}-k^2\lambda t}d^3\mathbf{k} \quad \rightarrow \quad T(\mathbf{r},0) = \frac{1}{(2\pi)^{3/2}}\int_{-\infty}^{+\infty} f(\mathbf{k})e^{i\mathbf{k}\cdot\mathbf{r}}d^3\mathbf{k}$$

Thus,

$$(3) \qquad f(\mathbf{k}) = \frac{1}{(2\pi)^{3/2}}\int_{-\infty}^{+\infty} T(\mathbf{r},0)e^{-i\mathbf{k}\cdot\mathbf{r}}d^3\mathbf{r},$$

so that knowledge of $T(\mathbf{r},0)$ yields knowledge of $f(\mathbf{k})$ and therefore $T(\mathbf{r},t)$ for all time.

Green's function (G_o)

If $T(\mathbf{r},0) = \delta^3(\mathbf{r}-\mathbf{r}_o)$ at $t = 0$, then for $t > 0$, $T(\mathbf{r},t) \equiv G_o(\mathbf{r}-\mathbf{r}_o,t)$. From (3) above,

$$f(\mathbf{k}) = \frac{1}{(2\pi)^{3/2}}\int_{-\infty}^{+\infty}\delta^3(\mathbf{r}-\mathbf{r}_o)e^{-i\mathbf{k}\cdot\mathbf{r}}d^3\mathbf{r} = \frac{e^{-i\mathbf{k}\cdot\mathbf{r}_o}}{(2\pi)^{3/2}}.$$

Substituting into (2):

$$T(\mathbf{r},t) = \frac{1}{(2\pi)^3}\int_{-\infty}^{+\infty} e^{i\mathbf{k}\cdot(\mathbf{r}-\mathbf{r}_o)-k^2\lambda t}d^3\mathbf{k}$$

$$= \frac{1}{(2\pi)^3}\left\{\int_{-\infty}^{+\infty} e^{ik_1(x-x_o)-k_1^2\lambda t}dk_1 \int_{-\infty}^{+\infty} e^{ik_2(y-y_o)-k_2^2\lambda t}dk_2 \int_{-\infty}^{+\infty} e^{ik_3(z-z_o)-k_3^2\lambda t}dk_3\right\}$$

Evaluating

$$\int_{-\infty}^{+\infty} e^{ik_1(x-x_o)-k_1^2\lambda t}dk_1 = \int_{-\infty}^{+\infty}\exp\left[-\lambda t\left(k_1 - \frac{i(x-x_o)}{2\lambda t}\right)^2 - \frac{(x-x_o)^2}{4\lambda t}\right]dk_1$$

$$= (\pi/\lambda t)^{1/2}e^{-(x-x_o)^2/4\lambda t},$$

and substituting, we have

$$T(\mathbf{r},t) = \frac{1}{(2\pi)^3}(\pi/\lambda t)^{3/2}e^{-\frac{|\mathbf{r}-\mathbf{r}_o|^2}{4\lambda t}}.$$

Thus,

$$G_o(|\mathbf{r}-\mathbf{r}_o|,t) = \frac{1}{(4\pi\lambda t)^{3/2}}e^{-\frac{|\mathbf{r}-\mathbf{r}_o|^2}{4\lambda t}} \text{ in 3 dimensions.}$$

(4)

$$G_o(x-x_o,t) = \frac{1}{(4\pi\lambda t)^{1/2}}e^{-\frac{(x-x_o)^2}{4\lambda t}} \text{ in 1 dimension.}$$

Properties of $G_o(|\mathbf{r}-\mathbf{r}_o|,t)$ for $t>0$

1. $\dfrac{\partial G_o}{\partial t} = \lambda\nabla^2 G_o$, i.e. G_o satisfies the homogeneous equation (1).

Proof: For $t>0$, $T(\mathbf{r},t) = G_o$. Thus, since $T(\mathbf{r},t)$ satisfies the homogeneous equation, so does G_o.

2. $\int_{-\infty}^{+\infty} G_o(|\mathbf{r}-\mathbf{r}_o|,t)d\tau = 1.$

Proof: $$\int_{-\infty}^{+\infty} G_o(|\mathbf{r}-\mathbf{r}_o|,t)d\tau = \iiint_{-\infty}^{+\infty}\frac{1}{(4\pi\lambda t)^{3/2}}e^{-\frac{(x-x_o)^2}{4\lambda t}-\frac{(y-y_o)^2}{4\lambda t}-\frac{(z-z_o)^2}{4\lambda t}}dxdydz$$

$$= \frac{1}{(4\pi\lambda t)^{1/2}}\left[(4\pi\lambda t)^{3/2}\right]^3 = 1.$$

3. As $t \to 0+$, $G_o \to \delta^3(\mathbf{r}-\mathbf{r}_o)$.

4. If $T(\mathbf{r},0)$ is known at $t = 0$, then for $t \geq 0$, $T(\mathbf{r},t) = \int G_o(\mathbf{r}-\mathbf{r}',t)T(\mathbf{r}',0)d\tau'$.

Proof:

(a) $\left(\frac{\partial}{\partial t} - \lambda\nabla^2\right)T(\mathbf{r},t) = \int\left(\frac{\partial}{\partial t} - \lambda\nabla^2\right)G_o(\mathbf{r}-\mathbf{r}',t)T(\mathbf{r}',0)d\tau' = 0$ from property 1 above. Thus, $T(\mathbf{r},t)$ satisfies the homogeneous equation.

(b) As $t \to 0$, $T(\mathbf{r},t) \to T(\mathbf{r},0)$ since $G_o(\mathbf{r}-\mathbf{r}',t) \to \delta^3(\mathbf{r}-\mathbf{r}')$, so that the boundary conditions at $t = 0$ are satisfied.

(c) As $r \to \infty$, $T(\mathbf{r},t) \to 0$ since $G_o(\mathbf{r}-\mathbf{r}',t) \to 0$, so that the boundary conditions as $r \to \infty$ are satisfied.

(B) INHOMOGENEOUS SOLUTION ($s \neq 0$)

The procedure, in general, is to find a particular solution of the inhomogeneous equation and the general solution of the homogeneous equation, and then combine to form the most general solution.

Particular solution

Theorem: If one can find an inhomogeneous Green's function $\mathcal{G}_o(\mathbf{r}-\mathbf{r}_o,t-t_o)$ such that

$$\left(\frac{\partial}{\partial t} - \lambda\nabla^2\right)\mathcal{G}_o(\mathbf{r}-\mathbf{r}_o,t-t_o) = \delta^3(\mathbf{r}-\mathbf{r}_o)\delta(t-t_o), \tag{5}$$

then a particular solution can always be written as:

$$T(\mathbf{r},t) = \iint \mathcal{G}_o(\mathbf{r}-\mathbf{r}',t-t')s(\mathbf{r}',t')d^3\mathbf{r}'dt'.$$

Proof: Define $\mathcal{G}_o(\mathbf{r}-\mathbf{r}_o,t-t_o) = G_o(\mathbf{r}-\mathbf{r}_o,t-t_o)$ for $t - t_o > 0$ and $= 0$ for $t - t_o < 0$, where G_o is the homogeneous Green's function.

Making the transformation $t' \equiv t - t_o$:

$\mathcal{G}_o = 0$ for $t' < 0$, so that the left- and right-hand sides of (5) are zero (the right-hand side by virtue of the δ-function property).

$\mathcal{G}_o = G_o$ for $t' > 0$, so that the left-hand side of (5) vanishes, while the right-hand side again vanishes by virtue of the δ-function property.

We now seek the behavior as $t' \to 0+$. Considering equation (5), from property 3 above, $\mathcal{G}_o = G_o \to \delta^3(\mathbf{r}-\mathbf{r}_o)$, while the right side ia another delta function that blows up as $\delta^3(\mathbf{r}-\mathbf{r}_o)\delta(t' \to 0+)$. Question: Do both sides have the same singularity? Consider, therefore, the following:

$$\int_{-\varepsilon}^{+\varepsilon} dt' \left(\frac{\partial}{\partial t'} - \lambda\nabla^2\right)\mathcal{G}_o(\mathbf{r}-\mathbf{r}_o,t') \underset{\varepsilon\to 0}{\overset{?}{=}} \int_{-\varepsilon}^{+\varepsilon} dt'\, \delta^3(\mathbf{r}-\mathbf{r}_o)\delta(t') = \delta^3(\mathbf{r}-\mathbf{r}_o).$$

From property 3 above, $\lim_{t'\to 0} \mathcal{G}_o(\mathbf{r}-\mathbf{r}_o,t') = \delta^3(\mathbf{r}-\mathbf{r}_o)$, so that the left side of the equation becomes

$$\int_{-\varepsilon}^{+\varepsilon} dt' \frac{\partial}{\partial t'} \mathcal{G}_o(\mathbf{r}-\mathbf{r}_o,t') - \lambda\int_{-\varepsilon}^{+\varepsilon} dt'\, \nabla^2 \mathcal{G}_o(\mathbf{r}-\mathbf{r}_o,t') \equiv \mathrm{I} + \mathrm{II}.$$

$$\text{Term I} = \left|\mathcal{G}_o(\mathbf{r}-\mathbf{r}_o,t')\right|_{-\varepsilon}^{+\varepsilon} = G_o(\mathbf{r}-\mathbf{r}_o,+\varepsilon) \to \delta^3(\mathbf{r}-\mathbf{r}_o) \text{ as } \varepsilon \to 0.$$

$$\text{Term II} = -\lambda\varepsilon(\nabla^2\mathcal{G}_o)_{t=0+} \to 0 \text{ as } \varepsilon \to 0.$$

Hence, both sides approach $\delta^3(\mathbf{r}-\mathbf{r}_o)$ as $t' \to 0+$, and the proof is complete. We have shown (5) to be satisfied for $t < 0$, for $t > 0$ and for $t' \to 0+$. Note the dependence of the inhomogeneous Green's function ($\mathcal{G}_o$) upon the homogeneous Green's function (G_o). This dependence hinges on the fact that there is a first-degree derivative in time. This will be seen again when the wave equation is analyzed.

Given an initial temperature distribution $T(\mathbf{r},0)$, one can consider this as being built up of many temperature pips, i.e. many $\delta^3(\mathbf{r}-\mathbf{r}_o)$. The resultant $T(\mathbf{r},t)$ for all $t > 0$ can then be found by summing up the diffusion of these pips throughout the infinite medium, i.e.

$$(6) \qquad T(\mathbf{r},t) = \int_{\tau'} G_o(\mathbf{r}-\mathbf{r}',t)T(\mathbf{r}',0)d\tau'.$$

If, instead, one is given an external supply of energy $s(\mathbf{r},t)$ at $t > 0$, then the diffusion of heat is expressed by

$$(7) \qquad T(\mathbf{r},t) = \iint \mathcal{G}_o(\mathbf{r}-\mathbf{r}',t-t')s(\mathbf{r}',t')d^3\mathbf{r}'dt',$$

i.e. the diffusion takes place in a manner similar to that of a temperature distribution.

General solution

Theorem: If $s(\mathbf{r},t)$ for $t > 0$ and $T(\mathbf{r},t=0)$ are given, then for $t \geq 0$,

$$(8) \qquad T(\mathbf{r},t) = \int_{-\infty}^{+\infty} G_o(\mathbf{r}-\mathbf{r}',t)T(\mathbf{r}',0)d^3\mathbf{r}' + \int_0^{\infty} dt' \int_{-\infty}^{+\infty} \mathcal{G}_o(\mathbf{r}-\mathbf{r}',t-t')s(\mathbf{r}',t')d^3\mathbf{r}',$$

where the first term on the right is the general solution of the homogeneous equation, and the second term is a particular solution of the inhomogeneous equation.

Proof:

(i) $T(\mathbf{r},t)$ satisfies the inhomogeneous equation

$$\left(\frac{\partial}{\partial t} - \lambda\nabla^2\right)T(\mathbf{r},t) = s(\mathbf{r},t).$$

(ii) At $r = \infty$, $G_o(\mathbf{r}-\mathbf{r}',t) \to 0$, while $\mathcal{G}_o(\mathbf{r}-\mathbf{r}',t) = G_o(\mathbf{r}-\mathbf{r}',t) \to 0$ for $t > 0$ and $\mathcal{G}_o = 0$ for $t < 0$. Hence, $T(\mathbf{r},t) \to 0$ at $r = \infty$.

(iii) If $T(\mathbf{r},0) = 0$ and $s(\mathbf{r},t) = 0$, then $T(\mathbf{r},t) = 0$ everywhere and for all t.

(iv) As $t \to 0, T(\mathbf{r},t) \to T(\mathbf{r},0)$.

Hence, $T(\mathbf{r},t)$ satisfies all the necessary boundary conditions.

Consider the expression

$$\int_0^{\infty} dt' \int_{-\infty}^{+\infty} \mathcal{G}_o(\mathbf{r}-\mathbf{r}',t-t')s(\mathbf{r}',t')d^3\mathbf{r}'.$$

Since $\mathcal{G}_o(\mathbf{r}-\mathbf{r}',t-t') = 0$ for $t' > t$, we can rewrite equation (8) as

$$(9) \qquad T(\mathbf{r},t) = \int_{-\infty}^{+\infty} G_o(\mathbf{r}-\mathbf{r}',t)T(\mathbf{r}',0)d^3\mathbf{r}' + \int_0^{t} dt' \int_{-\infty}^{+\infty} \mathcal{G}_o(\mathbf{r}-\mathbf{r}',t-t')s(\mathbf{r}',t')d^3\mathbf{r}'.$$

We consider next the problem of the semi-infinite medium.

Question: Why should one get, or even expect to get a new Green's function?

Answer: Reflections may occur at the boundaries, and the Green's function should incorporate these boundary properties in the description.

9.3 Semi-Infinite Medium

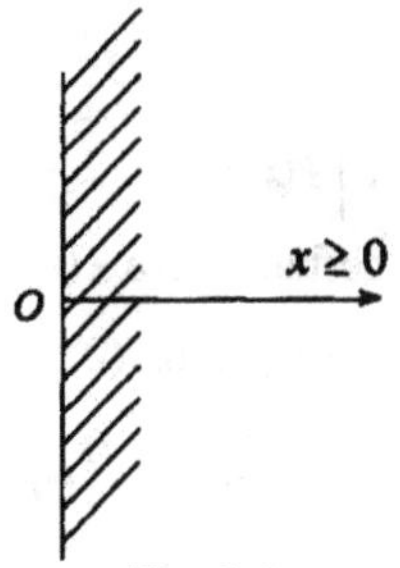

Fig. 9-1

Boundary conditions (Fig. 9-1):

(i) For all time $t \geq 0$, $T(x=0,t) = 0$ (i.e. in contact with a heat reservoir at temperature $T_0 = 0$ at $x = 0$), and $T(x=\infty,t) = 0$.

(ii) For $x \geq 0$, $\frac{\partial T(x,t)}{\partial t} - \lambda\nabla^2 T(x,t) = s(x,t)$ *given.*

(iii) $T(x,t=0)$ *given.*

Consider the following related diffusion problem for the case of an infinite medium (essentially an image problem). Assume that the above temperature distribution and source $s(x,t)$ are reflected in the $-x$ plane, with negative signs for $s(x,t)$ and $T(x,t)$. Then if $s'(x,t)$ and $T'(x,t)$ refer to an infinite medium,

$$T'(x,t=0) = T(x,t=0) \text{ for } x > 0 \text{ and } = -T(|x|,t=0) \text{ for } x < 0.$$

$$s'(x,t) = s(x,t) \text{ for } x > 0 \text{ and } = -s(|x|,t) \text{ for } x < 0.$$

From equation (9), Section 9.2, for $t \geq 0$:

$$(1) \qquad T'(x,t) = \int_{-\infty}^{+\infty} G_0(x-x',t)T'(x',0)dx' + \int_0^t dt' \int_{-\infty}^{+\infty} G_0(x-x',t-t')s'(x',t')dx'.$$

Since $G_0(x-x',t')$ and $s'(x',t')$ are even functions in $(x - x')$ while $T'(x',0)$ is an odd function in x':

$$T'(x,t) = -T'(-x,t) \quad \rightarrow \quad T(0,t) = 0.$$

Thus, the solution of the semi-infinite medium problem is the solution of the infinite medium problem for the $+x$ domain. In other words, for a semi-infinite medium, for $x \geq 0$:

$$(2) \qquad T(x,t) = \int_0^{+\infty} \left[G_0(x-x',t) - G_0(x+x',t)\right]T'(x',t=0)dx'$$

$$+ \int_0^t dt' \int_0^{+\infty} \left[G_0(x-x',t-t') - G_0(x+x',t-t')\right]s'(x',t')dx'.$$

Semi-infinite Green's function

Let $\left|T(x',t)\right|_{t=0} = \delta(x_o - x')$ for $x \geq 0$ and $x' > 0$. The Green's function $G(x,x_o,t)$ is defined from $T(x,t) \equiv G(x,x_o,t) = G_o(x-x_o,t) - G_o(x+x_o,t)$.

From equation (2):

$$(3) \qquad T(x,t) = \int_0^{+\infty} G(x-x',t)T(x',t=0)dx' + \int_0^t dt' \int_0^{+\infty} G(x-x',t-t')s'(x',t')dx'.$$

[Note here that $G(x,x',t)$, a function in 3 variables, x, x' and t, goes over into a function of 2 variables, $(x-x')$ and t.] Note that equation (3) has the same form as equation (1) for the case of an infinite medium, except for the domain of integration and the form of the Green's function.

Define the semi-infinite Green's function $\mathcal{G}(x,x',t) \equiv G(x,x',t)$ for $t > 0$ and $\equiv 0$ for $t < 0$.

Properties of $G(x,x',t)$

(i) $\left(\dfrac{\partial}{\partial t} - \lambda\nabla^2\right)G(x,x',t) = 0$. This follows since $T(x,t)$ satisfies the homogeneous equation.

(ii) $\left(\dfrac{\partial}{\partial t} - \lambda\nabla^2\right)\mathcal{G}(x,x',t) = \delta(x-x')\delta(t)$. For $t < 0$, both sides vanish, while for $t > 0$, $\mathcal{G} = G$ and the left side vanishes from (i) above, while the right side vanishes since $\delta(t \neq 0) = 0$.

(iii) As $t \to 0+$, $G(x,x',t) \to \delta(x-x')$.

(iv) $T(x,t) = \int_0^{+\infty} G(x-x',t)T(x',t=0)dx' + \int_0^t dt' \int_0^{+\infty} G_o(x-x',t-t')s'(x',t')dx'$.

(v) $\left|G(x-x',t)\right|_{x=0} = 0$ since $G_o(-x',t) - G_o(+x',t) = 0$, and $\left|G(x-x',t)\right|_{x=\infty} = 0$.

Thus, the Green's function $G(x-x',t)$ obeys the same boundary conditions as does the temperature.

Importance of the Green's function: The Green's function is completely determined, in principle, by the particular type of boundary in the problem, independent of initial temperature distribution and independent of heat source. Once the form of the Green's function is known, one can plug any given $s(x,t)$ and $T(x,0)$ into the problem in order to solve for $T(x,t)$ at some later time $t > 0$.

9.3.1 Age of the Earth (Treatment due to Lord Kelvin)

Assumptions

(i) At time $t = 0$ the earth starts to condense.

(ii) Non-curvature of the earth, i.e. a semi-infinite medium.

(iii) $T = 0$ at the surface of the earth today. (A fair assumption, since it is small compared with $T(t = 0) = 1200°C$.)

(iv) Uniform temperature of $1200°C$ as the melting point of rock at $t = 0$.

Given

(i) $T(t = 0) = 1200°C$. (Actually somewhat greater due to pressure.)

(ii) $\lambda = k/c = 0.005$ cm^2/sec.

(iii) $s = S/c \approx \dfrac{10^{-6}\text{ ergs/gm·sec}}{10^{7}\text{ ergs/gm·°K}} = 10^{-13}$ °K/sec.

(iv) At present time, $\left|\dfrac{\partial T}{\partial x}\right|_{x=0} = 32°/\text{km} = 32°/10^5\text{cm}.$

From equation (3), Section 9.3,

$$(1)\qquad T(x,t) = \int_0^{+\infty} G(x-x',t)T'(x',t=0)dx' + \int_0^t dt'\int_0^{+\infty} G(x-x',t-t')s'(x',t')dx'.$$

Assuming $T(x',0) \equiv T_0 = 1200°\text{C}$ and constant, and $s(x',t') \equiv s_0 = \text{constant}$,

$$T(x,t) = T_0\int_0^{+\infty} G(x-x',t)dx' + s_0\int_0^t dt'\int_0^{+\infty} G(x-x',t-t')dx'.$$

Differentiating, and evaluating at $x = 0$:

$$(2)\qquad \left|\frac{\partial T}{\partial x}\right|_{x=0} = T_0\int_0^{+\infty}\left|\frac{\partial G(x-x',t)}{\partial x}dx'\right|_{x=0} + s_0\int_0^t dt'\int_0^{+\infty}\left|\frac{\partial G(x-x',t)}{\partial x}dx'\right|_{x=0}.$$

Using the properties of $G(x,x',t)$:

$$\left|\frac{\partial G(x-x',t)}{\partial x}\right|_{x=0} = \left|\frac{\partial G_o(x-x',t)}{\partial x}\right|_{x=0} - \left|\frac{\partial G_o(x+x',t)}{\partial x}\right|_{x=0}$$

$$= \left|-\frac{\partial G_o(x-x',t)}{\partial x'} - \frac{\partial G_o(x+x',t)}{\partial x'}\right|_{x=0} = -2\frac{\partial G_o(x',t)}{\partial x'}.$$

Thus,

$$\int_0^\infty \frac{\partial G(x-x',t)}{\partial x}dx' = -2\int_0^\infty \frac{\partial G_o(x',t)}{\partial x'}dx' = -\frac{2}{(4\pi\lambda t)^{1/2}}\left|e^{-\frac{(x')^2}{4\lambda t}}\right|_{x'=0}^{\infty}$$

$$= \frac{1}{(\pi\lambda t)^{1/2}}.$$

Substituting into (2),

$$\left|\frac{\partial T}{\partial x}\right|_{x=0} = \frac{T_o}{(\pi\lambda t)^{1/2}} + 2s_o\int_0^t \frac{dt'}{\sqrt{4\pi\lambda(t-t')}}.$$

$$(3) \qquad \left|\frac{\partial T}{\partial x}\right|_{x=0} = \frac{T_o}{\sqrt{\pi\lambda t}} + \frac{2s_o\sqrt{t}}{\sqrt{\pi\lambda}} = \frac{1}{\sqrt{\pi\lambda t}}(T_o + 2s_o t).$$

One now compares the value of $\left|\frac{\partial T}{\partial x}\right|_{x=0}$ as calculated from (3) using known constants with the known current value of $\left|\frac{\partial T}{\partial x}\right|_{x=0}$. Expecting a value of $t \approx 10^9$ years $= 3 \times 10^{16}$sec:

$$\sqrt{\pi\lambda t} = (\pi\cdot 5\cdot 10^{-3}\cdot 3\cdot 10^{16})^{1/2} \approx (3^2\cdot 50\cdot 10^{12})^{1/2} \approx 3\cdot 7\cdot 10^6,$$

so that calculations from (3) yield

$$\left|\frac{\partial T}{\partial x}\right|_{x=0} = \frac{1200 + 6\cdot 10^{16}\cdot 10^{-13}}{21\cdot 10^6} = \frac{72}{21\cdot 10^4} \sim \frac{1}{3000},$$

while the current known value is

$$\left|\frac{\partial T}{\partial x}\right|_{x=0} \sim \frac{32°}{\text{km}} = \frac{32°}{10^5\text{cm}} \approx \frac{1}{300}.$$

When Lord Kelvin first computed t, he neglected s_0 which we see contributes the greater portion to $\left|\frac{\partial T}{\partial x}\right|_{x=0}$ in the ratio $\frac{6000}{1200} = 5$ to 1.

9.3.2 Temperature Variation of the Earth's Surface

In contrast with the previous problem for which the gradient $\partial T/\partial x$ at the surface of the earth extends through a distance of kilometers, we are here concerned with a gradient that extends through centimeters, or meters, i.e. we are seeking a daily or yearly variation in temperature, with boundary conditions that are functions of time.

Boundary conditions

(i) $T = A\sin\omega t$ at $x = 0$. (If not a sine function itself, then at least a Fourier series in sines.)

(ii) $T(x,t=0)$ given at $t = 0$.

(iii) $S(x,t)$ known for $t \geq 0$.

Whenever the boundary temperature is a function of time, as in boundary condition (i), it is only necessary to find a particular solution T_p that satisfies the homogeneous equation $\partial T_p/\partial t = \lambda\nabla^2 T_p$. Since T_p is given at $x = 0$, if we define $T' \equiv T - T_p$, then conditions for T' are:

(iv) $T' = 0$ at $x = 0$.

(v) T' will satisfy $\partial T'/\partial t = \lambda\nabla^2 T' + S$.

(vi) T' is known at $t = 0$.

In this manner we have reduced an inhomogeneous boundary condition to its corresponding homogeneous boundary condition for a semi-infinite medium.

We seek, then, a solution of the homogeneous equation:

$$(1) \qquad \frac{\partial T_p}{\partial t} = \lambda\nabla^2 T_p.$$

Let $T(x=0,t) = Ae^{i\omega t}$, the imaginary part of which we seek in order to satisfy our boundary conditions at $x = 0$.

Assuming a particular solution $T_p(x,t) = F(x)e^{i\omega t}$ and substituting into (1), we have $i\omega F(x)e^{i\omega t} = \lambda\nabla^2 F(x)e^{i\omega t}$, or

$$(2) \qquad \frac{d^2F(x)}{dx^2} = \frac{i\omega}{\lambda}F(x) = \frac{e^{i\pi/2}\omega}{\lambda}F(x).$$

Choosing $\omega > 0$, solutions to (2) are given by

$$(3) \qquad F(x) = B\exp\left[(i\pi/4)(\omega/\lambda)^{1/2}x\right] + C\exp\left[-(i\pi/4)(\omega/\lambda)^{1/2}x\right],$$

where B and C have yet to be determined. Since $e^{i\pi/4} = (1+i)/\sqrt{2}$, and since at $x = \infty$, $\exp\left[(i\pi/4)(\omega/\lambda)^{1/2}x\right] = \infty$, B must $= 0$, since $F(\infty) = 0$. Also, $F(0) = C = A$, so that $C = A$. We finally have for T_p:

$$\begin{aligned} T_p(x,t) &= A\exp\left[-(i\pi/4)(\omega/\lambda)^{1/2}x\right]\exp\left[i\omega t\right] \\ &= A\exp\left[-(\omega/2\lambda)^{1/2}(1+i)x + i\omega t\right] \\ &= A\exp\left[-(\omega/2\lambda)^{1/2}x\right]\exp\left[i(\omega t-(\omega/2\lambda)^{1/2}x\right]. \end{aligned}$$

Since we seek the imaginary part as the physical solution,

$$(4) \qquad T_p(x,t) = A\exp\left[-(\omega/2\lambda)^{1/2}x\right]\sin\left[\omega t-(\omega/2\lambda)^{1/2}x\right].$$

[Note that $T_p(x,t)$ has a part that decays in the $+x$ direction, and includes phase changes that accompany varying changes in x.

Note also that the equation $\partial T_p(x,t)/\partial t = \lambda\nabla^2 T_p(x,t)$ is a real equation when λ is real. If T_p is complex, one can always equate real and imaginary parts. The Schrödinger equation, however, is not real since λ is complex.]

Yearly variation

For soil,

$$\lambda = \frac{0.002\ \text{cm}^2}{\text{sec}}. \quad \omega_y = \frac{2\pi}{T} = \frac{2\pi}{3\cdot 10^7\ \text{sec}}.$$

$$(\omega_y/2\lambda)^{1/2} = [2\pi/(3\cdot 10^7\cdot 4\cdot 10^{-3})]^{1/2}\ \text{cm}^{-1} \approx (1/2\cdot 10^4)^{1/2} \approx 7\cdot 10^{-3}\ \text{cm}^{-1}.$$

At $x = 0$, $T_p = A\sin\omega_y t$ (Fig. 9-2a).

At $x = 1$ meter,

$$(\omega_y/2\lambda)^{1/2}x = 0.7 \approx \pi/4.$$

$$e^{-0.7} \approx 1/2.$$

$T_p(x,t) = (A/2)\sin(\omega_y t - \pi/4)$ (Fig. 9-2b).

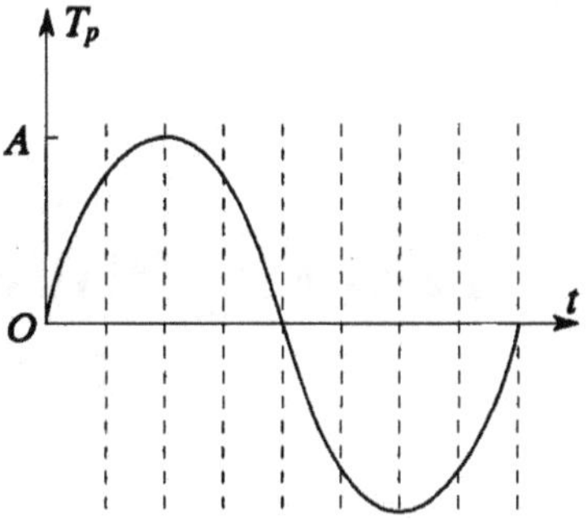

Fig. 9-2a

At $x = 4$ meter,

$$(\omega_y/2\lambda)^{1/2}x \approx \pi.$$

$$\exp\left[-(\omega_y/2\lambda)^{1/2}x\right] \approx 1/2^4.$$

$T_p(x,t) = (A/2)\sin(\omega_y t - \pi)$.

Thus, at a depth of 4 meters, the earth is coldest in summer and hottest in winter (Fig. 9-2c).

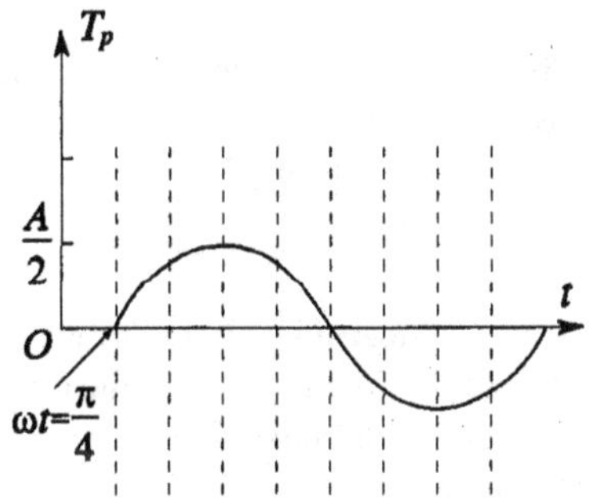

Fig. 9-2b

Daily variation

Yearly and daily phase changes will be equal when $\sqrt{\omega_d}x_d = \sqrt{\omega_y}x_y$ or when

$$\frac{x_d}{x_y} = (\omega_d/\omega_y)^{-1/2} \approx \frac{1}{20}.$$

Thus, $x = 1$ meter (yearly) $\Rightarrow$ $x = 5$ cm (daily) to get $\pi/4$ phase change, and $x = 4$ meters (yearly) $\Rightarrow$ $x = 20$ cm (daily) to get π phase change.

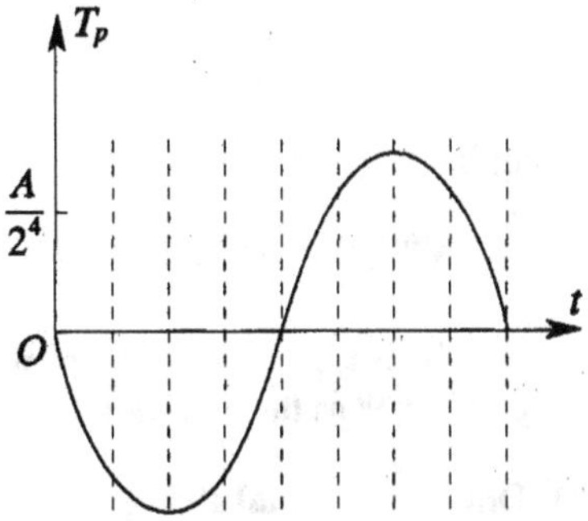

Fig. 9-2c

Chapter 10

The Eigenvalue Problem

10.1 Eigenvalues and Eigenfunctions

In Section 8.1 we considered the particular example of the wave equation type of differential equation,

$$(1) \qquad \nabla^2\Phi - \frac{1}{\upsilon^2}\frac{\partial^2\Phi}{\partial t^2} = 0.$$

For a one-dimensional vibrating string, Φ is the displacement and $\upsilon^2 = T\,(\text{tension})/\rho\,(\text{density})$. For the case of 2 dimensions, we are dealing with a vibrating membrane. We seek the characteristic frequency of the string/membrane.

Assuming $\Phi(\mathbf{r},t) = \Psi(\mathbf{r})e^{i\omega t}$ and substituting into (1), we have

$$(\nabla^2\Psi)e^{i\omega t} + \frac{\omega^2}{\upsilon^2}\frac{\partial^2\Psi}{\partial t^2}e^{i\omega t} = 0.$$

Setting $\lambda = \omega^2/\upsilon^2$, we get:

$$(2) \qquad -\nabla^2\Psi(\mathbf{r}) = \lambda\Psi(\mathbf{r})$$

Equation (2) is typical of an eigenvalue problem, where one regards $-\nabla^2$ as the *operator*, Ψ as the *eigenfunction*, and λ as the *eigenvalue*.

(A) One-dimensional string

For the one-dimensional vibrating string, equation (2) becomes

$$-\frac{d^2\Psi(x)}{dx^2} = \lambda\Psi(x).$$

Solutions are given by $\Psi(x) = A\sin kx + B\cos kx$, where $\lambda = k^2$. Since boundary conditions demand that $\Psi = 0$ at $x = 0$, we see that $B = 0$.
In addition, since boundary conditions also demand that $\Psi = 0$ at $x = L$, it follows that $A\sin kL = 0 \Rightarrow k = n\pi/L$ $(n = 0,1\ldots\infty)$. Thus, due to the boundary conditions, $\lambda = k^2 = (n^2\pi^2)/L^2$ has certain discrete values, namely the eigenvalues.

Substituting for k, we have $\Psi(x) = A\sin n\pi x/L$.

Note that $\int_0^L \Psi_n^2(x)dx = \int_0^L A^2\sin^2\frac{n\pi x}{L}\,dx = \frac{A^2}{2}\int_0^L(1 - \cos 2n\pi x/L)dx = \frac{A^2L}{2}$.

Thus, since we want to normalize the Ψ_n, we choose $A = (2/L)^{1/2}$, yielding for $\Psi_n(x)$:

(3) $\quad \Psi_n(x) = (2/L)^{1/2}\sin n\pi x/L,$

where $\lambda_n = k^2 = (n^2\pi^2)/L^2$ $(n = 1,2\ldots\infty)$, satisfying $-\nabla^2\Psi_n(x) = \lambda_n\Psi_n(x)$.

Characteristics of the eigenvalue problem

1. $-d^2/dx^2$ is a good Hermitian operator for all functions with boundary conditions $\Psi_n = 0$ at $x = 0$ and $x = L$, i.e.

$$\int_0^L \Psi^*\left(-\frac{d^2}{dx^2}\right)\Phi dx = \left\{\int_0^L \Phi^*\left(-\frac{d^2}{dx^2}\right)\Psi dx\right\}^*,$$

where Φ and Ψ are in the class of functions admitted by the boundary conditions of the problem.

Proof:

$$\int_0^L \Psi^*\frac{d^2\Phi}{dx^2}dx = \int_0^L \frac{d}{dx}\left[\Psi^*\frac{d\Phi}{dx}\right]dx - \int_0^L \frac{d\Psi^*}{dx}\frac{d\Phi}{dx}dx$$

$$= \left[\Psi^*\frac{d\Phi}{dx}\right]_0^L - \int_0^L \frac{d\Psi^*}{dx}\frac{d\Phi}{dx}dx = 0 - \int_0^L \frac{d\Psi^*}{dx}\frac{d\Phi}{dx}dx$$

$$= -\int_0^L \frac{d}{dx}\left[\Phi\frac{d\Psi^*}{dx}\right]dx + \int_0^L \frac{d^2\Psi^*}{dx^2}\Phi\,dx = 0 + \int_0^L \frac{d^2\Psi^*}{dx^2}\Phi\,dx.$$

2. $\int_0^L \Psi_m(x)\Psi_n(x)dx = \delta_{mn}$.

3. The $\Psi_n(x)$ ($n = 1,2\ldots\infty$) form a complete and orthonormal set for any function $\Psi(x)$ satisfying $\Psi(x) = 0$ at $x = 0$ and $x = L$.

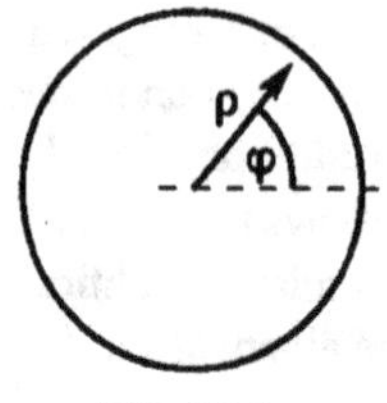

Fig. 10-1

(B) Two-dimensional circular membrane

The eigenvalue equation [from (2)] is given by

$$-\nabla^2\Psi(\rho,\varphi) = \lambda\Psi(\rho,\varphi),$$

where $\lambda = \omega^2/\upsilon^2$ under the condition that $\Psi = 0$ at $\rho = \rho_0$ (radius of the membrane) (Fig. 10-1). To determine the eigenvalues and the eigenfunctions, we can expand $\Psi(\rho,\varphi)$ as $\Psi(\rho,\varphi) = \sum_{m=-\infty}^{+\infty} f_m(\rho)e^{im\varphi}$, since on a circle the $e^{im\varphi}$, or equivalently, the Fourier series, form a complete set, and any function $g(\varphi)$ can be expanded in a Fourier series. Substituting into the eigenvalue equation:

$$\sum_{m=-\infty}^{+\infty}\left[\frac{1}{\rho}\frac{d}{d\rho}\left(\frac{\rho\, df(\rho)}{d\rho}\right) - \frac{m^2 f(\rho)}{\rho^2} + \lambda f(\rho)\right]e^{im\varphi} = 0.$$

Multiplying by $e^{-im\varphi}$ and integrating over the circle:

$$\int_0^{2\pi} e^{-im\varphi}d\varphi \sum_{m=-\infty}^{+\infty} e^{im\varphi}\left[\frac{d^2f(\rho)}{d\rho^2} + \frac{1}{\rho}\frac{df(\rho)}{d\rho} + \left(\lambda - \frac{m^2}{\rho^2}\right)f(\rho)\right] = 0.$$

Using the orthonormality properties of the $e^{im\varphi}$ yields

$$(4) \qquad \frac{d^2f(\rho)}{d\rho^2} + \frac{1}{\rho}\frac{df(\rho)}{d\rho} + \left(\lambda - \frac{m^2}{\rho^2}\right)f(\rho) = 0.$$

Set $\sqrt{\lambda}\rho = z$, so that $\frac{d}{d\rho} = \sqrt{\lambda}\frac{d}{dz} \rightarrow \frac{d^2}{d\rho^2} = \lambda\frac{d^2}{dz^2}$, and (4) becomes

$$\lambda\frac{d^2f(z)}{dz^2} + \frac{\lambda}{z}\frac{df(z)}{dz} + \left(\lambda - \frac{m^2\lambda}{z^2}\right)f(z) = 0.$$

Dividing through by λ we arrive at Bessel's equation with integral m values:

$$\text{(5)} \qquad \frac{d^2 f_m(z)}{dz^2} + \frac{1}{z}\frac{df_m(z)}{dz} + \left(1 - \frac{m^2}{z^2}\right) f_m(z) = 0.$$

The m are necessarily integers since they arose from expanding $\Psi(\rho,\varphi)$ into a Fourier series. In Section 7.5.4 we discussed the case for m being integer-valued, for which the second solution of Bessel's equation is given by the Hankel functions H_m^1 or H_m^2. Thus, the most general solution of the above vibrating membrane differential equation (5) is

$$f_m(z) = AJ_m(z) + BH_m^1(z).$$

Since we want $f_m(z)$ to be finite at $\rho = 0$ (i.e. at $z = 0$), while on the other hand $H_m^1(z)$ possesses a logarithmic singularity at $z = 0$, we must have $B = 0$. Thus,

$$\text{(6)} \qquad f_m(z) = AJ_m(z).$$

Since $\Psi(\rho_o,\varphi) = 0$ at $\rho = \rho_o$ for all φ, it follows that $J_m(\sqrt{\lambda}\rho_o) = 0$. It is this boundary condition which gives λ its discreteness property. We now seek the roots of $J_m(z) = 0$.

For $m = 0$ (Fig. 10-2), roots are z_{01}, z_{02}, z_{03}, etc.

For $m \neq 0$ (Fig. 10-3), roots are z_{mn}, i.e. z_{m1}, z_{m2}, z_{m3}, etc . The origin is not included as a root since at the origin, $z = 0$, or $\sqrt{\lambda}\rho = 0$. If $\rho \neq 0$, then λ must $= 0$, which, however, means no vibration. Hence, roots are z_{mn}, where $n = 1, 2 \ldots \infty$.

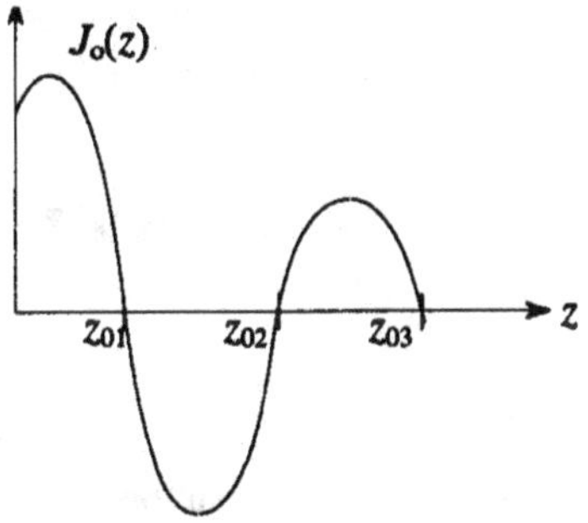

Fig. 10-2

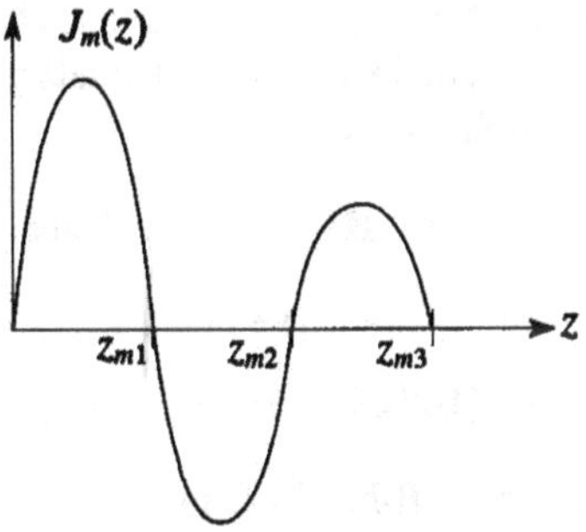

Fig. 10-3

Corresponding to roots z_{mn}, there are the eigenvalues

(7) $\lambda_{mn} = (z_{mn}/\rho_o)^2$,

and the corresponding eigenfunctions

(8) $\Psi_{mn}(\rho,\varphi) = AJ_m(\sqrt{\lambda_{mn}}\,\rho)e^{im\varphi}$.

Degeneracy

For the case of $+m$ and $-m$ (but $\neq 0$), if we recall that $J_{+m}(z)$ differs from $J_{-m}(z)$ only by the factor $(-1)^m$, then it follows that both $J_{+m}(z)$ and $J_{-m}(z)$ will have identical roots, giving rise to identical eigenvalues. The $\Psi_{mn}(\rho,\varphi)$, however, do not depend only on $J_m(z)$, so that we have different eigenfunctions that give rise to the same eigenvalue, i.e. *degeneracy*.

Tabulating the roots in increasing order, we have:

$z_{01} = 2.405$; $z_{11} = 3.832$; $z_{21} = 5.135$; $z_{02} = 5.520$; $z_{31} = 6.379$; etc.

From $\lambda = \omega^2/\upsilon^2$, we note that λ_{mn} has a frequency, $\omega_{mn} = \sqrt{\lambda_{mn}}\,\upsilon = z_{mn}\upsilon/\rho_o$, associated with it. Thus, the eigenvalues of the circular membrane are given by

(9) $$\omega_{mn} = \frac{z_{mn}\upsilon}{\rho_o}.$$

We now seek the vibration patterns for the above characteristic frequencies.

m = 0, n = 1 This is a non-degenerate case, with only one eigenfunction, $\Psi_{01} = AJ_o(z_{01}\rho/\rho_o)$. Note that for $\rho < \rho_o$ (i.e. within the boundary of the problem), $z_{01}\rho/\rho_o < z_{01}$, and $J_o(z)$ has no roots for $z < z_{01}$ (Fig. 10.2). Hence there is a node only at the boundary, and the amplitude of the oscillation is the amplitude of $J_o(z)$.

m = 1, n = 1 Here we have degeneracy ($m = \pm1$). $\Psi_{\pm1,1} = AJ_1(z_{11}\rho/\rho_o)e^{\pm i\varphi}$.

$$(1/2)(\Psi_{+1,1}) + (1/2)(\Psi_{-1,1}) = AJ_1(z_{11}\rho/\rho_o)\cos\varphi.$$
$$(1/2i)(\Psi_{+1,1}) - (1/2i)(\Psi_{-1,1}) = AJ_1(z_{11}\rho/\rho_o)\sin\varphi.$$

Since the left-hand sides of both equations are separately eigensolutions, $J_1(z_{11}\rho/\rho_o)\cos\varphi$ and $J_1(z_{11}\rho/\rho_o)\sin\varphi$ are also eigensolutions.

In fact, any linear combination of these degenerate solutions will also be eigenfunctions. Hence, for $m = \pm 1, n = 1$:

$$\Psi_{\pm 1,1}(\rho,\varphi) \propto J_1(z_{11}\rho/\rho_o)\sin(\varphi-\theta),$$

where θ is an arbitrary constant.

Question: What will the vibration patterns look like, and where else but at the boundary will $\Psi = 0$?

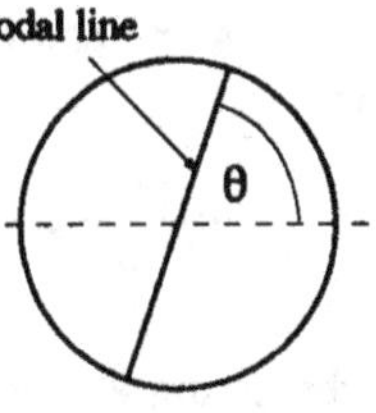

Fig. 10-4

From the form of Ψ above, we see that $\Psi = 0$ at $\varphi = \theta$. Since $J_1(z)$ is zero at the origin (Fig. 10.3) and at the root z_{11} (i.e. when $\rho = \rho_o$), nodes occur at $\rho = \rho_o$ for all φ, and along the nodal line $\varphi = \theta$ (Fig. 10-4).

m = ±2, n = 1 In this case we have degeneracy again. $\Psi_{\pm 2,1} = AJ_2(z_{21}\rho/\rho_o)e^{\pm 2i\varphi}$. Writing as before:

$$(1/2)[\Psi_{+2,1} + \Psi_{-2,1}] = AJ_2(z_{21}\rho/\rho_o)\cos 2\varphi.$$

$$(1/2i)[\Psi_{+2,1} - \Psi_{-2,1}] = AJ_2(z_{21}\rho/\rho_o)\sin 2\varphi.$$

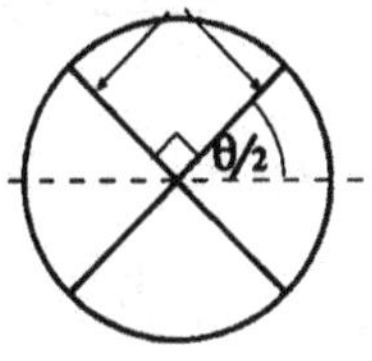

Fig. 10-5

Since $J_2(z_{21}\rho/\rho_o)\cos 2\varphi$ and $J_2(z_{21}\rho/\rho_o)\sin 2\varphi$ are eigensolutions, a linear combination of the two will also be eigensolutions.

Thus,

$$\Psi_{\pm 2,1}(\rho,\varphi) \propto J_2(z_{21}\rho/\rho_o)\sin(2\varphi-\theta),$$

so that $\Psi_{\pm 2,1}(\rho,\varphi) = 0$ at $\varphi = \theta/2$, or $\varphi = \theta/2 + \pi/2$, which results in the two nodal lines shown in Fig. 10-5.

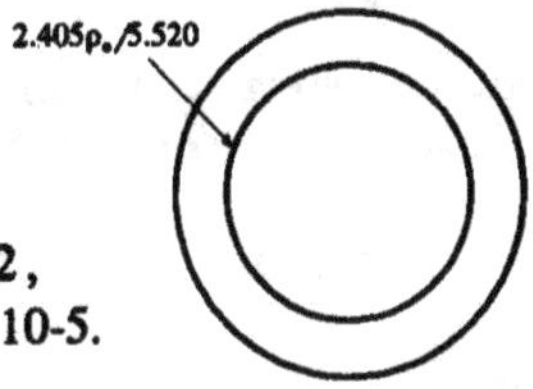

Fig. 10-6

m = 0, n = 2 As was the case for $m = 0$, there is degeneracy.

$\Psi_{02} = AJ_o(z_{02}\rho/\rho_o) \Rightarrow J_o(z_{02}\rho/\rho_o) = 0$ at $z = z_{01}, z_{02}$, etc. (Fig. 10-2), i.e. $\Psi_{02} = 0$ when $z_{02}\rho/\rho_o = z_{02} \Rightarrow \rho = \rho_o$, and $\Psi_{02} = 0$ when $z_{02}\rho/\rho_o = z_{01}$, or when $\rho = z_{01}\rho_o/z_{02} = 2.405\rho_o/5.520$. The nodal lines for all φ are shown in Fig. 10-6.

Nodal lines for $m \neq 0$ appear as a result of degeneracy. The location of the nodal line in these cases is extremely arbitrary, due to cylindrical symmetry.

10.2 Harmonic Oscillator/Free Particle in a Sphere

(A) HARMONIC OSCILLATOR

In the harmonic oscillator treatment (Section 7.3.1), we saw that by setting $p \rightarrow -i\hbar \frac{\partial}{\partial x}$ and $E \rightarrow i\hbar \frac{\partial}{\partial t}$ in the Schrödinger equation, $E = \frac{p^2}{2m} + V$, we get

$$i\hbar \frac{\partial \Psi}{\partial t} = -\frac{\hbar^2}{2m}\frac{\partial^2 \Psi}{\partial x^2} + V\Psi \rightarrow E\Psi = \left(-\frac{\hbar^2}{2m}\nabla^2 + V\right)\Psi.$$

Substituting $V = kx^2/2$, multiplying through by $2/\hbar\omega = 2(m/k)^{1/2}/\hbar$, and letting $z = (\sqrt{mk}/\hbar)^{1/2}\, x$, we arrive at

$$\frac{d^2\Psi}{dz^2} - z^2\Psi = -\frac{2E\Psi}{\hbar\omega}.$$

Setting $f(z) = \Psi(z)e^{z^2/2}$ results in the Hermite equation

$$\text{(1)} \qquad \frac{d^2 f(z)}{dz^2} - 2z\frac{df(z)}{dz} + 2\lambda f(z) = 0,$$

where $2E/\hbar\omega = (2\lambda + 1) \rightarrow E = (\lambda + 1/2)\hbar\omega$. Solutions of the Hermite equation were given as $f(z) = H_n(z)$, so that $\Psi_n(z) = e^{-z^2/2}H_n(z)$, and unnormalized solutions of the harmonic oscillator are given by

$$\text{(2)} \qquad \Psi_n(\alpha x) = e^{-\alpha^2 x^2/2}H_n(\alpha x).$$

Since

$$\int_{-\infty}^{+\infty} H_n(z)H_m(z)e^{-z^2}dz = \delta_{nm}2^n n!\sqrt{\pi},$$

we have as normalized solutions:

$$\text{(3)} \qquad \Psi_{on}(\alpha x) = \left[\frac{\alpha}{\sqrt{\pi}2^n n!}\right]^{1/2} e^{-\alpha^2 x^2/2}H_n(\alpha x),$$

where $z = \alpha x = x\left[\frac{\sqrt{km}}{\hbar}\right]^{1/2}$.

The eigenvalue equation is given by

(4) $$\left(-\frac{\hbar^2}{2m}\nabla^2 + V\right)\Psi_n = E_n\Psi_n,$$

where

(5) $$E_n = (n + 1/2)\hbar(k/m)^{1/2}.$$

(6) $$\int_{-\infty}^{+\infty}\Psi_{on}(\alpha x)\Psi_{om}(\alpha x)d(\alpha x) = \delta_{nm}.$$

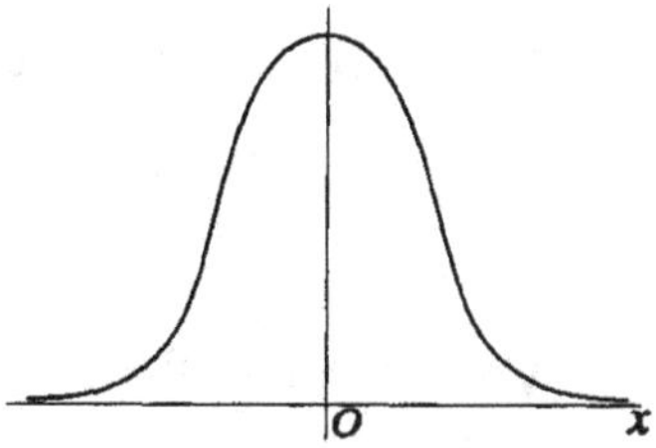

Fig. 10-7

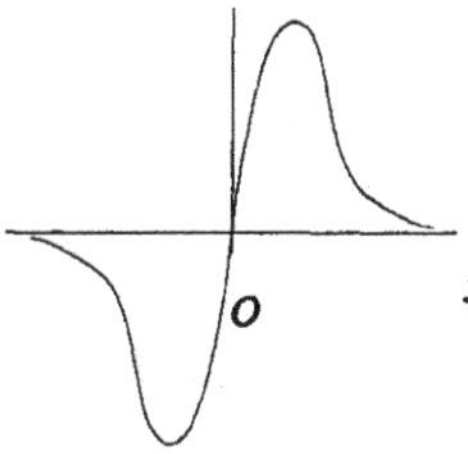

Fig. 10-8

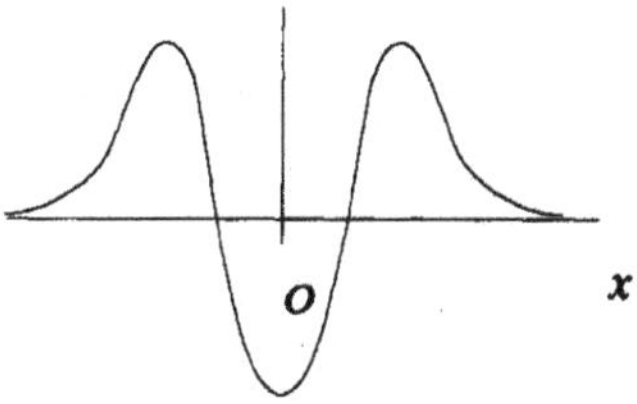

Fig. 10-9

For $n = 0$: $\Psi_{oo}(\alpha x) = Ae^{-\alpha^2x^2/2}$.

For $n = 1$: $\Psi_{o1}(\alpha x) = Bxe^{-\alpha^2x^2/2}$.

For $n = 2$: $\Psi_{o2}(\alpha x) = Ce^{-\alpha^2x^2/2}(Dx^2 - F)$.

Note that the lowest eigenvalue, E_0, corresponds to the case of no node (Fig. 10-7), while the next highest eigenvalue, E_1, corresponds to the case of 1 node (Fig. 10-8), and the next highest eigenvalue, E_2, corresponds to the case of 2 nodes (Fig. 10-9). These are to be compared with the case of the circular membrane.

(B) FREE PARTICLE IN A SPHERE

Since $V = 0$ for a free particle, Schrödinger's equation becomes $-\frac{\hbar^2}{2m}\nabla^2\Psi(\mathbf{r}) = E\Psi(\mathbf{r})$, where $\Psi(\mathbf{r}) = 0$ at $r = a$ (radius of sphere) and is finite for $r \leq a$. If we set $E = \frac{\hbar^2k^2}{2m}$, we have for Schrödinger's equation:

$$-\nabla^2\Psi(\mathbf{r}) = k^2\Psi(\mathbf{r}) = \lambda\Psi(\mathbf{r}).$$

Expanding $\Psi(r,\theta,\varphi)$ in terms of the complete set of spherical harmonics:

$$\Psi(r,\theta,\varphi) = \sum_{l=0}^{\infty}\sum_{m_l=-l}^{+l} R_l(r) Y_{l,m_l}(\theta,\varphi).$$

$$\nabla^2\Psi(r,\theta,\varphi) + k^2\Psi(r,\theta,\varphi) = \sum_{lm_l}\left[\frac{1}{r^2}\frac{d}{dr}\left(r^2\frac{dR_l(r)}{dr}\right)\right]Y_{l,m_l}(\theta,\varphi)$$

$$+\sum_{lm_l}\left\{R_l(r)\left[\frac{1}{r^2\sin\theta}\frac{\partial}{\partial\theta}\sin\theta\frac{\partial Y_{l,m_l}(\theta,\varphi)}{\partial\theta} + \frac{1}{r^2\sin^2\theta}\frac{\partial^2 Y_{l,m_l}(\theta,\varphi)}{\partial\varphi^2}\right]\right\}$$

$$+\sum_{lm_l} k^2 Y_{l,m_l}(\theta,\varphi) R_l(r) = 0.$$

Substituting

$$\frac{1}{\sin\theta}\frac{\partial}{\partial\theta}\sin\theta\frac{\partial Y_{l,m_l}(\theta,\varphi)}{\partial\theta} + \frac{1}{\sin^2\theta}\frac{\partial^2 Y_{l,m_l}(\theta,\varphi)}{\partial\varphi^2} = -l(l+1)Y_{l,m_l}(\theta,\varphi):$$

$$\sum_{lm_l}\left[\frac{1}{r^2}\frac{d}{dr}r^2\frac{dR_l(r)}{dr} + \left(k^2 - \frac{l(l+1)}{r^2}\right)R_l(r)\right]Y_{l,m_l}(\theta,\varphi) = 0.$$

Employing the orthonormality relations for $Y_{l,m_l}(\theta,\varphi)$, we have

$$(7)\qquad \frac{1}{r^2}\frac{d}{dr}r^2\frac{dR_l(r)}{dr} + \left(k^2 - \frac{l(l+1)}{r^2}\right)R_l(r) = 0.$$

Setting $R_l(r) = \dfrac{f_l(r)}{\sqrt{r}}$,

$$\frac{dR_l(r)}{dr} = \frac{1}{\sqrt{r}}\frac{df(r)}{dr} - \frac{1}{2}\frac{f(r)}{r^{3/2}} \Rightarrow \frac{d^2R_l(r)}{dr^2} = \frac{-1}{r^{3/2}}\frac{df(r)}{dr} + \frac{1}{\sqrt{r}}\frac{d^2f(r)}{dr^2} + \frac{3f(r)}{4r^{5/2}}.$$

Substituting in (7):

$$\frac{d^2f(r)}{dr^2} + \frac{1}{r}\frac{df(r)}{dr} - \frac{f(r)}{4r^2} + k^2 f(r) - \frac{l(l+1)f(r)}{r^2} = 0.$$

$$\Downarrow$$

$$\frac{d^2f(r)}{dr^2} + \frac{1}{r}\frac{df(r)}{dr} + k^2 f(r) - \frac{(l+1/2)^2 f(r)}{r^2} = 0.$$

If we now let $z = kr$, we have

$$(8) \qquad \frac{d^2f(z)}{dz^2} + \frac{1}{z}\frac{df(z)}{dz} + \left[1 - \frac{(l+1/2)^2}{z^2}\right] f(z) = 0,$$

which we recognize as Bessel's equation with $\lambda = l + 1/2$. Solutions are given by

$$f_l(z) = \text{const.}\, J_{l+1/2}(z) \quad \rightarrow \quad R_l(r) = \text{const.}\, \frac{J_{l+1/2}(kr)}{\sqrt{r}}.$$

Roots of $J_{l+1/2}(kr) = 0$ are given by $k_{l,n}$, where $n = 1, 2 \ldots \infty$ $\rightarrow$ $\lambda_{l,n} = k_{l,n}^2$.
Finally then, we have for the eigenfunctions:

$$(9) \qquad \Psi_{l,m,n}(r,\theta,\varphi) = \text{const.}\, \frac{J_{l+1/2}(k_{l,n}r)}{\sqrt{r}}\, Y_{l,m}(\theta,\varphi).$$

10.3 The Variational Principle

In Section 5.3.2 it was pointed out that a Hermitian operator H satisfies

$$\int \Psi^* H \Phi d\tau = \left\{\int \Phi^* H \Psi d\tau\right\}^*,$$

where $d\tau$ includes the range of the particular problem.

Eigenvalue characteristics

For $H\Psi = \lambda\Psi$:

(A) All the eigenvalues, λ, are real.

Proof: $\int \Psi^* H \Psi d\tau = \lambda \int \Psi^* \Psi d\tau$, from which

$$\lambda = \frac{\int \Psi^* H \Psi d\tau}{\int \Psi^* \Psi d\tau} \quad \rightarrow \quad \lambda^* = \frac{\left(\int \Psi^* H \Psi d\tau\right)^*}{\left(\int \Psi^* \Psi d\tau\right)^*} = \frac{\int \Psi^* H \Psi d\tau}{\int \Psi^* \Psi d\tau} = \lambda.$$

(B) If (a): $H\Psi = \lambda\Psi$ and (b): $H\Psi' = \lambda'\Psi'$, then $\int \Psi'^* \Psi d\tau = 0$, which says that eigenfunctions Ψ' and Ψ are orthogonal.

(i) For $\lambda \neq \lambda'$:

Proof: Multiplying (a) by Ψ'^* and integrating:

(2) $$\int \Psi'^* H \Psi d\tau = \lambda \int \Psi'^* \Psi d\tau.$$

Multiplying (b) by Ψ^* and integrating:

(4) $$\int \Psi^* H \Psi' d\tau = \lambda' \int \Psi^* \Psi' d\tau \rightarrow \left(\int \Psi^* H \Psi' d\tau\right)^* = \lambda' \int \Psi'^* \Psi d\tau.$$

Substituting $\left(\int \Psi^* H \Psi' d\tau\right)^* = \int \Psi'^* H \Psi d\tau$ into (4):

(5) $$\int \Psi'^* H \Psi d\tau = \lambda' \int \Psi'^* \Psi d\tau.$$

Subtracting (5) from (2), we have $(\lambda - \lambda')\int \Psi'^* \Psi d\tau = 0$, so that $\int \Psi'^* \Psi d\tau = 0$ if $\lambda \neq \lambda'$,

(ii) For $\lambda = \lambda'$:

In this case we have degeneracy, since two eigenfunctions Ψ and Ψ' give rise to the same eigenvalues $\lambda = \lambda'$, and Ψ and Ψ' are not necessarily orthogonal. However, one can use the Schmidt orthogonalization method, discussed in Section 6.4.1, to make them orthogonal. Since the characteristics above hold for H real or complex, if H is a real Hermitian operator, and if $H\Psi = \lambda\Psi$, then $H\Psi^* = \lambda\Psi^*$. Since this equation is satisfied for the real and imaginary parts, separately, of any complex eigenfunction, we will always choose Ψ to be real.

Functionals $F(\Psi)$

Definition: For every given function Ψ, there is a unique value for the functional $F(\Psi)$.

An example of a functional is the expression $\langle\Psi|H|\Psi\rangle \equiv \dfrac{\int \Psi H \Psi d\tau}{\int \Psi^2 d\tau}$.

Functional Variation $\delta F(\Psi)$

Definition: $\delta F(\Psi) \equiv F(\Psi + \delta\Psi) - F(\Psi)$, where terms of order $O(\delta\Psi)^2$ are neglected.

Example (i): If $F(\Psi) = \int \Psi^2 d\tau$, then

$$\delta F(\Psi) = \int (\Psi + \delta\Psi)^2 d\tau - \int \Psi^2 d\tau = \int \left[2\Psi\delta\Psi + (\delta\Psi)^2\right] d\tau \rightarrow 2\int \Psi\delta\Psi d\tau.$$

Example (ii): If $F(\Psi) = -\int \Psi\nabla^2\Psi d\tau$.

$$F(\Psi) = -\int \nabla\cdot(\Psi\nabla\Psi)d\tau + \int(\nabla\Psi)^2 d\tau = -\int(\Psi\nabla\Psi)\cdot n_o dS + \int(\nabla\Psi)^2 d\tau.$$

We saw in Section 10.1 that ∇^2 will be a good Hermitian operator if we set $\Psi = 0$ at the surface boundary. Thus,

$$F(\Psi) = \underset{(a)}{-\int \Psi\nabla^2\Psi d\tau} = \underset{(b)}{\int(\nabla\Psi)^2 d\tau}$$

where both expressions (a) and (b) represent the same functional.

(a) $\delta F(\Psi) = -\int \delta\Psi\nabla^2\Psi d\tau - \int \Psi\nabla^2\delta\Psi d\tau$.

Due to the Hermitian property of ∇^2, $\int \Psi\nabla^2\delta\Psi d\tau = \int \delta\Psi\nabla^2\Psi d\tau$, so that $\delta F(\Psi) = -2\int \delta\Psi\nabla^2\Psi d\tau$.

$$\text{(b)}\quad \delta F(\Psi) = 2\int \nabla\Psi\cdot\nabla(\delta\Psi)d\tau = 2\int \nabla\cdot(\delta\Psi\nabla\Psi)d\tau - 2\int \delta\Psi\cdot\nabla^2\Psi d\tau$$
$$= 2\int(\delta\Psi\nabla\Psi)\cdot n_o dS - 2\int \delta\Psi\cdot\nabla^2\Psi d\tau.$$

As above, if $\delta\Psi = 0$ on the boundary, then $\delta F(\Psi) = -2\int \delta\Psi\nabla^2\Psi d\tau$, in agreement with (a) above.

10.3.1 Distribution of the Eigenvalues

Let $H\Psi_i = \lambda_i\Psi_i$, where H is the Hermitian operator for the particular Ψ_i that satisfies the boundary conditions $\Psi_i = 0$ on the boundary. Let us arrange the λ_i such that $\lambda_1 \le \lambda_2 \le \lambda_3 \le \lambda_4 \le \ldots$, i.e. a spectrum of eigenvalues.

Define $F(\Psi)$, the functional in question, as

$$(1)\qquad \langle\Psi|H|\Psi\rangle = \frac{\int \Psi H\Psi d\tau}{\int \Psi^2 d\tau},$$

where the domain of integration is the allowed domain of the problem (such as the length of a string, area of a membrane, etc.), and where Ψ satisfies the proper boundary conditions.

Theorem (i): If $H\Psi_i = \lambda_i\Psi_i$ and if $\Psi = \Psi_i$, then $\langle\Psi|H|\Psi\rangle = \lambda_i$.

Proof:

$$\langle\Psi|H|\Psi\rangle = \langle\Psi_i|H|\Psi_i\rangle = \frac{\int\Psi_i H\Psi_i d\tau}{\int\Psi_i^2 d\tau} = \frac{\lambda_i\int\Psi_i^2 d\tau}{\int\Psi_i^2 d\tau} = \lambda_i .$$

Theorem (ii): If $\delta\langle\Psi|H|\Psi\rangle = 0$, then $H\Psi = \langle\Psi|H|\Psi\rangle\Psi$, i.e. Ψ is an eigenfunction (since $\langle\Psi|H|\Psi\rangle$ is just a number), and $\langle\Psi|H|\Psi\rangle$ is an eigenvalue.

Proof:

$$\delta\langle\Psi|H|\Psi\rangle = \frac{\int[\delta\Psi H\Psi + \Psi H\delta\Psi]d\tau}{\int\Psi^2 d\tau} - \frac{2\int\Psi H\Psi d\tau}{\int\Psi^2 d\tau}\frac{\int\Psi\delta\Psi d\tau}{\int\Psi^2 d\tau} = 0.$$

Since for Hermitian operators, $\int\delta\Psi H\Psi d\tau = \int\Psi H\delta\Psi d\tau$,

$$\delta\langle\Psi|H|\Psi\rangle = \frac{2\int\delta\Psi H\Psi d\tau}{\int\Psi^2 d\tau} - \frac{2\int\Psi H\Psi d\tau}{\int\Psi^2 d\tau}\frac{\int\Psi\delta\Psi d\tau}{\int\Psi^2 d\tau}$$

$$(2) \qquad \delta\langle\Psi|H|\Psi\rangle = \frac{2}{\int\Psi^2 d\tau}\left[\int\delta\Psi\left(H\Psi - \langle\Psi|H|\Psi\rangle\Psi\right)d\tau\right] = 0.$$

Choosing a point $\mathbf{r}_p$ within the domain, such that $\delta\Psi = \delta^3(\mathbf{r} - \mathbf{r}_p)$, we have $H\Psi = \langle\Psi|H|\Psi\rangle\Psi$ within the domain.

Theorem (iii): If $\Psi = \Psi_i + \delta\varphi$, where $\delta\varphi$ is an infinitesimal, it follows that

$$\langle\Psi|H|\Psi\rangle = \lambda_i + O(\delta\varphi)^2,$$

implying that $\langle\Psi|H|\Psi\rangle$ is stationary at Ψ_i. (For example, if $f = f(x)$ and $x = x_0 + \delta x$, and if $f = f(x_0) + O(\delta x)^2$, then f is stationary at $x = x_0$.)

Theorem (iv): The minimum value of $\langle\Psi|H|\Psi\rangle$ is λ_1 (the lowest eigenvalue) for all Ψ satisfying the boundary conditions.

Proof: If a minimum, then $\delta\langle\Psi|H|\Psi\rangle = 0$ and $\langle\Psi|H|\Psi\rangle = \lambda_i$. Hence, choose the lowest $\lambda_i = \lambda_1$.

Consider the following cases:

(1) Vibrating membrane with a fixed boundary at S.

(2) Same vibrating membrane as in (1), except that constraints are added at certain areas S_1, S_2, S_3, etc., so that $\Psi = 0$ there.

For both (1) and (2) we have $H\Psi = \lambda\Psi$, where $\lambda = \omega^2/\upsilon^2$ and $H = -\nabla^2$.

Theorem (v): Regarding cases (1) and (2) above, $(\lambda_1)_{(1)} \le (\lambda_1)_{(2)}$, where we are dealing with different wavefunctions, since boundary conditions for (1) and (2) are different..

Proof: Since the boundary conditions of (2) include, and therefore satisfy the boundary conditions of (1):

$(\lambda_1)_{(2)}$ is a minimum of $\langle\Psi|H|\Psi\rangle$ for all Ψ satisfying the boundary conditions of (2), thereby also satisfying the boundary conditions of (1).

$(\lambda_1)_{(1)}$ is a minimum of $\langle\Psi|H|\Psi\rangle$ for all Ψ satisfying the boundary conditions of (1) only.

[In Section 10.1 we discussed the vibrating circular membrane and showed that the roots z_{mn} for $m \neq 0$, i.e. equivalent to case (2) above, are greater than z_{on} for $m = 0$, i.e. equivalent to case (1) above, and $\lambda_{mn} \propto \omega_{mn}^2 \propto z_{mn}^2$.]

Thus, the effect of constraints is to raise the minimum λ_1, and as a result, $(\lambda_1)_{(1)} \le (\lambda_1)_{(2)}$.

Theorem (vi): Let $H\Psi_1 = \lambda_1\Psi_1$, where λ_1 is the lowest eigenvalue. Then the lowest value of $\langle\Psi|H|\Psi\rangle$ is λ_2 for all Ψ satisfying the boundary conditions and obeying the constraint condition $\int \Psi\Psi_1 d\tau = 0$, i.e. orthogonal to Ψ_1.

Proof: We use the method of Lagrangian multipliers, i.e.

$$(3) \qquad \delta\langle\Psi|H|\Psi\rangle + m\delta\int \Psi\Psi_1 d\tau = 0,$$

where Ψ is regarded as arbitrary and Ψ_1 is fixed, i.e. $\delta\Psi_1 = 0$. Substituting for $\delta\langle\Psi|H|\Psi\rangle$ from equation (2):

$$\frac{2}{\int \Psi^2 d\tau}\left[\int \delta\Psi\left(H\Psi - \langle\Psi|H|\Psi\rangle\Psi\right)d\tau\right] + m\delta\int(\Psi\delta\Psi_1 + \Psi_1\delta\Psi)d\tau = 0.$$

By hypothesis $\delta\Psi_1 = 0$, so that after setting $m' = \frac{m}{2}\int \Psi^2 d\tau$, we get

$$\frac{2}{\int \Psi^2 d\tau}\left[\int \delta\Psi\left(H\Psi - \langle\Psi|H|\Psi\rangle\Psi + m'\Psi_1\right)d\tau\right] = 0.$$

Again choosing a point $\mathbf{r}_p$ within the domain such that $\delta\Psi = \delta^3(\mathbf{r} - \mathbf{r}_p)$:

(4) $\quad H\Psi = \langle\Psi|H|\Psi\rangle\Psi - m'\Psi_1.$

We impose the constraint condition $\int \Psi\Psi_1 d\tau = 0$ by multiplying by Ψ_1 and integrating. Thus, the left-hand side of (4) becomes

$$\int \Psi_1 H\Psi d\tau = \int \Psi H\Psi_1 d\tau = \lambda_1 \int \Psi\Psi_1 d\tau = 0,$$

while the first term on the right-hand side of (4) becomes

$$\int \langle\Psi|H|\Psi\rangle\Psi_1\Psi d\tau = \langle\Psi|H|\Psi\rangle \int \Psi_1\Psi d\tau = 0.$$

Thus, $m'\int \Psi_1^2 d\tau = 0 \quad \rightarrow \quad m' = 0 \quad \rightarrow \quad H\Psi = \langle\Psi|H|\Psi\rangle\Psi.$

The λ_i that yields the lowest $\langle\Psi|H|\Psi\rangle$ cannot be λ_1 since the corresponding Ψ_1 does not satisfy $\int \Psi_1^2 d\tau = 0$. For all other λ_i, however, Ψ_i can be found which satisfy $\int \Psi_i\Psi_1 d\tau = 0$. Thus, the next lowest $\langle\Psi|H|\Psi\rangle$ is λ_2.

Theorem (vii): [Generalization of Theorem (vi)]

The lowest $\langle\Psi|H|\Psi\rangle$ is λ_n for all Ψ satisfying the boundary conditions and $\int \Psi\Psi_i d\tau = 0$, where $i = 1, 2 \ldots n-1$. For example, for $\Psi_1, i = 1$ and $n = 2$, implying λ_2; for $\Psi_2, i = 2$ and $n = 3$, implying λ_3, etc.

Consider any set of functions $f_1 \cdots f_n$ which may or may not satisfy the boundary conditions. Define $m(f_1 \cdots f_n) \equiv \text{minimum}\langle\Psi|H|\Psi\rangle$ for Ψ satisfying the boundary conditions and $\int \Psi f_i d\tau = 0,\ i = 1, 2 \ldots n$.

Theorem (viii): The maximum value of $m(f_1 \cdots f_n)$ for all $f_1 \cdots f_n$ is λ_{n+1}.

Proof:

(i) If $f_1 = \Psi_1,\ m(\Psi_1) = \lambda_2 \underset{\text{Theorem (vi)}}{\rightarrow} \text{maximum}\left[m(\Psi_1)\right] = \lambda_2.$

(ii) For any f_1, choose $\Psi = a_1\Psi_1 + a_2\Psi_2$ (where Ψ_1 and Ψ_2 are eigenfunctions of H with eigenvalues λ_1, λ_2) to be orthogonal to f_1, i.e.

$$\int f_1\Psi d\tau = \int f_1 a_1\Psi_1 d\tau + \int f_1 a_2\Psi_2 d\tau = 0 \quad \rightarrow \quad \frac{a_1}{a_2} = -\frac{\int f_1\Psi_2 d\tau}{\int f_1\Psi_1 d\tau}.$$

We have

$$\langle\Psi|H|\Psi\rangle = \frac{\int (a_1\Psi_1 + a_2\Psi_2)H(a_1\Psi_1 + a_2\Psi_2)d\tau}{\int (a_1\Psi_1 + a_2\Psi_2)^2 d\tau},$$

so that, substituting $\int \Psi_i H\Psi_j d\tau = \lambda_j \int \Psi_i\Psi_j d\tau = \lambda_i\delta_{ij}$,

$$\langle\Psi|H|\Psi\rangle = \frac{a_1^2\lambda_1 + a_2^2\lambda_2}{a_1^2 + a_2^2},$$

and $m(f_1) \equiv \min \langle\Psi|H|\Psi\rangle = \min \dfrac{a_1^2\lambda_1 + a_2^2\lambda_2}{a_1^2 + a_2^2} \leq \lambda_2$ for arbitrary f_1, and maximum $m(f_1) = \lambda_2$.

Theorem (ix): For the resonator problem discussed above, in which (2) refers to the addition of constraints, $(\lambda_2)_{(1)} \leq (\lambda_2)_{(2)}$, i.e. *all* frequencies are raised due to the constraints.

Proof: For any f, we seek min $\langle\Psi|H|\Psi\rangle$ for Ψ satisfying the boundary conditions for (1) and for (2), and $\int \Psi f d\tau = 0$. Any Ψ satisfying the boundary conditions for (2) and $\int \Psi f d\tau = 0$ will also satisfy the boundary conditions for (1) and $\int \Psi f d\tau = 0$. Hence,

$$\left[m(f)\right]_{(2)} \geq \left[m(f)\right]_{(1)} \rightarrow \max\left[m(f_1)\right]_{(1)} \leq \max\left[m(f_1)\right]_{(2)} \rightarrow (\lambda_2)_{(1)} \leq (\lambda_2)_{(2)}.$$

Thus, for any sequence $f_1 f_2 \ldots f_n$, $(\lambda_{n+1})_{(1)} \leq (\lambda_{n+1})_{(2)}$. The purpose in introducing the set of functions f_i is so that one can compare two different problems, i.e. although for problem (1), $\int \Psi f_i d\tau = 0$, it is not necessarily so for problem (2), since the Ψ are two different functions obeying different boundary conditions. We have finally seen that by placing constraints upon resonators, all higher frequencies are increased as well.

10.3.2 Upper Bound to the Eigenvalues

Theorem: If $-\nabla^2\Psi_i = \lambda_i\Psi_i$ (where $H = -\nabla^2$), and $\Psi_i = 0$ on a closed surface S, then $\lim\limits_{n\to\infty} \lambda_n = \infty$, i.e. the eigenvalues have no bound.

Proof:

(i) Let S first be the surface of a cube of side L. We construct a Ψ that will vanish on the surface of such a cube, e.g. choose

$$\Psi_{n_1n_2n_3}(xyz) = \sin\frac{n_1\pi x}{L}\sin\frac{n_2\pi y}{L}\sin\frac{n_3\pi z}{L},$$

so that $\Psi = 0$ at $x = 0$ and L, at $y = 0$ and L, and at $z = 0$ and L. Thus,

$$\nabla^2\Psi_{n_1n_2n_3}(xyz) = -\frac{(n_1^2+n_2^2+n_3^2)\pi^2}{L^2}\left\{\sin\frac{n_1\pi x}{L}\sin\frac{n_2\pi y}{L}\sin\frac{n_3\pi z}{L}\right\}$$

$$= -\frac{\pi^2}{L^2}\left(\sum_{i=1}^{3} n_i^2\right)\Psi_{n_1n_2n_3}(xyz) \;\rightarrow\; -\lambda_{n_1n_2n_3}\Psi_{n_1n_2n_3}(xyz).$$

Hence

$$\lambda_{n_1n_2n_3} = \frac{\pi^2}{L^2}\sum_{i=1}^{3} n_i^2 \;\rightarrow\; \lim_{n\to\infty}\lambda_n = \infty.$$

(ii) Consider a cube having an enclosed surface S. We require that $\Psi = 0$ on S and in the region between S and the edges of the cube (Fig. 10-10).

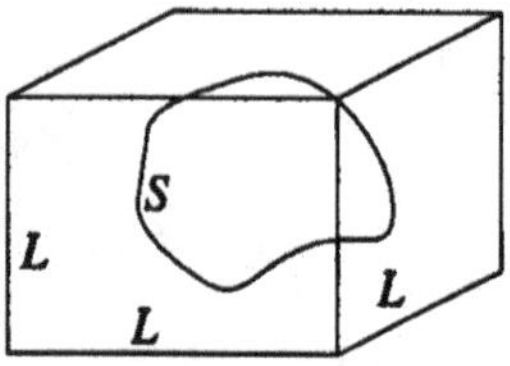

Fig. 10-10

$$\langle\Psi|H|\Psi\rangle_{\text{cube}} = \frac{\int_{\text{cube}}\Psi H\Psi d\tau}{\int_{\text{cube}}\Psi^2 d\tau} = \frac{\int_{V_s}\Psi H\Psi d\tau}{\int_{V_s}\Psi^2 d\tau}.$$

Thus, $\langle\Psi|H|\Psi\rangle_{\text{cube}} = \langle\Psi|H|\Psi\rangle_{V_S}$ integrating over a volume V_S of surface S, for Ψ to satisfy (ii). Since Ψ satisfying (ii) constitutes a more restrictive boundary condition than Ψ satisfying (i), we have, denoting $\lambda_{(2)}$ as λ_S corresponding to min $\langle\Psi|H|\Psi\rangle_{V_S}$, and denoting $\lambda_{(1)}$ as λ_{cube} corresponding to min $\langle\Psi|H|\Psi\rangle_{\text{cube}}$ for Ψ satisfying boundary conditions (i):

$$\lambda_{(2)} \geq \lambda_{(1)} \;\rightarrow\; \lambda_{(2)} = \lambda_S \rightarrow \infty, \text{ since } \lambda_{(1)} \rightarrow \infty.$$

Theorem: If $-\nabla^2\Psi_n + V\Psi_n = \lambda_n\Psi_n$, where $H = -\nabla^2 + V$, V is a function of space only, $\Psi_n = 0$ on a closed surface S, and $V \geq V_0$ (a finite constant) within S, then $\lim_{n\to\infty} \lambda_n = \infty$.

Proof:

(i) If $V = V_0$ (const.), then $-\nabla^2\Psi_n = (\lambda_n - V_0)\Psi_n$. From our previous theorem, however, $\lim_{n\to\infty} (\lambda_n - V_0) = \infty$, and since V_0 is a constant, we have $\lim_{n\to\infty} \lambda_n = \infty$.

(ii) For $V \geq V_0$ (inside S). Define $H(\xi) = -\nabla^2 + \xi(V - V_0) + V_0$, so that $dH/d\xi = V - V_0$. We also have that $H(+1) = -\nabla^2 + V$ and $H(0) = -\nabla^2 + V_0$.

Consider $H(\xi)\Psi_n(\xi) = \lambda_n(\xi)\Psi_n(\xi)$, where ξ is regarded as a parameter.

$$\underset{(1)}{\frac{dH(\xi)}{d\xi}\Psi_n(\xi)} + \underset{(2)}{H(\xi)\frac{d\Psi_n(\xi)}{d\xi}} = \underset{(3)}{\frac{d\lambda_n(\xi)}{d\xi}\Psi_n(\xi)} + \underset{(4)}{\lambda_n(\xi)\frac{d\Psi_n(\xi)}{d\xi}}.$$

Multiplying by $\Psi_n(\xi)$ and integrating term (2) over the domain (taking into account that H is Hermitian),

$$\int \Psi_n(\xi) H(\xi) \frac{d\Psi_n(\xi)}{d\xi} d\tau = \int \frac{d\Psi_n(\xi)}{d\xi} H(\xi)\, \Psi_n(\xi)\, d\tau$$

$$= \lambda_n(\xi) \int \frac{d\Psi_n(\xi)}{d\xi} \Psi_n(\xi)\, d\tau.$$

Thus, since terms (2) and (4) are equal, term (1) must equal term (3), i.e.

$$\int \Psi_n^2(\xi)(V - V_0) d\tau = \int \frac{d\lambda_n(\xi)}{d\xi} \Psi_n^2(\xi) d\tau = \frac{d\lambda_n(\xi)}{d\xi} \int \Psi_n^2(\xi),$$

and

$$\frac{d\lambda_n(\xi)}{d\xi} = \frac{\int \Psi_n^2(\xi)(V - V_0) d\tau}{\int \Psi_n^2(\xi)} \geq 0.$$

Since $\frac{d\lambda_n(\xi)}{d\xi} \geq 0$, it follows that $\lambda_n(\xi{=}1) \geq \lambda_n(\xi{=}0)$; but $\lambda_n(\xi{=}0)$ (i.e. $V = V_0$) has no upper bound, so that $\lambda_n(\xi{=}1)$ (i.e. $V > V_0$) has no upper bound.

Importance of there being no upper bound to the eigenvalues

Theorem: If $H\Psi_n = \lambda_n\Psi_n$ and if $\lim_{n\to\infty}\lambda_n = \infty$, then the orthonormal set $\Psi_1\Psi_2\Psi_3\ldots$ forms a *complete* set for functions satisfying the boundary conditions.

Proof: Define $\overline{H} \equiv H - \lambda_1$, so that $\overline{H}\Psi_n = (\lambda_n - \lambda_1)\Psi_n$. Let Ψ be any function satisfying the boundary conditions, and define

$$F_n \equiv \Psi - \sum_{m=1}^{n} a_m\Psi_m,$$

where $a_m = \int \Psi\Psi_m d\tau$. It then follows that

$$\int F_n\Psi_{m'}d\tau = \int \Psi\Psi_{m'}d\tau - \sum_{m=1}^{n} a_m \int \Psi_m\Psi_{m'}d\tau = \int \Psi\Psi_{m'}d\tau - \sum_{m=1}^{n} a_m\delta_{mm'}$$

$$= \int \Psi\Psi_{m'}d\tau - \sum_{m=1}^{n} a_{m'} = \int \Psi\Psi_{m'}d\tau - \int \Psi\Psi_{m'}d\tau \equiv 0.$$

Hence, $\int F_n\Psi_{m'}d\tau = 0$, $m = 1, 2 \ldots n$, i.e. F_n is orthogonal to the first n eigenfunctions.

Consider now $\int F_n\overline{H}F_n d\tau = \int F_n\overline{H}\Psi d\tau - \int F_n\overline{H}\sum_{m=1}^{n} a_m\Psi_m d\tau$.

Since F_n is orthogonal to Ψ_m, the second term on the right vanishes, so that

$$\int F_n\overline{H}F_n d\tau = \int F_n\overline{H}\Psi d\tau = \int \left[\Psi - \sum_{m=1}^{n} a_m\Psi_m\right]\overline{H}\Psi d\tau$$

$$= \int \Psi\overline{H}\Psi d\tau - \sum_{m=1}^{n} a_m \int \Psi_m\overline{H}\Psi d\tau.$$

Since H is Hermitian,

$$\sum_{m=1}^{n} a_m \int \Psi_m\overline{H}\Psi d\tau = \sum_{m=1}^{n} a_m \int \Psi\overline{H}\Psi_m d\tau = \sum_{m=1}^{n} a_m(\lambda_m - \lambda_1)\int \Psi_m\Psi d\tau$$

$$= \sum_{m=1}^{n} a_m^2(\lambda_m - \lambda_1).$$

Thus

$$\int F_n\overline{H}F_n d\tau = \int \Psi\overline{H}\Psi d\tau - \sum_{m=1}^{n} a_m^2(\lambda_m - \lambda_1).$$

Recalling (Sect. 10.3.1) that $\langle F_n|\overline{H}|F_n\rangle \geq \lambda_{n+1} - \lambda_1$, we have

$$\langle F_n|\overline{H}|F_n\rangle = \frac{\int F_n\overline{H}F_n d\tau}{\int F_n^2 d\tau} = \frac{\int \Psi\overline{H}\Psi d\tau - \sum_{m=1}^{n} a_m^2(\lambda_m - \lambda_1)}{\int F_n^2 d\tau} \geq \lambda_{n+1} - \lambda_1$$

$$\Downarrow$$

$$\frac{\int \Psi\overline{H}\Psi d\tau}{\int F_n^2 d\tau} \geq \lambda_{n+1} - \lambda_1 \quad \rightarrow \quad \int F_n^2 d\tau \leq \frac{\int \Psi\overline{H}\Psi d\tau}{\lambda_{n+1} - \lambda_1}.$$

Since $\int \Psi\overline{H}\Psi d\tau$ is independent of n $\left[\text{given a } \Psi, \text{ one calculates } \int \Psi\overline{H}\Psi d\tau\right]$, $\lim_{n\to\infty} \int F_n^2 d\tau = 0 \quad \Rightarrow \quad \lim_{n\to\infty} F_n = 0$. Hence,

(1) $$F_{n\to\infty} = \Psi - \sum_{m=1}^{\infty} a_m\Psi_m = 0,$$

and the Ψ_m form a complete set.

<u>Short review plus additional remarks</u> (If $H\Psi_n = \lambda_n\Psi_n$)

(i) The Ψ_n constitute an orthonormal set.

(ii) The Ψ_n constitute a complete set if $\lim_{n\to\infty} \lambda_n = \infty$.

(iii) For $H = -\frac{\hbar^2}{2m}\frac{d^2}{dx^2} + \frac{kx^2}{2}$ (a particular example of $H = -\nabla^2 + V$), solutions were shown to be

(2) $$\Psi_n(\alpha x) = H_n(\alpha x)e^{-\alpha^2 x^2/2},$$

where $\alpha = (\sqrt{mk}/\hbar)^{1/2}$. The eigenvalues are given by

(3) $$E_n = \lambda_n = (n + 1/2)\hbar\omega = (n + 1/2)\hbar(k/m)^{1/2} \rightarrow \infty \text{ as } n \rightarrow \infty.$$

Thus, for z real,

(4) $$\Psi(z) = \sum_{n=0}^{\infty} A_n H_n(z)e^{-z^2/2},$$

for any function Ψ that vanishes at $x = \pm\infty$, thereby making H Hermitian.

(iv) For $H = -\dfrac{d^2}{d\varphi^2}$, the $\Psi_m(\varphi)$ are $\sin m\varphi$ and $\cos m\varphi$, with eigenvalues $\lambda_m = +m^2$, and as $m \to \infty$, $\lambda_m \to \infty$. Hence, $\sin\varphi$ and $\cos\varphi$ form a complete set, so that for any function Ψ:

(5) $$\Psi(\varphi) = \sum_{m=0}^{\infty}(A_m \sin m\varphi + B_m \cos m\varphi).$$

(v) If $H = -\dfrac{1}{\sin\theta}\dfrac{\partial}{\partial\theta}\sin\theta\dfrac{\partial}{\partial\theta} - \dfrac{1}{\sin^2\theta}\dfrac{\partial^2}{\partial\varphi^2}$, the corresponding eigenfunctions are $Y_{lm_l}(\theta,\varphi)$ with eigenvalues $l(l+1)$ which $\to \infty$ as $l \to \infty$. Hence, the spherical harmonics form a complete set, and any function of θ and φ can be written:

(6) $$\Psi(\theta,\varphi) = \sum_{lm_l} A_{lm_l} Y_{lm_l}(\theta,\varphi).$$

(vi) If $H = -\nabla^2$ inside a sphere, then the eigenfunctions are given by

(7) $$\Psi_{lm_ln}(r,\theta,\varphi) = \text{const.}\,\frac{J_{l+1/2}(k_{ln}r)}{\sqrt{r}}\,Y_{lm_l}(\theta,\varphi),$$

and

(8) $$\Psi(r,\theta,\varphi) = \sum_{lm_ln} A_{lm_ln}\frac{J_{l+1/2}(k_{ln}r)}{\sqrt{r}}\,Y_{lm_l}(\theta,\varphi).$$

Chapter 11

Wave Equations

11.1 Infinite Medium

Consider the time-dependent wave equation

(1) $$\Delta\Phi - \frac{1}{c^2}\ddot{\Phi} = 0,$$

where Φ may be the component of a magnetic vector potential, the displacement of a vibrating solid, etc.

Boundary conditions

At $t = 0$: $\Phi(\mathbf{r}, t=0) = u(\mathbf{r})$ and $\dot{\Phi}(\mathbf{r}, t=0) = \upsilon(\mathbf{r})$ are known. Carrying out a Fourier analysis,

(2) $$\Phi(\mathbf{r},t) = \frac{1}{(2\pi)^{3/2}}\int_{V_k} F_{\mathbf{k}}(t)e^{i\mathbf{k}\cdot\mathbf{r}}d^3\mathbf{k}.$$

Substituting into (1):

$$\Delta\Phi - \frac{1}{c^2}\ddot{\Phi} = \frac{1}{(2\pi)^{3/2}}\int_{V_k}\left[-k^2 - \frac{1}{c^2}\frac{d^2}{dt^2}\right]F_{\mathbf{k}}(t)e^{i\mathbf{k}\cdot\mathbf{r}}d^3\mathbf{k} = 0.$$

Multiplying by $e^{-i\mathbf{k}_o\cdot\mathbf{r}}$ and integrating over all space:

$$\frac{1}{(2\pi)^{3/2}}\int_{V_k}\left[-k^2 - \frac{1}{c^2}\frac{d^2}{dt^2}\right]F_{\mathbf{k}}(t)d^3\mathbf{k}\int_{V_r} e^{i\mathbf{r}\cdot(\mathbf{k}-\mathbf{k}_o)}d^3\mathbf{r}$$

$$= C\int_{V_k}\left[-k^2 - \frac{1}{c^2}\frac{d^2}{dt^2}\right]F_{\mathbf{k}}(t)d^3\mathbf{k}\,\delta^3(\mathbf{k}-\mathbf{k}_o) \rightarrow \left[k_o^2 + \frac{1}{c^2}\frac{d^2}{dt^2}\right]F_{\mathbf{k}_o}(t) = 0.$$

Since $\mathbf{k}_o$ is arbitrary, the last equation becomes

$$\frac{d^2F_{\mathbf{k}}(t)}{dt^2} = -k^2c^2F_{\mathbf{k}}(t),$$

solutions of which are given by

$$(3) \qquad F_{\mathbf{k}}(t) = f_{\mathbf{k}}e^{ikct} + g_{\mathbf{k}}e^{-ikct},$$

so that

$$(4) \qquad \Phi(\mathbf{r},t) = \frac{1}{(2\pi)^{3/2}}\int_{V_k}\left[f_{\mathbf{k}}e^{i(\mathbf{k}\cdot\mathbf{r}+kct)} + g_{\mathbf{k}}e^{i(\mathbf{k}\cdot\mathbf{r}-kct)}\right]d^3\mathbf{k}.$$

Applying boundary conditions at $t = 0$:

$$(5) \qquad \begin{aligned} \Phi(\mathbf{r},0) &= u(\mathbf{r}) = \frac{1}{(2\pi)^{3/2}}\int_{V_k}\left[f_{\mathbf{k}} + g_{\mathbf{k}}\right]e^{i\mathbf{k}\cdot\mathbf{r}}d^3\mathbf{k}. \\ \dot{\Phi}(\mathbf{r},0) &= \upsilon(\mathbf{r}) = \frac{1}{(2\pi)^{3/2}}\int_{V_k}ikc\left[f_{\mathbf{k}} - g_{\mathbf{k}}\right]e^{i\mathbf{k}\cdot\mathbf{r}}d^3\mathbf{k} \end{aligned}$$

Taking inverse Fourier transforms of (5):

$$(6) \qquad \begin{aligned} f_{\mathbf{k}} + g_{\mathbf{k}} &= \frac{1}{(2\pi)^{3/2}}\int_{V_r}u(\mathbf{r})e^{-i\mathbf{k}\cdot\mathbf{r}}d^3\mathbf{r}. \\ ikc(f_{\mathbf{k}} - g_{\mathbf{k}}) &= \frac{1}{(2\pi)^{3/2}}\int_{V_r}\upsilon(\mathbf{r})e^{-i\mathbf{k}\cdot\mathbf{r}}d^3\mathbf{r}. \end{aligned}$$

Thus, given $u(\mathbf{r})$ and $\upsilon(\mathbf{r})$, $f_{\mathbf{k}}$ and $g_{\mathbf{k}}$ are known from (6), $F_{\mathbf{k}}(t)$ is known from (3), and $\Phi(\mathbf{r},t)$ is known from (2) for all time.

<u>Green's function (D-function) for the infinite medium</u>

(i) If at $t = 0$, $\Phi(\mathbf{r},0) = u(\mathbf{r}) = 0$ and $\dot{\Phi}(\mathbf{r},0) = \upsilon(\mathbf{r}) = \delta^3(\mathbf{r})$, then for all time, $\Phi(\mathbf{r},t) \equiv D(\mathbf{r},t)$ defines the D-function.

Given the above boundary conditions, we have from (6):

$$f_{\mathbf{k}} + g_{\mathbf{k}} = 0. \quad (f_{\mathbf{k}} - g_{\mathbf{k}}) = \frac{1}{ikc(2\pi)^{3/2}}. \quad \Rightarrow f_{\mathbf{k}} = -g_{\mathbf{k}} = \frac{1}{2ikc(2\pi)^{3/2}}.$$

Substituting for $f_{\mathbf{k}}$ and $g_{\mathbf{k}}$ into (3) and then for $F_{\mathbf{k}}(t)$ into (2):

$$\Phi(\mathbf{r},t) = \frac{1}{(2\pi)^{3/2}}\int_{V_k}\frac{1}{2ikc(2\pi)^{3/2}}\left(e^{ikct}-e^{-ikct}\right)e^{i\mathbf{k}\cdot\mathbf{r}}d^3\mathbf{k},$$

yielding the D-function

(7) $$D(\mathbf{r},t) = \frac{1}{(8\pi)^3}\int_{V_k}\frac{\sin kct\, e^{i\mathbf{k}\cdot\mathbf{r}}}{kc}d^3\mathbf{k}.$$

(ii) If at $t = 0, \Phi(\mathbf{r},0) = u(\mathbf{r}) = \delta^3(\mathbf{r})$ and $\dot{\Phi}(\mathbf{r},0) = \upsilon(\mathbf{r}) = 0$, then from (6):

$$f_{\mathbf{k}} + g_{\mathbf{k}} = \frac{1}{(2\pi)^{3/2}}. \quad (f_{\mathbf{k}} - g_{\mathbf{k}}) = 0. \quad \Rightarrow \quad f_{\mathbf{k}} = g_{\mathbf{k}} = \frac{1}{2(2\pi)^{3/2}}.$$

Then for all time,

$$\Phi(\mathbf{r},t) = \frac{1}{(2\pi)^{3/2}}\frac{1}{2(2\pi)^{3/2}}\int_{V_k}\left(e^{ikct}+e^{-ikct}\right)e^{i\mathbf{k}\cdot\mathbf{r}}d^3\mathbf{k}.$$

$$\Downarrow$$

(8) $$\Phi(\mathbf{r},t) = \frac{1}{(8\pi)^3}\int_{V_k}\cos kct\, e^{i\mathbf{k}\cdot\mathbf{r}}d^3\mathbf{k} = \frac{\partial D(\mathbf{r},t)}{\partial t}.$$

Thus, for any arbitrary boundary conditions at $t = 0$, i.e. $\Phi(\mathbf{r}) = u(\mathbf{r})$ and $\dot{\Phi}(\mathbf{r}) = \upsilon(\mathbf{r})$, we have for any time t:

(9) $$\Phi(\mathbf{r},t) = \int_{V_{r'}}\left[D(\mathbf{r}-\mathbf{r}',t)\upsilon(\mathbf{r}') + \frac{\partial D(\mathbf{r}-\mathbf{r}',t)}{\partial t}u(\mathbf{r}')\right]d^3\mathbf{r}'.$$

Proof: We have already shown that $D(\mathbf{r},t)$ and $\dfrac{\partial D(\mathbf{r},t)}{\partial t}$ are particular solutions of the wave equation, under the boundary conditions specified above. Thus,

(i) $\Delta\Phi - \ddot{\Phi}/c^2 = 0.$

(ii) As $t \to 0$, it follows from (9) and the definitions of D and $\partial D/\partial t$:

$$\Phi(\mathbf{r},t\to 0) = \int\delta^3(\mathbf{r}-\mathbf{r}')u(\mathbf{r}')d^3\mathbf{r}' = u(\mathbf{r}). \quad \dot{\Phi}(\mathbf{r},t\to 0) = \upsilon(\mathbf{r}).$$

These last equations follow since

$$\frac{\partial D(\mathbf{r},t\to 0)}{\partial t} = \delta^3(\mathbf{r}).$$

$$\frac{\partial^2 D(\mathbf{r},t)}{\partial t^2} = -\frac{k^2c^2}{8\pi^3}\int \frac{\sin kct}{kc}e^{i\mathbf{k}\cdot\mathbf{r}}d^3\mathbf{k} \ \to\ 0 \ \text{as}\ t\to 0.$$

Hence, Φ satisfies the wave equation and the boundary conditions at $t = 0$.

For the wave equation $\Delta\Phi - \ddot{\Phi}/c^2 = 0$, if $\Phi(\mathbf{r},t_o)$ and $\dot{\Phi}(\mathbf{r},t_o)$ are known at $t = t_o$, then since the equation is of 2nd-order in time, Φ should be known at all other times. In fact, for $t > t_o$,

$$(10)\qquad \Phi(\mathbf{r},t) = \int_{V_{r'}}\left[D(\mathbf{r}-\mathbf{r}',t-t_o)\dot{\Phi}(\mathbf{r}',t_o) + \frac{\partial D(\mathbf{r}-\mathbf{r}',t-t_o)}{\partial t}\Phi(\mathbf{r}',t_o)\right]d^3\mathbf{r}',$$

where for the case of an infinite medium,

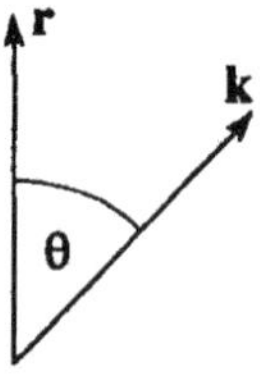

Fig. 11-1

$$(7)\qquad D(\mathbf{r},t) = \frac{1}{(8\pi)^3}\int_{V_k}\frac{\sin kct\, e^{i\mathbf{k}\cdot\mathbf{r}}}{kc}d^3\mathbf{k}.$$

Since $d^3\mathbf{k} = k^2 d\Omega_k dk$, where $d\Omega_k$ is the solid angle surrounding the **k** vector in k-space, if we choose the axes so that **r** lies along z (Fig. 11-1), then

$$\int e^{i\mathbf{k}\cdot\mathbf{r}}d\Omega_k = \int_0^{2\pi}d\varphi\int_0^{\pi}e^{ikr\cos\theta}\sin\theta d\theta = \frac{2\pi}{ikr}\left(e^{ikr} - e^{-ikr}\right).$$

Substituting into (7):

$$D(\mathbf{r},t) = \frac{2\pi}{8\pi^3}\int_0^{\infty}\frac{\left(e^{ikr} - e^{-ikr}\right)}{ikr}\frac{\left(e^{ikct} - e^{-ikct}\right)}{2ikc}k^2dk$$

$$= -\frac{1}{8\pi^2}\int_0^{\infty}\frac{e^{ik(r+ct)} + e^{-ik(r+ct)} - e^{ik(r-ct)} - e^{-ik(r-ct)}}{rc}dk$$

$$= -\frac{1}{8\pi^2}\int_{-\infty}^{\infty}\frac{e^{ik(r+ct)} - e^{ik(r-ct)}}{rc}dk.$$

This last equation can be expressed as

$$(11)\qquad D(\mathbf{r},t) = \frac{1}{4\pi cr}\left[\delta(\mathbf{r}-ct)-\delta(\mathbf{r}+ct)\right],$$

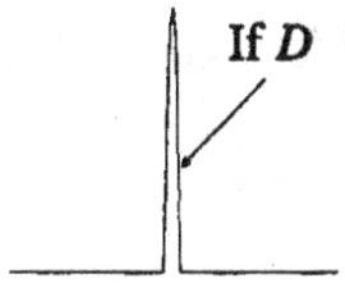

Fig. 11-2

which implies that the D-function vanishes everywhere but at the points $t = \mathbf{r}/c$ and $t = -\mathbf{r}/c$, which in turn means that $\partial D/\partial t$ is zero everywhere but infinitesimally near $t = \pm\mathbf{r}/c$. If D is as shown in Fig. 11-2, then $\partial D/\partial t$ will be as shown in Fig. 11-3.

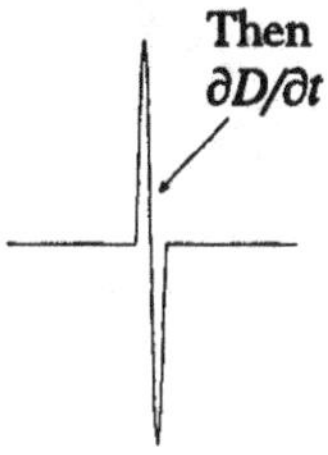

Fig. 11-3

The D-function can be pictured as a function that exists only on the surface of the cone $r = \pm ct$ (Fig. 11-4).

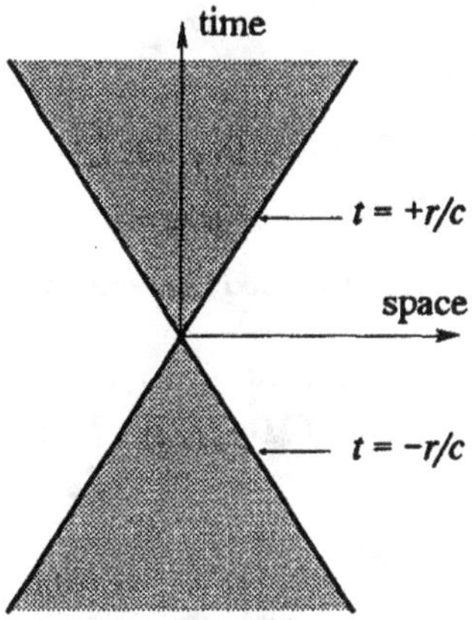

Fig. 11-4

<u>Summarizing $D(\mathbf{r},t)$ characteristics</u>

(i) $D(\mathbf{r},t) = \dfrac{1}{8\pi^3}\displaystyle\int_{V_k}\frac{\sin kct\, e^{i\mathbf{k}\cdot\mathbf{r}}}{kc}\,d^3\mathbf{k}.$

(ii) $D(\mathbf{r},t) = \dfrac{1}{4\pi cr}\left[\delta(\mathbf{r}-ct)-\delta(\mathbf{r}+ct)\right].$

(iii) $D(\mathbf{r},0) = 0;\quad \dfrac{\partial^2 D(\mathbf{r},0)}{\partial t^2} = 0.$

(iv) $\dfrac{\partial D(\mathbf{r},0)}{\partial t} = \delta^3(\mathbf{r}).$

(v) $\Delta D(\mathbf{r},t) - \dfrac{1}{c^2}\dfrac{\partial^2 D(\mathbf{r},t)}{\partial t^2} = 0$, since $D(\mathbf{r},t)$ was constructed to be a particular solution of the wave equation.

(vi) $\Phi(\mathbf{r},t) = \displaystyle\int_{V_r}\left[D(\mathbf{r}-\mathbf{r}',t-t_0)\dot{\Phi}(\mathbf{r}',t_0) + \frac{\partial D(\mathbf{r}-\mathbf{r}',t-t_0)}{\partial t}\Phi(\mathbf{r}',t_0)\right]d^3\mathbf{r}'.$

11.2 Retarded and Advanced D-Functions

Let $\rho(\mathbf{r},t)$ be the source function, so that the inhomogeneous wave equation reads

(1) $$\Delta\Phi - \frac{1}{c^2}\ddot{\Phi} = -4\pi\rho(\mathbf{r},t).$$

We define the retarded D-function, $D_{\text{ret}}(\mathbf{r},t)$ as:

(2) $$D_{\text{ret}}(\mathbf{r},t) = \begin{cases} 4\pi c^2 D(\mathbf{r},t) \text{ for } t > 0. \\ 0 \text{ for } t < 0. \end{cases}$$

Theorem:

(3) $$\Box D_{\text{ret}}(\mathbf{r},t) = -4\pi\delta^3(\mathbf{r})\,\delta(t).$$

Proof:

(a) For $t < 0$: $D_{\text{ret}} = 0$, $\Box D_{\text{ret}} = 0$ and $\delta(t) = 0$ $\rightarrow$ true for $t < 0$.

(b) For $t > 0$: $\Box D_{\text{ret}} = \Box D = 0$ and $\delta(t) = 0$ $\rightarrow$ true for $t > 0$.

(c) We now need to investigate the singularities of both sides of (3) at $t = 0$.

$$\int_{-\varepsilon}^{+\varepsilon>0} dt\,\Box D_{\text{ret}}(\mathbf{r},t) = \int_{-\varepsilon}^{+\varepsilon>0} dt\,\Delta D_{\text{ret}}(\mathbf{r},t) - \frac{1}{c^2}\int_{-\varepsilon}^{+\varepsilon>0} dt \frac{\partial}{\partial t}\frac{\partial D_{\text{ret}}(\mathbf{r},t)}{\partial t}$$

$$= \varepsilon\Delta D(\mathbf{r},t)4\pi c^2 - \frac{1}{c^2}\frac{\partial}{\partial t}\left|D_{\text{ret}}(\mathbf{r},t)\right|_{-\varepsilon}^{+\varepsilon} \rightarrow -4\pi\delta^3(\mathbf{r}).$$

Thus, as $\varepsilon \rightarrow 0$, the left-hand side of (3) approaches $-4\pi\delta^3(\mathbf{r})$. Since from (3) the right-hand side satisfies

$$\int_{-\varepsilon}^{+\varepsilon>0} -\delta^3(\mathbf{r})\,\delta(t) 4\pi dt = -4\pi\delta^3(\mathbf{r}),$$

the theorem is proved.

From equation (2) and equation (11) Section 11.1, for $t > 0$:

$$D_{\text{ret}}(\mathbf{r},t) = \frac{4\pi c^2}{4\pi cr}\left[\delta(r - ct) - \delta(r + ct)\right],$$

Since for $t > 0$, $\delta(r + ct) = 0$ always, we have

$$(4) \qquad D_{\text{ret}}(\mathbf{r},t) = \frac{c}{r}\delta(r - ct).$$

Theorem: If at $t = t_o$, $\Phi(\mathbf{r},t_o)$ and $\dot{\Phi}(\mathbf{r},t_o)$ are known, and if $\rho(\mathbf{r},t)$ is given, then for $t > t_o$:

$$(5) \qquad \Phi(\mathbf{r},t) = \int_{V_{r'}}\left[D(\mathbf{r}-\mathbf{r}',t-t_o)\dot{\Phi}(\mathbf{r}',t_o) + \frac{\partial D(\mathbf{r}-\mathbf{r}',t-t_o)}{\partial t}\Phi(\mathbf{r}',t_o)\right]d^3\mathbf{r}'$$

$$+\int_{t_o}^{\infty}dt'\int D_{\text{ret}}(\mathbf{r}-\mathbf{r}',t-t')\rho(\mathbf{r}',t')d^3\mathbf{r}'.$$

Proof:

(i) Since $\Box D = 0$ and $\Box\frac{\partial D}{\partial t} = \frac{\partial}{\partial t}\Box D = 0$ [(v), Sect. 11.1]:

$$\Box\Phi = \int_{t_o}^{\infty}dt'\int d^3\mathbf{r}'\left[-4\pi\delta^3(\mathbf{r}-\mathbf{r}')\,\delta(t-t')\rho(\mathbf{r}',t')\right] = -4\pi\rho(\mathbf{r},t)$$

for $t > t_o$, where we have set $\delta^3(\mathbf{r}) = \delta^3(\mathbf{r}-\mathbf{r}')$ and $\delta^3(t) = \delta^3(t-t')$.

For $t < t_o$: $\delta(t - t') = 0$, since t' runs from t_o to ∞.

Thus, $\Phi(\mathbf{r},t)$ satisfies the inhomogeneous wave equation for all t.

(ii) As $t \to t_{o+}$, $\Phi(\mathbf{r},t) \to$?

Since $D(\mathbf{r}-\mathbf{r}',t \to t_o) = 0$ and $\frac{\partial D(\mathbf{r}-\mathbf{r}',t \to t_o)}{\partial t} = \delta^3(\mathbf{r}-\mathbf{r}')$, then as $t \to t_{o+}$:

$$\int_{t_o}^{\infty}dt'\int D_{\text{ret}}(\mathbf{r}-\mathbf{r}',t-t')\rho(\mathbf{r}',t')d^3\mathbf{r}' = \int_{t_o}^{\infty}dt'\int D_{\text{ret}}(\mathbf{r}-\mathbf{r}',t<0)\rho(\mathbf{r}',t')d^3\mathbf{r}' \to 0,$$

since $D_{\text{det}} = 0$ for time < 0. Thus, as $t \to t_{o+}$,

$$\Phi(\mathbf{r},t) \to \int_{V_{r'}}\delta^3(\mathbf{r}-\mathbf{r}')\Phi(\mathbf{r}',t_o)d^3\mathbf{r}' = \Phi(\mathbf{r},t_o).$$

(iii) As $t \to t_{0+}$, $\dot{\Phi}(\mathbf{r},t) \to ?$

Since $\dfrac{\partial D(\mathbf{r}-\mathbf{r}',t \to t_0)}{\partial t} = \delta^3(\mathbf{r}-\mathbf{r}')$ and $\dfrac{\partial^2 D(\mathbf{r}-\mathbf{r}',t \to t_0)}{\partial t^2} = 0$, from (5):

$$\lim_{t \to t_{0+}} \int_{t_0}^{\infty} dt' \int \frac{\partial D_{\text{ret}}(\mathbf{r}-\mathbf{r}',t' < 0)}{\partial t} \rho(\mathbf{r}',t') d^3\mathbf{r}' = 0.$$

Hence, as $t \to t_{0+}$, $\dot{\Phi}(\mathbf{r},t) \to \int_{V_{r'}} \delta^3(\mathbf{r}-\mathbf{r}') \dot{\Phi}(\mathbf{r}',t_0) d^3\mathbf{r}' = \dot{\Phi}(\mathbf{r},t_0)$.

If we seek an expression for $\Phi(\mathbf{r},t)$ for $t < t_0$, we cannot use the D_{ret} function since the boundary conditions were stated for $t = t_0$. Instead we define a new function, the advanced D-function, D_{adv}, as follows:

$$(6) \qquad D_{\text{adv}}(\mathbf{r},t) = \begin{cases} 0 \text{ for } t > 0. \\ -4\pi c^2 D(\mathbf{r},t) \text{ for } t < 0. \end{cases}$$

Recalling from (ii) Sect. 11.1 that $D(\mathbf{r},t) = \dfrac{1}{4\pi c r}\left[\delta(\mathbf{r}-ct) - \delta(\mathbf{r}+ct)\right]$, and that for $t < 0, \delta(\mathbf{r}-ct) = 0$, we have $D(\mathbf{r},t) = -\delta(\mathbf{r}+ct)/4\pi c r$. Thus,

$$(7) \qquad D_{\text{adv}}(\mathbf{r},t) = \frac{c\,\delta(\mathbf{r}+ct)}{r} \quad \text{for } t < 0.$$

Theorem:

$$(8) \qquad \Box D_{\text{adv}}(\mathbf{r},t) = -4\pi\delta^3(\mathbf{r})\,\delta(t).$$

Proof:

(a) For $t < 0$: $\Box D_{\text{adv}} = -4\pi c^2 \Box D = 0$ and $\delta(t) = 0$ $\to$ true for $t < 0$.

(b) For $t > 0$: $\Box D_{\text{adv}} = 0$ and $\delta(t) = 0$ $\to$ true for $t > 0$.

(c) We now need to investigate the singularities of both sides of (8) at $t = 0$.

$$\int_{-\varepsilon}^{+\varepsilon>0} dt\, \Box D_{\text{adv}}(\mathbf{r},t) = \int_{-\varepsilon}^{+\varepsilon>0} dt\, \Delta D_{\text{adv}}(\mathbf{r},t) - \frac{1}{c^2}\int_{-\varepsilon}^{+\varepsilon>0} dt \frac{\partial}{\partial t}\frac{\partial D_{\text{adv}}(\mathbf{r},t)}{\partial t}$$

$$= -\varepsilon \Delta D_{\text{adv}}(\mathbf{r},t) - \frac{1}{c^2}\frac{\partial}{\partial t}\left|D_{\text{adv}}(\mathbf{r},t)\right|_{-\varepsilon}^{+\varepsilon} \to -4\pi\delta^3(\mathbf{r}).$$

Thus, as $\varepsilon \to 0$, the left-hand side of (8) approaches $-4\pi\delta^3(\mathbf{r})$. Since from (8) the right-hand side satisfies

$$\int_{-\varepsilon}^{+\varepsilon>0} -\delta^3(\mathbf{r})\,\delta(t)4\pi dt = -4\pi\delta^3(\mathbf{r}),$$

the theorem is proved.

Theorem: If at $t = t_0$, $\Phi(\mathbf{r},t_0)$ and $\dot{\Phi}(\mathbf{r},t_0)$ are known, and if $\rho(\mathbf{r},t)$ is also given, then for $t < t_0$:

$$(9) \qquad \Phi(\mathbf{r},t) = \int_{V_{r'}}\left[D(\mathbf{r}-\mathbf{r}',t-t_0)\dot{\Phi}(\mathbf{r}',t_0) + \frac{\partial D(\mathbf{r}-\mathbf{r}',t-t_0)}{\partial t}\Phi(\mathbf{r}',t_0)\right]d^3\mathbf{r}'$$

$$+\int_{-\infty}^{t_0} dt' \int_{V_{r'}} D_{adv}(\mathbf{r}-\mathbf{r}',t-t')\rho(\mathbf{r}',t')d^3\mathbf{r}'.$$

Proof:

(i) Since $\Box D = 0$ and $\Box\dfrac{\partial D}{\partial t} = \dfrac{\partial}{\partial t}\Box D = 0$ [(v), Sect. 11.1]:

$$\Box\Phi = \int_{-\infty}^{t_0} dt'\left[-4\pi\delta(t-t')\rho(\mathbf{r}',t')\right] = -4\pi\rho(\mathbf{r},t)$$

for $t < t_0$, where again we have set $\delta^3(\mathbf{r}) = \delta^3(\mathbf{r}-\mathbf{r}')$ and $\delta^3(t) = \delta^3(t-t')$.

For $t > t_0$: $\delta(t-t') = 0$, since t' runs from $-\infty$ to t_0. Thus, $\Phi(\mathbf{r},t)$ satisfies the inhomogeneous wave equation for all t.

(ii) As $t \to t_{0+}$, $\Phi(\mathbf{r},t) \to$?

Since $D(\mathbf{r}-\mathbf{r}',t \to t_0) = 0$ and $\dfrac{\partial D(\mathbf{r}-\mathbf{r}',t \to t_0)}{\partial t} = \delta^3(\mathbf{r}-\mathbf{r}')$, we have from (9) as $t \to t_{0+}$:

$$\int_{-\infty}^{t_0} dt' \int D_{adv}(\mathbf{r}-\mathbf{r}',t-t')\rho(\mathbf{r}',t')d^3\mathbf{r}' = \int_{-\infty}^{t_0} dt' \int D_{adv}(\mathbf{r}-\mathbf{r}',t>0)\rho(\mathbf{r}',t')d^3\mathbf{r}' \to 0,$$

since $D_{adv} = 0$ for time > 0. Thus, as $t \to t_{0+}$,

$$\Phi(\mathbf{r},t) \to \int_{V_{r'}} \delta^3(\mathbf{r}-\mathbf{r}')\Phi(\mathbf{r}',t_0)d^3\mathbf{r}' = \Phi(\mathbf{r},t_0).$$

(iii) As $t \to t_{0+}$, $\dot{\Phi}(\mathbf{r},t) \to ?$

$\frac{\partial D(\mathbf{r}-\mathbf{r}',t \to t_0)}{\partial t} = \delta^3(\mathbf{r}-\mathbf{r}')$ and $\frac{\partial^2 D(\mathbf{r}-\mathbf{r}',t \to t_0)}{\partial t^2} = 0$. Thus, from (9) we have:

$$\lim_{t \to t_{0+}} \int_{t_0}^{\infty} dt' \int \frac{\partial D_{\text{ret}}(\mathbf{r}-\mathbf{r}',t-t')}{\partial t} \rho(\mathbf{r}',t')d^3\mathbf{r}' = 0.$$

Hence, as $t \to t_{0+}$, $\dot{\Phi}(\mathbf{r},t) \to \int_{V_{r'}} \delta^3(\mathbf{r}-\mathbf{r}')\dot{\Phi}(\mathbf{r}',t_0)d^3\mathbf{r}' = \dot{\Phi}(\mathbf{r},t_0)$.

11.3 Field Due to a Moving Point Charge

Let $t_0 = -\infty$ and $\Phi(\mathbf{r},t_0) = \dot{\Phi}(\mathbf{r},t_0) = 0$, i.e. let there be no initial disturbance. Then at any later time t, we have from equation (5), Sect. 11.2,

$$\text{(1)} \qquad \Phi(\mathbf{r},t) = \int_{-\infty}^{\infty} dt' \int_{V_{r'}} D_{\text{ret}}(\mathbf{r}-\mathbf{r}',t-t')\rho(\mathbf{r}',t')d^3\mathbf{r}',$$

which says that the disturbance at any later time t is due only to the source function $\rho(\mathbf{r},t)$. We rewrite (1), using $D_{\text{ret}} = c\,\delta(r - ct)/r$.

$$\Phi(\mathbf{r},t) = \int_{V_{r'}} d^3\mathbf{r}' \int_{-\infty}^{\infty} dt' \frac{c}{r}\delta(r - ct)\rho(\mathbf{r}',t'),$$

where we have set $\delta(r-ct) \equiv \delta\left[|\mathbf{r}-\mathbf{r}'| - c(t-t')\right]$.

Define $\eta \equiv |\mathbf{r}-\mathbf{r}'| - c(t-t')$, so that in the integration over time, $d\eta = c\,dt'$. Substituting and integrating:

$$\text{(2)} \qquad \Phi(\mathbf{r},t) = \int_{V_{r'}} d^3\mathbf{r}' \int_{-\infty}^{\infty} d\eta \frac{\rho(\mathbf{r}',t')}{|\mathbf{r}-\mathbf{r}'|}\delta(\eta) = \int_{V_{r'}} d^3\mathbf{r}' \frac{\rho\left(\mathbf{r}', t - \frac{|\mathbf{r}-\mathbf{r}'|}{c}\right)}{|\mathbf{r}-\mathbf{r}'|}.$$

Note that the form $\int_{V_{r'}} d^3\mathbf{r}' \frac{\rho(\mathbf{r}',t')}{|\mathbf{r}-\mathbf{r}'|}$, where $t' = t - \frac{|\mathbf{r}-\mathbf{r}'|}{c}$, suggests the Poisson problem, i.e. although one investigates Φ at a point $\mathbf{r}$ and at a time t, the source is at point $\mathbf{r}'$ at a time $t' = t - \frac{|\mathbf{r}-\mathbf{r}'|}{c}$, which is the difference between time t and the time it takes the functional disturbance to travel a distance $|\mathbf{r}-\mathbf{r}'|$ with velocity c (Fig. 11-5).

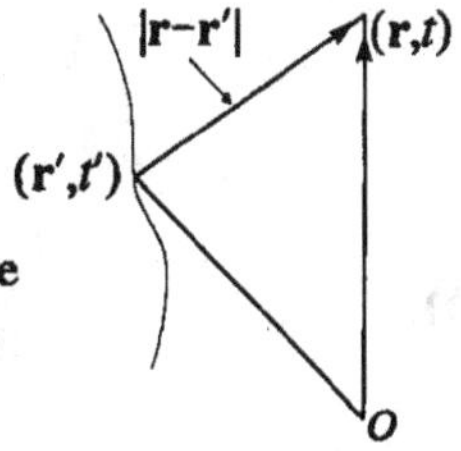

Fig. 11-5

As $c \to \infty$, $t' \to t$, resulting in Poisson's equation $\Delta\Phi = -4\pi\rho$, i.e. instantaneous propagation.

Returning to our problem, let $\rho(\mathbf{r}',t') = e\,\delta^3[\mathbf{r}'-\mathbf{r}_e(t')]$ refer to a moving source located at $\mathbf{r}_e(t)$ (specifying the trajectory of the source particle), such that for a fixed time t', $\int \rho(\mathbf{r}',t')d^3\mathbf{r}' = e$.

Substituting into equation (2):

$$(3) \qquad \Phi(\mathbf{r},t) = \int_{V_{r'}} d^3\mathbf{r}' \left| \frac{e\,\delta^3[\mathbf{r}'-\mathbf{r}_e(t')]}{|\mathbf{r}-\mathbf{r}'|} \right|_{t'=t-\frac{|\mathbf{r}-\mathbf{r}'|}{c}} .$$

If we define the vector $\bar{\eta} \equiv \mathbf{r}' - \mathbf{r}_e(t')$, equation (3) becomes

$$(4) \qquad \Phi(\mathbf{r},t) = \int_{V_\eta} d^3\bar{\eta}\, \frac{e\,\delta^3(\bar{\eta})J_{\bar{\eta}}}{|\mathbf{r}-\mathbf{r}'|} = \left| \frac{e\,J_{\bar{\eta}}}{|\mathbf{r}-\mathbf{r}'|} \right|_{\bar{\eta}=0} ,$$

where we have used the expression for the transformation of volume elements from $\mathbf{r}'$-space to $\bar{\eta}$-space, e.g.

$$d^3\mathbf{r}' = J_{\bar{\eta}}\left(\frac{x'\ y'\ z'}{\eta_1\ \eta_2\ \eta_3}\right) d\eta_1\, d\eta_2\, d\eta_3 ,$$

where $J_{\bar{\eta}}$ is the Jacobian of the transformation, i.e.

$$J_{\bar{\eta}}\left(\frac{x'\ y'\ z'}{\eta_1\ \eta_2\ \eta_3}\right) = \begin{vmatrix} \frac{\partial x'}{\partial \eta_1} & \frac{\partial y'}{\partial \eta_1} & \frac{\partial z'}{\partial \eta_1} \\ \frac{\partial x'}{\partial \eta_2} & \frac{\partial y'}{\partial \eta_2} & \frac{\partial z'}{\partial \eta_2} \\ \frac{\partial x'}{\partial \eta_3} & \frac{\partial y'}{\partial \eta_3} & \frac{\partial z'}{\partial \eta_3} \end{vmatrix} .$$

By definition, $\bar{\eta} = \mathbf{r}' - \mathbf{r}_e(t') \Rightarrow d\bar{\eta} = d\mathbf{r}' - \dot{\mathbf{r}}_e(t')dt'$. In addition,

$$t' = t - \frac{|\mathbf{r}-\mathbf{r}'|}{c} = t - \frac{[(\mathbf{r}-\mathbf{r}')\cdot(\mathbf{r}-\mathbf{r}')]^{1/2}}{c} .$$

$$dt' = -\frac{1}{2}\frac{[(\mathbf{r}-\mathbf{r}')\cdot(\mathbf{r}-\mathbf{r}')]^{-1/2}\,[-2d\mathbf{r}'\cdot(\mathbf{r}-\mathbf{r}')]}{c} = \frac{d\mathbf{r}'\cdot(\mathbf{r}-\mathbf{r}')}{c|(\mathbf{r}-\mathbf{r}')|}.$$

Thus,

$$(5) \qquad d\bar{\eta} = d\mathbf{r}' - \frac{\bar{\upsilon}_e(\mathbf{r}-\mathbf{r}')\cdot d\mathbf{r}'}{c|(\mathbf{r}-\mathbf{r}')|}.$$

Since we want J evaluated at a particular time t and location $\mathbf{r}$, we choose the x-axis parallel to $\dot{\mathbf{r}}_e = \bar{\upsilon}_e$. In component form,

$$d\eta_1 = dx' - \frac{\upsilon_e(x-x')dx'}{c|(\mathbf{r}-\mathbf{r}')|} - \frac{\upsilon_e(y-y')dy'}{c|(\mathbf{r}-\mathbf{r}')|} - \frac{\upsilon_e(z-z')dz'}{c|(\mathbf{r}-\mathbf{r}')|}.$$

$$d\eta_2 = dy'.$$

$$d\eta_3 = dz'.$$

setting up as simultaneous equations:

$$d\eta_1 = dx'\left[1 - \frac{\upsilon_e(x-x')}{c|(\mathbf{r}-\mathbf{r}')|}\right] + dy'\left[-\frac{\upsilon_e(y-y')}{c|(\mathbf{r}-\mathbf{r}')|}\right] + dz'\left[-\frac{\upsilon_e(z-z')}{c|(\mathbf{r}-\mathbf{r}')|}\right].$$

$$d\eta_2 = 0 + dy' + 0.$$

$$d\eta_3 = 0 + 0 + dz'.$$

Hence the det of the transformation $J_{r'}\left(\frac{\eta}{r'}\right)$ is given by

$$J_{r'}\left(\frac{\eta}{r'}\right) = 1 - \frac{\upsilon_e(x-x')}{c|(\mathbf{r}-\mathbf{r}')|} = 1 - \frac{\vec{\upsilon}_e\cdot(\mathbf{r}-\mathbf{r}')}{c|(\mathbf{r}-\mathbf{r}')|}.$$

Since $J_{r'}\left(\frac{\eta}{r'}\right) = \left[J_{\bar{\eta}}\left(\frac{r'}{\eta}\right)\right]^{-1}$,

$$J_{\bar{\eta}}\left(\frac{r'}{\eta}\right) = \frac{1}{1 - \dfrac{\vec{\upsilon}_e\cdot(\mathbf{r}-\mathbf{r}')}{c|(\mathbf{r}-\mathbf{r}')|}}.$$

Thus, from (4):

$$\Phi(\mathbf{r},t) = \int_{V_\eta} d^3\bar{\eta}\, \frac{e\,\delta^3(\bar{\eta})}{|\mathbf{r}-\mathbf{r}'|} \frac{1}{1 - \dfrac{\bar{\upsilon}_e\cdot(\mathbf{r}-\mathbf{r}')}{c|\mathbf{r}-\mathbf{r}'|}} = \left| \frac{e}{|\mathbf{r}-\mathbf{r}'| - \dfrac{\bar{\upsilon}_e\cdot(\mathbf{r}-\mathbf{r}')}{c}} \right|_{\bar{\eta}=0,t'} .$$

Since $\bar{\eta} = 0$ implies $\mathbf{r}' = \mathbf{r}_e(t')$, we have

$$(5) \qquad \Phi(\mathbf{r},t) = \frac{e}{|\mathbf{r}-\mathbf{r}_e(t')| - \dfrac{\bar{\upsilon}_e\cdot(\mathbf{r}-\mathbf{r}_e(t'))}{c}},$$

<u>Physical interpretation of (5)</u>

A source at $\mathbf{r}_e(t')$ moves along some path C, and at any instant has a velocity $\bar{\upsilon}_e$. At a point $\mathbf{r}$ and at time t the disturbance is given by $\dfrac{e}{|\mathbf{r}-\mathbf{r}_e|}$, where an additional term appears in the denominator. In order for the disturbance to be felt at time t, it must have been produced at $\mathbf{r}_e$ at some previous (retarded) time $t' = t - \dfrac{|\mathbf{r}-\mathbf{r}_e|}{c}$, where $\dfrac{|\mathbf{r}-\mathbf{r}_e|}{c}$ is the time it takes to go from $\mathbf{r}_e$ to $\mathbf{r}$.

Note: If the wave equation is given by

$$\Delta\mathbf{A} - \frac{1}{c^2}\ddot{\mathbf{A}} = -\frac{4\pi\mathbf{j}}{c},$$

and if

$$\mathbf{j}(\mathbf{r}',t') = e\bar{\upsilon}_e\delta^3[\mathbf{r}' - \mathbf{r}_e(t')],$$

then at point $\mathbf{r}$ and at time t, we have

$$\mathbf{A}(\mathbf{r},t) = \left| \frac{e\bar{\upsilon}_e(t')}{|\mathbf{r}-\mathbf{r}_e(t')| - \dfrac{\bar{\upsilon}_e\cdot(\mathbf{r}-\mathbf{r}_e(t'))}{c}} \right|_{\text{at } t' = \text{retarded time}},$$

which are the so-called Liénard-Wiechart potentials.

11.4 Finite Boundary Medium

Program:

Try to find the Green's function for the homogeneous case, $\nabla^2\Phi - \ddot{\Phi}/c^2 = 0$, and then deduce the Green's function for the inhomogeneous case. Then, knowing Φ and $\dot{\Phi}$ at time $t = t_0$, Φ will be known at all other times.

One does not perform a Fourier analysis, since the boundary in this case is not at infinity, and might be quite complicated.

(A) Homogeneous Case

Consider $H\Psi_n = -\nabla_n^2\Psi_n = \lambda_n\Psi_n$, where $\Psi_n = 0$ on surface S, and imagine that this eigenvalue problem has been solved, i.e. that we know the entire set of eigensolutions and their respective corresponding eigenvalues. We will obtain the Green's function from this eigenvalue problem.

$$\int\Psi_n H\Psi_n d\tau = \int\lambda_n\Psi_n^2 d\tau.$$

$$\Downarrow$$

$$\lambda_n = \frac{\int\Psi_n(-\nabla_n^2)\Psi_n d\tau}{\int\Psi_n^2 d\tau} = \frac{-\int\nabla\cdot(\Psi_n\nabla\Psi_n)d\tau + \int(\nabla\Psi_n)^2 d\tau}{\int\Psi_n^2 d\tau}$$

$$= \frac{-\int(\Psi_n\nabla\Psi_n)\cdot\mathbf{n}_o dS + \int(\nabla\Psi_n)^2 d\tau}{\int\Psi_n^2 d\tau} = \frac{\int(\nabla\Psi_n)^2 d\tau}{\int\Psi_n^2 d\tau}.$$

The last step follows since by hypothesis $\Psi_n = 0$ on S. Hence,

$$\lambda_n = \frac{\int(\nabla\Psi_n)^2 d\tau}{\int\Psi_n^2 d\tau} \geq 0.$$

(The λ_n correspond to ω_n^2, the frequencies of vibration, and must be ≥ 0.)

Define the Green's function $G(\mathbf{r},\mathbf{r}',t)$:

(1) $$G(\mathbf{r},\mathbf{r}',t) = \sum_n \Psi_n(\mathbf{r})\Psi_n(\mathbf{r}')\frac{\sin\sqrt{\lambda_n}ct}{c\sqrt{\lambda_n}}.$$

Note that if one sets $\Psi_n(\mathbf{r})\Psi_n(\mathbf{r}') = e^{i\mathbf{k}\cdot(\mathbf{r}-\mathbf{r}')}$, sets $\sqrt{\lambda_n} = k$ and lets the $\Sigma \to \int$, then $G(\mathbf{r},\mathbf{r}',t)$ for finite media becomes similar to $D(\mathbf{r},\mathbf{r}',t)$ for infinite media [equation (7), Section 11.1].

Properties of $G(\mathbf{r},\mathbf{r}'t)$

(i) $\nabla^2 G(\mathbf{r},\mathbf{r}',t) - \dfrac{1}{c^2}\dfrac{\partial^2 G(\mathbf{r},\mathbf{r}',t)}{\partial t^2} = 0.$

Proof:

$$\nabla^2 G = \sum_n -\lambda_n \Psi_n(\mathbf{r})\Psi_n(\mathbf{r}')\frac{\sin\sqrt{\lambda_n}ct}{c\sqrt{\lambda_n}}.$$

$$\frac{1}{c^2}\frac{\partial^2 G}{\partial t^2} = \sum_n -\lambda_n \Psi_n(\mathbf{r})\Psi_n(\mathbf{r}')\frac{\sin\sqrt{\lambda_n}ct}{c\sqrt{\lambda_n}}.$$

(ii) $\nabla^2 \dfrac{\partial G(\mathbf{r},\mathbf{r}',t)}{\partial t} - \dfrac{1}{c^2}\dfrac{\partial^2}{\partial t^2}\dfrac{\partial G(\mathbf{r},\mathbf{r}',t)}{\partial t} = 0.$

Proof:

$$\frac{\partial G}{\partial t} = \sum_n \Psi_n(\mathbf{r})\Psi_n(\mathbf{r}')\cos\sqrt{\lambda_n}ct.$$

$$\nabla^2\frac{\partial G}{\partial t} = \sum_n -\lambda_n \Psi_n(\mathbf{r})\Psi_n(\mathbf{r}')\cos\sqrt{\lambda_n}ct.$$

$$\frac{1}{c^2}\frac{\partial^2}{\partial t^2}\frac{\partial G}{\partial t} = \sum_n -\lambda_n \Psi_n(\mathbf{r})\Psi_n(\mathbf{r}')\cos\sqrt{\lambda_n}ct.$$

(iii) $G(\mathbf{r},\mathbf{r}',t) = 0$ and $\dfrac{\partial G(\mathbf{r},\mathbf{r}',t)}{\partial t} = 0$ for $\mathbf{r}$ on S or for $\mathbf{r}'$ on S, since by hypothesis, $\Psi_n(\mathbf{r}) = 0$ for $\mathbf{r}$ on S, and $\Psi_n(\mathbf{r}') = 0$ for $\mathbf{r}'$ on S.

(iv) $G(\mathbf{r},\mathbf{r}',0) = 0$ and $\dfrac{\partial^2 G(\mathbf{r},\mathbf{r}',0)}{\partial t^2} = 0$ by virtue of term $\sin\sqrt{\lambda_n}ct$.

(v) $\dfrac{\partial G(\mathbf{r},\mathbf{r}',0)}{\partial t} = \delta^3(\mathbf{r}-\mathbf{r}')$.

Proof: $\dfrac{\partial G(\mathbf{r},\mathbf{r}',0)}{\partial t} = \sum_n \Psi_n(\mathbf{r})\Psi_n(\mathbf{r}')$. Since the Ψ_n form a complete (as well as orthonormal) set, any function that satisfies the boundary conditions can be expanded in a set of Ψ_n. Thus, $\delta^3(\mathbf{r}-\mathbf{r}') = \sum_n a_n\Psi_n(\mathbf{r})$, from which

$$\int \Psi_m(\mathbf{r})\delta^3(\mathbf{r}-\mathbf{r}')d\tau = \int \sum_n a_n\Psi_n(\mathbf{r})\Psi_m(\mathbf{r})d\tau = \sum_n a_n\delta_{nm} = a_m .$$

Hence, $a_m = \Psi_m(\mathbf{r}') \rightarrow \delta^3(\mathbf{r}-\mathbf{r}') = \dfrac{\partial G(\mathbf{r},\mathbf{r}',0)}{\partial t}$.

(vi) If $\Phi(\mathbf{r},t_o)$ and $\dot{\Phi}(\mathbf{r},t_o)$ are given, then at any later time t:

$$(2) \qquad \Phi(\mathbf{r},t) = \int \left[G(\mathbf{r},\mathbf{r}',t-t_o)\dot{\Phi}(\mathbf{r}',t_o) + \frac{\partial G(\mathbf{r},\mathbf{r}',t-t_o)}{\partial t}\Phi(\mathbf{r}',t_o)\right] d^3\mathbf{r}' .$$

Proof:

(a) Since $\nabla^2 G(\mathbf{r},\mathbf{r}',t-t_o) - \dfrac{1}{c^2}\dfrac{\partial^2 G(\mathbf{r},\mathbf{r}',t-t_o)}{\partial t^2} = 0$ from (i) above, and since $\nabla^2 \dfrac{\partial G(\mathbf{r},\mathbf{r}',t-t_o)}{\partial t} - \dfrac{1}{c^2}\dfrac{\partial^2}{\partial t^2}\dfrac{\partial G(\mathbf{r},\mathbf{r}',t-t_o)}{\partial t} = 0$ from (ii) above, we have

$$\nabla^2\Phi(\mathbf{r},t) - \frac{1}{c^2}\frac{\partial^2\Phi(\mathbf{r},t)}{\partial t^2} = 0.$$

(b) As $t \rightarrow t_o$, $\Phi(\mathbf{r},t) \rightarrow \Phi(\mathbf{r},t_o)$ and $\dot{\Phi}(\mathbf{r},t) \rightarrow \dot{\Phi}(\mathbf{r},t_o)$ from (iv) and (v) above.

(c) For $\mathbf{r}$ on S, $\Phi(\mathbf{r},t) = 0$ from (iii) above.

Thus, $\Phi(\mathbf{r},t)$ satisfies the wave equation and the boundary conditions.

(B) Inhomogeneous Case

For the inhomogeneous case, where a source $\rho(\mathbf{r},t)$ is present, we define:

$$\mathcal{G}_{ret}(\mathbf{r},\mathbf{r}',t) = \begin{cases} 4\pi c^2 G(\mathbf{r},\mathbf{r}',t) \text{ for } t > 0. \\ 0 \text{ for } t < 0. \end{cases}$$

$$\mathcal{G}_{adv}(\mathbf{r},\mathbf{r}',t) = \begin{cases} 0 \text{ for } t > 0. \\ -4\pi c^2 G(\mathbf{r},\mathbf{r}',t) \text{ for } t < 0. \end{cases}$$

<u>Properties of $\mathcal{G}_{adv}(\mathbf{r},\mathbf{r}',t)$ and $\mathcal{G}_{ret}(\mathbf{r},\mathbf{r}',t)$</u>

(3) $$\left(\nabla^2 - \frac{1}{c^2}\frac{\partial^2}{\partial t^2}\right)\mathcal{G}_{adv/ret}(\mathbf{r},\mathbf{r}',t) = -4\pi\delta^3(\mathbf{r}-\mathbf{r}')(\delta t).$$

The method of proof follows what was done previously for $D_{ret/adv}$ (Section 11.2): we show it to be true for $t < 0$ and for $t > 0$, and then show that the singularities on both sides of (3) are the same as $t \to 0$.

It follows from (3) above, that when a source $\rho(\mathbf{r},t)$ is present,

$$\left(\nabla^2 - \frac{1}{c^2}\frac{\partial^2}{\partial t^2}\right)\Phi(\mathbf{r},\mathbf{r}',t) = -4\pi\rho(\mathbf{r},t),$$

and for $t > t_o$ and $t < t_o$,

(4) $$\Phi(\mathbf{r},t) = \Phi_o(\mathbf{r},t) + \Phi_1(\mathbf{r},t),$$

where

(5) $$\Phi_o(\mathbf{r},t) = \int\left[G(\mathbf{r},\mathbf{r}',t-t_o)\dot{\Phi}(\mathbf{r}',t_o) + \frac{\partial G(\mathbf{r},\mathbf{r}',t-t_o)}{\partial t}\Phi(\mathbf{r}',t_o)\right]d^3\mathbf{r}'.$$

For $t > t_o$,

(6) $$\Phi_1(\mathbf{r},t) = \int_{t_o}^{\infty}dt'\int d^3\mathbf{r}'\,\mathcal{G}_{ret}(\mathbf{r},\mathbf{r}',t-t')\rho(\mathbf{r}',t'),$$

while for $t < t_o$,

(7) $$\Phi_1(\mathbf{r},t) = \int_{-\infty}^{t_o}dt'\int d^3\mathbf{r}'\,\mathcal{G}_{adv}(\mathbf{r},\mathbf{r}',t-t')\rho(\mathbf{r}',t').$$

(These results are to be compared with equations (5) and (9), Section 11.2.)

11.5 Green's Function Method Applied to Schrödinger's Equation and to Heat Conduction

Consider the eigenvalue equation

$$H\Psi = \xi\frac{\partial\Psi}{\partial t}.$$

For the case of Schrödinger's equation, $H = -\frac{\hbar^2}{2m}\Delta + V$ and $\xi = -\frac{\hbar}{i}$ (eigenvalue ξ is imaginary).

For the case of heat conduction, $H = -\Delta$ and $\xi = -\frac{1}{\lambda}$ (eigenvalue ξ is real).

In both cases, operator H is real and Hermitian.

Problem: Given $\Psi = \Psi(\mathbf{r},0)$ at $t = 0$ in both cases. Find $\Psi(\mathbf{r},t)$ at all other times.

One supposes the particular eigenvalue problem $H\Psi_n = \lambda_n\Psi_n$ to be solved, where the Ψ_n satisfy certain boundary conditions. The Ψ_n then form a complete set for the particular operator H for any $\Psi(\mathbf{r},t)$ satisfying the boundary conditions, i.e.

$$(1) \qquad \Psi(\mathbf{r},t) = \sum_n f_n(t)\Psi_n(\mathbf{r}),$$

so that

$$H\Psi(\mathbf{r},t) = \sum_n f_n(t)H\Psi_n(\mathbf{r}) = \sum_n f_n(t)\lambda_n\Psi_n(\mathbf{r}).$$

$$\Downarrow$$

$$H\Psi - \xi\dot{\Psi} = \sum_n \Psi_n(\mathbf{r})\left[f_n(t)\lambda_n - \xi\dot{f}_n\right] = 0,$$

and since the Ψ_n form an orthonormal set,

$$\frac{df_n(t)}{dt} = \frac{\lambda_n}{\xi}f_n(t) \quad\Rightarrow\quad f_n(t) = a_n\exp\left(\frac{\lambda_n t}{\xi}\right),$$

and

$$(2) \qquad \Psi(\mathbf{r},t) = \sum_n a_n\exp\left(\frac{\lambda_n t}{\xi}\right)\Psi_n(\mathbf{r}).$$

We seek the Green's function for the problem.

Boundary conditions require that $G(\mathbf{r},\mathbf{r}',t)$ = δ–function at $t = 0$, and at a later time t, $G(\mathbf{r},\mathbf{r}',t) = \Psi(\mathbf{r},t)$.

At $t = 0$, $\Psi(\mathbf{r},0) = \sum_n a_n\Psi_n(\mathbf{r}) = \delta^3(\mathbf{r}-\mathbf{r}')$. Since $a_n = \Psi_n(\mathbf{r}')$ [Sect. 11.4], it follows that

(3) $$G(\mathbf{r},\mathbf{r}',t) = \sum_n \Psi_n(\mathbf{r})\Psi_n(\mathbf{r}')\exp\left(\frac{\lambda_n t}{\xi}\right).$$

is the appropriate Green's function.

Properties of $G(\mathbf{r},\mathbf{r}',t)$

(i) $HG = \xi\dot{G}$.

Proof:

$$HG(\mathbf{r},\mathbf{r}',t) = \sum_n \lambda_n\Psi(\mathbf{r})\Psi(\mathbf{r}')\exp\left(\frac{\lambda_n t}{\xi}\right).$$

$$\xi\dot{G}(\mathbf{r},\mathbf{r}',t) = \sum_n \lambda_n\Psi_n(\mathbf{r})\Psi_n(\mathbf{r}')\exp\left(\frac{\lambda_n t}{\xi}\right).$$

(ii) As $t \to 0$, $G(\mathbf{r},\mathbf{r}',t) \to \delta^3(\mathbf{r}-\mathbf{r}')$.

(iii) For all time t:

(4) $$\Psi(\mathbf{r},t) = \int G(\mathbf{r},\mathbf{r}',t-t_0)\Psi(\mathbf{r}',t_0)d^3\mathbf{r}'.$$

(Note that this is of the required form for the homogeneous equation of heat conduction in the infinite medium, as discussed in Section 9.2, Green's function property 4.)

Real versus imaginary eigenvalues

For the *Schrödinger equation*, $\exp\left(\lambda_n t/\xi\right) \to \exp\left(-i\lambda_n t/\hbar\right)$ represents an oscillation in time, and since Ψ represents matter density which must be conserved, $\int\Psi^2 d\tau$ = finite.

For the case of *heat conduction*, $\exp\left(\lambda_n t/\xi\right) \to \exp\left(-\lambda\lambda_n t\right)$. Since $\lambda\lambda_n$ is positive, this represents an exponential decay in time for $t > 0$ and an exponential increase in time for $t < 0$, which agrees with physical interpretations.

Appendix B

Problems and Solutions

Problem Set A

I. If $f(z)$ is analytic and $f(z) = \varphi + i\psi$, show that both φ and ψ obey Laplace's equation, i.e. show that $\Delta\varphi = 0$ and $\Delta\psi = 0$.

Solution: Since $f(z)$ is analytic and is expressible as $f(z) = \varphi + i\psi$, then φ and ψ must obey the Cauchy-Riemann equations

$$\frac{\partial\varphi}{\partial x} = \frac{\partial\psi}{\partial y}, \qquad \frac{\partial\varphi}{\partial y} = -\frac{\partial\psi}{\partial x}.$$

Evaluating $\Delta\varphi$ and $\Delta\psi$:

$$\Delta\varphi = \frac{\partial^2\varphi}{\partial x^2} + \frac{\partial^2\varphi}{\partial y^2} = \frac{\partial}{\partial x}\left(\frac{\partial\psi}{\partial y}\right) + \frac{\partial}{\partial y}\left(-\frac{\partial\psi}{\partial x}\right) = \frac{\partial^2\psi}{\partial x\partial y} - \frac{\partial^2\psi}{\partial y\partial x} = 0.$$

$$\Delta\psi = \frac{\partial^2\psi}{\partial x^2} + \frac{\partial^2\psi}{\partial y^2} = \frac{\partial}{\partial x}\left(-\frac{\partial\varphi}{\partial y}\right) + \frac{\partial}{\partial y}\left(\frac{\partial\varphi}{\partial x}\right) = -\frac{\partial^2\varphi}{\partial x\partial y} + \frac{\partial^2\varphi}{\partial y\partial x} = 0.$$

II. Given a conducting circular cylinder of radius r_o, whose potential φ on the surface has the value A.

(a) Find $f(z)$, a solution to the problem (trial and error).

(b) Show that $f(z)$ satisfies the boundary conditions.

(c) Sketch lines of **E** and φ = constant.

Solution: Observe that we want φ to be a constant on the surface and lines of constant φ to be concentric with the cylinder. Hence $\varphi = f(\rho)$ only. Also, since lines of constant ψ should be perpendicular to lines of constant φ, we see that $\psi = g(\theta)$ only. This suggests the form $f(z) = B\ln z$.

(a) Let $f(z) = B\ln z = B\ln\rho + iB\theta$. Then, $\varphi = B\ln\rho$ and $\psi = B\theta$. At $\rho = r_o$, $\varphi = A$, so that $B = A/\ln r_o$. Thus,

$$\begin{pmatrix} \varphi = \dfrac{A}{\ln r_o}\ln\rho. \\ \psi = \dfrac{A}{\ln r_o}\theta. \end{pmatrix} \quad \rightarrow \quad f(z) = \frac{A}{\ln r_o}\ln z.$$

(b) In cylindrical coordinates,

$$\nabla^2\varphi = \frac{1}{\rho}\frac{\partial}{\partial\rho}\left(\rho\frac{\partial\varphi}{\partial\rho}\right),$$

or

$$\frac{1}{\rho}\frac{\partial}{\partial\rho}\left(\rho\frac{A}{\rho\ln r_o}\right) = 0 \quad \rightarrow \quad \nabla^2\varphi = 0,\ \varphi = A \text{ on the surface.}$$

Thus, boundary conditions are satisfied.

(c) Lines of **E** are parallel to lines of ψ = constant or θ = constant.

$$\mathbf{E} = -\nabla\varphi = -\frac{A}{\ln r_o}\nabla\ln\rho$$

$$= -\frac{A}{\ln r_o}\frac{1}{\rho}\vec{\rho_o},$$

and the mapping is as shown in Fig. B.1.

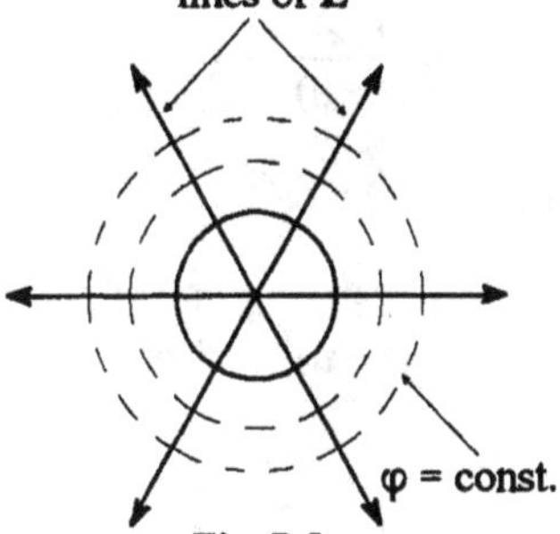

Fig. B.1

III. Show that if $f(z)$ is analytic on and within C (Fig. B.2), and a is a point within C_1, then

$$f'(a) = \frac{1}{2\pi i}\oint_C \frac{f(z)}{(z-a)^2}dz.$$

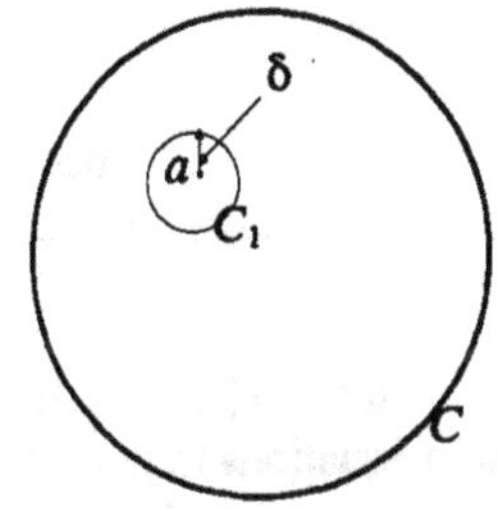

Fig. B.2

Solution: Define a function $g(z) = \frac{f(z) - f(a)}{(z-a)} - f'(a)$, which is defined on the circle C_1, of radius δ, and is analytic on and within C_1. Then, as $\delta \to 0$, $g(z) \to 0$ since $\lim_{\delta \to 0}\left[\frac{f(z)-f(a)}{(z-a)} - f'(a)\right] = 0$.

We have shown that if $f(z)$ is analytic on and within C, then

$$\oint_C \frac{f(z)dz}{(z-a)^2} = \oint_{C_1} \frac{f(z)dz}{(z-a)^2},$$

where C_1 is the circle of radius δ surrounding point a. Thus,

$$\oint_{C_1} \frac{f(z)dz}{(z-a)^2} = \oint_{C_1} \frac{g(z)dz}{(z-a)} + f'(a)\oint_{C_1} \frac{dz}{(z-a)} + f(a)\oint_{C_1} \frac{dz}{(z-a)^2}.$$

Since $(z-a)$ is finite over C_1, while $g(z)$ is analytic on and within C_1, we have, by Cauchy's integral theorem,

$$\oint_{C_1} \frac{g(z)dz}{(z-a)} = 0.$$

Since $(z-a) = \delta e^{i\theta}$,

$$\oint_{C_1} \frac{dz}{(z-a)^2} = \int_0^{2\pi} \frac{\delta i e^{i\theta}}{\delta^2 e^{2i\theta}} = \frac{i}{\delta}\int_0^{2\pi} e^{-i\theta} d\theta = 0.$$

Hence,

$$\oint_C \frac{f(z)dz}{(z-a)^2} = \oint_{C_1} \frac{f(z)dz}{(z-a)^2} = f'(a)\int_0^{2\pi} \frac{\delta i e^{i\theta} d\theta}{\delta e^{i\theta}} = 2\pi i f'(a).$$

Thus,

$$f'(a) = \frac{1}{2\pi i}\oint_C \frac{f(z)dz}{(z-a)^2}.$$

IV. Show that if $f(z)$ is analytic and if $f(z) = \varphi + i\psi$, then the Cauchy-Riemann equations in polar form are satisfied, i.e.

$$\frac{\partial \varphi}{\partial \rho} = \frac{1}{\rho}\frac{\partial \psi}{\partial \theta} \quad \text{and} \quad \frac{1}{\rho}\frac{\partial \varphi}{\partial \theta} = -\frac{\partial \psi}{\partial \rho}.$$

Solution: Since $f(z)$ is analytic and $f(z) = \varphi + i\psi$, φ and ψ obey the Cauchy-Riemann equations:

$$\frac{\partial\varphi}{\partial x} = \frac{\partial\psi}{\partial y} \quad \text{and} \quad \frac{\partial\varphi}{\partial y} = -\frac{\partial\psi}{\partial x}.$$

If we regard $f(z) = f(\rho,\theta)$, then $\varphi = \varphi(\rho,\theta)$ and $\psi = \psi(\rho,\theta)$. The Cauchy-Riemann equations then become:

$$\frac{\partial\varphi}{\partial\rho}\frac{\partial\rho}{\partial x} + \frac{\partial\varphi}{\partial\theta}\frac{\partial\theta}{\partial x} = \frac{\partial\psi}{\partial\rho}\frac{\partial\rho}{\partial y} + \frac{\partial\psi}{\partial\theta}\frac{\partial\theta}{\partial y}$$

and

$$\frac{\partial\varphi}{\partial\rho}\frac{\partial\rho}{\partial y} + \frac{\partial\varphi}{\partial\theta}\frac{\partial\theta}{\partial y} = -\left[\frac{\partial\psi}{\partial\rho}\frac{\partial\rho}{\partial x} + \frac{\partial\psi}{\partial\theta}\frac{\partial\theta}{\partial x}\right].$$

But from $\rho = (x^2 + y^2)^{1/2}$ and $\theta = \tan^{-1}\frac{y}{x}$, we have

$$\frac{\partial\rho}{\partial x} = \frac{x}{\rho} = \cos\theta \quad \text{and} \quad \frac{\partial\rho}{\partial y} = \frac{y}{\rho} = \sin\theta.$$

$$\frac{\partial\theta}{\partial x} = -\frac{y}{\rho^2} = -\frac{\sin\theta}{\rho} \quad \text{and} \quad \frac{\partial\theta}{\partial y} = \frac{x}{\rho^2} = \frac{\cos\theta}{\rho}.$$

Thus we have

(1) $$\frac{\partial\varphi}{\partial\rho}\cos\theta - \frac{\partial\varphi}{\partial\theta}\frac{\sin\theta}{\rho} = \frac{\partial\psi}{\partial\rho}\sin\theta + \frac{\partial\psi}{\partial\theta}\frac{\cos\theta}{\rho}.$$

(2) $$\frac{\partial\varphi}{\partial\rho}\sin\theta + \frac{\partial\varphi}{\partial\theta}\frac{\cos\theta}{\rho} = -\frac{\partial\psi}{\partial\rho}\cos\theta + \frac{\partial\psi}{\partial\theta}\frac{\sin\theta}{\rho}.$$

Multiply (1) by $\cos\theta$ and (2) by $\sin\theta$ and add to get:

$$\frac{\partial\varphi}{\partial\rho}(\cos^2\theta + \sin^2\theta) = \frac{\partial\psi}{\partial\theta}\left(\frac{\cos^2\theta + \sin^2\theta)}{\rho}\right) \rightarrow \frac{\partial\varphi}{\partial\rho} = \frac{1}{\rho}\frac{\partial\psi}{\partial\theta}.$$

Multiply (1) by $-\sin\theta$ and (2) by $\cos\theta$ and add to get:

$$\frac{\partial\varphi}{\partial\theta}\left(\frac{\cos^2\theta+\sin^2\theta}{\rho}\right) = -\frac{\partial\psi}{\partial\rho}(\cos^2\theta+\sin^2\theta) \quad\rightarrow\quad \frac{1}{\rho}\frac{\partial\varphi}{\partial\theta} = -\frac{\partial\psi}{\partial\rho}.$$

V. If $f(z)$ is analytic when $|z-a|<R$, then for $0<r<R$, show that

$$f'(a) = \frac{1}{\pi r}\int_0^{2\pi}\varphi e^{-i\theta}d\theta,$$

where $f(z)=\varphi+i\psi$.

Solution: Expanding $f(z)$ in a Taylor series about point a,

$$f(z) = f(a)+f'(a)(z-a)+\frac{f''(a)}{2!}(z-a)^2+\cdots+\frac{f^n(a)}{n!}(z-a)^n+\cdots$$

Writing $z-a=re^{i\theta}$, we have

$$\int_0^{2\pi} f(z)e^{i\theta}d\theta = \int_0^{2\pi}\left[f(a)e^{i\theta}+f'(a)re^{i2\theta}+\frac{f''(a)}{2!}r^2e^{i3\theta}+\cdots\right]d\theta$$

$$+\cdots+\int_0^{2\pi}\left[\frac{f^n(a)}{n!}r^ne^{i(n+1)\theta}+\cdots\right]d\theta.$$

$$= f(a)\int_0^{2\pi}e^{i\theta}d\theta+f'(a)r\int_0^{2\pi}e^{i2\theta}d\theta+\frac{f''(a)}{2!}r^2\int_0^{2\pi}e^{i3\theta}d\theta+\cdots$$

$$+\cdots+\frac{f^n(a)}{n!}r^n\int_0^{2\pi}e^{i(n+1)\theta}d\theta+\cdots = 0.$$

Hence $\int_0^{2\pi} f(z)e^{i\theta}d\theta = 0$. It follows that $\left\{\int_0^{2\pi} f(z)e^{i\theta}d\theta\right\}^* = \int_0^{2\pi} f^*(z)e^{-i\theta}d\theta = 0$,

or, $\int_0^{2\pi}(\varphi-i\psi)e^{-i\theta}d\theta = 0 \quad\rightarrow\quad \int_0^{2\pi}\varphi e^{-i\theta}d\theta = i\int_0^{2\pi}\psi e^{-i\theta}d\theta.$

We have shown in Problem 3 above that $f'(a) = \frac{1}{2\pi i}\oint_C \frac{f(\xi)}{(\xi - a)^2} d\xi$. Since $(\xi - a)^2 = r^2 e^{i2\theta}$ and $d\xi = ire^{i\theta}d\theta$, we have

$$f'(a) = \frac{1}{2\pi i}\int_0^{2\pi} \frac{(\varphi + i\psi)ire^{i\theta}}{r^2 e^{i2\theta}} d\theta = \frac{1}{2\pi r}\left[\int_0^{2\pi} \varphi e^{-i\theta}d\theta + i\int_0^{2\pi} \psi e^{-i\theta}d\theta\right].$$

Since, as we have just shown, these two integrals on the right side are equal,

$$f'(a) = \frac{1}{\pi r}\int_0^{2\pi} \varphi e^{-i\theta}d\theta.$$

Problem Set B

I. How many singularities do the following functions have? Discuss the nature of these singularities.

(a) $f(z) = e^z$

(b) $f(z) = \cos z$

(c) $f(z) = (z-a)^{1/3}(z-b)^{2/5}$

(d) $f(z) = e^{-z^2}$

(e) $f(z) = \ln(z-a)$

Solution:

(a) $f(z) = e^z = 1 + z + \frac{z^2}{2!} + \frac{z^3}{3!} + \cdots = \sum_{n=0}^{\infty} \frac{z^n}{n!}$.

Setting $w = 1/z$,

$$g(w) = f(z) = e^{1/w} = 1 + \frac{1}{w} + \frac{1}{2w^2} + \frac{1}{3!w^3} + \cdots = \sum_{n=0}^{\infty} \frac{1}{n!w^n},$$

and we note that $g(w)$ has an essential singularity at $w = 0$; hence, e^z has an essential singularity at z = infinity.

(b) $f(z) = \cos z$.

Setting $w = 1/z$,

$$\cos z = \cos\frac{1}{w} = 1 - \frac{1}{2w^2} + \frac{1}{4!w^4} - \cdots,$$

which has an essential singularity at $w = 0$. Hence, $\cos z$ has an essential singularity at z = infinity.

(c) $f(z) = (z-a)^{1/3}(z-b)^{2/5}$

Noting that there are *branch points* at a and b, we run a cut from a to ∞ and from b to ∞ (Fig. B.3).

Setting $z = 1/w$,

$$g(w) = \left(\frac{1}{w} - a\right)^{1/3}\left(\frac{1}{w} - b\right)^{2/5}$$

$$= \frac{(1 - aw)^{1/3}(1 - bw)^{2/5}}{w^{1/3 + 2/5}}.$$

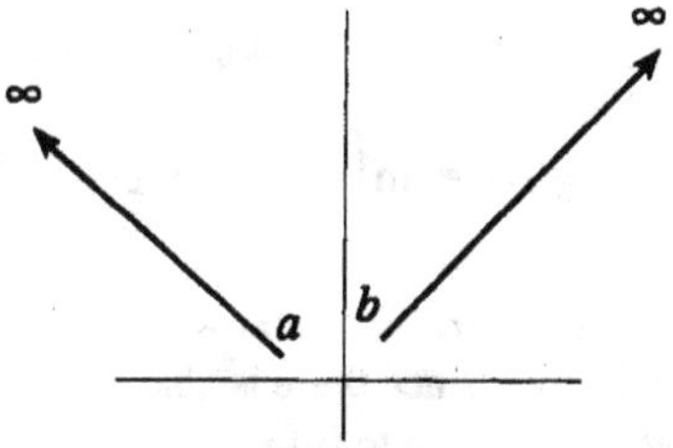

Fig. B.3

We see that we have a singularity at $w = 0$,
hence $f(z)$ has a singularity at z = infinity.
Note also that we have a branch point at z = infinity.

To determine the required number of Riemann sheets, note that $(z - a)^{1/3} = \rho^{1/3}e^{i\theta/3}$, and not until $\theta = 6\pi$ is $(z - a)^{1/3}$ at $(\theta = 6\pi) = (z - a)^{1/3}$ at $\theta = 0$.

The best procedure is to choose a particular value: e.g. $f(z) = (z - a)^{1/3}$.
Setting $(z - a) = -1$,

$$(-1) = e^{i\pi} = e^{3i\pi} = e^{5i\pi} = e^{7i\pi} = e^{9i\pi}, \text{ etc.}$$

$$(-1)^{1/3} = e^{i\pi/3} = e^{3i\pi/3} = e^{5i\pi/3} = e^{7i\pi/3} = e^{9i\pi/3} = \cdots$$

Note that the series starts to repeat at the 4th term — where $e^{i\pi/3}$ occurs again — i.e. the series is triple valued, and one needs 3 Riemann sheets.

Setting $(z - b) = -1$,

$$(-1) = e^{i\pi} = e^{3i\pi} = e^{5i\pi} = e^{7i\pi} = e^{9i\pi} = e^{11i\pi}, \text{ etc.}$$

$$(-1)^{2/5} = e^{2i\pi/5} = e^{6i\pi/5} = e^{10i\pi/5} = e^{14i\pi/5} = e^{18i\pi/5} = e^{22i\pi/5} \cdots$$

Note that the series starts to repeat at the 6th term — where $e^{2i\pi/5}$ occurs again — i.e. the series is quintuple valued, and one needs 5 Riemann sheets.

(d) $f(z) = e^{-z^2}$. Setting $z = 1/w$,

$$g(w) = e^{-1/w^2} = 1 - \frac{1}{w^2} + \frac{1}{2w^4} - \cdots,$$

which has an essential singularity at $w = 0$; hence e^{-z^2} has an essential singularity at z = infinity.

(e) $f(z) = \ln(z - a)$. Setting $z = 1/w$,

$$g(w) = \ln\left(\frac{1}{w} - a\right) = \ln\frac{1 - aw)}{w} = \ln(1 - aw) - \ln w.$$

$g(w)$ has a branch point at $w = 0$; hence $\ln(z - a)$ has a branch point at z = infinity. Since there is also a branch point at $z = a$, we introduce a cut extending from a to infinity.

II. A function which is analytic everywhere, except maybe at infinity, is defined to be an entire function (or an integral function). Prove that if an entire function is not a polynomial, then it must have an essential singularity at infinity.

Solution: If a function $F(z)$ is an integral function, then it can be expanded in a Taylor series about the point $z = 0$, i.e. $F(z) = \sum_{n=o}^{\infty} a_n z^n$.
If $F(z)$ is not a polynomial, then it must have a singularity at z = infinity, in fact an essential singularity, since the series goes to infinity with increasing powers of z.

III. Find the poles and residues of the functions

(a) $\dfrac{1}{\sinh z}$. (b) $\dfrac{1}{\cos z}$. (c) $z \cot z$.

Solution:

(a) The zeros of sinhz lie at values of z for which $e^z - e^{-z} = 0$, or $e^{2z} = 1 = e^{2i\pi n}$, i.e. zeros at $z = i\pi n$. Hence we are interested in points along the imaginary axis, namely 0, $i\pi$, $2i\pi$, $3i\pi$, etc.

For $z = 0$: Setting $z = iw$, $\dfrac{1}{\sinh z} = \dfrac{1}{\sinh(iw)} = \dfrac{1}{i \sin w}$.

As $w \to 0$, $\sin w \to w$, so that $\lim\limits_{w \to o} \dfrac{1}{\sinh z} = \dfrac{1}{iw} = \dfrac{1}{z}$. Hence the residue at $z = 0$ is 1.

For $z = in\pi$: Setting $z = i(n\pi + w)$,

$$\sinh z = \sinh i(n\pi + w) = i\sin(n\pi + w) = i\,[\sin n\pi \cos w + \cos n\pi \sin w]$$
$$= i\sin w(-1)^n.$$

It follows that $\lim_{w\to o} \frac{1}{\sinh z} = \frac{1}{(-1)^n iw} = \frac{1}{(-1)^n\,[z - in\pi]}$. Hence the residue at $z = in\pi = \frac{1}{(-1)^n}$.

At

$$\begin{array}{lll} z = i\pi & n = 1 & R = -1 \\ z = 2i\pi & n = 2 & R = +1 \\ z = 3i\pi & n = 3 & R = -1 \end{array} \quad \Rightarrow \quad \begin{array}{l} R = +1 \text{ for even } \pi. \\ R = -1 \text{ for odd } \pi. \end{array}$$

(b) Since $\cos z = \frac{e^{iz} + e^{-iz}}{2}$, we see that zeros occur when $e^{2iz} + 1 = 0$, or when $(e^{iz} - i)(e^{iz} + i) = 0 \Rightarrow e^{iz} = \pm i$, or $e^{iz} = e^{i[(\pi/2) + n\pi]}$. Hence we have zeros when $z = (\pi/2) + n\pi$. Letting $z = n\pi + (\pi/2) + w$,

$$\begin{aligned} \cos z &= \cos[n\pi + (\pi/2) + w] \\ &= \cos[n\pi + (\pi/2)]\cos w - \sin[n\pi + (\pi/2)]\sin w \\ &= (-1)^{n+1}\sin w. \end{aligned}$$

$$\lim_{w\to o} \frac{1}{\cos z} = \frac{1}{(-1)^{n+1}w} = \frac{1}{(-1)^{n+1}[z - (n\pi + \pi/2)]}.$$

Hence, the residue at $z = n\pi + \pi/2$ is $\frac{1}{(-1)^{n+1}}$.

At

$$\begin{array}{lll} z = \pi/2 & n = 0 & R = -1 \\ z = \pi + \pi/2 & n = 1 & R = +1 \quad \text{etc.} \\ z = 2\pi + \pi/2 & n = 2 & R = -1 \end{array}$$

(c) $z\cot z = (z\cos z)/\sin z$ and zeros of $\sin z$ occur at $z = n\pi$ (except $n = 0$). Setting $z = n\pi + w$ $(n \neq 0)$,

$$z\cot z = \frac{(n\pi + w)\cos(n\pi + w)}{\sin(n\pi + w)} = = \frac{(n\pi + w)\cos w}{\sin w}.$$

$$\lim_{w\to o} z\cot z = \frac{n\pi}{w} = \frac{n\pi}{(z - n\pi)} \quad \Rightarrow \quad \text{Residue at } z = n\pi \text{ is } n\pi.$$

IV. Use contour integration to prove that

(a) $$\int_{-\infty}^{+\infty} \frac{dx}{(1+x^2)^{n+1}} = \frac{\pi(2n)!}{4^n(n!)^2}.$$

(b) $$\int_{-\infty}^{+\infty} \frac{x^{2m}dx}{1+x^{2n}} = \frac{\pi}{n}\csc\left(\frac{2m+1}{2n}\pi\right),$$

where m and n are positive integers and $n > m$.

Solution:

(a) $$\int_{-\infty}^{+\infty} \frac{dx}{(1+x^2)^{n+1}} = 2\int_{0}^{+\infty} \frac{dx}{(1+x^2)^{n+1}},$$

since the integrand is an even function of x.

Setting $x = \tan\theta$, $dx = \sec^2\theta d\theta$, $(1+x^2)^{n+1} = (1+\tan^2\theta)^{n+1} = \sec^{2n+2}\theta$. Thus,

$$2\int_{0}^{+\infty} \frac{dx}{(1+x^2)^{n+1}} = 2\int_{0}^{\pi/2} \frac{\sec^2\theta d\theta}{\sec^{2n+2}\theta} = 2\int_{0}^{\pi/2} \cos^{2n}\theta d\theta = \frac{1}{2}\int_{0}^{2\pi} \cos^{2n}\theta d\theta.$$

On the unit circle, $z = e^{i\theta}$, $dz = ie^{i\theta}d\theta = izd\theta$ and $d\theta = dz/iz$. We also have that

$$\cos\theta = \frac{e^{i\theta}+e^{-i\theta}}{2} = \frac{z+1/z}{2} \rightarrow \cos^{2n}\theta = \frac{1}{4^n}(z+1/z)^{2n}.$$

Thus,

$$2\int_{0}^{+\infty} \frac{dx}{(1+x^2)^{n+1}} = \frac{1}{2}\int_{0}^{2\pi} \frac{1}{4^n}\left(z+\frac{1}{z}\right)^{2n} \frac{dz}{iz} = \frac{1}{2\cdot 4^n\cdot i}\oint \frac{(z+1/z)^{2n}}{z}dz$$

$$= \frac{\pi}{4^n} \text{ residue of } \frac{(z+1/z)^{2n}}{z}.$$

Evaluating the residue:

$$\frac{(z+1/z)^{2n}}{z} = \frac{1}{z}\left[z^{2n} + 2nz^{2n-1}\frac{1}{z} + \frac{2n(2n-1)}{2!}z^{2n-2}\frac{1}{z^2} + \cdots\right]$$

$$\cdots + \frac{1}{z}\left[\frac{2n(2n-1)(2n-2)\cdots(n+1)z^n}{n!}\frac{1}{z^n} + \cdots\right]$$

Note that the coefficient of $1/z$ in this *entire* expression is that term within the bracket which contains no z term, i.e.

$$\frac{2n(2n-1)(2n-2)\cdots(n+1)}{n!} = \frac{(2n)!}{(n!)^2}.$$

Hence

$$\int_{-\infty}^{+\infty}\frac{dx}{(1+x^2)^{n+1}} = \frac{\pi}{4^n}\frac{(2n)!}{(n!)^2}.$$

(b) $\displaystyle\int_{-\infty}^{+\infty}\frac{x^{2m}dx}{1+x^{2n}}.$

Initial method: Setting $y = x^{2n}$,

$$dy = 2nx^{2n-1}dx \rightarrow dx = \frac{1}{2n}x^{1-2n}dy = \frac{y^{(1-2n)/2n}}{2n}dy,\ x^{2m} = y^{2m/2n}.$$

Thus,

$$\int_{-\infty}^{+\infty}\frac{x^{2m}dx}{1+x^{2n}} = 2\int_0^{\infty}\frac{y^{2m/2n}y^{(1-2n)/2n}dy}{2n(1+y)} = \frac{1}{n}\int_0^{\infty}\frac{y^{\frac{2m+1}{2n}-1}dy}{1+y}.$$

If we now set $\alpha = (2m+1)/2n$, then this integral is of the form $\displaystyle\int_0^{\infty}\frac{y^{\alpha-1}dy}{1+y}$, where α satisfies the condition that it is not an integer; in fact $\alpha < 1$.

Also note that $\dfrac{y^{\alpha}}{1+y} \rightarrow 0$ as $\begin{cases} y \rightarrow 0 \\ y \rightarrow \infty \end{cases}$ and $\dfrac{1}{1+z}$ is analytic on the real axis.

We have considered this type of integral in our notes dealing with complex integration, Section 6.10.1, type III, and have there shown that the proper contour is as shown in Fig. B.4, and that

$$\int_0^{\infty} y^{\alpha-1}F(y)dy = \frac{1}{1-e^{2\pi i\alpha}}\oint_C y^{\alpha-1}F(y)dy.$$

We have shown (see text reference above) that

$$\text{Residue}\ \frac{y^{\alpha-1}}{(1+y)} = (-1)^{\alpha-1} = e^{i\pi(\alpha-1)}$$
$$= e^{i\pi\alpha}e^{-i\pi} = -e^{i\pi\alpha}.$$

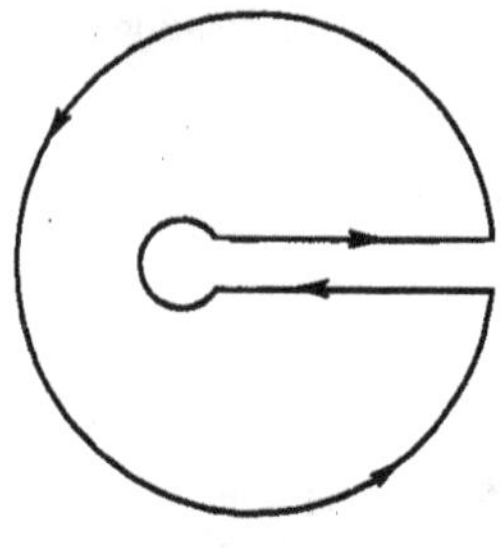

Fig. B.4

Hence,

$$\int_0^{\infty}\frac{y^{\alpha-1}}{(1+y)}\,dy = \frac{1}{1-e^{2\pi i\alpha}}\,2\pi i\,(-e^{i\pi\alpha})$$

$$= \frac{2i\pi}{e^{i\pi\alpha}-e^{-i\pi\alpha}} = \frac{\pi}{\sin\alpha\pi} = \pi\csc\alpha\pi.$$

Finally,

$$\int_{-\infty}^{+\infty}\frac{x^{2m}\,dx}{1+x^{2n}} = \frac{1}{n}\int_0^{\infty}\frac{y^{\alpha-1}}{(1+y)}\,dy = \frac{\pi}{n}\csc\left[\frac{2m+1}{2n}\,\pi\right].$$

Alternative method: Consider

$$\int_{-\infty}^{+\infty}\frac{z^{2m}\,dz}{1+z^{2n}} + \int_{+\infty\ \text{arc}}^{-\infty}\frac{z^{2m}\,dz}{1+z^{2n}},$$

where the integration in the second term is over the arc shown in Fig. B.5. From B to $-B$, $\int_B^{-B}\frac{z^{2m}dz}{1+z^{2n}} \to 0$ as $z \to \infty$ for $n > m$ and both integers. Hence,

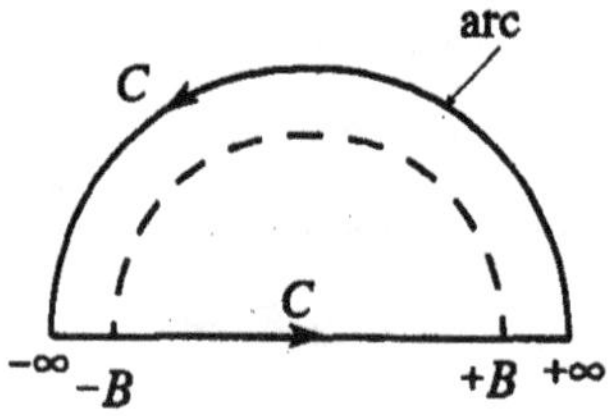

Fig. B.5

$$\int_{-\infty}^{+\infty}\frac{x^{2m}\,dx}{1+x^{2n}} = \oint_C\frac{z^{2m}\,dz}{1+z^{2n}} = 2\pi i\sum\text{Residues (R)},$$

where C is the complete upper half contour shown in Fig. B.5.

At $z^{2n} = -1$, roots are $z^{2n} = e^{i\pi k} \Rightarrow z = e^{i\pi k/2n}$, where $k = 1, 3, 5 \cdots (2n-1)$, yielding n roots in the upper half-plane. (For $k = -1, -3, -5 \cdots -(2n-1)$, we have n roots in the lower half-plane, but these are neglected as we are only concerned with the upper half-plane.)

$$\sum R = R_{z_1} + R_{z_2} + \cdots + R_{z_n}$$

$$= \lim_{z\to z_1} \frac{z_1^{2m}(z - z_1)}{(1 + z^{2n})|_{z_1}} + \lim_{z\to z_2} \frac{z_2^{2m}(z - z_2)}{(1 + z^{2n})|_{z_2}} + \cdots + \lim_{z\to z_n} \frac{z_n^{2m}(z - z_n)}{(1 + z^{2n})|_{z_n}}.$$

Using L'Hopital's rule, i.e. if $\frac{f_1}{g_1} = \frac{0}{0}$ then $\lim_{x\to x_1} \frac{f}{g} = \frac{f_1'}{g_1'}$, we have

$$\sum R = \left[\frac{z_1^{2m}}{2nz_1^{2n-1}} + \frac{z_2^{2m}}{2nz_2^{2n-1}} + \cdots + \frac{z_n^{2m}}{2nz_n^{2n-1}}\right].$$

$$z_1 = e^{i\pi\cdot 1/2n}. \quad z_2 = e^{i\pi\cdot 3/2n}. \quad z_3 = e^{i\pi\cdot 5/2n}. \quad \ldots \quad z_n = e^{i\pi\cdot(2n-1)/2n}.$$

$$z_1^{2m} = e^{i\pi\cdot m/n}. \quad z_2^{2m} = e^{i\pi\cdot 3m/n}. \quad z_3^{2m} = e^{i\pi\cdot 5m/n}. \quad \ldots \quad z_n^{2m} = e^{i\pi\cdot(2n-1)m/n}.$$

$$2n\sum R = \frac{e^{i\pi\cdot 1\cdot 2m/2n}}{e^{i\pi(2n-1)/2n}} + \frac{e^{i\pi\cdot 3\cdot 2m/2n}}{e^{i\pi(2n-1)\cdot 3/2n}} + \cdots + \frac{e^{i\pi\cdot(2n-1)\cdot 2m/2n}}{e^{i\pi(2n-1)(2n-1)/2n}}$$

$$= e^{i\pi\cdot(2m-2n+1)\cdot 1/2n} + e^{i\pi\cdot(2m-2n+1)\cdot 3/2n} + \cdots$$

$$+ \cdots + e^{i\pi\cdot(2m-2n+1)\cdot(2n-1)/2n}$$

$$2n\sum R = \sum_{r=0}^{n-1} e^{i\pi\cdot(2r+1)\cdot(2m-2n+1)/2n} = e^{i\pi\cdot(2m-2n+1)/2n} \sum_{r=0}^{n-1} e^{2i\pi r\cdot(2m-2n+1)/2n}.$$

But since

$$\sum_{r=0}^{n-1}\left(e^{2i\pi(2m-2n+1)/2n}\right)^r = \frac{1 - e^{i\pi\cdot 2n\cdot(2m-2n+1)/2n}}{1 - e^{2i\pi\cdot(2m-2n+1)/2n}} = \frac{1 - e^{i\pi\cdot 2n\cdot(2m-2n+1)/2n}}{1 - e^{2i\pi\cdot(2m+1)/2n}},$$

we have

$$2n\sum R = -e^{i\pi\cdot(2m+1)/2n} \cdot \frac{1 - e^{i\pi\cdot 2n\cdot(2m-2n+1)/2n}}{1 - e^{2i\pi\cdot(2m+1)/2n}}$$

$$= \frac{1 - e^{i\pi\cdot(2m-2n+1)}}{e^{i\pi\cdot(2m+1)/2n} - e^{-i\pi\cdot(2m+1)/2n}}.$$

Since $2n - 2m - 1$ is an odd number,

$$e^{-i\pi(2n-2m-1)} = e^{+i\pi(2n-2m-1)} = -1,$$

so that

$$2n\sum \mathrm{R} = \frac{2}{2i \sin\left[\frac{(2m+1)\pi}{2n}\right]}.$$

Finally then,

$$\int_{-\infty}^{+\infty} \frac{x^{2m}dx}{1 + x^{2n}} = 2\pi i \sum \mathrm{R} = \frac{2\pi i \cdot 2}{4ni} \csc\left[\frac{(2m+1)\pi}{2n}\right] = \frac{\pi}{n} \csc\left[\frac{(2m+1)\pi}{2n}\right].$$

V. Prove, using $\int_0^\infty e^{-x^2} dx = \frac{\sqrt{\pi}}{2}$, that

(a) $$\int_0^\infty \cos x^2 dx = \int_0^\infty \sin x^2 dx = \frac{\sqrt{\pi/2}}{2}.$$

(b) $$\int_0^\infty \cos(ax) e^{-x^2} dx = \frac{\sqrt{\pi}}{2} e^{-a^2/4}.$$

Solution:

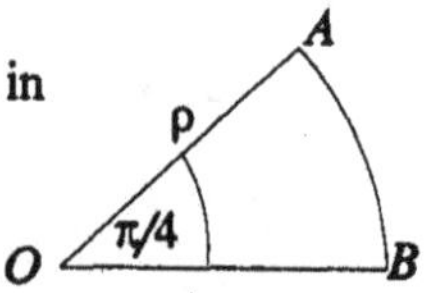

Fig. B.6

(a) Consider $\oint e^{iz^2} dz$ over the closed path OBA shown in Fig. B.6. Since e^{iz^2} is analytic within this closed region, $\oint e^{iz^2} dz = 0$. [We use the fact that along AO, $z = \rho e^{i\theta}$ and $dz = d\rho e^{i\theta}$].

(i) $$\int_0^B e^{iz^2} dz = \int_0^B e^{ix^2} dx.$$

(ii) $$\int_A^0 e^{iz^2} dz = \int_A^0 e^{i\rho^2 e^{2i\theta}} e^{i\theta} d\rho \underset{\theta = \pi/4}{\rightarrow} \int_A^0 e^{i\rho^2 \cdot i} e^{i\pi/4} d\rho = e^{i\pi/4} \int_A^0 e^{-\rho^2} d\rho.$$

(iii) $\int_{B\,\text{arc}}^{A} e^{iz^2}dz:\; z = \rho e^{i\theta};\; dz = i\rho e^{i\theta}d\theta;\; z^2 = \rho^2 e^{2i\theta} = \rho^2\cos 2\theta + i\,\rho^2\sin 2\theta.$

$$\left|e^{iz^2}dz\right| = \left|e^{i\rho^2 e^{2i\theta}}\,i\rho e^{i\theta}d\theta\right| = \left|e^{i\rho^2\cos 2\theta - \rho^2\sin 2\theta}\,i\rho e^{i\theta}d\theta\right| = \left|e^{-\rho^2\sin 2\theta}\rho d\theta\right|$$

However, in the interval between $\theta = 0$ and $\theta = \pi/4$, $\sin 2\theta > \theta$, so that $\left|e^{iz^2}dz\right| < e^{-\rho^2\theta}\rho d\theta$ and

$$\left|\int_{B\,\text{arc}}^{A} e^{iz^2}dz\right| < \int_0^{\pi/4} e^{-\rho^2\theta}\rho d\theta = \left|-\frac{1}{\rho}e^{-\rho^2\theta}\right|_0^{\pi/4} = \frac{1}{\rho}\left(1 - e^{-\rho^2\pi/4}\right) \underset{\rho\to\infty}{\to} 0.$$

Hence, as $\rho \to \infty$,

$$\oint_{OBA} e^{iz^2}dz = \int_0^\infty e^{ix^2}dx - \int_0^\infty e^{i\pi/4}e^{-x^2}dx = 0,$$

so that

$$\int_0^\infty e^{ix^2}dx = e^{i\pi/4}\int_0^\infty e^{-x^2}dx = \left(\cos\frac{\pi}{4} + i\sin\frac{\pi}{4}\right)\frac{\sqrt{\pi}}{2}.$$

Finally, then,

$$\int_0^\infty \cos x^2 dx = \text{Real part of} \int_0^\infty e^{ix^2}dx = \frac{1}{\sqrt{2}}\frac{\sqrt{\pi}}{2} = \frac{\sqrt{\pi/2}}{2}.$$

$$\int_0^\infty \sin x^2 dx = \text{Imaginary part of} \int_0^\infty e^{ix^2}dx = \frac{1}{\sqrt{2}}\frac{\sqrt{\pi}}{2} = \frac{\sqrt{\pi/2}}{2}.$$

(b) Consider $\oint e^{iaz}e^{-z^2}dz$ over the contour shown in Fig. B.7.

(i) $\int_{-B}^{+B} e^{iaz}e^{-z^2}dz = \int_{-B}^{+B} e^{ia\rho}e^{-\rho^2}d\rho,$

since $z = \rho e^{i\theta}$, $dz = d\rho$ and $z^2 = \rho^2$ at $\theta = 0$.

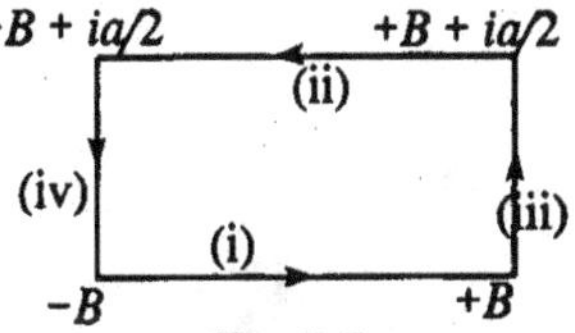

Fig. B.7

(ii) $\int_{+B+ia/2}^{-B+ia/2} e^{iaz}e^{-z^2}dz = \int_B^{-B} e^{ia(x+ia/2)}e^{-(x^2+iax-a^2/4)}dx,$

$$= \int_B^{-B} e^{-(x^2+a^2)/4}dx.$$

The last step follows since along this path, $z = x + iy$, $dz = dx$, $dy = 0$. Also, $z^2 = x^2 + 2ixy - y^2$, and for $y = a/2$, $z^2 = x^2 + iax - a^2/4$.

(iii) Along path (iii), $dx = 0$, $dz = idy$. Also, $z^2 = x^2 + 2ixy - y^2$, $iaz = ia(x + iy)$, $e^{iaz} = e^{-ay}e^{iax}$, and $e^{-z^2} = e^{y^2-x^2}e^{-2ixy}$. Therefore, $\left|e^{iaz}e^{-z^2}dz\right| = \left|e^{-ay}e^{iax}e^{y^2-x^2}e^{-2ixy}idy\right| = \left|e^{-ay}e^{y^2-x^2}dy\right|$. Hence,

$$\left|\int_B^{B+ia/2} e^{iaz}e^{-z^2}dz\right| = \left|\int_0^{a/2} e^{-ay}e^{y^2-x^2}dy\right| = \left|e^{-x^2}\int_0^{a/2} e^{y^2-ay}dy\right| \to 0 \text{ as } x\to\infty.$$

(iv) Similarly, for path (iv), $\left|\int_{-B+ia/2}^{-B} e^{iaz}e^{-z^2}dz\right| \to 0$ as $x\to\infty$.

Hence, as $x\to\infty$,

$$\oint e^{iaz}e^{-z^2}dz = \int_{-\infty}^{+\infty} e^{iax}e^{-x^2}dx + \int_{+\infty}^{-\infty} e^{-x^2}e^{-a^2/4}dx = 0,$$

since by Cauchy's theorem, the function e^{iaz-z^2} is analytic within the region. Thus,

$$\int_{-\infty}^{+\infty} e^{iax}e^{-x^2}dx = e^{-a^2/4}\int_{-\infty}^{+\infty} e^{-x^2}dx = \sqrt{\pi}e^{-a^2/4}.$$

$$\text{Real part of } \int_{-\infty}^{+\infty} e^{iax}e^{-x^2}dx = \int_{-\infty}^{+\infty} \cos(ax)e^{-x^2}dx$$

$$= 2\int_0^{+\infty} \cos(ax)e^{-x^2}dx = \sqrt{\pi}e^{-a^2/4}.$$

Hence,

$$\int_0^{+\infty} \cos(ax)e^{-x^2}dx = \frac{\sqrt{\pi}e^{-a^2/4}}{2}.$$

Problem Set C

I. Prove that if $\mathrm{Re}(z) > 0$, then

$$\Gamma(z) = s^z \int_0^\infty e^{-st} t^{z-1} dt,$$

where s is any complex number whose real part > 0. $\Gamma(z)$ is defined as

$$\Gamma(z) \equiv \int_0^\infty e^{-t} t^{z-1} dt.$$

Solution: Define a new variable $\upsilon \equiv st$, necessarily complex by virtue of s being complex. Then $d\upsilon = s\,dt$ and $d\upsilon/st = dt/t = d\upsilon/\upsilon$. It follows then that

$$f(z) = s^z \int_0^\infty e^{-st} t^{z-1} dt = \int_0^\infty e^{-\upsilon} \frac{\upsilon^z}{t} dt = \int_0^\infty e^{-\upsilon} \upsilon^{z-1} d\upsilon.$$

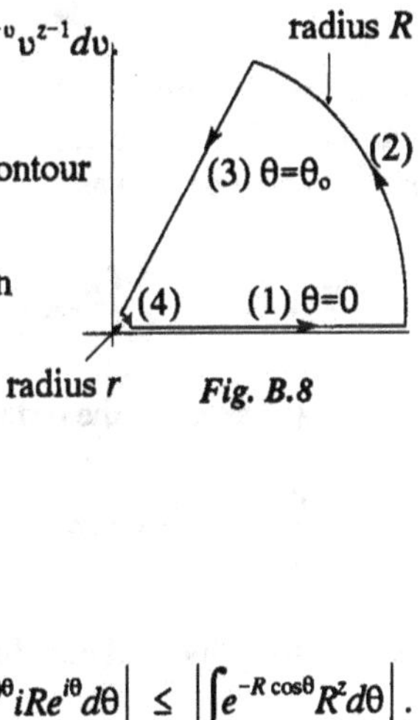

Fig. B.8

Consider the integral $\oint e^{-\upsilon} \upsilon^{z-1} d\upsilon$, where the contour is taken in the complex υ plane (Fig. B.8).

Since the integrand is analytic within the region shown, $\oint e^{-\upsilon} \upsilon^{z-1} d\upsilon = 0$. Hence,

$$\int_{(1)} + \int_{(2)} + \int_{(3)} + \int_{(4)} = 0.$$

(a) Consider $\int_{(2)} e^{-\upsilon} \upsilon^{z-1} d\upsilon$:

$$\left| \int_{(2)} e^{-\upsilon} \upsilon^{z-1} d\upsilon \right| \le \left| \int_{(2)} e^{-R\cos\theta - iR\sin\theta} R^{z-1} e^{i(z-1)\theta} iRe^{i\theta} d\theta \right| \le \left| \int_{(2)} e^{-R\cos\theta} R^z d\theta \right|.$$

For $-\pi/2 < \theta_0 < \pi/2$, $\cos\theta > 0$, and as $R \to \infty$, $e^{-R\cos\theta} R^z \to 0$.

(b) Consider $\int_{(4)} e^{-\upsilon} \upsilon^{z-1} d\upsilon$:

$$\left| \int_{(4)} e^{-\upsilon} \upsilon^{z-1} d\upsilon \right| \le \left| \int_{(4)} e^{-r\cos\theta - ir\sin\theta} r^{z-1} e^{i(z-1)\theta} ire^{i\theta} d\theta \right| < \left| \int_{(4)} e^{-r\cos\theta} r^z d\theta \right|.$$

As $r \to 0$, for Re $z > 0$, $e^{-r\cos\theta}r^z \to 0$.

(c) Consider $\int_{(1)} e^{-\upsilon}\upsilon^{z-1}d\upsilon = \int_0^\infty e^{-t}t^{z-1}dt = \Gamma(z)$.

(d) Consider $\int_{(3)} e^{-\upsilon}\upsilon^{z-1}d\upsilon = \int_\infty^0 s^z e^{-st}t^{z-1}dt = -\int_0^\infty s^z e^{-st}t^{z-1}dt$, since along $\theta = \theta_0$, $\upsilon = \upsilon_0 e^{i\theta_0} = ts_0 e^{i\theta_0}$ (i.e. real part of $s > 0$). Hence,

$$\int_{(1)} e^{-\upsilon}\upsilon^{z-1}d\upsilon + \int_{(3)} e^{-\upsilon}\upsilon^{z-1}d\upsilon = 0,$$

or,

$$\Gamma(z) = -\int_{(3)} e^{-\upsilon}\upsilon^{z-1}d\upsilon = s^z\int_0^\infty e^{-st}t^{z-1}dt.$$

II. Consider the equation $\Delta\Psi + \upsilon(\rho)\Psi = 0$, where υ is a function of the cylindrical radial variable, ρ, only. Show that the general solution of this equation can be written as

$$\Psi(\rho\theta z) = \sum_{m=-\infty}^{+\infty}\int \Phi_{mk}(\rho)e^{im\theta+ikz}dk,$$

where $\Phi_{mk}(\rho)$ satisfies the ordinary differential equation

$$\left[\frac{d^2}{d\rho^2} + \frac{1}{\rho}\frac{d}{d\rho} + \upsilon(\rho) - k^2 - \frac{m^2}{\rho^2}\right]\Phi_{mk} = 0.$$

Solution: In cylindrical coordinates, we have $x = \rho\cos\theta$, $y = \rho\sin\theta$, $z = z$.

$$h_1 = \sqrt{g_{11}} = \left[\left(\frac{\partial x}{\partial\rho}\right)^2 + \left(\frac{\partial y}{\partial\rho}\right)^2 + \left(\frac{\partial z}{\partial\rho}\right)^2\right]^{1/2} = \left(\cos^2\theta + \sin^2\theta\right)^{1/2} = 1.$$

$$h_2 = \sqrt{g_{22}} = \left[\left(\frac{\partial x}{\partial\theta}\right)^2 + \left(\frac{\partial y}{\partial\theta}\right)^2 + \left(\frac{\partial z}{\partial\theta}\right)^2\right]^{1/2} = \left(\rho^2\cos^2\theta + \rho^2\sin^2\theta\right)^{1/2} = \rho.$$

$$h_3 = \sqrt{g_{33}} = \left[\left(\frac{\partial x}{\partial z}\right)^2 + \left(\frac{\partial y}{\partial z}\right)^2 + \left(\frac{\partial z}{\partial z}\right)^2\right]^{1/2} = 1.$$

Hence,

$$\Delta\Psi = \frac{1}{\rho}\left[\frac{\partial}{\partial\rho}\left(\rho\frac{\partial\Psi}{\partial\rho}\right)+\frac{\partial}{\partial\theta}\left(\frac{1}{\rho}\frac{\partial\Psi}{\partial\theta}\right)+\frac{\partial}{\partial z}\left(\rho\frac{\partial\Psi}{\partial z}\right)\right]$$

$$= \frac{1}{\rho}\frac{\partial}{\partial\rho}\left(\rho\frac{\partial\Psi}{\partial\rho}\right)+\frac{1}{\rho^2}\frac{\partial^2\Psi}{\partial\theta^2}+\frac{\partial^2\Psi}{\partial z^2}.$$

$\Psi = \Psi(\rho\theta z)$ and can be written as $\Psi_1(\rho z, \theta)$, a function of θ, with ρ and z regarded as parameters. We know that any function of θ can be expanded in a Fourier series:

$$\Psi(\rho\theta z) = \sum_{m=-\infty}^{+\infty}\Phi_m(\rho z)e^{im\theta}.$$

Employing the Fourier transform, regarding ρ as parameter:

$$\Phi_m(\rho z) = \frac{1}{\sqrt{2\pi}}\int_{-\infty}^{+\infty}\varphi_{m'k}(\rho)e^{ikz}dk.$$

Setting $\varphi_{m'k}(\rho) = \sqrt{2\pi}\,\varphi_{mk}(\rho)$, we have

$$\Psi(\rho\theta z) = \sum_{m=-\infty}^{+\infty}\int_{-\infty}^{+\infty}\varphi_{mk}(\rho)e^{im\theta+ikz}dk.$$

$$\Delta\Psi(\rho\theta z) = \sum_{m=-\infty}^{+\infty}\int_{-\infty}^{+\infty}e^{im\theta+ikz}\left[\frac{1}{\rho}\frac{\partial}{\partial\rho}\left(\rho\frac{\partial\varphi_{mk}(\rho)}{\partial\rho}\right)\right]dk$$

$$-\frac{1}{\rho^2}\sum_{m=-\infty}^{+\infty}\int_{-\infty}^{+\infty}m^2\varphi_{mk}(\rho)e^{im\theta+ikz}dk-\sum_{m=-\infty}^{+\infty}\int_{-\infty}^{+\infty}k^2\varphi_{mk}(\rho)e^{im\theta+ikz}dk,$$

so that

$$\Delta\Psi + \upsilon(\rho)\Psi = \sum_{m=-\infty}^{+\infty}\int_{-\infty}^{+\infty}e^{im\theta+ikz}\left[\frac{1}{\rho}\frac{\partial}{\partial\rho}\left(\rho\frac{\partial\varphi_{mk}(\rho)}{\partial\rho}\right)\right]dk$$

$$-\frac{m^2}{\rho^2}\sum_{m=-\infty}^{+\infty}\int_{-\infty}^{+\infty}\varphi_{mk}(\rho)e^{im\theta+ikz}dk-k^2\sum_{m=-\infty}^{+\infty}\int_{-\infty}^{+\infty}\varphi_{mk}(\rho)e^{im\theta+ikz}dk$$

$$+\sum_{m=-\infty}^{+\infty}\int_{-\infty}^{+\infty}\varphi_{mk}(\rho)\upsilon e^{im\theta+ikz}dk = 0.$$

$\Psi(\rho\theta z)$ will be a solution to $\Delta\Psi(\rho\theta z) + \upsilon(\rho)\Psi(\rho\theta z) = 0$ when

$$\left[\frac{1}{\rho}\frac{d}{d\rho}\left(\rho\frac{d}{d\rho}\right) - \frac{m^2}{\rho^2} - k^2 + \upsilon(\rho)\right]\varphi_{mk}(\rho) = 0,$$

where the partial derivatives have been replaced by total derivatives, i.e. $\varphi_{mk}(\rho)$ satisfies

$$\left[\frac{d^2}{d\rho^2} + \frac{1}{\rho}\frac{d}{d\rho} - \frac{m^2}{\rho^2} - k^2 + \upsilon(\rho)\right]\varphi_{mk}(\rho) = 0.$$

III. The Schrödinger equation for an electron in a hydrogen-like atom and in an external electric field in the z-direction may be written as

$$\Delta\Psi(\mathbf{r}) + u\Psi(\mathbf{r}) = 0 \quad \text{(Stark effect)},$$

where $u = a + b/r + cz$, with a, b and c as constants, and where r is the spherical radial variable. [Due to the term cz, u is not a function of r alone.]

(a) Show that, by using parabolic coordinates, $(\xi_1\ \xi_2\ \varphi)$, defined as

$$x = \xi_1\xi_2\cos\varphi, \quad y = \xi_1\xi_2\sin\varphi, \quad z = (\xi_1^2 - \xi_2^2)/2,$$

the differential equation may be written as

$$\sum_{i=1}^{2}\left\{\frac{1}{\xi_i}\frac{\partial}{\partial\xi_i}\left(\xi_i\frac{\partial}{\partial\xi_i}\right) + \frac{1}{\xi_i^2}\frac{\partial^2}{\partial\varphi^2} + a\xi_i^2 + b - (-1)^i\frac{c}{2}\xi_i^4\right\}\Psi(\mathbf{r}) = 0.$$

(b) Expand $\Psi(\mathbf{r})$ into the Fourier series, $\Psi(\mathbf{r}) = \sum_m \Phi_m(\xi_1\xi_2)e^{im\varphi}$. Give the explicit form of the differential equation satisfied by Φ_m.

(c) Show that if the technique of separation of variables is used, i.e.

$$\Phi_m = f_1(\xi_1)\cdot f_2(\xi_2),$$

then f_1 and f_2 satisfy the two differential equations:

$$\frac{1}{\xi_i}\frac{d}{d\xi_i}\left(\xi_i\frac{df_i}{d\xi_i}\right) + \left[a\xi_i^2 + b - \frac{m^2}{\xi_i^2} - (-1)^i\left(\frac{c\xi_i^4}{2} + d\right)\right]f_i = 0, \quad (i = 1, 2)$$

where d is a new constant.

Solution:

(a)
$$dx = \xi_2\cos\varphi d\xi_1 + \xi_1\cos\varphi d\xi_2 - \xi_1\xi_2\sin\varphi d\varphi.$$

$$dy = \xi_2\sin\varphi d\xi_1 + \xi_1\sin\varphi d\xi_2 + \xi_1\xi_2\cos\varphi d\varphi.$$

$$dz = \xi_1 d\xi_1 - \xi_2 d\xi_2.$$

$$\begin{aligned} ds^2 &= dx^2 + dy^2 + dz^2 = \left(\cos\varphi[\xi_2 d\xi_1 + \xi_1 d\xi_2] - \xi_1\xi_2\sin\varphi d\varphi\right)^2 \\ &\quad + \left(\sin\varphi[\xi_2 d\xi_1 + \xi_1 d\xi_2] + \xi_1\xi_2\cos\varphi d\varphi\right)^2 + \left(\xi_1 d\xi_1 - \xi_2 d\xi_2\right)^2 \\ &= \left(\xi_2 d\xi_1 + \xi_1 d\xi_2\right)^2 + \xi_1^2\xi_2^2 d\varphi^2 + \left(\xi_1 d\xi_1 - \xi_2 d\xi_2\right)^2. \end{aligned}$$

Rewriting,

$$ds^2 = d\xi_1^2\left[\xi_1^2 + \xi_2^2\right] + d\xi_2^2\left[\xi_1^2 + \xi_2^2\right] + d\varphi^2\left[\xi_1^2\xi_2^2\right].$$

Thus, the new parabolic coordinates are seen to be an orthogonal set, for which

$$h_1 = \sqrt{\xi_1^2 + \xi_2^2};\ h_2 = \sqrt{\xi_1^2 + \xi_2^2};\ h_3 = \sqrt{\xi_1^2\xi_2^2} = \xi_1\xi_2.$$

One can now use the expression for $\Delta\Psi$ in generalized coordinates, given by

$$\begin{aligned} \Delta\Psi &= \frac{1}{h_1h_2h_3}\left(\frac{\partial}{\partial\xi_1}\left[\frac{1}{h_1}\frac{\partial\Psi}{\partial\xi_1}h_2h_3\right] + \frac{\partial}{\partial\xi_2}\left[\frac{1}{h_2}\frac{\partial\Psi}{\partial\xi_2}h_1h_3\right]\right) \\ &\quad + \frac{1}{h_1h_2h_3}\left(\frac{\partial}{\partial\xi_3}\left[\frac{1}{h_3}\frac{\partial\Psi}{\partial\xi_3}h_1h_2\right]\right). \end{aligned}$$

Substituting,

$$\begin{aligned} \Delta\Psi &= \frac{1}{\xi_1\xi_2\left(\xi_1^2 + \xi_2^2\right)}\left(\frac{\partial}{\partial\xi_1}\left[\frac{1}{\sqrt{\xi_1^2 + \xi_2^2}}\frac{\partial\Psi}{\partial\xi_1}\xi_1\xi_2\sqrt{\xi_1^2 + \xi_2^2}\right]\right) \\ &\quad + \frac{1}{\xi_1\xi_2\left(\xi_1^2 + \xi_2^2\right)}\left(\frac{\partial}{\partial\xi_2}\left[\frac{1}{\sqrt{\xi_1^2 + \xi_2^2}}\frac{\partial\Psi}{\partial\xi_2}\xi_1\xi_2\sqrt{\xi_1^2 + \xi_2^2}\right]\right) \\ &\quad + \frac{1}{\xi_1\xi_2\left(\xi_1^2 + \xi_2^2\right)}\left(\frac{\partial}{\partial\varphi}\left[\frac{1}{\xi_1\xi_2}\frac{\partial\Psi}{\partial\varphi}\left(\xi_1^2 + \xi_2^2\right)\right]\right). \end{aligned}$$

Rearranging,

$$\Delta\Psi = \frac{1}{\left(\xi_1^2+\xi_2^2\right)}\left[\frac{1}{\xi_1}\frac{\partial}{\partial\xi_1}\left(\xi_1\frac{\partial\Psi}{\partial\xi_1}\right)+\frac{1}{\xi_2}\frac{\partial}{\partial\xi_2}\left(\xi_2\frac{\partial\Psi}{\partial\xi_2}\right)\right]$$

$$+\frac{1}{\left(\xi_1^2+\xi_2^2\right)}\left(\frac{1}{\xi_1^2}+\frac{1}{\xi_2^2}\right)\frac{\partial^2\Psi}{\partial\varphi^2}.$$

$$r = \sqrt{x^2+y^2+z^2} = \left(\xi_1^2\xi_2^2+\frac{1}{4}\left[\xi_1^4-2\xi_1^2\xi_2^2+\xi_2^4\right]\right)^{1/2} = \frac{1}{2}\left(\xi_1^2+\xi_2^2\right).$$

Thus, $u = a + 2b\left(\xi_1^2+\xi_2^2\right)^{-1} + \frac{c}{2}\left(\xi_1^2-\xi_2^2\right)$. Substituting into $\Delta\Psi + u\Psi = 0$:

$$\Delta\Psi = \frac{1}{\left(\xi_1^2+\xi_2^2\right)}\left[\frac{1}{\xi_1}\frac{\partial}{\partial\xi_1}\left(\xi_1\frac{\partial\Psi}{\partial\xi_1}\right)+\frac{1}{\xi_2}\frac{\partial}{\partial\xi_2}\left(\xi_2\frac{\partial\Psi}{\partial\xi_2}\right)\right]$$

$$+\frac{1}{\left(\xi_1^2+\xi_2^2\right)}\left[\left(\frac{1}{\xi_1^2}+\frac{1}{\xi_2^2}\right)\frac{\partial^2\Psi}{\partial\varphi^2}+\left(\xi_1^2+\xi_2^2\right)a\Psi+2b\Psi\right]$$

$$+\frac{1}{\left(\xi_1^2+\xi_2^2\right)}\frac{c}{2}\left(\xi_1^4-\xi_2^4\right)\Psi = 0.$$

Rewriting,

$$\sum_{i=1}^{2}\left\{\frac{1}{\xi_i}\frac{\partial}{\partial\xi_i}\left(\xi_i\frac{\partial}{\partial\xi_i}\right)+\frac{1}{\xi_i^2}\frac{\partial^2}{\partial\varphi^2}+a\xi_i^2+b-(-1)^i\frac{c}{2}\xi_i^4\right]\Psi = 0.$$

(b) Expand Ψ in a Fourier series, $\Psi = \sum_m \Phi_m(\xi_1\xi_2)e^{im\varphi}$, and substitute into $\Delta\Psi + u\Psi = 0$ to get:

$$\sum_m e^{im\varphi}\left[\frac{1}{\xi_1}\frac{\partial}{\partial\xi_1}\left(\xi_1\frac{\partial\Phi_m}{\partial\xi_1}\right)+\frac{1}{\xi_2}\frac{\partial}{\partial\xi_2}\left(\xi_2\frac{\partial\Phi_m}{\partial\xi_2}\right)\right]$$

$$+\sum_m e^{im\varphi}\left[\left(\frac{-m^2}{\xi_1^2}-\frac{m^2}{\xi_2^2}\right)\frac{\partial^2\Phi_m}{\partial\varphi^2}+\left(\xi_1^2+\xi_2^2\right)a\Phi_m+2b\Phi_m\right]$$

$$+\sum_m e^{im\varphi}\left[+\frac{c}{2}\left(\xi_1^4-\xi_2^4\right)\Phi_m\right] = 0.$$

Since this must hold for all Φ, and since the $e^{im\varphi}$ form an orthogonal set, we have

$$\frac{1}{\xi_1}\frac{\partial}{\partial \xi_1}\left(\xi_1\frac{\partial \Phi_m}{\partial \xi_1}\right)+\frac{1}{\xi_2}\frac{\partial}{\partial \xi_2}\left(\xi_2\frac{\partial \Phi_m}{\partial \xi_2}\right)-m^2\left(\frac{1}{\xi_1^2}+\frac{1}{\xi_2^2}\right)\Phi_m$$

$$+\left(\xi_1^2+\xi_2^2\right)a\Phi_m+2b\Phi_m+\frac{c}{2}\left(\xi_1^4-\xi_2^4\right)\Phi_m = 0.$$

Rewriting,

$$\sum_{i=1}^{2}\left\{\frac{1}{\xi_i}\frac{\partial}{\partial \xi_i}\left(\xi_i\frac{\partial}{\partial \xi_i}\right)-\frac{m^2}{\xi_i^2}+a\xi_i^2+b-(-1)^i\frac{c}{2}\xi_i^4\right\}\Phi_m(\xi_1\xi_2) = 0.$$

(c) We try for a solution of the form $\Phi_m(\xi_1\xi_2)=f_1(\xi_1)\cdot f_2(\xi_2)$ and substitute into the last equation of (b) above. After dividing by $f_1(\xi_1)\cdot f_2(\xi_2)$:

$$\frac{1}{f_1}\frac{1}{\xi_1}\frac{\partial}{\partial \xi_1}\left(\xi_1\frac{\partial f_1}{\partial \xi_1}\right)+\frac{1}{f_2}\frac{1}{\xi_2}\frac{\partial}{\partial \xi_2}\left(\xi_1\frac{\partial f_2}{\partial \xi_2}\right)-m^2\left(\frac{1}{\xi_1^2}+\frac{1}{\xi_2^2}\right)$$

$$+a\left(\xi_1^2+\xi_2^2\right)+2b+\frac{c}{2}\left(\xi_1^4-\xi_2^4\right) = 0.$$

Rewriting, noting that the above partial derivatives are now total derivatives:

$$\left\{\frac{1}{f_1}\frac{1}{\xi_1}\frac{d}{d\xi_1}\left(\xi_1\frac{df_1}{d\xi_1}\right)-\frac{m^2}{\xi_1^2}+a\xi_1^2+b+\frac{c}{2}\xi_1^4\right\}$$

$$+\left\{\frac{1}{f_2}\frac{1}{\xi_2}\frac{d}{d\xi_2}\left(\xi_2\frac{df_2}{d\xi_2}\right)-\frac{m^2}{\xi_2^2}+a\xi_2^2+b-\frac{c}{2}\xi_2^4\right\} = 0.$$

This equation can be written as $G(\xi_1)+H(\xi_2) = 0$, which can hold only if $G(\xi_1)=-H(\xi_2)=-d$, a constant. Thus,

$$\frac{1}{\xi_i}\frac{d}{d\xi_i}\left(\xi_i\frac{df_i}{d\xi_i}\right)+\left(a\xi_i^2+b-\frac{m^2}{\xi_i^2}-(-1)^i\left[\frac{c}{2}\xi_i^4+d\right]\right)f_i = 0.$$

IV. Any inhomogeneous second-order linear differential equation can be written as $\frac{d^2f}{dz^2} + p(z)\frac{df}{dz} + q(z)f = F(z)$, where p, q and F are known functions of z.

(a) With the transformations

$$f(z) = g(z)\exp\left(-\frac{1}{2}\int_{z_0}^{z} p(z')dz'\right), \qquad F(z) = G(z)\exp\left(-\frac{1}{2}\int_{z_0}^{z} p(z')dz'\right),$$

show that the differential equation can be written as

$$\frac{d^2g}{dz^2} + J(z)g = G(z),$$

and find the expression for $J(z)$.

(b) Define $g(z) \equiv \sum_n g_n(z)$, where

$$g_0(z) = \int_{z_0}^{z} (z - z')G(z')dz'$$

$$g_n(z) = -\int_{z_0}^{z} (z - z')J(z')g_{n-1}(z')dz' \quad \text{for } n \geq 1.$$

Prove that $g(z)$ defined above is a formal solution of $d^2g/dz^2 + J(z)g = G(z)$.

(c) Prove that at a point z where $J(z)$ and $G(z)$ are analytic, by suitable definition of the line integrals appearing in $g_0(z)$ and $g_n(z)$ expressions above, the sum $\sum_n g_n(z)$ must exist. (In fact, it converges uniformly.)

Solution:

(a) Setting $s = -\frac{1}{2}\int_{z_0}^{z} p(z')dz'$, we have $f(z) = g(z)e^s$ and $F(z) = G(z)e^s$.

$$\frac{df}{dz} = e^s\frac{dg}{dz} - \frac{g}{2}pe^s = e^s\frac{dg}{dz} - \frac{pf}{2}.$$

$$\frac{d^2f}{dz^2} = e^s\frac{d^2g}{dz^2} - \frac{1}{2}\frac{dg}{dz}pe^s - \frac{f}{2}\frac{dp}{dz} - \frac{p}{2}\frac{df}{dz}$$

Rearranging,

$$\frac{d^2f}{dz^2} = e^s\frac{d^2g}{dz^2} - \frac{p}{2}\left(\frac{df}{dz} + \frac{pf}{2}\right) - \frac{f}{2}\frac{dp}{dz} - \frac{p}{2}\frac{df}{dz}$$

$$= e^s\frac{d^2g}{dz^2} - p\frac{df}{dz} - \frac{p^2f}{4} - \frac{f}{2}\frac{dp}{dz}.$$

After rearranging again and substituting into $\frac{d^2f}{dz^2} + p(z)\frac{df}{dz} + q(z)f = F(z)$, we have

$$e^s\left[\frac{d^2g}{dz^2} - \frac{p^2g}{4} - \frac{g}{2}\frac{dp}{dz} + qg\right] = G(z)e^s \quad \Rightarrow \quad \frac{d^2g}{dz^2} + J(z)g = G(z),$$

where $J(z) = -\frac{p^2}{4} - \frac{1}{2}\frac{dp}{dz} + q(z)$.

(b) Define $g(z) \equiv \sum_{n=0}^{\infty} g_n(z)$, where

$$g_0(z) = \int_{z_0}^{z} (z - z')G(z')dz'$$

$$g_n(z) = -\int_{z_0}^{z} (z - z')J(z')g_{n-1}(z')dz' \quad \text{for } n \geq 1.$$

Assuming that one can differentiate inside the sum:

$$\frac{d^2g}{dz^2} = \frac{d^2g_0}{dz^2} + \sum_{n=1}^{\infty}\frac{d^2g_n}{dz^2}.$$

$$\frac{dg_0}{dz} = \int_{z_0}^{z} G(z')dz' \quad \Rightarrow \quad \frac{d^2g_0}{dz^2} = G(z).$$

$$\frac{dg_n}{dz} = -\int_{z_0}^{z}\frac{\partial}{\partial z}(z - z')J(z')g_{n-1}(z')dz' = -\int_{z_0}^{z} J(z')g_{n-1}(z')dz'.$$

Thus,

$$\frac{d^2g_n}{dz^2} = -J(z)g_{n-1}(z) \Rightarrow \sum_{n=1}^{\infty}\frac{d^2g_n}{dz^2} = -J(z)\sum_{n=1}^{\infty}g_{n-1}(z) = -J(z)g(z)$$

by hypothesis.

Thus,

$$\frac{d^2g}{dz^2} = G(z) - J(z)g,$$

and $g(z) = \sum_{n=0}^{\infty} g_n(z)$ is a formal solution to the equation

$$\frac{d^2g(z)}{dz^2} + J(z)g(z) = G(z),$$

provided, of course that $\sum_{n=0}^{\infty} g_n(z)$ converges uniformly.

(c) Let z_0 be the point for which $J(z)$ and $G(z)$ are analytic, i.e. let z_0 be an ordinary point. Then there exists a region about z_0 wherein $J(z)$ and $G(z)$ are analytic. Over the path from z to z_0, as shown in Fig. B.9, $J(z)$ and $G(z)$ are analytic. It follows then that

$$\int_{z_0}^{z} (z - z')G(z')dz' = g_0(z)$$

and

$$g_n(z) = -\int_{z_0}^{z} (z - z')J(z')g_{n-1}(z')dz'$$

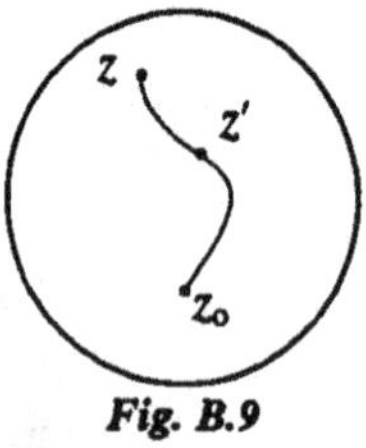

Fig. B.9

are also analytic. [Note that if g_0 is analytic, then g_1 is analytic, from which it follows that g_2 is analytic, etc.] We now prove the following theorem:

Theorem: $|g_n(z)| \le \frac{VU^n}{n!}|z - z_0|^{2n}$.

Proof: Let $|J(z)| \le U$ over the line integral from z to z_0 and let $|g_0(z)| \le V$.

For $n = 0$, the theorem is true by hypothesis. We prove by the method of induction, i.e. assume true for $n - 1$, then show it to be true for n.

$$|g_{n-1}(z')| \le \frac{VU^{n-1}}{(n-1)!}|z' - z_0|^{2n-2}.$$

$$|g_n(z)| < \int_{z_0}^{z} |z - z'||J(z')||g_{n-1}(z')||dz'|.$$

Since $g_n(z)$ is analytic in the region shown, the path from z_o to z is immaterial; hence take the straight path z_o to z. For this path, $|z - z'| \le |z - z_o|$, so that

$$|g_n(z)| < \frac{VU^{n-1}U}{(n-1)!}|z - z_o|\int_{z_o}^{z}|z' - z_o|^{2n-2}|dz'|.$$

Letting $s = |z' - z_o|$, $ds = |dz'|$, so that

$$|g_n(z)| < \frac{VU^n}{(n-1)!}|z - z_o|\int_0^{|z-z_o|}s^{2n-2}ds = \frac{VU^n|z - z_o|^{2n}}{(n-1)!(2n-1)}.$$

For $n \ge 1$, $2n - 1 \ge n$, so that $|g_n(z)| < \dfrac{VU^n|z - z_o|^{2n}}{n!}$.

Setting $U|z - z_o|^2 = \xi$, $g(z) = \sum_{n=o}^{\infty} g_n(z)$ forms a series of terms, each one of which is less than the corresponding term of the power series for e^{ξ}, i.e. $e^{\xi} = \sum_{n=o}^{\infty}\frac{\xi^n}{n!}$, which converges for all ξ. Hence, $g(z) = \sum_{n=o}^{\infty} g_n(z)$ converges and therefore exists.

Problem Set D

I. Let $H_n(z)$ be the Hermite polynomials. Prove that

(a) $$\int_{-\infty}^{+\infty} H_n H_m e^{-z^2} z\,dz = (n+1)!2^n\sqrt{\pi}\,\delta_{m,n+1} + (m+1)!2^m\sqrt{\pi}\,\delta_{m,n-1}.$$

(b) $$\int_{-\infty}^{+\infty} H_n^2 e^{-z^2} z^2 dz = \left(\frac{2n+1}{2}\right) 2^n n!\sqrt{\pi}.$$

(c) $$\int_{-\infty}^{+\infty} \left[\frac{d}{dz}\left(H_n e^{-z^2/2}\right)\right]^2 dz = \left(\frac{2n+1}{2}\right) 2^n n!\sqrt{\pi}.$$

Solution:

(a) The generating function for the Hermite polynomials is given by

$$e^{-t^2+2zt} = \sum_{n=0}^{\infty} \frac{H_n(z)t^n}{n!}. \qquad e^{-s^2+2zs} = \sum_{m=0}^{\infty} \frac{H_m(z)s^m}{m!}.$$

$$\sum_{m,n=0}^{\infty} \frac{t^n s^m}{n!m!} \int_{-\infty}^{+\infty} H_n(z)H_m(z)e^{-z^2} z\,dz = \int_{-\infty}^{+\infty} e^{-t^2-s^2+2sz+2tz-z^2} z\,dz$$

$$= e^{2st}\int_{-\infty}^{+\infty} e^{-(z-s-t)^2} z\,dz = e^{2st}\int_{-\infty}^{+\infty} e^{-(z-s-t)^2}[(z-s-t)+(s+t)]dz$$

$$= (s+t)e^{2st}\sqrt{\pi} = (s+t)\sqrt{\pi}\sum_{n=0}^{\infty}\frac{(2st)^n}{n!}$$

$$= \sqrt{\pi}\sum_{n=0}^{\infty}\frac{2^n s^{n+1} t^n}{n!} + \sqrt{\pi}\sum_{m=0}^{\infty}\frac{2^m s^m t^{m+1}}{m!}$$

$$= \sqrt{\pi}\sum_{m,n=0}^{\infty}\frac{2^n s^m t^n}{n!}\,\delta_{m,n+1} + \sqrt{\pi}\sum_{m,n=0}^{\infty}\frac{2^m s^m t^n}{m!}\,\delta_{n,m+1}.$$

Equating coefficients of $t^n s^m$:

$$\frac{1}{n!m!}\int_{-\infty}^{+\infty} H_n H_m e^{-z^2} z\,dz = \sqrt{\pi}\,\frac{2^n}{n!}\,\delta_{m,n+1} + \sqrt{\pi}\,\frac{2^m}{m!}\,\delta_{n,m+1},$$

so that

$$\int_{-\infty}^{+\infty} H_n(z)H_m(z)e^{-z^2} z\,dz = \sqrt{\pi}\,2^n(n+1)!\delta_{m,n+1} + \sqrt{\pi}\,2^m(m+1)!\delta_{n,m+1}.$$

(b) As above,

$$e^{-t^2+2zt} = \sum_{n=0}^{\infty}\frac{H_n(z)t^n}{n!}. \qquad e^{-s^2+2zs} = \sum_{m=0}^{\infty}\frac{H_m(z)s^m}{m!}.$$

$$\sum_{m,n=0}^{\infty}\frac{t^n s^m}{n!m!}\int_{-\infty}^{+\infty} H_n(z)H_m(z)e^{-z^2} z^2\,dz = \int_{-\infty}^{+\infty} e^{-t^2-s^2+2sz+2tz-z^2} z^2\,dz$$

$$= e^{2st}\int_{-\infty}^{+\infty} e^{-(z-s-t)^2} z^2\,dz = e^{2st}\int_{-\infty}^{+\infty} e^{-(z-s-t)^2}\left[(z-s-t)^2 + 2(s+t)z - (s+t)^2\right]dz.$$

But

$$\int_{-\infty}^{+\infty} e^{-(z-s-t)^2}(z-s-t)^2\,dz = -\frac{1}{2}\left| e^{-(z-s-t)^2}(z-s-t)\right|_{-\infty}^{+\infty} + \frac{1}{2}\int_{-\infty}^{+\infty} e^{-(z-s-t)^2}\,dz = \frac{\sqrt{\pi}}{2},$$

and

$$\int_{-\infty}^{+\infty} e^{-(z-s-t)^2} z(s+t)\,dz = (s+t)\left[\int_{-\infty}^{+\infty} e^{-(z-s-t)^2}[(z-s-t)+(s+t)]\,dz\right] = (s+t)^2\sqrt{\pi}.$$

Summing, we have

$$\sum_{m,n=0}^{\infty}\frac{t^n s^m}{n!m!}\int_{-\infty}^{+\infty} H_n(z)H_m(z)e^{-z^2} z^2\,dz = e^{2st}\left[\frac{\sqrt{\pi}}{2} + (s+t)^2\sqrt{\pi}\right]$$

$$= \sum_{n=0}^{\infty}\frac{(2st)^n}{n!}\frac{\sqrt{\pi}}{2} + \sum_{n=0}^{\infty}(s+t)^2\frac{(2st)^n}{n!}\sqrt{\pi}.$$

In the first term on the left, s and t occur with equal powers, namely for $n = m$. Comparing coefficients of $(st)^n$ on both sides of the equation:

$$\frac{1}{n!^2}\int_{-\infty}^{+\infty} H_n^2(z)e^{-z^2}z^2dz = \frac{2^n}{n!}\frac{\sqrt{\pi}}{2} + \frac{2^n\sqrt{\pi}}{(n-1)!} = \frac{(2n+1)}{2}\frac{2^n\sqrt{\pi}}{n!}.$$

Thus,

$$\int_{-\infty}^{+\infty} H_n^2(z)e^{-z^2}z^2dz = \frac{(2n+1)}{2}2^n n!\sqrt{\pi}.$$

(c)

$$\int_{-\infty}^{+\infty}\left[\frac{d}{dz}\left(H_n e^{-z^2/2}\right)\right]^2 dz = \int_{-\infty}^{+\infty}\left[\frac{dH_n}{dz}e^{-z^2/2} - zH_n e^{-z^2/2}\right]^2 dz$$

$$= \int_{-\infty}^{+\infty} e^{-z^2}\left[\frac{dH_n}{dz} - zH_n\right]^2 dz = \int_{-\infty}^{+\infty} e^{-z^2}\left[2nH_{n-1} - zH_n\right]^2 dz$$

$$= \int_{-\infty}^{+\infty} e^{-z^2}\left[2zH_n - H_{n+1} - zH_n\right]^2 dz = \int_{-\infty}^{+\infty} e^{-z^2}\left[zH_n - H_{n+1}\right]^2 dz$$

$$= \int_{-\infty}^{+\infty} H_n^2 e^{-z^2}z^2dz - 2\int_{-\infty}^{+\infty} zH_nH_{n+1}e^{-z^2}dz + \int_{-\infty}^{+\infty} H_{n+1}^2 e^{-z^2}dz$$

$$= \sqrt{\pi}\left\{\frac{2n+1}{2}2^n n! - 2(n+1)!2^n + 2^{n+1}(n+1)!\right\} = \sqrt{\pi}2^n n!\frac{2n+1}{2}.$$

II. The matrix elements of any operator O for a harmonic oscillator in quantum mechanics are defined to be

$$<n|O|m> = \int_{-\infty}^{+\infty}\Psi_n(x)O\Psi_m(x)dx,$$

where

$$\Psi_n(x) = \left[\frac{\alpha}{\sqrt{\pi}2^n n!}\right]^{1/2} H_n(\alpha x)e^{-(\alpha x)^2/2}, \qquad \text{where } \alpha = \left(\frac{m_0 k}{\hbar^2}\right)^{1/4},$$

and where k is the force constant of the harmonic oscillator.

(a) Show that $\langle n|x^2|n\rangle = \sum_{m=o}^{\infty} \langle n|x|m\rangle\langle m|x|n\rangle$, i.e. matrix behavior.

(b) If $(\Delta x)_n$ and $(\Delta p)_n$ are defined as

$$(\Delta x)_n = \left[\langle n|x^2|n\rangle - \langle n|x|n\rangle^2\right]^{1/2}, \quad (\Delta p)_n = \left[\langle n|p^2|n\rangle - \langle n|p|n\rangle^2\right]^{1/2},$$

prove that $(\Delta x)_n\,(\Delta p)_n = \dfrac{2n+1}{2}\hbar \geq \dfrac{\hbar}{2}$.

Solution:

(a)
$$\langle n|x^2|n\rangle = \int_{-\infty}^{+\infty} \Psi_n(x)\Psi_n(x)x^2 dx = \int_{-\infty}^{+\infty} \frac{\alpha x^2}{\sqrt{\pi}\,2^n n!} H_n^2(\alpha x) e^{-(\alpha x)^2} dx$$

$$= \int_{-\infty}^{+\infty} \frac{\alpha^2 x^2}{\alpha^2\sqrt{\pi}\,2^n n!} H_n^2(\alpha x) e^{-(\alpha x)^2} d(\alpha x)$$

$$= \frac{1}{\alpha^2\sqrt{\pi}\,2^n n!}\left(\frac{2n+1}{2}\,2^n n!\sqrt{\pi}\right) = \frac{2n+1}{2\alpha^2}.$$

$$\langle n|x|m\rangle = \int_{-\infty}^{+\infty} \Psi_n(x)\Psi_m(x)x\,dx$$

$$= \int_{-\infty}^{+\infty} \frac{\alpha x}{\alpha\sqrt{\pi}}\,\frac{1}{\sqrt{2^n 2^m n!m!}} H_n(\alpha x)H_m(\alpha x) e^{-(\alpha x)^2} d(\alpha x).$$

$$\sum_{m=o}^{\infty} \langle n|x|m\rangle\langle m|x|n\rangle = \sum_{m=o}^{\infty} \pi\left(\frac{(n+1)!2^n\delta_{m,n+1} + (m+1)!2^m\delta_{m,n-1}}{\alpha\sqrt{\pi}\sqrt{2^n 2^m n!m!}}\right)^2$$

$$= \sum_{m=o}^{\infty} \pi\left(\frac{[(n+1)!2^n\delta_{m,n+1}]^2 + [(m+1)!2^m\delta_{m,n-1}]^2}{\alpha^2\pi 2^n 2^m n!m!}\right)$$

$$= \frac{(n+1)!^2 2^{2n}}{\alpha^2 2^n 2^{n+1} n!(n+1)!} + \frac{n!^2 2^{2n-2}}{\alpha^2 2^n 2^{n-1} n!(n-1)!}$$

$$= \frac{n+1}{2\alpha^2} + \frac{n}{2\alpha^2} = \frac{2n+1}{2\alpha^2} = \langle n|x^2|n\rangle.$$

Hence, $\langle n|x^2|n\rangle = \sum_{m=o}^{\infty} \langle n|x|m\rangle\langle m|x|n\rangle$.

(b) From (a) above, $<n|x^2|n> = \dfrac{2n+1}{2\alpha^2}$. We also have:

$$<n|x|n> = \int_{-\infty}^{+\infty}\Psi_n^2(x)x\,dx = \int_{-\infty}^{+\infty}\frac{\alpha x}{\alpha\sqrt{\pi}2^n n!}H_n^2(\alpha x)e^{-(\alpha x)^2}d(\alpha x) = 0.$$

Hence,

$$(\Delta x)_n = \left[<n|x^2|n> - <n|x|n>^2\right]^{1/2} = \left[<n|x^2|n>\right]^{1/2} = \left(\frac{2n+1}{2}\right)^{1/2}\frac{1}{\alpha}.$$

From $p = \dfrac{\hbar}{i}\dfrac{\partial}{\partial x} \;\Rightarrow\; p^2 = -\hbar^2\dfrac{\partial^2}{\partial x^2}$,

$$<n|p^2|n> = \int_{-\infty}^{+\infty}\frac{-\alpha\hbar^2}{\sqrt{\pi}2^n n!}\left[H_n(\alpha x)e^{-(\alpha x)^2/2}\frac{d^2}{dx^2}\left\{H_n(\alpha x)e^{-(\alpha x)^2/2}\right\}\right]dx.$$

Multiplying numerator and denominator by α^2:

$$<n|p^2|n> = \int_{-\infty}^{+\infty}\frac{-\alpha^2\hbar^2}{\sqrt{\pi}2^n n!}\left[H_n(\alpha x)e^{-(\alpha x)^2/2}\frac{d^2}{d(\alpha x)^2}\left\{H_n(\alpha x)e^{-(\alpha x)^2/2}\right\}\right]d(\alpha x).$$

Setting $\omega = \alpha x$,

$$\frac{d^2}{d\omega^2}\left[H_n(\omega)e^{-\omega^2/2}\right] = \frac{d}{d\omega}\left[e^{-\omega^2/2}\left(-\omega H_n + \frac{dH_n}{d\omega}\right)\right]$$

$$= \frac{d}{d\omega}\left[e^{-\omega^2/2}(\omega H_n - H_{n+1})\right]$$

$$= e^{-\omega^2/2}\left[\omega\frac{dH_n}{d\omega} + H_n - \frac{dH_{n+1}}{d\omega}\right] - \omega e^{-\omega^2/2}[\omega H_n - H_{n+1}]$$

$$= e^{-\omega^2/2}\left[H_n(1-\omega^2) + \omega H_{n+1} + \omega(2\omega H_n - H_{n+1}) - 2(n+1)H_n\right]$$

$$= e^{-\omega^2/2}\left[H_n(\omega^2 - 2n - 1) + H_{n+1}(\omega - \omega)\right] = e^{-\omega^2/2}H_n(\omega^2 - 2n - 1).$$

Hence,

$$<n|p^2|n> = \int_{-\infty}^{+\infty} \frac{-\alpha^2\hbar^2}{\sqrt{\pi}2^n n!}\left[H_n(\alpha x)e^{-(\alpha x)^2} H_n(\alpha^2 x^2 - 2n - 1)\right]d(\alpha x)$$

$$= \frac{-\alpha^2\hbar^2}{\sqrt{\pi}2^n n!}\int_{-\infty}^{+\infty} H_n^2(\alpha x)e^{-(\alpha x)^2}(\alpha x)^2 d(\alpha x)$$

$$+ \frac{(2n+1)\alpha^2\hbar^2}{\sqrt{\pi}2^n n!}\int_{-\infty}^{+\infty} H_n^2(\alpha x)e^{-(\alpha x)^2} d(\alpha x)$$

$$= \frac{-\alpha^2\hbar^2}{\sqrt{\pi}2^n n!}\left[\frac{2n+1}{2}2^n n!\sqrt{\pi} - (2n+1)2^n n!\sqrt{\pi}\right] = \frac{2n+1}{2}\alpha^2\hbar^2.$$

$$<n|p|n> = \int_{-\infty}^{+\infty} \frac{\alpha\hbar}{i\sqrt{\pi}2^n n!} H_n(\alpha x)e^{-(\alpha x)^2/2}\frac{d}{dx} H_n(\alpha x)e^{-(\alpha x)^2/2}dx$$

$$= \int_{-\infty}^{+\infty} \frac{\alpha\hbar}{i\sqrt{\pi}2^n n!} H_n(\alpha x)e^{-(\alpha x)^2/2}\frac{d}{d(\alpha x)} H_n(\alpha x)e^{-(\alpha x)^2/2}d(\alpha x)$$

$$= \frac{\alpha\hbar}{i\sqrt{\pi}2^n n!}\int_{-\infty}^{+\infty} H_n(\alpha x)e^{-(\alpha x)^2}\left[(\alpha x)H_n(\alpha x) - H_{n+1}(\alpha x)\right]d(\alpha x)$$

$$= \frac{\alpha\hbar}{i\sqrt{\pi}2^n n!}\int_{-\infty}^{+\infty} H_n^2(\alpha x)e^{-(\alpha x)^2}(\alpha x)d(\alpha x)$$

$$- \frac{\alpha\hbar}{i\sqrt{\pi}2^n n!}\int_{-\infty}^{+\infty} H_n(\alpha x)H_{n+1}(\alpha x)e^{-(\alpha x)^2}d(\alpha x).$$

From the results of Problem I (a), the first term vanishes for $m = n$, while the second term vanishes for $n \neq n + 1$. Hence, $<n|p|n> = 0$ and

$$(\Delta p)_n = \left[<n|p^2|n>\right]^{1/2} = \left[\frac{2n+1}{2}\right]^{1/2}\alpha\hbar.$$

Finally, then:

$$(\Delta x)_n(\Delta p)_n = \left(\frac{2n+1}{2}\right)^{1/2}\frac{1}{\alpha}\left[\frac{2n+1}{2}\right]^{1/2}\alpha\hbar = \frac{2n+1}{2}\hbar \geq \frac{\hbar}{2}.$$

III. The Laguerre's equation is given by

$$\frac{d^2f}{dz^2} + \frac{1-z}{z}\frac{df}{dz} + \frac{n}{z}f = 0.$$

(a) What are the regular and irregular singular points?

(b) Prove that one of the two independent solutions is an entire function and that the other one has a branch point at $z = 0$.

(c) Show that if n is a positive integer, then one of the solutions can be written as

$$f(z) = L_n(z) = \sum_{m=0}^{N}(-1)^m\left(\frac{n!}{m!}\right)^2\frac{1}{(n-m)!}z^m,$$

which is defined to be the Laguerre polynomial, with $n = 0, 1, 2, \ldots$

Solution: (refer to Section 7.2)

(a) Since $p(z) = \frac{1}{z} - 1$ and $q(z) = \frac{n}{z}$, we have $p_{-1} = 1$, $q_{-1} = n$, and $q_{-2} = 0$. Hence, $pz = 1 - z$ and $qz^2 = nz$ are analytic at $z = 0$, so that $z = 0$ is a regular singular point of the Laguerre equation.

To investigate $z = \infty$, we consider $2z - pz^2 = 2z - z + z^2 = z^2 + z$ and $q(z)z^4 = nz^3$, both of which are not analytic at $z = \infty$. Hence, $z = \infty$ is a singular point, and therefore an irregular point, since both pz and qz^2 are not analytic at $z = \infty$.

(b) The indicial equation has the form

$$s(s-1) + sp_{-1} + q_{-2} = 0 \;\rightarrow\; s^2 - s + s = 0 \;\rightarrow\; s^2 = 0,$$

where for this problem $q_{-2} = 0$ and $p_{-1} = 1$. There thus results only one solution, e.g. $s = 0$. In the vicinity of $z = 0$, one independent solution is given by

$$f_1(z) = \sum_{m=0}^{\infty} a_m z^m.$$

Since $f_1(z)$ is regular at all points except $z = \infty$, $f_1(z)$ is an entire function with an infinite radius of convergence.

Furthermore, for p having a pole of order 1 and q a pole of order 1, ($m \le 1, n \le 2$), we have at worst a function with a branch point at $z = 0$; this function constitutes the second independent solution.

(c) $$f_1(z) = \sum_{m=0}^{\infty} a_m z^m. \quad \frac{df_1(z)}{dz} = \sum_{m=0}^{\infty} m a_m z^{m-1}. \quad \frac{d^2 f_1(z)}{dz^2} = \sum_{m=0}^{\infty} m(m-1) a_m z^{m-2}.$$

$$\frac{1-z}{z}\frac{df_1(z)}{dz} = \sum_{m=0}^{\infty}\left(m a_m z^{m-1} - m a_m z^{m-1}\right).$$

$$\frac{n}{z} f_1(z) = \sum_{m=0}^{\infty} n a_m z^{m-1}.$$

Substituting into the Laguerre equation:

$$\sum_{m=0}^{\infty} z^m \left[(m+2)(m+1)a_{m+2} + (m+2)a_{m+2} - (m+1)a_{m+1} + n a_{m+1}\right] = 0.$$

$$\Downarrow$$

$$(m+2)^2 a_{m+2} = a_{m+1}(m+1-n) \quad \rightarrow \quad a_{m+2} = \frac{m+1-n}{(m+2)^2} a_{m+1}.$$

$$\Downarrow$$

$$a_{m+1} = \frac{m-n}{(m+1)^2} a_m .$$

If n is a positive integer the series will stop when $m = n$.

$$f_1(z) = a_o + a_1 z + a_2 z^2 + \ldots + a_{m-1} z^{m-1} + a_m z^m + a_{m+1} z^{m+1} + \ldots$$

$$= a_o\left[1 - nz + \frac{n(n-1)z^2}{2^2} - \frac{n(n-1)(n-2)z^3}{3!^2} + \ldots\right]$$

$$+ a_o\left[\ldots + \frac{(-1)^r n(n-1)\ldots(n-r+1)z^r}{r!^2}\right].$$

For $m = n$, the last term will be that for which the power of z is m, namely $a_m z^m$:

$$f_1(z) = a_0\left[1 - nz + \frac{n(n-1)z^2}{2!^2} - \frac{n(n-1)(n-2)z^3}{3!^2} + \frac{(-1)^n n! z^n}{n!^2}\right]$$

Choosing $a_0 = (-1)^n n!$:

$$f_1(z) = (-1)^n n! + \frac{(-1)^{n+1} n n! z}{1!} + \frac{(-1)^{n+2} n(n-1) n! z^2}{2!^2} + \ldots + \frac{(-1)^{2n}(n!)^2 z^n}{n!^2}$$

$$= \frac{(-1)^n (n!)^2}{(0!)^2(n-0)!} + \frac{(-1)^{n+1}(n!)^2 z}{(1!)^2(n-1)!} + \frac{(-1)^{n+2}(n!)^2 z^2}{(2!)^2(n-2)!} + \ldots$$

$$\ldots + \frac{(-1)^{2n}(n!)^2 z^n}{n!^2(n-n)!}$$

This can be expressed as:

$$f_1(z) = \sum_{m=0}^{n} (-1)^{n+m}\left(\frac{n!}{m!}\right)^2 \frac{z^m}{(n-m)!}.$$

If one now chooses $a_0 = n!$ rather than $(-1)^n n!$, one obtains, after dividing through by $(-1)^n$:

$$f_1(z) = \sum_{m=0}^{n} (-1)^m \left(\frac{n!}{m!}\right)^2 \frac{z^m}{(n-m)!} = \mathcal{L}(z),$$

which is the Laguerre polynomial.

IV. Show that the Laguerre polynomials satisfy the following equations:

(a) $$\mathcal{L}_n(z) = \frac{n!}{2\pi i}\oint\left[t^{n+1}(1-t)\right]^{-1} \exp\left(\frac{-tz}{1-t}\right) dt,$$

where the closed contour is taken to include the origin in the t plane.

(b) $$(1-t)^{-1}\exp\left(\frac{-tz}{1-t}\right) = \sum_{n=0}^{\infty} \frac{\mathcal{L}_n(z)}{n!} t^n.$$

(c) $$(1 + 2n - z)\mathcal{L}_n - n^2\mathcal{L}_{n-1} - \mathcal{L}_{n+1} = 0.$$

Solution:

(a) Define $y_n(z) = \frac{1}{2\pi i}\oint \frac{t^{-n-1}}{(1-t)}\exp\left(\frac{-tz}{1-t}\right)dt$, so that

$$\frac{dy_n(z)}{dz} = -\frac{1}{2\pi i}\oint \frac{t\cdot t^{-n-1}}{(1-t)^2}\exp\left(\frac{-tz}{1-t}\right)dt.$$

$$\frac{d^2y_n(z)}{dz^2} = \frac{1}{2\pi i}\oint \frac{t^2\cdot t^{-n-1}}{(1-t)^3}\exp\left(\frac{-tz}{1-t}\right)dt.$$

$$z\frac{d^2y_n(z)}{dz^2} + (1-z)\frac{dy_n(z)}{dz} + ny_n(z) = \frac{1}{2\pi i}\oint \frac{t^{-n-1}}{(1-t)}\exp\left(\frac{-tz}{1-t}\right)\frac{zt^2}{(1-t)^2}dt$$

$$-\frac{1}{2\pi i}\oint \frac{t^{-n-1}}{(1-t)}\exp\left(\frac{-tz}{1-t}\right)\left(\frac{(1-z)t}{(1-t)} - n\right)dt.$$

Since

$$-\frac{d}{dt}\left[\frac{t^{-n}}{(1-t)}\exp\left(\frac{-tz}{1-t}\right)\right] = \exp\left(\frac{-tz}{1-t}\right)\frac{t^{-n-1}}{1-t}\left[n - \frac{t}{1-t} + \frac{zt}{1-t} + \frac{zt^2}{(1-t)^2}\right],$$

$$\frac{z\,d^2y_n(z)}{dz^2} + (1-z)\frac{dy_n(z)}{dz} + ny_n(z) = -\frac{1}{2\pi i}\oint \frac{d}{dt}\left[\frac{t^{-n}}{(1-t)}\exp\left(\frac{-tz}{1-t}\right)\right]dt = 0,$$

i.e., $y_n(z)$ satisfies the Laguerre equation. From residue theory,

$$y_n(0) = \frac{1}{2\pi i}\oint \frac{1+t+t^2+\ldots}{t^{n+1}}dt = 1,$$

while $\mathcal{L}_n(0) = n!$, so that $\mathcal{L}_n(0) = n!y_n(0)$. Hence, $\mathcal{L}_n(z) = n!y_n(z)$, and

$$\mathcal{L}_n(z) = \frac{n!}{2\pi i}\oint \frac{t^{-n-1}}{(1-t)}\exp\left(\frac{-tz}{1-t}\right)dt.$$

(b) Since from residue theory, $\frac{1}{2\pi i}\oint t^{-n-1}\sum_{n=0}^{\infty} y_n(z)t^n dt = y_n(z)$, it follows that

$$\frac{\exp\left(\frac{-tz}{1-t}\right)}{1-t} = \sum_{n=0}^{\infty} y_n(z)t^n = \sum_{n=0}^{\infty}\frac{\mathcal{L}_n(z)t^n}{n!}.$$

(c)

$$\frac{\exp\left(\frac{-tz}{1-t}\right)}{1-t} = \sum_{n=0}^{\infty}\frac{y_n(z)t^n}{n!} = \sum_{n=0}^{\infty}\frac{\mathcal{L}_n(z)t^n}{n!}.$$

Taking derivatives with respect to t:

$$\exp\left(\frac{-tz}{1-t}\right)\left[\frac{1}{(1-t)^2}-\frac{z}{(1-t)^3}\right] = \frac{\exp\left(\frac{-tz}{1-t}\right)}{(1-t)}\left[\frac{1-t-z}{(1-t)^2}\right] = \sum_{n=0}^{\infty}\frac{\mathcal{L}_n(z)t^{n-1}}{(n-1)!}.$$

$$\sum_{n=0}^{\infty}\frac{\mathcal{L}_n(z)t^n}{n!}(1-t-z) = (1-2t+t^2)\sum_{n=0}^{\infty}\frac{\mathcal{L}_n(z)t^{n-1}}{(n-1)!}.$$

Equating coefficients of t^n:

$$\frac{(1-z)\mathcal{L}_n(z)}{n!}-\frac{\mathcal{L}_{n-1}(z)}{(n-1)!} = \frac{\mathcal{L}_{n+1}(z)}{n!}-\frac{2\mathcal{L}_n(z)}{(n-1)!}+\frac{\mathcal{L}_{n-1}(z)}{(n-2)!},$$

and multiplying through by $n!$:

$$(1-z+2n)\mathcal{L}_n(z)-n^2\mathcal{L}_{n-1}(z)-\mathcal{L}_{n+1}(z) = 0.$$

V. The associated Laguerre polynomials are defined as

$$\mathcal{L}_n^k(z) = \frac{d^k\mathcal{L}_n(z)}{dz^k}.$$

Show that

(a) $$\left(\frac{d^2}{dz^2}+\frac{k+1-z}{z}\frac{d}{dz}+\frac{n-k}{z}\right)\mathcal{L}_n^k(z) = 0.$$

(b) If $f(z) = z^{(k-1)/2}e^{-z/2}\mathcal{L}_n^k(z)$, show that

$$\frac{d^2f(z)}{dz^2}+\frac{2}{z}\frac{df(z)}{dz}+\frac{1}{4z^2}\left[-k^2+1+(4n-2k+2)z-z^2\right]f(z) = 0.$$

Solution:

(a) The Laguerre equation reads $z\dfrac{d^2y}{dz^2}+(1-z)\dfrac{dy}{dz}+ny = 0.$

Differentiating k times with respect to z:

$$(k=1)\qquad z\frac{d^3y}{dz^3}+\frac{d^2y}{dz^2}+\frac{d^2y}{dz^2}-z\frac{d^2y}{dz^2}-\frac{dy}{dz}+n\frac{dy}{dz} = 0.$$

$$(k=2)\qquad z\frac{d^4y}{dz^4}+\frac{d^3y}{dz^3}+2\frac{d^3y}{dz^3}-z\frac{d^3y}{dz^3}-\frac{d^2y}{dz^2}-\frac{d^2y}{dz^2}+n\frac{d^2y}{dz^2} = 0.$$

$$(k=k)\qquad z\frac{d^{k+2}y}{dz^{k+2}}+(k+1-z)\frac{d^{k+1}y}{dz^{k+1}}+(n-k)\frac{d^ky}{dz^k} = 0.$$

However, since $\dfrac{d^{k+j}y}{dz^{k+j}} = \dfrac{d^j}{dz^j}\dfrac{d^ky}{dz^k} = \dfrac{d^j}{dz^j}\mathcal{L}_n^k(z)$, we have

$$\left[z\frac{d^2}{dz^2}+(k+1-z)\frac{d}{dz}+(n-k)\right]\mathcal{L}_n^k(z) = 0.$$

(b) Let $y^k \equiv \mathcal{L}_n^k(z)$, so that if $f(z) = z^{(k-1)/2}e^{-z/2}\mathcal{L}_n^k(z) = z^{(k-1)/2}e^{-z/2}y^k$, then

$$\frac{df}{dz} = y^{k+1}e^{-z/2}z^{(k-1)/2}+\frac{(k-1)}{2}y^ke^{-z/2}z^{(k-3)/2}-\frac{1}{2}y^ke^{-z/2}z^{(k-1)/2}.$$

$$\begin{aligned}\frac{d^2f}{dz^2} = {} & y^{k+2}e^{-z/2}z^{(k-1)/2}+\frac{(k-1)}{2}y^{k+1}e^{-z/2}z^{(k-3)/2}-\frac{1}{2}y^{k+1}e^{-z/2}z^{(k-1)/2}\\ & -\frac{1}{2}y^{k+1}e^{-z/2}z^{(k-1)/2}-\frac{(k-1)}{4}y^ke^{-z/2}z^{(k-3)/2}+\frac{1}{4}y^ke^{-z/2}z^{(k-1)/2}\\ & +\frac{(k-1)}{2}y^{k+1}e^{-z/2}z^{(k-3)/2}+\frac{(k-1)}{2}\frac{(k-3)}{2}y^ke^{-z/2}z^{(k-3)/2}\\ & -\frac{(k-1)}{4}y^ke^{-z/2}z^{(k-3)/2}.\end{aligned}$$

$$z\frac{d^2f}{dz^2} = y^{k+2}e^{-z/2}z^{(k+1)/2} + (k-1)y^{k+1}e^{-z/2}z^{(k-1)/2} - y^{k+1}e^{-z/2}z^{(k+1)/2}$$

$$-\frac{(k-1)}{2}y^ke^{-z/2}z^{(k-1)/2} + \frac{1}{4}y^ke^{-z/2}z^{(k+1)/2} + \frac{(k-1)(k-3)}{4}y^ke^{-z/2}z^{(k-3)/2}.$$

$$2\frac{df}{dz} = 2y^{k+1}e^{-z/2}z^{(k-1)/2} + (k-1)y^ke^{-z/2}z^{(k-3)/2} - y^ke^{-z/2}z^{(k-1)/2}.$$

$$z\frac{d^2f}{dz^2} + 2\frac{df}{dz} = e^{-z/2}z^{(k-1)/2}\left[zy^{k+2} + (k+1-z)y^{k+1} - \frac{(1+k)y^k}{2}\right]$$

$$+ e^{-z/2}z^{(k-1)/2}\left[\frac{z}{4}y^k + \frac{(k^2-1)}{4z}y^k\right].$$

Since from part (a) above,

$$zy^{k+2} + (k+1-z)y^{k+1} + (n-k)y^k = 0,$$

we have

$$z\frac{d^2f}{dz^2} + 2\frac{df}{dz} = e^{-z/2}z^{(k-1)/2}\left[(k-n)y^k - \frac{(k+1)}{2}y^k + \frac{z}{4}y^k + \frac{(k^2-1)}{4z}y^k\right]$$

$$= -e^{-z/2}z^{(k-1)/2}y^k\,\frac{(1-k^2-z^2-2zk+2z+4zn)}{4z}.$$

$$z\frac{d^2f}{dz^2} + 2\frac{df}{dz} + \frac{1}{4z}\left[(1-k^2) + (4n-2k+2)z - z^2\right]f = 0.$$

VI. Prove that for any values of λ and μ: (Refer to Sections 6.11.1 and 7.5.1)

(a) $$\sum_{m=0}^{n}\frac{\Gamma(\lambda+n+1)\Gamma(\mu+n+1)}{(m!)(n-m)!\Gamma(\lambda+n-m+1)\Gamma(\mu+m+1)} = \frac{\Gamma(\lambda+\mu+2n+1)}{(n!)\Gamma(\lambda+\mu+n+1)}.$$

(*Hint:* Compare coefficients of x^n in $(1+x)^{\lambda+n}(1+x)^{\mu+n}$ with those in $(1+x)^{\lambda+\mu+2n}$.)

(b) $$J_\lambda(z)\,J_\mu(z) = \sum_{m=o}^{\infty} \frac{(-1)^n\Gamma(\lambda+\mu+2n+1)(z/2)^{\lambda+\mu+2n}}{n!\,\Gamma(\lambda+\mu+n+1)\Gamma(\lambda+n+1)} \cdot \frac{1}{\Gamma(\mu+n+1)}.$$

(c) $$J_{-\lambda}(z)\,J_{\lambda-1}(z) + J_{-\lambda+1}(z)\,J_\lambda(z) = \frac{2\sin\lambda\pi}{\pi z}.$$

Solution:

(a) $$\sum_{m=o}^{n} \frac{\Gamma(\lambda+n+1)\Gamma(\mu+n+1)}{(m!)(n-m)!\,\Gamma(\lambda+n-m+1)\Gamma(\mu+m+1)}$$

$$= \frac{1}{0!n!}\left\{\underset{m=0}{\frac{\Gamma(n+\mu+1)}{\Gamma(\mu+1)}} + \underset{m=n}{\frac{\Gamma(n+\lambda+1)}{\Gamma(\lambda+1)}}\right\} + \frac{1}{1!(n-1)!}\left\{\underset{m=1}{\frac{\Gamma(n+\lambda+1)\Gamma(n+\mu+1)}{\Gamma(\mu+2)\Gamma(n+\lambda)}}\right\}$$

$$+ \frac{1}{1!(n-1)!}\left\{\underset{m=n-1}{\frac{\Gamma(n+\lambda+1)\Gamma(n+\mu+1)}{\Gamma(\mu+2)\Gamma(n+\lambda)}}\right\} + \ldots$$

$$= \frac{1}{n!}\left\{\frac{(n+\mu)!}{\mu!} + \frac{(n+\lambda)!}{\lambda!}\right\} + \frac{1}{(n-1)!}\left\{\frac{(n+\lambda)!(n+\mu)!}{(n+\lambda-1)!(\mu+1)!}\right\}$$

$$+ \frac{1}{(n-1)!}\left\{\frac{(n+\mu)!(n+\lambda)!}{(n+\mu-1)!(\lambda+1)!}\right\} + \ldots$$

$$= \frac{1}{n!}\Big[(n+\mu)(n+\mu-1)\ldots(\mu+1) + (\lambda+n)(\lambda+n-1)\ldots(\lambda+1)\Big]$$

$$+ \frac{1}{(n-1)!}\Big[(n+\lambda)\Big((\mu+n)(\mu+n-1)\ldots(\mu+2)\Big)\Big]$$

$$+ \frac{1}{(n-1)!}\Big[(n+\mu)\Big((\lambda+n)(\lambda+n-1)\ldots(\lambda+2)\Big)\Big] + \ldots$$

The product $(1+x)^{\lambda+n}(1+x)^{\mu+n}$ can be expressed as $[A][B]$, where

$$[A] = \left[1 + (\lambda+n)x + \frac{(\lambda+n)(\lambda+n-1)x^2}{2!}\right] + \ldots$$

$$\ldots + \left[\frac{(\lambda+n)(\lambda+n-1)\ldots(\lambda+1)x^n}{n!} + \ldots\right].$$

$[B]$ is identical to $[A]$, except that λ is replaced by μ. The coefficient of x^n in this expansion product is seen to be:

$$\frac{1}{n!}\Big[(n+\mu)(n+\mu-1)\ldots(\mu+1)+(\lambda+n)(\lambda+n-1)\ldots(\lambda+1)\Big]$$

$$+\frac{1}{(n-1)!}\Big[(n+\lambda)\big((\mu+n)(\mu+n-1)\ldots(\mu+2)\big)\Big]$$

$$+\frac{1}{(n-1)!}\Big[(n+\mu)\big((\lambda+n)(\lambda+n-1)\ldots(\lambda+2)\big)\Big]$$

$$=\sum_{m=0}^{n}\frac{\Gamma(\lambda+n+1)\Gamma(\mu+n+1)}{(m!)(n-m)!\Gamma(\lambda+n-m+1)\Gamma(\mu+m+1)}\,.$$

Consider the expression

$$\frac{\Gamma(\lambda+\mu+2n+1)}{(n!)\Gamma(\lambda+\mu+n+1)}=\frac{1}{n!}\frac{(\lambda+\mu+2n)(\lambda+\mu+2n-1)\ldots(\lambda+\mu+1)\Gamma(\lambda+\mu+1)}{(\lambda+\mu+n)(\lambda+\mu+n-1)\ldots(\lambda+\mu+1)\Gamma(\lambda+\mu+1)}$$

$$=\frac{(\lambda+\mu+2n)(\lambda+\mu+2n-1)\ldots(\lambda+\mu+n+1)}{n!}\,.$$

The coefficient of x^n in the product expansion of $(1+x)^{\mu+\lambda+2n}$ is given by:

$$\frac{(2n+\lambda+\mu)(2n+\lambda+\mu-1)\ldots(2n+\lambda+\mu-[n-1])}{n!}=\frac{\Gamma(\lambda+\mu+2n+1)}{(n!)\Gamma(\lambda+\mu+n+1)}\,.$$

Hence, since the expansions on both sides of (a) are for the same quantity, namely, $(1+x)^{\lambda+n}(1+x)^{\mu+n}=(1+x)^{\mu+\lambda+2n}$, we have that

$$\sum_{m=0}^{n}\frac{\Gamma(\lambda+n+1)\Gamma(\mu+n+1)}{(m!)(n-m)!\Gamma(\lambda+n-m+1)\Gamma(\mu+m+1)}=\frac{\Gamma(\lambda+\mu+2n-1)}{(n!)\Gamma(\lambda+\mu+n+1)}\,.$$

(b) Since

$$J_\lambda(z)=\sum_{k=0}^{\infty}\left(\frac{z}{2}\right)^{2k+\lambda}\frac{(-1)^k}{k!\Gamma(\lambda+k+1)}\quad\text{and}\quad J_\mu(z)=\sum_{l=0}^{\infty}\left(\frac{z}{2}\right)^{2l+\mu}\frac{(-1)^l}{l!\Gamma(\mu+l+1)},$$

$$J_\lambda(z)\,J_\mu(z)=\sum_{k=0}^{\infty}\sum_{l=0}^{\infty}\left(\frac{z}{2}\right)^{2(k+l)+\lambda+\mu}\frac{(-1)^{k+l}}{k!l!\Gamma(\lambda+k+1)\Gamma(\mu+l+1)}\,.$$

If $n = k + l$, then as n runs from 0 to infinity, l also runs from 0 to infinity, but at best equal to n and never greater. Thus,

$$J_\lambda(z)\,J_\mu(z) = \sum_{n=0}^{\infty}\left(\frac{z}{2}\right)^{2n+\lambda+\mu}(-1)^n\sum_{l=0}^{n}\frac{1}{l!(n-l)!\Gamma(\lambda+n-l+1)\Gamma(\mu+l+1)}.$$

However, from part (a) above,

$$\sum_{l=0}^{n}\frac{1}{l!(n-l)!\Gamma(\lambda+n-l+1)\Gamma(\mu+l+1)}$$

$$= \frac{\Gamma(\lambda+\mu+2n-1)}{(n!)\Gamma(\lambda+\mu+n+1)\Gamma(\lambda+n+1)\Gamma(\mu+n+1)},$$

so that

$$J_\lambda(z)\,J_\mu(z) = \sum_{m=0}^{n}\frac{(-1)^n\Gamma(\lambda+\mu+2n+1)(z/2)^{\lambda+\mu+2n}}{n!\Gamma(\lambda+\mu+n+1)\Gamma(\lambda+n+1)}\cdot\frac{1}{\Gamma(\mu+n+1)}.$$

(c) Setting $\mu = \lambda-1$ and $\lambda = -\lambda$ and then $\mu = -\lambda+1$ and $\lambda = +\lambda$ in the results of part (b) above, we have

$$J_{-\lambda}\,J_{\lambda-1} + J_{-\lambda+1}\,J_\lambda = \sum_{n=0}^{\infty}\frac{(-1)^n\Gamma(2n)(z/2)^{2n-1}}{n!\Gamma(n)\Gamma(n-\lambda+1)\Gamma(\lambda+n)}$$

$$+\sum_{m=0}^{\infty}\frac{(-1)^m\Gamma(2m+2)(z/2)^{2m+1}}{(m)!\Gamma(m+2)\Gamma(m+\lambda+1)\Gamma(m-\lambda+2)}.$$

Setting $n = m + 1$:

$$J_{-\lambda}\,J_{\lambda-1} + J_{-\lambda+1}\,J_\lambda = \frac{(-1)^0\Gamma(0)(z/2)^{-1}}{0!\Gamma(0)\Gamma(1-\lambda)\Gamma(\lambda)}$$

$$+\sum_{m=0}^{\infty}\frac{(-1)^{m+1}\Gamma(2m+2)(z/2)^{2m+1}}{(m+1)!\Gamma(m+1)\Gamma(m-\lambda+2)\Gamma(1+\lambda+m)}$$

$$+\sum_{m=0}^{\infty}\frac{(-1)^m\Gamma(2m+2)(z/2)^{2m+1}}{(m)!\Gamma(m+2)\Gamma(m-\lambda+2)\Gamma(1+\lambda+m)}.$$

Combining,

$$J_{-\lambda} J_{\lambda-1} + J_{-\lambda+1} J_\lambda = \frac{\Gamma(0)(2/z)}{\Gamma(0)\Gamma(1-\lambda)\Gamma(\lambda)}$$

$$+ \sum_{m=0}^{\infty} \frac{(-1)^m \Gamma(2m+2)(z/2)^{2m+1}}{(m)!\Gamma(m+1)\Gamma(m-\lambda+2)\Gamma(1+\lambda+m)} \left[\frac{1}{m+1} - \frac{1}{m+1}\right].$$

Hence,

$$J_{-\lambda} J_{\lambda-1} + J_{-\lambda+1} J_\lambda = \frac{2}{z\,\Gamma(1-\lambda)\Gamma(\lambda)}.$$

But $\Gamma(\lambda)\Gamma(1-\lambda) = \dfrac{\pi}{\sin\pi\lambda}$ (Section 7.5.2) $\rightarrow$ $J_{-\lambda} J_{\lambda-1} + J_{-\lambda+1} J_\lambda = \dfrac{2\sin\pi z}{\pi z}$.

VII. Let $Z_\lambda(z) = AJ_\lambda(z) + BH^1_\lambda(z)$, where A and B are arbitrary constants. Prove that the general solutions of the following differential equations are:

(a) $$\frac{d^2 f}{dz^2} + \frac{1}{z}\frac{df}{dz} - \left[\frac{1}{z} + \left(\frac{\lambda}{2z}\right)^2\right] f = 0 \quad \rightarrow \quad f = Z_\lambda(2i\sqrt{z})$$

(b) $$\frac{d^2 f}{dz^2} + bz^m f = 0 \quad \rightarrow \quad f = \sqrt{z}\, Z_{1/m+2}\left(\frac{2\sqrt{b}}{m+2} z^{(m+2)/2}\right)$$

Solution:

(a) Let $y = 2i\sqrt{z}$, so that $J_\lambda(2i\sqrt{z}) \rightarrow J_\lambda(y)$.

$$\frac{dJ_\lambda(y)}{dz} = \frac{dJ_\lambda(y)}{dy}\frac{dy}{dz} = \frac{dJ_\lambda(y)}{dy}\frac{i}{\sqrt{z}} = \frac{-2}{y}\frac{dJ_\lambda(y)}{dy}.$$

$$\frac{d^2 J_\lambda(y)}{dz^2} = \frac{-2}{y}\frac{d}{dy}\left[\frac{-2}{y}\frac{dJ_\lambda(y)}{dy}\right] = \frac{-4}{y^3}\frac{dJ_\lambda(y)}{dy} + \frac{4}{y^2}\frac{d^2 J_\lambda(y)}{dy^2}.$$

$$\frac{1}{z}\frac{dJ_\lambda(y)}{dz} = \frac{8}{y^3}\frac{dJ_\lambda(y)}{dy}. \qquad \left[\frac{1}{z} + \left(\frac{\lambda}{2z}\right)^2\right] J_\lambda(y) = \frac{-4J_\lambda(y)}{y^2} + \frac{4\lambda^2 J_\lambda(y)}{y^4}.$$

$$\frac{d^2J_\lambda}{dz^2}+\frac{1}{z}\frac{dJ_\lambda}{dz}-\left[\frac{1}{z}+\left(\frac{\lambda}{2z}\right)^2\right]J_\lambda = \frac{4}{y^2}\left[\frac{d^2J_\lambda}{dy^2}+\frac{1}{y}\frac{dJ_\lambda}{dy}+\left(1-\frac{\lambda^2}{y^2}\right)J_\lambda\right] = 0,$$

since $J_\lambda(y)$ is a solution of Bessel's equation.

Thus we have shown that $J_\lambda(2i\sqrt{z})$ is a solution of equation (a). Since the Hankel function H^1_λ (Section 7.5.4) is a combination of $J_\lambda(2i\sqrt{z})$ and $J_{-\lambda}(2i\sqrt{z})$, it too is a solution of (a). Hence the most general solution of equation (a) is $Z_\lambda(z) = AJ_\lambda(z) + BH^1_\lambda(z)$.

(b) $\dfrac{d^2f}{dz^2} + bz^m f = 0.$

Let $y = \dfrac{2\sqrt{b}}{m+2}z^{(m+2)/2}$, so that $\dfrac{dy}{dz} = \sqrt{b}z^{m/2}$. Letting $w = \dfrac{y(m+2)}{2\sqrt{b}}$, we have $z = w^{2/(m+2)}$, $z^{m/2} = w^{m/(m+2)}$, $z^{(m/2)-1} = w^{-2/(m+2)}$, and $\dfrac{dw}{dy} = \dfrac{m+2}{2\sqrt{b}}$.

Consider $f = \sqrt{z}\, Z_{1/m+2}\left(\dfrac{2\sqrt{b}}{m+2}z^{(m+2)/2}\right) \to \sqrt{z}J_\lambda(y).$

$$\frac{df}{dz} = \frac{z^{-1/2}J_\lambda}{2} + z^{1/2}\frac{dJ_\lambda}{dy}\sqrt{b}w^{m/(m+2)}$$

$$= \frac{w^{-1/(m+2)}J_\lambda}{2} + w^{(m+1)/(m+2)}\sqrt{b}\frac{dJ_\lambda}{dy}.$$

$$\frac{d^2f}{dz^2} = \frac{d^2J_\lambda}{dy^2}\left[bw^{(2m+1)/(m+2)}\right] + \frac{dJ_\lambda}{dy}\left[\frac{\sqrt{b}(m+2)}{2}w^{(m-1)/(m+2)}\right]$$
$$+ J_\lambda\left[\frac{(-1)}{4}w^{-3/(m+2)}\right].$$

$$\frac{d^2f}{dz^2} + bz^m f = \frac{d^2J_\lambda}{dy^2}\left[bw^{(2m+1)/(m+2)}\right] + \frac{dJ_\lambda}{dy}\left[\frac{\sqrt{b}(m+2)}{2}w^{(m-1)/(m+2)}\right]$$
$$+ J_\lambda\left[bw^{(2m+1)/(m+2)} - \frac{1}{4}w^{-3/(m+2)}\right].$$

Rearranging:

$$\frac{d^2f}{dz^2} + bz^m f = bw^{\frac{(2m+1)}{(m+2)}} \left\{ \frac{d^2J_\lambda}{dy^2} + \frac{dJ_\lambda}{dy}\left(\frac{(m+2)w^{-1}}{2\sqrt{b}} \right) + J_\lambda \left(1 - \frac{w^{-2}}{4b} \right) \right\}.$$

Substituting $w = \frac{(m+2)y}{2\sqrt{b}}$,

$$\frac{d^2f}{dz^2} + bz^m f = bw^{\frac{(2m+1)}{(m+2)}} \left\{ \frac{d^2J_\lambda}{dy^2} + \frac{dJ_\lambda}{dy}\frac{1}{y} + J_\lambda \left(1 - \frac{1}{y^2(m+2)^2} \right) \right\}.$$

But if $\lambda = \frac{1}{m+2}$,

$$\frac{d^2f}{dz^2} + bz^m f = bw^{\frac{(2m+1)}{(m+2)}} \left\{ \frac{d^2J_{1/m+2}}{dy^2} + \frac{dJ_{1/m+2}}{dy}\frac{1}{y} + J_{1/m+2} \left(1 - \frac{\lambda^2}{y^2} \right) \right\} = 0,$$

since $J_{1/(m+2)}(y)$ is a solution of Bessel's equation. We have thus shown that

$$\sqrt{z}\, Z_{1/m+2}\left(\frac{2\sqrt{b}}{m+2} z^{(m+2)/2} \right)$$

is a solution of $\frac{d^2f}{dz^2} + bz^m f = 0$, and therefore that

$$\sqrt{z} H^1_\lambda(y) = \frac{\sqrt{z}\, i}{\sin\lambda\pi}\left[e^{-\lambda\pi i} J_\lambda(y) - J_{-\lambda}(y) \right]$$

$$= \frac{i}{\sin\lambda\pi}\sqrt{z}\, e^{-\lambda\pi i} J_\lambda(y) - \frac{i}{\sin\lambda\pi}\sqrt{z} J_{-\lambda}(y)$$

is also a solution of $\frac{d^2f}{dz^2} + bz^m f = 0$. Hence, the most general solution of $\frac{d^2f}{dz^2} + bz^m f = 0$ is

$$f = \sqrt{z}\, Z_{1/m+2}\left(\frac{2\sqrt{b}}{m+2} z^{(m+2)/2} \right).$$

(c) Show that $f = z^{\alpha}Z_{\lambda}(\beta z)$ is the most general solution of the differential equation

$$\text{[1]} \qquad \frac{d^2f}{dz^2} + \frac{1-2\alpha}{z}\frac{df}{dz} + \left(\beta^2 + \frac{\alpha^2-\lambda^2}{z^2}\right)f = 0.$$

Solution:

Let $y = \beta z$, so that $z = y/\beta$ and $J_{\lambda}(\beta z) \to J_{\lambda}(y)$.

$$\frac{dJ_{\lambda}(y)}{dz} = \frac{dJ_{\lambda}(y)}{dy}\frac{dy}{dz} = \beta\frac{dJ_{\lambda}(y)}{dy}. \qquad \frac{d^2J_{\lambda}(y)}{dz^2} = \beta^2\frac{d^2J_{\lambda}(y)}{dy^2}.$$

Consider $f = z^{\alpha}Z_{\lambda}(\beta z)$.

$$\frac{df}{dz} = \alpha z^{\alpha-1}J_{\lambda} + z^{\alpha}\beta\frac{dJ_{\lambda}}{dy} = \frac{1}{\beta^{\alpha-1}}\left(\alpha y^{\alpha-1}J_{\lambda} + y^{\alpha}\frac{dJ_{\lambda}}{dy}\right).$$

$$\frac{d^2f}{dz^2} = \frac{1}{\beta^{\alpha-2}}\left(\alpha(\alpha-1)y^{\alpha-2}J_{\lambda} + 2\alpha y^{\alpha-1}\frac{dJ_{\lambda}}{dy} + y^{\alpha}\frac{d^2J_{\lambda}}{dy^2}\right)$$

$$= \frac{y^{\alpha}}{\beta^{\alpha-2}}\left(\frac{d^2J_{\lambda}}{dy^2} + \frac{2\alpha}{y}\frac{dJ_{\lambda}}{dy} + \frac{\alpha(\alpha-1)}{y^2}J_{\lambda}\right).$$

$$\frac{1-2\alpha}{z}\frac{df}{dz} = \frac{(1-2\alpha)}{\beta^{\alpha-2}y}\left(\alpha y^{\alpha-1}J_{\lambda} + y^{\alpha}\frac{dJ_{\lambda}}{dy}\right).$$

$$\left(\beta^2 + \frac{\alpha^2-\lambda^2}{z^2}\right)f = \left(\beta^2 + \frac{(\alpha^2-\lambda^2)\beta^2}{y^2}\right)\frac{y^{\alpha}}{\beta^{\alpha}}J_{\lambda} = \frac{y^{\alpha}}{\beta^{\alpha-2}}\left(1 + \frac{(\alpha^2-\lambda^2)}{y^2}\right)J_{\lambda}.$$

$$\frac{d^2f}{dz^2} + \frac{1-2\alpha}{z}\frac{df}{dz} + \left(\beta^2 + \frac{\alpha^2-\lambda^2}{z^2}\right)f$$

$$= \frac{y^{\alpha}}{\beta^{\alpha-2}}\left\{\frac{d^2J_{\lambda}}{dy^2} + \frac{1}{y}\frac{dJ_{\lambda}}{dy} + J_{\lambda}\left(1 - \frac{\lambda^2}{y^2}\right)\right\} = 0,$$

since $J_{\lambda}(y)$ satisfies Bessel;'s equation. Hence, $z^{\alpha}J_{\lambda}(\beta z)$ is a solution of equation [1], and therefore $z^{\alpha}H_{\lambda}^{1}(\beta z)$ is a solution as well. Hence, the most general solution of equation [1] is $z^{\alpha}Z_{\lambda}(\beta z)$.

Problem Set E

I. Consider the differential equation for heat conduction (Chapter 9), $\frac{\partial T}{\partial t} = \lambda \Delta T + S$, for a one-dimensional medium ($x \geq 0$). Let the boundary condition at $x = 0$ be $(\partial T)/(\partial x) = 0$ for all time (i.e. the medium is surrounded by an insulating wall at $x = 0$.

(a) Show that the Green's function for the homogeneous equation ($S = 0$) is given by

$$G(xx';t) = \frac{1}{\sqrt{4\pi\lambda t}}\left[e^{-\frac{(x-x')^2}{4\lambda t}} + e^{-\frac{(x+x')^2}{4\lambda t}}\right].$$

(b) Find the Green's function for the inhomogeneous equation.

Solution:

(a) If $G(xx';t) = \frac{1}{\sqrt{4\pi\lambda t}}\left[e^{-\frac{(x-x')^2}{4\lambda t}} + e^{-\frac{(x+x')^2}{4\lambda t}}\right] = G_o(x-x',t) + G_o(x+x',t)$, we would be able to say that for all time t and for $x \geq 0$,

$$T(x,t) = \int_0^\infty G(xx';t)T(x',0)dx'.$$

We must show that $T(x,t)$ satisfies proper boundary conditions.

(i) $$\left|\frac{\partial T(xt)}{\partial x}\right|_{x=0} = \int_0^\infty \left|\frac{\partial G(xx';t)}{\partial x}\right|_{x=0} T(x'0)dx'.$$

But

$$\left|\frac{\partial G(xx';t)}{\partial x}\right|_{x=0} = \left|\frac{\partial G_o(x-x',t)}{\partial x}\right|_{x=0} + \left|\frac{\partial G_o(x+x',t)}{\partial x}\right|_{x=0}$$

$$= \left|-\frac{\partial G_o(x-x',t)}{\partial x'}\right|_{x=0} + \left|\frac{\partial G_o(x+x',t)}{\partial x'}\right|_{x=0}$$

$$= \frac{1}{\sqrt{4\pi\lambda t}} e^{-\frac{(x')^2}{4\lambda t}}\left[\frac{2x'}{4\lambda t} - \frac{2x'}{4\lambda t}\right] = 0.$$

Thus, $\left|\dfrac{\partial T(xt)}{\partial x}\right|_{x=0} = 0$, and the boundary condition at $x = 0$ is satisfied.

(ii) $$\frac{\partial T(xt)}{\partial t} - \lambda\Delta T(xt) = \int_0^\infty \left(\frac{\partial}{\partial t} - \lambda\Delta\right) G(xx';t)T(x'0)dx'.$$

$$\sqrt{4\pi\lambda t}\,\Delta G(x,x';t) = \frac{\partial^2}{\partial x^2}\left[e^{-\frac{(x-x')^2}{4\lambda t}} + e^{-\frac{(x+x')^2}{4\lambda t}}\right]$$

$$= \frac{\partial}{\partial x}\left[-\frac{(x-x')}{2\lambda t}e^{-\frac{(x-x')^2}{4\lambda t}} - \frac{(x+x')}{2\lambda t}e^{-\frac{(x+x')^2}{4\lambda t}}\right]$$

$$= -\frac{1}{2\lambda t}e^{-\frac{(x-x')^2}{4\lambda t}} + \frac{(x-x')^2}{4\lambda^2 t^2}e^{-\frac{(x-x')^2}{4\lambda t}}$$

$$-\frac{1}{2\lambda t}e^{-\frac{(x+x')^2}{4\lambda t}} + \frac{(x+x')^2}{4\lambda^2 t^2}e^{-\frac{(x+x')^2}{4\lambda t}}.$$

$$\lambda\sqrt{4\pi\lambda t}\Delta G(xx';t) = e^{-\frac{(x-x')^2}{4\lambda t}}\left[\frac{(x-x')^2}{4\lambda t^2} - \frac{1}{2t}\right] + e^{-\frac{(x+x')^2}{4\lambda t}}\left[\frac{(x+x')^2}{4\lambda t^2} - \frac{1}{2t}\right].$$

$$\frac{\partial G(xx';t)}{\partial t} = \frac{\partial}{\partial t}\left[\frac{e^{-\frac{(x-x')^2}{4\lambda t}} + e^{-\frac{(x+x')^2}{4\lambda t}}}{\sqrt{4\pi\lambda t}}\right]$$

$$= \frac{1}{\sqrt{4\pi\lambda t}}\left[\frac{(x-x')^2}{4\lambda t^2}e^{-\frac{(x-x')^2}{4\lambda t}} + \frac{(x+x')^2}{4\lambda t^2}e^{-\frac{(x+x')^2}{4\lambda t}}\right]$$

$$-\frac{1}{t\sqrt{4\pi\lambda t}}\left[e^{-\frac{(x-x')^2}{4\lambda t}} + e^{-\frac{(x+x')^2}{4\lambda t}}\right]$$

$$\sqrt{4\pi\lambda t}\,\frac{\partial G(xx';t)}{\partial t} = e^{-\frac{(x-x')^2}{4\lambda t}}\left[\frac{(x-x')^2}{4\lambda t^2} - \frac{1}{2t}\right] + e^{-\frac{(x+x')^2}{4\lambda t}}\left[\frac{(x+x')^2}{4\lambda t^2} - \frac{1}{2t}\right].$$

$$= \lambda\sqrt{4\pi\lambda t}\Delta G(x,x';t).$$

Hence we see that $\frac{\partial G(xx';t)}{\partial t} - \lambda\Delta G(xx';t) = 0$, so that

$$\frac{\partial T(xt)}{\partial t} - \lambda\Delta T(xt) = 0 \quad \Rightarrow \quad T(xt) = \int_0^\infty G(xx';t)T(x'0)dx'$$

satisfies the heat conduction equation.

(iii) As $t \to 0+$, $G(xx';t) = G_0(x-x',t) + G_0(x+x',t) \to \delta(x-x')$ for $x' > 0$.

Hence,

$$\underset{t\to 0+}{T(xt)} = \int_0^\infty \delta(x-x')T(x'0)dx' = T(x0),$$

and a second boundary condition is satisfied.

(iv) As $x \to +\infty$, $G(xx';t) = \frac{1}{\sqrt{4\pi\lambda t}}\left[e^{-\frac{(x-x')^2}{4\lambda t}} + e^{-\frac{(x+x')^2}{4\lambda t}}\right] \to 0$, so that $\underset{x\to\infty}{T(xt)} \to 0$.

Thus, all boundary conditions are satisfied,

(b) To find the Green's function for the inhomogeneous equation ($S \neq 0$), we must find a Green's function $\mathcal{G}(x-x_o,t-t_o)$ satisfying

$$[1] \quad \left(\frac{\partial}{\partial t} - \lambda\Delta\right)\mathcal{G}(x-x_o,t-t_o) = \delta(x-x_o)\delta(t-t_o),$$

so that a particular solution may be written as

$$T(xt) = \int_0^\infty dt' \int_0^\infty \mathcal{G}(x-x',t-t')S(x't')dx'.$$

Define

$$\mathcal{G}(x-x_o,t-t_o) \equiv G(xx',t') \text{ for } t' > 0 \text{ (i.e. } t > t_o).$$

$$\mathcal{G}(x-x_o,t-t_o) \equiv 0 \text{ for } t' < 0 \text{ (i.e. } t < t_o).$$

We must show that this $\mathcal{G}(x-x_o,t-t_o)$ satisfies equation [1].

(i) For $t' > 0$ $(t > t_o)$, $\mathcal{G} = G$ and $\left(\frac{\partial}{\partial t} - \lambda\Delta\right)G(xx',t) = 0$ from (ii) part (a), while $\delta(x-x_o)\delta(t-t_o) = 0$ for $t > t_o$. Hence the above condition is satisfied for $t' > 0$.

(ii) For $t' < 0$ $(t < t_o)$, $\mathcal{G} = 0$ while $\delta(x-x_o)\delta(t-t_o) = 0$ for $t < t_o$. Hence the condition is satisfied for $t' < 0$.

(iii) As $t \to t_o$, the right-hand side of equation [1] approaches $\delta(x-x_o)\delta(0)$, while the left-hand side approaches $\delta(x-x_o) + \delta(x+x_o) \to \delta(x-x_o)$ for $+x_o$.

We must investigate whether both sides have the same singularity at $t = t_o$.

$$\int_{-\varepsilon}^{+\varepsilon} dt' \left(\frac{\partial}{\partial t} - \lambda\Delta\right)\mathcal{G}(x-x_o,t') \overset{?}{=} \int_{-\varepsilon}^{+\varepsilon} dt'\delta(x-x_o)\delta(t') = \delta(x-x_o).$$

$$\lim_{\varepsilon\to 0}\int_{-\varepsilon}^{+\varepsilon} dt' \frac{\partial \mathcal{G}(x-x_o,t')}{\partial t} - \lim_{\varepsilon\to 0}\lambda\int_{-\varepsilon}^{+\varepsilon} dt'\Delta\mathcal{G}(x-x_o,t')$$

$$= \lim_{\varepsilon\to 0}\left|\mathcal{G}(x-x_o,t')\right|_{-\varepsilon}^{+\varepsilon} - \lim_{\varepsilon\to 0}\lambda\varepsilon\Delta\mathcal{G}(x-x_o,\varepsilon) = \lim_{\varepsilon\to 0}\mathcal{G}(x-x_o,\varepsilon) - 0$$

$$= \delta(x-x_o).$$

Hence for $t \to t_o$, both sides approach the same singularity, namely $\delta(x-x_o)$.

Summarizing, we have shown $\mathcal{G}(x-x_o,t-t_o)$ to satisfy equation [1], so that a particular solution of the inhomogeneous equation is given by

$$T(xt) = \int_0^\infty dt' \int_0^\infty \mathcal{G}(x-x',t-t')S(x't)dx'.$$

Since for $t' > t$, $t - t' < 0$ and $\mathcal{G} = 0$, we can rewrite the particular solution as

$$T(xt) = \int_0^t dt' \int_0^\infty G(x-x',t-t')S(x't')dx',$$

and the most general solution will then be given by

$$T(xt) = \int_0^\infty G(xx',t)T(x'0)dx' + \int_0^t dt' \int_0^\infty G(x-x',t-t')S(x't')dx'.$$

II. Consider problem I above, where instead of the adiabatic boundary condition, one has at $x = 0$ and for all time,

$$\frac{\partial T}{\partial x} + \alpha T = 0,$$

where α is a constant.

(a) Find the corresponding Green's function for the homogeneous and the inhomogeneous equation.

(b) Express, explicitly, the temperature distribution for all $t \geq 0$ in terms of its initial distribution at $t = 0$ and the source function $S(xt)$.

Solution:

(a) Suspecting that $G_0(x-x',t)$ and $G_0(x+x',t)$ will not suffice to give $(\partial T)/(\partial x)$, we try adding a continuous source function $a(\eta)$, i.e. we try

$$G(xx',t) = G_0(x-x',t) + CG_0(x+x',t) + \int_{-\infty}^{-x'} a(\eta) G_0(x-\eta,t) d\eta .$$

[1] $$G(xx',t) = \frac{1}{\sqrt{4\pi\lambda t}} \left[e^{-\frac{(x-x')^2}{4\lambda t}} + Ce^{-\frac{(x+x')^2}{4\lambda t}} + \int_{-\infty}^{-x'} a(\eta) e^{-\frac{(x-\eta)^2}{4\lambda t}} d\eta \right].$$

At $x = 0$:

[2] $$\sqrt{4\pi\lambda t}\, G(0x',t) = (1 + C) e^{-\frac{(x')^2}{4\lambda t}} + \int_{-\infty}^{-x'} a(\eta) e^{-\frac{\eta^2}{4\lambda t}} d\eta .$$

From equation [1],

$$\sqrt{4\pi\lambda t} \left| \frac{\partial G(xx',t)}{\partial x} \right|_{x=0} = (1 - C) x' \frac{e^{-\frac{(x')^2}{4\lambda t}}}{2\lambda t} + \frac{\partial}{\partial x} \int_{-\infty}^{-x'} \left| a(\eta) e^{-\frac{(x-\eta)^2}{4\lambda t}} d\eta \right|_{x=0} .$$

Note that

$$\frac{\partial}{\partial x} \left| e^{-\frac{(x-\eta)^2}{4\lambda t}} d\eta \right|_{x=0} = -\frac{\partial}{\partial \eta} \left| e^{-\frac{(x-\eta)^2}{4\lambda t}} d\eta \right|_{x=0} .$$

Thus,

$$\frac{\partial}{\partial x}\int_{-\infty}^{-x'}\left|a(\eta)e^{-\frac{(x-\eta)^2}{4\lambda t}}d\eta\right|_{x=0} = -\int_{-\infty}^{-x'}a(\eta)\frac{\partial}{\partial\eta}\left|e^{-\frac{(x-\eta)^2}{4\lambda t}}d\eta\right|_{x=0}$$

$$= -\left|a(\eta)e^{-\frac{\eta^2}{4\lambda t}}\right|_{-\infty}^{-x'} + \int_{-\infty}^{-x'}\left|e^{-\frac{(x-\eta)^2}{4\lambda t}}\right|_{x=0}\frac{\partial a(\eta)}{\partial\eta}d\eta$$

$$= -\left|a(\eta)e^{-\frac{\eta^2}{4\lambda t}}\right|_{-\infty}^{-x'} + \int_{-\infty}^{-x'}e^{-\frac{\eta^2}{4\lambda t}}\frac{\partial a(\eta)}{\partial\eta}d\eta$$

$$= -a(x')e^{-\frac{(x')^2}{4\lambda t}} + \int_{-\infty}^{-x'}e^{-\frac{\eta^2}{4\lambda t}}\frac{\partial a(\eta)}{\partial\eta}d\eta$$

Thus, at $x = 0$,

[3] $$\sqrt{4\pi\lambda t}\,\frac{\partial G(xx',t)}{\partial x} = e^{-\frac{(x')^2}{4\lambda t}}\left[\frac{(1-C)x'}{2\lambda t} - a(-x')\right] + \int_{-\infty}^{-x'}e^{-\frac{\eta^2}{4\lambda t}}\frac{\partial a(\eta)}{\partial\eta}d\eta.$$

Since we want, as boundary condition at $x = 0$, that $(\partial G)/(\partial x) + \alpha G = 0$, we have, combining equations [2] and [3]:

[4] $$e^{-\frac{(x')^2}{4\lambda t}}\left[\frac{(1-C)x'}{2\lambda t} - a(-x')\right] + (1+C)\alpha e^{-\frac{(x')^2}{4\lambda t}} + \int_{-\infty}^{-x'}e^{-\frac{\eta^2}{4\lambda t}}\frac{\partial a(\eta)}{\partial\eta}d\eta$$

$$+\alpha\int_{-\infty}^{-x'}a(\eta)e^{-\frac{\eta^2}{4\lambda t}}d\eta = 0.$$

Since equation [4] must be true for all $t > 0$, we collect terms having the same powers of t:

(i) $1 - C = 0 \;\rightarrow\; C = 1.$

(ii) $a(-x') - \alpha(1 + C) = 0 \;\rightarrow\; a(-x') = 2\alpha.$

(iii) $\dfrac{\partial a(\eta)}{\partial\eta} + \alpha a(\eta) = 0 \;\rightarrow\; a(\eta) = be^{-\alpha\eta}.$

At $\eta = -x'$, $a(-x') = 2\alpha = be^{\alpha x'} \;\rightarrow\; b = 2\alpha e^{-\alpha x'}$, so that $a(\eta) = 2\alpha e^{-\alpha(\eta+x')}$.

[5] $$\sqrt{4\pi\lambda t}\,G(xx',t) = e^{-\frac{(x-x')^2}{4\lambda t}} + e^{-\frac{(x+x')^2}{4\lambda t}} + 2\alpha e^{-\alpha x'}\int_{-\infty}^{-x'}e^{-\frac{(x-\eta)^2}{4\lambda t}}e^{-\alpha\eta}d\eta.$$

<u>Properties of $G(xx',t)$</u>

(i) $G(xx',t)$ satisfies the boundary condition that at $x = 0$, $\dfrac{\partial G}{\partial x} + \alpha G = 0$. (It was deliberately constructed so.)

(ii) $G(xx',t)$ satisfies the heat conduction equation, $\dfrac{\partial G(xx',t)}{\partial t} = \lambda \Delta G(xx',t)$.

Proof: By letting

$$G^{o}(xx',t) = G_o(x-x',t) + G_o(x+x',t)$$

and

$$G''(xx',t) = \frac{2\alpha e^{-\alpha x'}}{\sqrt{4\pi\lambda t}} \int_{-\infty}^{-x'} e^{-\frac{(x-\eta)^2}{4\lambda t}} e^{-\alpha\eta} d\eta,$$

so that $G(xx',t) = G^{o}(x-x',t) + G''(x+x',t)$, it then follows that

$$\frac{\partial G}{\partial t} = \frac{\partial G^{o}}{\partial t} + \frac{\partial G''}{\partial t}. \qquad \lambda G = \lambda G^{o} + \lambda G''.$$

Since we already know that $\dfrac{\partial G^{o}}{\partial t} = \lambda \Delta G^{o}$, it remains for us to show that $\dfrac{\partial G''}{\partial t} = \lambda \Delta G''$.

$$\frac{1}{2\alpha e^{-\alpha x'}} \frac{\partial G''}{\partial t} = \frac{\partial}{\partial t}\left[\frac{\int_{-\infty}^{-x'} e^{-\frac{(x-\eta)^2}{4\lambda t}} e^{-\alpha\eta} d\eta}{\sqrt{4\pi\lambda t}}\right]$$

$$= \frac{\int_{-\infty}^{-x'} \frac{(x-\eta)^2}{4\lambda t^2} e^{-\frac{(x-\eta)^2}{4\lambda t}} e^{-\alpha\eta} d\eta}{\sqrt{4\pi\lambda t}} - \frac{1}{2t\sqrt{4\pi\lambda t}} \int_{-\infty}^{-x'} e^{-\alpha\eta} e^{-\frac{(x-\eta)^2}{4\lambda t}} d\eta$$

$$= \frac{1}{\sqrt{4\pi\lambda t}}\left[\int_{-\infty}^{-x'} e^{-\alpha\eta} e^{-\frac{(x-\eta)^2}{4\lambda t}} \left(\frac{(x-\eta)^2}{4\lambda t^2} - \frac{1}{2t}\right) d\eta\right].$$

$$\frac{\lambda\Delta G''}{2\alpha e^{-\alpha x'}} = \lambda\frac{\partial^2}{\partial x^2}\left[\frac{\int_{-\infty}^{-x'} e^{-\frac{(x-\eta)^2}{4\lambda t}}e^{-\alpha\eta}d\eta}{\sqrt{4\pi\lambda t}}\right]$$

$$= \frac{\lambda}{\sqrt{4\pi\lambda t}}\int_{-\infty}^{-x'}\frac{\partial}{dx}\left[\frac{-(x-\eta)}{2\lambda t}e^{-\frac{(x-\eta)^2}{4\lambda t}}e^{-\alpha\eta}d\eta\right]$$

$$= \frac{\lambda}{\sqrt{4\pi\lambda t}}\int_{-\infty}^{-x'}\left[\frac{(x-\eta)^2}{4\lambda^2t^2}-\frac{1}{2\lambda t}\right]e^{-\frac{(x-\eta)^2}{4\lambda t}}e^{-\alpha\eta}d\eta = \frac{1}{2\alpha e^{-\alpha x'}}\frac{\partial G''}{\partial t}.$$

Thus,

$$\frac{\partial G''}{\partial t} = \lambda\Delta G'' \quad\Rightarrow\quad \frac{\partial G}{\partial t} = \lambda\Delta G.$$

(iii) As $x \to \infty$, $G \to 0$, as required.

(b) For $S \neq 0$, we must find a new Green's function $\mathcal{G}(xx',t)$ such that

[6] $$\left(\frac{\partial}{\partial t} - \lambda\Delta\right)\mathcal{G}(xx',t') = \delta(x-x')\delta(t').$$

Define

$$\mathcal{G}(x-x_o,t-t_o) = G(xx',t') \text{ for } t' > 0 \text{ (i.e. } t > t_o).$$

$$\mathcal{G}(x-x_o,t-t_o) = 0 \text{ for } t' < 0 \text{ (i.e. } t < t_o).$$

To show that the new $\mathcal{G}(x-x_o,t-t_o)$ satisfies [6] above, we show it to be true for the following cases:

(i) For $t' > 0$ $(t > t_o)$, $\mathcal{G} = G$. But, from part (a) for $S = 0$, we showed that $\left(\frac{\partial}{\partial t} - \lambda\Delta\right)G = 0$, while for $t' \neq 0$, $\delta(x-x')\delta(t') = 0$, thus satisfying both sides of [6] for $t' > 0$.

(ii) For $t' < 0$ $(t < t_0)$, $\mathcal{G} = 0$, while for $t' \neq t_0$, $\delta(x-x')\delta(t') = 0$, thus satisfying both sides of [6] for $t' < 0$.

(iii) As $t \to t_0$ $(t' \to 0)$, $\mathcal{G} \to G \underset{t'\to 0}{\to} G^0 + G''$.

To compare singularities on both sides of [6] as $t \to t_0$, consider:

$$\lim_{\varepsilon\to 0}\int_{-\varepsilon}^{+\varepsilon} dt\left(\frac{\partial}{\partial t} - \lambda\Delta\right)\mathcal{G}(x-x_0,t-t_0) = \lim_{\varepsilon\to 0}\left[\left|\mathcal{G}(x-x_0,t-t_0)\right|_{-\varepsilon}^{+\varepsilon} - \lambda\varepsilon\Delta\mathcal{G}\right]$$

$$= \lim_{\varepsilon\to 0}\left[\mathcal{G}(x-x_0,\varepsilon) - \lambda\varepsilon\Delta\mathcal{G}\right] \to \delta(x-x_0)$$

$$\lim_{\varepsilon\to 0}\int_{-\varepsilon}^{+\varepsilon}\delta(x-x_0)\delta(t)dt = \delta(x-x_0).$$

Hence, both sides of [6] approach the same singularity, namely, $\delta(x-x_0)$. We have thus shown that $\mathcal{G}$ satisfies [6], and can therefore say that for all $t \geq 0$,

$$T(x,t) = \int_0^\infty G(xx',t)T(x',0)dx' + \int_0^t dt'\int_0^\infty G(xx',t)S(x',t')dx'.$$

III. Consider a ring conductor of length a. The heat conduction equation is given by

$$\frac{\partial T}{\partial t} = \lambda\frac{\partial^2 T}{\partial x^2} + S,$$

where x is the length measured along the circumference. The boundary condition is given as $T(x,t) = T(x+a,t)$,

(a) Prove that the Green's function is

$$G(x-x',t) = \frac{1}{a}\left[1 + 2\sum_{n=1}^{\infty}\cos\frac{2n\pi(x-x')}{a}\,e^{-\frac{4n^2\pi^2\lambda t}{a^2}}\right].$$

Solution: We must show that $G(x-x',t)$ has the following properties:

(i) $G(x-x',t)$ satisfies the heat conduction equation for $S = 0$.

Proof:

$$\frac{\partial G}{\partial t} = -\frac{1}{a}\left[2\sum_{n=1}^{\infty}\cos\frac{2n\pi(x-x')}{a}\,e^{-\frac{4n^2\pi^2\lambda t}{a^2}}\left(\frac{4n^2\pi^2\lambda}{a^2}\right)\right].$$

$$\frac{\partial G}{\partial x} = -\frac{1}{a}\left[2\sum_{n=1}^{\infty}\frac{2n\pi}{a}\sin\frac{2n\pi(x-x')}{a}\,e^{-\frac{4n^2\pi^2\lambda t}{a^2}}\right].$$

$$\frac{\partial^2 G}{\partial x^2} = -\frac{1}{a}\left[2\sum_{n=1}^{\infty}\frac{4n^2\pi^2}{a^2}\cos\frac{2n\pi(x-x')}{a}\,e^{-\frac{4n^2\pi^2\lambda t}{a^2}}\right] = \frac{1}{\lambda}\frac{\partial G}{\partial t}.$$

(ii) $G(x-x',0)$ at $t=0$ is the delta function $\delta(x-x')$.

Proof:

As $t \to 0$ for $x = x'$:

$$G(0,t) \to \lim_{x\to x'}\frac{1}{a}\left[1+2\left(\cos\frac{2\pi(x-x')}{a}+\cos\frac{4\pi(x-x')}{a}+\ldots\right)\right]$$

$$= \frac{1}{a}\left[1+2(1+1+1+\ldots)\right] \Rightarrow \infty.$$

As $t \to 0$ for $x \neq x'$: Letting $y = x-x'$, we have

$$\int_{-a/2}^{+a/2} G(y,0)dy = \frac{1}{a}\int_{-a/2}^{+a/2}\left[1+2\sum_{n=1}^{\infty}\cos\frac{2n\pi y}{a}\right]dy$$

$$= \frac{1}{a}\int_{-a/2}^{+a/2}dy + \frac{2}{a}\int_{-a/2}^{+a/2}\sum_{n=1}^{\infty}\cos\frac{2n\pi y}{a}\,dy$$

$$= 1 + \left|\sum_{n=1}^{\infty}\frac{a}{2n\pi}\sin\frac{2n\pi y}{a}\right|_{-\frac{a}{2}}^{+\frac{a}{2}} = 1+0 = 1.$$

Thus, $\int_{-a/2}^{+a/2} G(y,0)dy = 1$ and $G(x-x',0)$ behaves like the delta function $\delta(x-x')$.

(iii) $G(x+a-x',t) = G(x,t)$.

Proof:

$$aG(x+a-x',t) = 1+2\sum_{n=1}^{\infty}\cos\frac{2n\pi(x+a-x')}{a}e^{-\frac{4n^2\pi^2\lambda t}{a^2}}.$$

$$= 1+2\sum_{n=1}^{\infty}\cos\left(\frac{2n\pi(x-x')}{a}+2n\pi\right)e^{-\frac{4n^2\pi^2\lambda t}{a^2}}$$

$$= 1+2e^{-\frac{4n^2\pi^2\lambda t}{a^2}}\left[\cos\left(\frac{2\pi(x-x')}{a}+2\pi\right)+\cos\left(\frac{4\pi(x-x')}{a}+4\pi\right)+\ldots\right]$$

$$= 1+2e^{-\frac{4n^2\pi^2\lambda t}{a^2}}\left[\cos\frac{2\pi(x-x')}{a}+\cos\frac{4\pi(x-x')}{a}+\ldots\right]$$

$$= 1+2\sum_{n=1}^{\infty}\cos\frac{2n\pi(x-x')}{a}e^{-\frac{4n^2\pi^2\lambda t}{a^2}} = aG(x-x',t).$$

IV. The four theta functions are defined as follows:

$$\theta_1(zq) = 2\sum_{n=0}^{\infty}(-1)^n q^{(n+1/2)^2}\sin(2n+1)z.$$

$$\theta_2(zq) = 2\sum_{n=0}^{\infty}q^{(n+1/2)^2}\cos(2n+1)z.$$

$$\theta_3(zq) = 1+2\sum_{n=1}^{\infty}q^{n^2}\cos(2nz).$$

$$\theta_4(zq) = 1+2\sum_{n=1}^{\infty}(-1)^n q^{n^2}\cos(2nz).$$

(a) Prove that the Green's function in Problem III is related to the theta functions through the relation

$$G(x-x',t) = \theta_3(zq)/a,$$

where $z = \frac{\pi(x-x')}{a}$ and $q = e^{-\frac{4\pi^2\lambda t}{a^2}}$.

(b) Show that the theta functions are related to each other by the following:

(i) $\theta_1(zq) = -\theta_2(z+\pi/2,q) = -iM\theta_3(z+\pi/2+\pi\tau/2,q) = -iM\theta_4(z+\pi\tau/2,q)$

(ii) $\theta_2(zq) = M\theta_3(z+\pi\tau/2,q) = M\theta_4(z+\pi/2+\pi\tau/2,q) = \theta_1(z+\pi/2,q)$

(iii) $\theta_3(zq) = \theta_4(z+\pi/2,q) = M\theta_1(z+\pi/2+\pi\tau/2,q) = M\theta_2(z+\pi\tau/2,q)$

(iv) $\theta_4(zq) = -iM\theta_1(z+\pi\tau/2,q) = iM\theta_2(z+\pi/2+\pi\tau/2,q) = \theta_3(z+\pi/2,q)$

where $q = e^{i\pi\tau}$ and $M = q^{1/4}e^{iz}$.

(c) Prove that $\left(\frac{\pi i}{4}\frac{\partial^2}{\partial z^2} + \frac{\partial}{\partial \tau}\right)\theta_\lambda(z,e^{i\pi\tau}) = 0$, for $\lambda = 1, 2, 3, 4$.

Solution:

(a) From Problem III: $aG(x-x',t) = 1 + 2\sum_{n=1}^{\infty}\cos\frac{2n\pi(x-x')}{a}\,e^{-\frac{4n^2\pi^2\lambda t}{a^2}}$.

$$aG(x-x',t) = 1 + 2\sum_{n=1}^{\infty}\cos\frac{2n\pi(x-x')}{a}\,e^{-\frac{4n^2\pi^2\lambda t}{a^2}}.$$

$$\theta_3(zq) = 1 + 2\sum_{n=1}^{\infty}q^{n^2}\cos(2nz) = 1 + 2\sum_{n=1}^{\infty}\cos\frac{2n\pi(x-x')}{a}\,e^{-\frac{4n^2\pi^2\lambda t}{a^2}}$$

$$= aG(x-x',t).$$

(b)

(i) $$\theta_2(z+\pi/2,q) = 2\sum_{n=0}^{\infty}q^{(n+1/2)^2}\cos(2n+1)(z+\pi/2).$$

$$= 2\sum_{n=0}^{\infty}q^{(n+1/2)^2}\left[-\sin z(2n+1)\sin(2n+1)\pi/2\right]$$

$$= 2\sum_{n=0}^{\infty}(-)(-1)^n q^{(n+1/2)^2}\sin(2n+1)z = -\theta_1(zq).$$

[1] $$\theta_2(z+\pi/2,q) = -\theta_1(zq).$$

$$-iM\theta_3(z+\pi/2+\pi\tau/2,q) = -ie^{iz}q^{1/4}\left[1+2\sum_{n=1}^{\infty}q^{n^2}\cos 2n(z+\pi/2+\pi\tau/2)\right]$$

$$= -ie^{iz}q^{1/4}\left[1+\sum_{n=1}^{\infty}q^{n^2}e^{2ni(z+\pi/2+\pi\tau/2)}+e^{-2ni(z+\pi/2+\pi\tau/2)}\right]$$

$$= -ie^{iz}q^{1/4}\sum_{n=-\infty}^{\infty}q^{n^2}e^{2ni(z+\pi/2+\pi\tau/2)}$$

$$= -i\sum_{n=-\infty}^{\infty}q^{(n^2+1/4)}e^{i(2n+1)z}(-1)^n q^n$$

$$= -i\sum_{n=-\infty}^{\infty}q^{(n^2+n+1/4)}e^{i(2n+1)z}(-1)^n = -i\sum_{n=-\infty}^{\infty}(-1)^n q^{(n+1/2)^2}e^{i(2n+1)z}$$

Consider the terms that arise from values $n = \pm 2$ and ± 3. We get results like $(-i)\left(q^{25/4}e^{5iz}+q^{9/4}e^{-3iz}-q^{49/4}e^{7iz}-q^{25/4}e^{-5iz}\right)$, which gives rise to pairs of terms like $(-i)q^{25/4}\left(e^{5iz}-e^{-5iz}\right) = (-i)2iq^{25/4}\sin 5z = 2q^{(n+1/2)^2}\sin(2n+1)z$ for $n = 2$. Hence, the sum in question is equivalent to

$$2\sum_{n=0}^{\infty}(-1)^n q^{(n+1/2)^2}\sin(2n+1)z = \theta_1(zq).$$

[2] $$-iM\theta_3(z+\pi/2+\pi\tau/2,q) = \theta_1(zq).$$

$$-iM\theta_4(z+\pi\tau/2,q) = -ie^{iz}q^{1/4}\left[1+2\sum_{n=1}^{\infty}(-1)^n q^{n^2}\cos 2n(z+\pi\tau/2)\right]$$

$$= -ie^{iz}q^{1/4}\left[1+\sum_{n=1}^{\infty}(-1)^n q^{n^2}\left(e^{2ni(z+\pi\tau/2)}+e^{-2ni(z+\pi\tau/2)}\right)\right]$$

$$= -ie^{iz}q^{1/4}\sum_{n=-\infty}^{\infty}(-1)^n q^{n^2}e^{2ni(z+\pi\tau/2)} = -i\sum_{n=-\infty}^{\infty}(-1)^n q^{(n+1/2)^2}e^{i(2n+1)z}.$$

Following the previous pair-analysis procedure:

[3] $$-iM\theta_4(z+\pi\tau/2,q) = 2\sum_{n=0}^{\infty}(-1)^n q^{(n+1/2)^2}\sin(2n+1)z = \theta_1(zq).$$

Combining results [1], [2] and [3], we have

$$\theta_1(zq) = -\theta_2(z+\pi/2,q) = -iM\theta_3(z+\pi/2+\pi\tau/2,q) = -iM\theta_4(z+\pi\tau/2,q).$$

(ii) $$M\theta_3(z+\pi\tau/2,q) = e^{iz}q^{1/4}\left[1+2\sum_{n=1}^{\infty}q^{n^2}\cos 2n(z+\pi\tau/2)\right]$$

$$= e^{iz}q^{1/4}\sum_{n=-\infty}^{\infty}q^{n^2}e^{2ni(z+\pi\tau/2)} = e^{iz}\sum_{n=-\infty}^{\infty}q^{(n+1/2)^2}e^{i2nz}$$

$$= \sum_{n=-\infty}^{\infty}q^{(n+1/2)^2}e^{i(2n+1)z}.$$

Consider the terms that arise from values $n = \pm 2$ and ± 3. We get results like $q^{25/4}e^{5iz} + q^{9/4}e^{-3iz} + q^{49/4}e^{7iz} + q^{25/4}e^{-5iz}$, which give rise to terms like $q^{25/4}\left(e^{5iz} + e^{-5iz}\right) = 2q^{25/4}\cos(5iz) = 2q^{(n+1/2)^2}\cos(2n+1)z$ for $n = 2$. Hence the sum in question is equivalent to

$$2\sum_{n=0}^{\infty}q^{(n+1/2)^2}\cos(2n+1)z = \theta_2(zq).$$

[4] $$M\theta_3(z+\pi\tau/2,q) = \theta_2(zq).$$

$$M\theta_4(z+\pi/2+\pi\tau/2,q) = e^{iz}q^{1/4}\left[1+2\sum_{n=1}^{\infty}(-1)^n q^{n^2}\cos 2n(z+\pi/2+\pi\tau/2)\right].$$

$$= e^{iz}q^{1/4}\sum_{n=1}^{\infty}(-1)^n q^{n^2}e^{2nz}e^{in\pi}e^{n\pi\tau}$$

$$= \sum_{n=1}^{\infty}(-1)^n(-1)^n q^{(n+1/2)^2}e^{2nz}e^{i(2n+1)z}$$

$$= 2\sum_{n=0}^{\infty}q^{(n+1/2)^2}\cos(2n+1)z = \theta_2(zq).$$

[5] $$M\theta_4(z+\pi/2+\pi\tau/2,q) = \theta_2(zq).$$

$$\theta_1(z+\pi/2,q) = 2\sum_{n=0}^{\infty}(-1)^n q^{(n+1/2)^2}\sin(2n+1)(z+\pi/2)$$

$$= 2\sum_{n=0}^{\infty}(-1)^n q^{(n+1/2)^2}\cos(2n+1)z\,\sin(n+1/2)\pi$$

$$= 2\sum_{n=0}^{\infty}(-1)^n(-1)^n q^{(n+1/2)^2}\cos(2n+1)z$$

$$= 2\sum_{n=0}^{\infty}q^{(n+1/2)^2}\cos(2n+1)z = \theta_2(zq).$$

[6] $\theta_1(z+\pi/2,q) = \theta_2(zq).$

Combining results [4], [5] and [6], we have

$$\theta_2(zq) = M\theta_3(z+\pi\tau/2,q) = M\theta_4(z+\pi/2+\pi\tau/2,q) = \theta_1(z+\pi/2,q).$$

(iii) $$\theta_4(z+\pi/2,q) = 1+2\sum_{n=1}^{\infty}(-1)^n q^{n^2}\cos 2n(z+\pi/2)$$

$$= 1+2\sum_{n=1}^{\infty}(-1)^n q^{n^2}\cos 2nz\,\cos n\pi$$

$$= 1+2\sum_{n=1}^{\infty}(-1)^n q^{n^2}(-1)^n\cos 2nz$$

$$= 1+2\sum_{n=1}^{\infty}q^{n^2}\cos 2nz = \theta_3(zq).$$

[7] $\theta_4(z+\pi/2,q) = \theta_3(zq).$

$$M\theta_1(z+\pi/2+\pi\tau/2,q) = e^{iz}q^{1/4}\sum_{n=0}^{\infty}2(-1)^n q^{(n+1/2)^2}\sin(2n+1)(z+\pi/2+\pi\tau/2)$$

$$= 2e^{iz}q^{1/4}\sum_{n=0}^{\infty}(-1)^n q^{(n+1/2)^2}\sin(2n+1)\pi/2\,\cos(2n+1)(z+\pi\tau/2)$$

$$= 2e^{iz}q^{1/4}\sum_{n=0}^{\infty}(-1)^n(-1)^n q^{(n+1/2)^2}\cos(2n+1)(z+\pi\tau/2)$$

$$= e^{iz}q^{1/4}\sum_{n=0}^{\infty}q^{(n+1/2)^2}\left(e^{i(2n+1)(z+\pi\tau/2)} + e^{-i(2n+1)(z+\pi\tau/2)}\right),$$

which, by virtue of pair-analysis,

$$= e^{iz}q^{1/4}\sum_{n=-\infty}^{\infty}q^{(n+1/2)^2}e^{-i(2n+1)(z+\pi\tau/2)} = \sum_{n=-\infty}^{\infty}q^{n^2}e^{-i2nz}$$

$$= 1 + 2\sum_{n=1}^{\infty}q^{n^2}\cos 2nz = \theta_3(zq).$$

[8] $M\theta_1(z+\pi/2+\pi\tau/2,q) = \theta_3(zq).$

$$M\theta_2(z+\pi\tau/2,q) = e^{iz}q^{1/4}\,2\sum_{n=0}^{\infty}q^{(n+1/2)^2}\cos(2n+1)(z+\pi\tau/2)$$

$$= e^{iz}q^{1/4}\sum_{n=0}^{\infty}q^{(n+1/2)^2}\left(e^{i(2n+1)(z+\pi\tau/2)} + e^{-i(2n+1)(z+\pi\tau/2)}\right)$$

$$= e^{iz}q^{1/4}\sum_{n=-\infty}^{\infty}q^{(n+1/2)^2}\,e^{-i(2n+1)(z+\pi\tau/2)}.$$

Note that if $n = -5$ in the above expression, we get a term like $q^{81/4}e^{9i(z+\pi\tau/2}$, while if $n = +4$ in the expression preceding the above, we get a term like $q^{81/4}\left(e^{9i(z+\pi\tau/2} + e^{-9i(z+\pi\tau/2}\right)$. Thus, the sum in question is equivalent to

$$e^{iz}q^{1/4}\sum_{n=-\infty}^{\infty}q^{(n^2+n+1/4)}e^{-iz}e^{-2niz}q^{-1/2}q^{-n} = \sum_{n=-\infty}^{\infty}q^{n^2}e^{-2niz}$$

$$= 1 + 2\sum_{n=1}^{\infty}q^{n^2}\cos(2nz) = \theta_3(zq).$$

[9] $M\theta_2(z+\pi\tau/2,q) = \theta_3(zq).$

Combining results [7, [8] and [9], we have

$$\theta_3(zq) = \theta_4(z+\pi/2,q) = M\theta_1(z+\pi/2+\pi\tau/2,q) = M\theta_2(z+\pi\tau/2,q).$$

$$\text{(iv)}\ -iM\theta_1(z+\pi\tau/2,q) = -ie^{iz}q^{1/4}2\sum_{n=0}^{\infty}(-1)^n q^{(n+1/2)^2}\sin(2n+1)(z+\pi\tau/2)$$

$$= -e^{iz}q^{1/4}\sum_{n=0}^{\infty}(-1)^n q^{(n+1/2)^2}\left(e^{i(2n+1)(z+\pi\tau/2)} - e^{-i(2n+1)(z+\pi\tau/2)}\right),$$

which, by virtue of pair-analysis, is equivalent to

$$= -e^{iz}q^{1/4}\sum_{n=-\infty}^{\infty}(-1)^n q^{(n+1/2)^2}e^{i(2n+1)(z+\pi\tau/2)}$$

$$= e^{iz}q^{1/4}\sum_{n=-\infty}^{\infty}(-1)^n q^{(n+1/2)^2}e^{-i(2n+1)(z+\pi\tau/2)} = \sum_{n=-\infty}^{\infty}(-1)^n q^{n^2}e^{-i2nz}.$$

For $n = \pm 2$ and ± 3 etc., we get terms like $q^4e^{-4iz} - q^4e^{4iz} + \ldots$, so that the sum in question is

$$1 + 2\sum_{n=1}^{\infty}(-1)^n q^{n^2}\cos(2nz) = \theta_4(zq).$$

[10] $\quad -iM\theta_1(z+\pi\tau/2,q) = \theta_4(zq).$

$$iM\theta_2(z+\pi/2+\pi\tau/2,q) = ie^{iz}q^{1/4}2\sum_{n=0}^{\infty}q^{(n+1/2)^2}\cos(2n+1)(z+\pi/2+\pi\tau/2)$$

$$= -ie^{iz}q^{1/4}2\sum_{n=0}^{\infty}q^{(n+1/2)^2}\sin(2n+1)(\pi/2)\sin(2n+1)(z+\pi\tau/2)$$

$$= -ie^{iz}q^{1/4}2\sum_{n=0}^{\infty}(-1)^n q^{(n+1/2)^2}\sin(2n+1)(z+\pi\tau/2)$$

$$= -iM\theta_1(z+\pi\tau/2,q) = \theta_4(zq).$$

[11] $\quad iM\theta_2(z+\pi/2+\pi\tau/2,q) = -iM\theta_1(z+\pi\tau/2,q) = \theta_4(zq).$

$$\theta_3(z+\pi/2,q) = 1+2\sum_{n=1}^{\infty} q^{n^2}\cos 2n(z+\pi/2) = 1+2\sum_{n=1}^{\infty} q^{n^2}\cos n\pi\,\cos 2nz$$

$$= 1+2\sum_{n=1}^{\infty} q^{n^2}(-1)^n\cos 2nz = \theta_4(zq).$$

[12] $\quad \theta_3(z+\pi/2,q) = \theta_4(zq).$

Combining results [10], [11] and [12], we have

$$\theta_4(zq) = -iM\theta_1(z+\pi\tau/2,q) = iM\theta_2(z+\pi/2+\pi\tau/2,q) = \theta_3(z+\pi/2,q).$$

(c) Show that $\left(\frac{\pi i}{4}\frac{\partial^2}{\partial z^2}+\frac{\partial}{\partial\tau}\right)\theta_\lambda(z,e^{i\pi\tau}) = 0$, for $\lambda = 1, 2, 3, 4$.

$[\theta_1]$: $\quad \dfrac{\partial\theta_1}{\partial\tau} = \dfrac{\partial\theta_1}{\partial q}\dfrac{dq}{d\tau} = 2\sum_{n=0}^{\infty}(-1)^n(n+1/2)^2 q^{(n+1/2)^2-1}\sin(2n+1)z\,(i\pi)q$

$$= 2\pi i\sum_{n=0}^{\infty}(-1)^n(n+1/2)^2 q^{(n+1/2)^2}\sin(2n+1)z.$$

$$\frac{\partial\theta_1}{\partial z} = 2\sum_{n=0}^{\infty}(-1)^n(2n+1)q^{(n+1/2)^2}\cos(2n+1)z.$$

$$\frac{\partial^2\theta_1}{\partial z^2} = -2\sum_{n=0}^{\infty}(-1)^n 4(n+1/2)^2 q^{(n+1/2)^2}\sin(2n+1)z.$$

$$\frac{\pi i}{4}\frac{\partial^2\theta_1}{\partial z^2} = -2\pi i\sum_{n=0}^{\infty}(-1)^n(n+1/2)^2 q^{(n+1/2)^2}\sin(2n+1)z = -\frac{\partial\theta_1}{\partial\tau}.$$

$[\theta_2]$: $\quad \dfrac{\partial\theta_2}{\partial\tau} = \dfrac{\partial\theta_2}{\partial q}\dfrac{dq}{d\tau} = 2\sum_{n=0}^{\infty}(n+1/2)^2 q^{(n+1/2)^2-1}\cos(2n+1)z\,(i\pi)q$

$$= 2\pi i\sum_{n=0}^{\infty}(n+1/2)^2 q^{(n+1/2)^2}\cos(2n+1)z.$$

$$\frac{\partial\theta_2}{\partial z} = -2\sum_{n=0}^{\infty}(2n+1)q^{(n+1/2)^2}\sin(2n+1)z.$$

$$\frac{\partial^2\theta_2}{\partial z^2} = -2\sum_{n=0}^{\infty}4(n+1/2)^2q^{(n+1/2)^2}\cos(2n+1)z.$$

$$\frac{\pi i}{4}\frac{\partial^2\theta_2}{\partial z^2} = -2\pi i\sum_{n=0}^{\infty}(n+1/2)^2q^{(n+1/2)^2}\cos(2n+1)z = -\frac{\partial\theta_2}{\partial\tau}.$$

$[\theta_3]$: $$\frac{\partial\theta_3}{\partial\tau} = \frac{\partial\theta_3}{\partial q}\frac{dq}{d\tau} = 2\sum_{n=1}^{\infty}(n)^2q^{n^2-1}\cos 2nz\,(i\pi)q$$

$$= 2\pi i\sum_{n=1}^{\infty}(n)^2q^{n^2}\cos 2nz.$$

$$\frac{\partial\theta_3}{\partial z} = -2\sum_{n=1}^{\infty}(2n)q^{n^2}\sin 2nz. \qquad \frac{\partial^2\theta_3}{\partial z^2} = -2\sum_{n=1}^{\infty}4n^2q^{n^2}\cos 2nz.$$

$$\frac{\pi i}{4}\frac{\partial^2\theta_3}{\partial z^2} = -2\pi i\sum_{n=1}^{\infty}n^2q^{n^2}\cos 2nz = -\frac{\partial\theta_3}{\partial\tau}.$$

$[\theta_4]$: $$\frac{\partial\theta_4}{\partial\tau} = \frac{\partial\theta_4}{\partial q}\frac{dq}{d\tau} = 2\sum_{n=1}^{\infty}(-1)^n n^2q^{n^2-1}\cos 2nz\,(i\pi)q$$

$$= 2\pi i\sum_{n=1}^{\infty}(-1)^n n^2q^{n^2}\cos 2nz.$$

$$\frac{\partial\theta_4}{\partial z} = -2\sum_{n=1}^{\infty}(-1)^n 2nq^{n^2}\sin 2nz.$$

$$\frac{\partial^2\theta_4}{\partial z^2} = -2\sum_{n=1}^{\infty}(-1)^n 4n^2q^{n^2}\cos 2nz.$$

$$\frac{\pi i}{4}\frac{\partial^2\theta_4}{\partial z^2} = -2\pi i\sum_{n=1}^{\infty}(-1)^n n^2q^{n^2}\cos 2nz = -\frac{\partial\theta_4}{\partial\tau}.$$

www.ingramcontent.com/pod-product-compliance
Lightning Source LLC
Chambersburg PA
CBHW050633190326
41458CB00008B/2246
* 9 7 8 9 8 1 0 2 2 7 6 0 9 *